The Great Power Competition Volume 4

Adib Farhadi · Anthony Masys
Editors

The Great Power Competition Volume 4

Lessons Learned in Afghanistan: America's Longest War

Springer

Editors
Adib Farhadi
University of South Florida
Tampa, FL, USA

Anthony Masys
College of Public Health
University of South Florida
Tampa, FL, USA

ISBN 978-3-031-22933-6 ISBN 978-3-031-22934-3 (eBook)
https://doi.org/10.1007/978-3-031-22934-3

This Springer imprint is published by the registered company Springer Nature Switzerland AG
The registered company address is: Gewerbestrasse 11, 6330 Cham, Switzerland

Preface

On August 20, 1998, in response to the U.S. embassy terror attacks in Kenya and Tanzania, President Clinton launched *Operation Infinite Reach* by deploying cruise missiles to destroy al-Qaeda positions in Afghanistan. The attack inaugurated the first official military action in U.S. history on this landlocked nation, situated at the heart of Asia. The event likely marked the first time most Americans had ever heard of al-Qaeda; yet a mere three years later, when President George W. Bush declared a Global War on Terror in response to the 9/11 terror attacks, the Taliban, al-Qaeda, and its leader, Osama bin Laden, had all become household names. The resulting 2001 *Operation Enduring Freedom*, aimed to root out and destroy al-Qaeda and any group harboring them; and for the first time in its history, NATO invoked Article 5—the principle of collective defense. This act of solidarity firmly united the U.S. and its allies against global terrorism. From this initial special operations mission, the war on terror morphed into "the longest war" in American history, a conflict that spanned twenty years and four presidencies—*five* including President Clinton's.

The War on Terror eventually expanded into the U.S. invasion of Iraq to defeat Saddam Hussain and stabilize the country for the Iraqi people. However, the mission degenerated into a quagmire of insurgency that ultimately thwarted U.S.-led nation-building efforts. In 2011, President Obama ended the war in Iraq to shift U.S. attention to the "good war"[1] in Afghanistan, which itself was on the verge of becoming a quagmire. President Obama deployed more than 100,000 troops (at its peak, there were 130,000 NATO and other allied forces from 43 countries)[2] to eradicate the Taliban insurgency and end the war. That same year, Osama bin Laden was killed by U.S. special operations forces at his compound in Abbottabad, Pakistan. This victory against terror, however, did not accomplish the end goal. Osama bin Laden was swiftly succeeded by his deputy, Ayman al-Zawahiri, and the Afghan War continued.

[1] Landler, Mark, "The Afghan War and the Evolution of Obama", January 1, 2017, https://www.nytimes.com/2017/01/01/world/asia/obama-afghanistan-war.html.

[2] Shadi Khan Saif, "9,592 troops of 36 countries serving in Afghanistan" *Anadolu Agency,* April 17, 2021, https://www.aa.com.tr/en/asia-pacific/9-592-troops-of-36-countries-serving-in-afghanistan/2212251.

When President Trump took office, he called the war in Afghanistan a "total disaster "[3] and, like his predecessors, vowed to destroy al-Qaeda. Yet even President Trump's deployment of the "mother of all bombs," considered one of the largest non-nuclear weapons in use by the U.S. military, did not achieve that end. Instead, in a surprise move in February 2020, President Trump directly negotiated with the Taliban and signed the Doha Agreement to end the Afghan war. The Doha Agreement called for the complete withdrawal of all American troops by August 30, 2021. When President Biden inherited the agreement in early 2021, he chose to honor the terms of the complete withdrawal. Both administrations seemed to understand that eradicating terror and establishing durable stability and peace in Afghanistan could not be achieved through tactical means alone. As if to underscore this point, on August 15, 2021, the same Taliban that had been so easily ousted in 2001, marched victoriously back into the capital city of Kabul and seized the entire country before American troops had even departed.

Thus, the U.S. war in Afghanistan ended tragically amid frenzied evacuation efforts at Kabul airport. The world watched in disbelief as American and NATO troops departed, leaving behind crowds of terrified men, women, and children desperate to accompany them. Scene after shocking scene emerged of individuals clinging to C-17 aircraft in an attempt to escape, only to fall to their deaths as the planes became airborne.[4] In the face of such unabated horror, the Afghan War was immediately deemed a "strategic failure,"[5] an admission that drew a painful parallel to another "unmitigated failure"[6] in U.S. history—the Vietnam War.

Both wars challenged the U.S. beyond the battlefield with issues such as "western-style" nation-building, insurgencies, regional geopolitics, and illegal narcotics trade. How did two separate U.S. wars and nation-building efforts yield equally unfortunate outcomes, as if no lessons had been learned and applied in the five decades that separated them, despite substantial technological advancements and other military successes? This, and other urgent questions regarding U.S. efforts in Afghanistan, must now be earnestly addressed.

How is it possible that after all the blood and treasure, a staggering loss of life, and nearly two trillion spent—the U.S. faces many of the same challenges to its national security and global standing as before 9/11? How could this U.S.-led, global mission fail to yield the desired outcome, despite protracted efforts to ensure victory? Could the twenty years of hard-fought gains have been preserved and a catastrophic withdrawal avoided through more strategic patience, planning, and execution? What about the impact on those left behind, who served, and died, to achieve the goal

[3] Diaz, Daniella, "A history of Trump's thoughts on Afghanistan" CNN, August 21, 2017, https://www.cnn.com/2017/08/21/politics/history-president-trump-remarks-afghanistan-tweets.

[4] Porter, Tom, "Photos and videos appear to show Afghans trying to flee the Taliban falling out of planes as they leave Kabul airport" Business Insider, August 16, 2021, https://www.businessinsider.com/kabul-airport-people-fall-from-planes-leaving-afghanistan-videos-photos-2021-8

[5] Lubold, G. and Youssef, N.A. "Gen. Milley calls Afghan withdrawal 'strategic failure' in heated senate hearing" *The Wall Street Journal*, September 28, 2021, WSJ.com

[6] Gentile, G.P. "Vietnam: Ending the Lost War", *Between War and Peace: How America Ends Its Wars*, Matthew Moten (ed) pg 260 (259–280)

of eradicating the terror responsible for 9/11? What of the millions of Afghans abandoned and left on the brink of starvation? Did we adequately plan for the safety of those who supported U.S. efforts and fulfill the promises made to them, especially the Afghan women? Why did the U.S. ultimately fail to "win the hearts and minds" of the Afghan people? Finally, has abandoning Afghanistan irrevocably damaged U.S. global credibility, influence, trust, and the equilibrium of Great Power Competition in the region?

These are just some of the compelling questions the authors address in this book. The long and tumultuous course of America's Longest War in Afghanistan demands further rigorous academic scrutiny to better inform future strategic engagements, missions, and post-conflict reconstruction efforts. We must especially understand the vital interplay between geopolitical, security, economic, religious, cultural, and ethnic dimensions that critically impacted Afghan War efforts and which continue to affect the modern Great Power Competition today. China and Russia are moving quickly to fill the region's void and shift the balance of power on the geopolitical and economic battlefield of the Central Region. *Lessons from Afghanistan: America's Longest War* offers forward-looking suggestions that incorporate key lessons learned over the past two decades, to better inform U.S. re-engagement in this vital region.

Only when these lessons are learned and applied in earnest to create a new way forward can original U.S. objectives in Afghanistan be achieved and present-day obligations to a nation in crisis be met. Only then can "the longest war" in American history be considered a success.

Dr. Adib Farhadi, Ph.D.
University of South Florida
Tampa, USA

Acknowledgments

The editors would like to extend their sincerest gratitude to the entire University of South Florida and National Defense University Near East South Asia Center for Strategic Strategies for their work on the Great Power Competition Initiative and the resulting edited volume.

A most heartfelt thanks to Arman Mahmoudian, Jeanie LeGendre, Kathleen Whitaker, Sarah T. White, Andrew Roberts, and the contributing authors for making this book possible.

Dr. Adib Farhadi would like to thank his wife, Elaha, children, Adam, Sophia, and mother Maliha, for their support, patience, and love, not only in regard to this project but in everything that life brings.

Contents

About the Editors

Dr. Adib Farhadi is Assistant Professor and Faculty Director of Executive Education at the University of South Florida. His research, which is at the intersection of religion, politics, economics, and conflict with a particular focus on the "Silk Road" Central and South Asia (CASA) Region, is the subject of his recent book, *Countering Violent Extremism by Winning Hearts and Minds*. He is Editor and Chief for The Great Power Competition book series and Director of the Great Power Competition Initiative. Formerly, Dr. Farhadi served in senior positions for Afghanistan and extensively advised the U.S. government and various other international organizations. Dr. Farhadi earned his Ph.D. in Economics from the University of Canberra, M.A. from New York University, and B.S. from East Carolina University.

Dr. Anthony Masys is Associate Professor and Director of Global Disaster Management, Humanitarian Assistance and Homeland Security. Former Senior Air Force Officer, Dr. Masys, has a B.Sc. in Physics and M.Sc. in Underwater Acoustics and Oceanography from the Royal Military College of Canada and a Ph.D. from the University of Leicester. He is Editor in Chief for Springer Publishing book series: Advanced Sciences and Technologies for Security Applications and holds various advisory board positions with academic journals and books series. Dr. Masys is Internationally Recognized Author, Speaker, and Facilitator and has held workshops on security, visual thinking, design thinking, and systems thinking in Europe, Canada, South America, West Africa, and Asia. He has published extensively in the domains of physics and the social sciences. Dr. Masys supports the University of Leicester (U.K.) as Associate Tutor in their Distance M.Sc. Program on Risk Crisis and Disaster Management.

Lessons Learned from Afghanistan: The Heavy Price of Treating Long-Term Strategic Issues with Short-Term Tactical Approaches

Adib Farhadi

Abstract How did the U.S. "war on terror" start with the swift defeat of the Taliban post-9/11, only to end with the very same Taliban recapturing Afghanistan, as if twenty years of global efforts, sacrifices, achievements, and countless lives lost had never happened? *Lessons from Afghanistan: America's Longest War* gathers prominent scholars, experts, leaders, and decision-makers from around the globe to answer this and other vital questions. The authors in this volume argue that lessons learned in Afghanistan reveal why U.S.-led global efforts yielded unwanted outcomes, despite all the precious lives lost and nearly two trillion spent to ensure victory. Based on decades of academic research and practitioner experience, the authors in this collection of essays assert that we can and must utilize these lessons to inform future U.S. foreign policy to effectively compete in the modern Great Power Competition. This chapter opens the discussion by suggesting that a lack of U.S. strategic patience, alongside missed opportunities to develop a sustainable economy based on natural resources in Afghanistan, ultimately undermined two decades of heartfelt U.S. military assistance and resulted in the failure to "win the hearts and minds" of ordinary Afghan people. This critical misstep effected the final blow to top-down "western-style" nation-building efforts—levying a heavy cost to the U.S. by weakening national security and influence in the region and positioning in the broader Great Power Competition. Further, the chaotic U.S. withdrawal from the Kabul airport and subsequent Afghan humanitarian crisis resurrected memories of similar failed "western-style" nation-building efforts in Iraq and Vietnam. Global perceptions of these unfortunate failures indicate a lack of strategic patience that perpetuates claims of declining U.S. influence. This not only impacts U.S. national security but its allies and all those around the world who still hold true to the values of "liberty" and "freedom." In an era of shifting geopolitics, in which China and Russia vie for global hegemony, the U.S. must swiftly re-assess the effectiveness of its foreign policy and operational strategies in the region. Such an honest examination offers realistic hope for the U.S. to strengthen its credibility and influence in the region through reinvigorated resource-based economic facilitation in Afghanistan.

A. Farhadi (✉)
University of South Florida, Tampa, FL 33620, USA
e-mail: Farhadi@usf.edu

A. Farhadi and A. Masys (eds.), *The Great Power Competition Volume 4*,
https://doi.org/10.1007/978-3-031-22934-3_1

Keywords Great power competition · Afghanistan · Taliban · Doha Agreement · War on terror · Central Asia · Insurgency · National security · Nation-building · United States

1 Introduction

The swift collapse of the U.S.-backed Afghan Government on August 15, 2021, stunned the international community as "Taliban" fighters ripped through the country with an unprecedented lightning offensive. Afghan troops crumbled with little resistance, and the capital city of Kabul fell within hours. The striking Taliban coup unfolded with an eerie efficiency, mirroring the U.S.-led expulsion of this very same group almost nineteen years earlier—except this time, no shots were fired by the withdrawing U.S. and allied forces. Incredibly, zero resistance was offered. As black and white flags were raised throughout the city, signaling the triumphant return of the Islamic Emirate of Afghanistan ("Taliban"), widespread panic erupted among Afghans and expatriates, many of whom had worked with allied forces during the war—now fearing retribution. The fall of Kabul heralded an explosive transition of power that was to become a full-blown international evacuation crisis.

The chaotic evacuation of U.S. troops and allied civilians from Kabul airport continued over several long days as the world watched in disbelief. When General Chris Donahue, the last U.S. soldier to leave Afghanistan, boarded an outbound C-17 cargo jet under the "dark of night," his departure unceremoniously ended the "longest war" in American history [6]. General Mark Milley, Chairman of the Joint Chiefs of Staff, called the war in Afghanistan a "strategic failure." After two decades of hard-fought gains, America's long-standing war in Afghanistan came to an ignominious end, leaving behind thousands of grief-stricken and frightened Afghans and allies, vulnerable to the vicissitudes of the Taliban government.

Evidence of how quickly and easily Afghanistan could revert to a haven for violent extremists surfaced just shy of the one-year anniversary of the U.S. evacuation of troops, when the leader of al-Qaeda, al-Zawahiri, was discovered in a private residence in Kabul, in defiance of the Doha Agreement. On July 31, 2022, President Biden announced that a U.S. drone strike had killed the al-Qaeda leader [2]. The Taliban denied knowledge of al-Zawahiri's presence in the capital city and countered with accusations of U.S. agreement violations.

Polarizing events such as these not only place additional stress on the Taliban interim government, as it struggles to stabilize the nation and establish diplomatic relations within the international community, but also highlight the dangers of a destabilized Afghanistan. Marjorie Eastman aptly describes the global fallout of the U.S. decision to leave Afghanistan when she writes, "The reflection of that epic failure leaves little room for exaggeration on the ripple effect that took place over the last year. With the Taliban's swift overthrow of the government, evil retribution ensued for American allies left behind enemy lines, to the loss of trust in American leadership and hegemony" [4].

The gravity of the events surrounding the U.S. invasion, occupation, and withdrawal from Afghanistan demand prudent and swift analysis of the strategic flaws and successes that powered the "longest war" and ended it so precariously. Were there, in fact, alternative approaches that could have preserved the "hard-fought" gains and achievements of two decades of American nation-building, while better demonstrating America's honor, humanity, and fundamental core values for which it stands? These and other hard questions must now be addressed to extract vital lessons that can inform U.S. policy in post-war Afghanistan and other conflict-affected countries. Identifying what went wrong in Afghanistan and what went right will be critical as well to situating future U.S. responses within the broader context of the modern Great Power Competition (GPC).

Lessons from Afghanistan: America's Longest War gathers prominent scholars, experts, leaders, and decision-makers from around the globe to address these challenging questions. From decades of research and experience, the authors in this collection of essays argue that the U.S. must utilize lessons from the War in Afghanistan to inform U.S. foreign policy; first, to honor our promises to the Afghan people and to the victim of 9/11; second, to compete more effectively in the modern Great Power Competition; and third, to assess the efficacy of top-down "western-style" nation-building. This chapter opens the discussion by suggesting that a lack of U.S. strategic patience and missed opportunities to develop a sustainable economy based on Afghanistan's vast natural resources, resulted in a failure to "win the hearts and minds" of the Afghan people—effecting the final blow to U.S. efforts in Afghanistan.

Specifically, a radically centralized and tactical economic development strategy, focused on short-term transactional aims and enacted mainly from within the "Kabul bubble," neglected the needs of ordinary Afghans living outside of urban centers [18]. As it became increasingly obvious to these populations, in desperate need of jobs, that economic development was aimed at urban-centric projects, a mistrust of "western intentions" spread, particularly among rural communities that comprise the bulk of the Afghan nation. This pervasive mistrust ultimately undermined the confidence of Afghan troops, who, in the end, were unwilling to fight for a foreign cause they could no longer believe in [19]. Short-term tactical solutions to strategic problems not only hindered the realization of long-term U.S. objectives and contributed to the present Afghan humanitarian crisis but left an opportune void for rival powers China and Russia to fill [8].

Thus, it was a lethal trifecta of top-down, centralized development strategies, missed opportunities for cooperation with other regional actors, and a lack of U.S. "strategic patience" that ultimately destroyed two decades of hard-fought gains in Afghanistan. According to the recent SIGAR report, *Why the Afghan Government Collapsed*, "The U.S. government consistently underestimated the amount of time required to rebuild Afghanistan and created unrealistic timelines and expectations that prioritized spending quickly. These choices increased corruption and reduced the effectiveness of programs" [19]. The last twenty years have only reinforced past lessons that such short-term, transactional strategies for "nation building" simply do not work in devastated conflict-affected regions like Afghanistan. Conversely, there is ample historical evidence for the efficacy of long-term constructivist approaches that employ strategic patience, such as the U.S. Marshall Plan enacted during post-WWII reconstruction.

As a fragile Afghanistan continues to suffer an unprecedented humanitarian crisis including mass starvation—the U.S. enters a new era of Great Power Competition in the Central Asia-South Asia (CASA) region. If the U.S. ultimately fails to meet its prior obligations to the Afghan people during this time of adversity, the resulting injury to U.S. global influence could be incalculable. China will continue to capitalize on fragile post-war conditions and anti-U.S. rhetoric in the region through its Belt and Road Initiative, leaving the U.S. significantly weakened and unable to compete successfully on the global playing field.

This chapter examines how twenty years of missed opportunities and short-term tactical solutions in Afghanistan undermined reconstruction efforts and cost the U.S. precious assets and positioning in the region and the wider GPC. While this chapter aims to highlight the specific flaws and missteps and their resulting consequences, it also seeks to offer realistic hope that these lessons can be parlayed into reinvigorated U.S. engagement in Afghanistan, through strategic economic cooperation, intervention, and facilitation.

2 Missed Opportunities: A Failure to Capitalize on Afghanistan's Vast Natural Resources

Throughout history, Afghanistan's geostrategic location has proven vital to great powers. Its wealth of natural resources, particularly critical minerals, should render this landlocked nation a natural focal point in the CASA region for U.S. strategic cooperation, consensus building, and resource sharing. Yet, the theory of the "resource curse," furthered by international development institutions, alongside logistical and security challenges, has deterred investors from developing the Afghan mining sector. During the last two decades, the U.S. failed to strategically develop and adequately capitalize on the Afghan mineral sector. According to a recent SIGAR report, "US efforts in Afghanistan's extractives industry presented one of the greatest opportunities for the Afghan government to generate royalty income, grow its economy, and provide meaningful employment" [17]. Further, "SIGAR found that U.S. agencies did not perform required oversight of the programs and that the programs did not meet their goals due, in part, to the same challenges that plagued previous U.S. efforts in the sector" [17]. The report also suggests that the "USAID, [U.S. Geological Survey] (USGS), and [Commercial Law Development Program Department of Commerce] (CLDP), did not conduct required program oversight for [Extractives Technical Assistance] (ETA) and [Multi-Dimensional Economic and Legal Reform Assistance] (MELRA). Specifically, USAID and USGS did not develop a Monitoring, Evaluation, and Learning (MEL) plan for the ETA program, including performance indicators, as USAID's Automated Directives System (ADS) required" [17]. According to economic expert Shabir Bashiri, this was exacerbated by the fact that, "in the 20 years of US presence, there was no developmental and infrastructural strategy or plan" [1].

These reports support the central claim of this chapter that though the U.S. invested ample resources in technical capacity building, it was unable to fully develop the Afghan economy due to its tactical and overly complicated bureaucratic approach. Alongside, or even *before*, tactical and technical aspects of Afghan resource development were addressed, essential transport infrastructure, required for regional market connectivity and trade facilitation, should have been developed. Thus, the U.S. missed vital opportunities to capitalize on Afghan natural resources—not due to a lack of technical capacity-building projects, but rather, a lack of strategic vision. Rest assured, Russia and China will not forgo these same opportunities now that the U.S. has left Afghanistan [3].

As the international community observed the harrowing events leading up to and during the U.S. withdrawal from Afghanistan in late 2021, China stepped in and acted, quickly filling the void left by the U.S. Continued lack of U.S. engagement in the region is detrimental, not only to Afghanistan but also to U.S. national security and geopolitical competition. Allowing China to gain preferential access to Afghan natural resources such as lithium, when it already possesses the critical and rare earth minerals near monopoly, will gravely shift geopolitical power in China's favor, given the current momentum of its existing Belt and Road Initiative (BRI).

Strategic minerals like lithium are essential to industry, military, and civilian uses and have no viable substitutes. For nearly two decades, U.S. intervention in Afghanistan kept the BRI from any serious attempts to claim the untapped resources of this landlocked nation. However, with the U.S. exempted and the region's geopolitical climate changing, Afghanistan, with its vast wealth of natural resources, is again within China's orbit. At this juncture, supplying weight to a rival power's almost complete monopoly of critical resources only impairs the GPC equilibrium further.

China, the largest hegemon in the region, is particularly eager to fill the vacuum left by the U.S. in Afghanistan. This eagerness may be seen in the speed with which China offered humanitarian assistance to the Taliban, the first foreign power to do so, thus establishing itself as a "principal partner" of the de facto Afghanistan government. The relationship continues to expand, with China's ambassador, Wang Yu, publicly emphasizing the great power's strategic approach, "as Afghans are good neighbors, brothers, and partners, China will stand firm on the concept of community with the shared future for mankind, guided by the global development initiative and global security initiative. China will substantially enhance bilateral cooperation in all areas" [5]. Only six days into 2023, China began to fulfill its recent promise as, "A Chinese company signed a $540 million deal with Afghanistan to develop an oil-and-gas field, as Beijing moves to secure access to the country's vast mineral wealth after the exit of American troops" [15].

Concrete evidence of such enhanced cooperation in the region is most apparent in China's BRI regional integration strategy that currently encompasses 146 nations. This transcontinental collaboration aims to connect the region in all directions to increase trade and economic growth. As involvement in the BRI continues to expand, the possibility of the U.S. being shut out of a region that possesses vast critical and rare earth minerals becomes an alarming reality. China already claims the overwhelming monopoly of these essential resources, which presents a challenge to our national security.

Further, in 2019, Chinese President Xi Jinping went on record to assure Russian President Vladimir Putin that China was "ready to go hand in hand" with Russia. Despite Russia's 2022 invasion of Ukraine, strong alliances between the two countries are likely to continue. News agencies for both countries reported that this alliance would "almost double their trade over the next five years, hitting $200 billion by 2024 compared to $107 billion in 2018, by implementing collaborative projects in energy, industry, and agriculture [in the region]" [16]. Like China, Russia recognizes an opportunity for a stronger foothold in CASA to reinforce its own geopolitical positioning and the vital need for critical resources. Russia currently seeks foreign investments of "1.5 billion to develop further and expand its own rare earth mineral industry to become the second largest rare-earth producer after China by 2030" [14].

Like China, Russia recognizes the geostrategic importance of Afghanistan. Following the fall of Kabul, President Putin quickly detailed terms for recognizing the legitimacy of the Taliban along with its resolve to "act regardless of what the United States and everybody else may think" [7]. Further, the U.S. must consider the regional impacts of Russia's invasion of Ukraine. The attack "sent a shock wave across all [of] Central Asia, as indeed it did to all of the new states that were formerly part of the USSR, which incited a response from regional powers that shifted the geopolitical landscape further" [20].

If there is any remaining doubt about the geostrategic value of Central Asian nations, the U.S. need only look at recent history. The countries that were once Soviet states have been targeted as "Russia, with steady persistence, has tried to lure them [the five CIS nations] back into its sphere of influence, if not of direct control, through economic and security alliances." Additionally, the U.S. and Europe have worked to develop them as market economies and implant civil society and democratic institutions there. China, meanwhile, has assigned them key roles in its Belt and Road Initiative and loaned them billions to develop economic strengths that complement China's own" [20].

From this perspective, the modern GPC in the CASA Region constitutes a competition for access to a finite supply of critical and rare earth minerals; with Afghanistan quickly becoming ground zero for its wealth of untapped natural resources. As rival powers China and Russia already have a distinct advantage, the U.S. could be summarily excluded from the massive economic development already underway in CASA. Yet, "strategic competition," as first put forth by President Biden in 2021, offers hope for renewed U.S. engagement in fragile and post-conflict regions like Afghanistan—that can be aimed squarely at resource development and economic integration through great power cooperation. The 2021 National Security Strategic Guidance expressly states, "strategic competition does not, and should not, preclude working with China when it is in our national interest to do so."

Terms such as "geopolitical competition" and "transformative cooperation" figure prominently in the latest National Security Strategy as well, committing the U.S. to outmaneuver China and Russia's expansionist ambitions in places such as Central Asia and Afghanistan. The 2022 NSS further emphasizes the role of strategic cooperation, iterating, "to preserve and increase international cooperation in an age of competition, we will pursue a dual-track approach. On one track, we will cooperate with any country, including our geopolitical rivals, that is willing to work constructively with us to address shared challenges. We will also fully engage with, and work to strengthen, international institutions."

In particular, the NSS commits the U.S. to seek these areas of cooperation with competitors such as China from a position of strength. To that end, the U.S. can most effectively compete by fashioning a foreign policy in the region based on soft power and trade-oriented economic strategies. Now that the U.S. has abandoned kinetics-based operations in Afghanistan, it can effectively apply lessons learned over the past twenty years to transition to strategic economic competition in the region. Further, the U.S. can play a vital role as a convener and peacebuilder in CASA to increase its national security and strengthen its overall positioning, specifically by facilitating public–private partnerships that support the extraction and commercialization of Afghan minerals. Critically, acting as such can regain the trust and participation of local Afghan populations, who are crucial to sustainable development and will benefit immediately from local mining industry job growth.

However, if the U.S. remains on its current policy trajectory in Afghanistan or is too slow to engage, alternative future scenarios are likely grim. The probable outcome is China's utter domination of the region, including preferential access to critical minerals such as lithium. Lessons learned from failed Afghanistan reconstruction efforts highlight the need for greater U.S. cooperation, economic intervention, and facilitation among and with regional stakeholders to produce and bring Afghan minerals to market most effectively. Over the past twenty years, failure to do so has resulted in a diminished position on the GPC playing field in the CASA region and new challenges to U.S. national security.

3 A Devastating Misstep: A Failure to Win Hearts and Minds

Since the withdrawal of U.S. troops in 2021 and the resulting powershift, Afghanistan has struggled with a collapsed economy, humanitarian crisis, and severe drought that has highlighted and amplified the lack of sustainable development in Afghanistan. Afghans lack sufficient food, education, healthcare, jobs, and essential infrastructure. These factors have rendered the country extremely vulnerable to violent extremist organizations (VEOs) and the proliferation of the illegal drug trade. Decades of traditional aid from the U.S. have not succeeded in ameliorating these woes, and the consequences radiate outward to affect global security and stability. In large part, this was due to radically centralized development strategies that spawned a lack of trust among local populations and failed to "win the hearts and minds" of ordinary Afghan constituents. As rural communities became increasingly aware that they would be left out of significant reconstruction projects, U.S. and NATO intentions were called into question, and the critical participation of local Afghan entities waned.

This section explores how missed opportunities to build trust and stabilize Afghanistan through regional trade facilitation jeopardized the future of Afghanistan and U.S. national security, by allowing rival powers to capitalize on vulnerabilities. The central lesson here, is that durable economic development in Afghanistan cannot take hold without the cooperation and participation of great powers and local entities alike. As this section highlights, Afghanistan has been dependent on foreign aid since the nineteenth century, when the British provided large subsidies to Afghan Emirs. Afghanistan's debilitating dependence on foreign aid has extended to the present day, with the U. S.'s expenditure of over a trillion dollars on overall aid to Afghanistan from 2002 to 2021. This dependence on foreign aid, with its influx of cash and a lack of "accountability to its citizens as well as a lack of oversight from its donors," created a rentier state that Fryland describes as "fueling a culture of corruption never previously seen in Afghanistan."

Missed opportunities to successfully integrate Afghanistan into the global economic ecosystem undermined the nation's potential to become the "Heart of Asia." Conversely, the successful framing of a new narrative in Afghanistan as the heart of a modern silk road ecosystem could have eased the economic impact of U.S. military withdrawal and allowed for the gradual decrease of international aid. Such an occurrence would have allowed the U.S. to shift its vital resources to geopolitical competition with Russia and China. Instead, following the 2021 U.S. withdrawal, the nation has devolved into the "black hole" of Asia, where narcotics trafficking and opium production increasingly dominate its landscape—threatening the entire region and global community once again.

While modern Afghanistan's many challenges can be traced back to its landlocked geography, central location, and isolation, these very features might have become the basis for this nation's transformation into the nexus of a budding Silk Road trade ecosystem. Instead, the U.S. seems to have hit replay on its unsavory legacy in Vietnam. Did the U.S. earnestly compare twenty-first-century nation-building goals in Afghanistan with the unwanted historical outcomes of Vietnam? Did the U.S. take those hard-won experiences into Afghanistan? Surely there were lessons learned from strategies that served the U.S. well and those that did not. Such defining questions must be honestly addressed, as any hope of redirecting the present U.S. legacy in Afghanistan is quickly passing, while the Central Region recollects itself and great powers and regional actors seize new opportunities to benefit their individual interests.

Any reasonable strategic response based on sound foreign policy demands understanding what went wrong with the approach to nation-building in Afghanistan. Most critically, we must ask if top-down "western-style" nation-building in a landlocked country of such diverse peoples was sufficiently aimed at serving the Afghan people. Was there a genuine understanding of the fundamental needs of local populations, or were reconstruction efforts modeled on a particular idea of what this nation should look like? Were sufficient efforts made to elevate this nation from its failed status by connecting it to its neighbors? Finally, did the U.S. forge enduring relationships that will sustain us as China and Russia move in to fill the vacuum? The U.S. must engage these pressing questions from a strategic and humanitarian perspective equally. Answers to such questions about the War in Afghanistan will not only critically impact U.S. future foreign policy but influence U.S. future diplomatic relations on the world stage.

If the U.S. approaches the war in Afghanistan from a broader vantage, it may discover that the key is in fact *perspective.* Afghanistan was not necessarily approached as a strategic long-ranging initiative but principally a *tactical* mission, essentially fought one year at a time. As with Vietnam, the U.S. relied primarily on superior military might and resources to accomplish mission objectives, and, as with Vietnam, it did not fully achieve its intended goals. Like Vietnam, the chaotic exit from Afghanistan represents the unsettling reality that the ideals we value so highly in the West may not always translate to other cultures. According to McNamara and VanDeMark, "in Vietnam, we acted according to what we thought were the principles and traditions of this nation … we were wrong, terribly wrong" [11]. Can the same be said about Afghanistan? This reality should serve as a reminder that the world cannot be viewed through a singular ideological lens. In today's interconnected and increasingly complex global landscape, we must be especially cognizant of the larger context.

Indeed, the U.S. entered Afghanistan with the noblest of intentions to help the Afghan people by providing generous humanitarian and military assistance. The U.S. staved off imminent mass starvation, built clinics, hospitals, and schools for children throughout the country, and bolstered rights for Afghan women. In the first years of the war alone, the U.S. vaccinated 4.26 million children against measles and polio, likely preventing 20,000 deaths. It rehabilitated and built highways, roads, bridges, airports, irrigation networks, and courthouses. It initiated potable-water supply projects and municipal water systems in major cities. It brought electricity to rural villages and cities alike. It entirely rebuilt the Afghan national security forces and funded local and national elections. The list of aid, progress, and achievements goes on. Yet, despite these efforts, the U.S. still did not manage to entirely win over the ordinary Afghan people. Why? Why did these efforts fail to transform the Afghan nation into the western vision of a stabilized way of life? The U.S. must ask if it fully understood the practical needs and goals of the Afghan people, and if it was sufficiently committed to seeing the task of nation-building in such a challenging and devastated country through to full fruition.

U.S. efforts in Afghanistan focused largely on various beneficial short-term projects without first creating a strong foundation for economic durability. Initiatives for essential infrastructure projects, which were, and still are, desperately required for market connectivity, were not implemented sufficiently. According to Shah, "American experts a decade ago estimated the value of Afghanistan's mineral resources at $1 trillion, which include rare-earth minerals now used in electric cars. While war raged, this potential wealth was never [effectively] exploited [15]. More critically, most development projects were oriented toward urban centers and failed to benefit ordinary Afghans nationwide with much-needed jobs. In truth, the U.S. vision of victory, namely, the acculturation of "western" ideals, is only one version of what victory in Afghanistan might have looked like.

True nation-building is never a straightforward affair. Winning the hearts and minds of a diverse population of people requires nuanced strategic vision and a bird's eye perspective of needs and wants. Vietnam and Afghanistan both offer ample proof of this. The U.S. largely overlooked that Afghanistan is a landlocked country, a geographic feature that makes the nation dependent upon its neighbors. Could facilitating regionalist economic alliances have better strengthened Afghanistan against insurgency? The key question now is how the U.S. move forward in Afghanistan. More, how does the U.S. productively re-engage to fulfill its original, vital mission there?

Currently, China offers the greatest prospect for Afghan regional integration through its Belt and Road Initiative. However, the U.S. can still compete and cooperate effectively in the region by utilizing its "convening power" to develop sustainable resource development and economic alliances between Afghanistan and its neighbors. In so doing, the U.S. can ostensibly safeguard preferential access to critical Afghan resources, maintain a balance of power in the region, and capitalize on the fact that Afghanistan is likely "the safest it has been in decades to develop Afghan mining and oil projects" [15]. Finally, such positive re-engagement, framed by lessons learned over the past two decades, could win back the hearts and minds of the Afghan people—who, against all odds, still hope for a future that offers employment, healthcare, education opportunities, and sustainable economic autonomy.

The following from George Marshall's famous speech in 1947 denotes the heart of durable reconstruction strategies—"the remedy lies in breaking the vicious circle and restoring the confidence of the European people in the economic future of their own countries and of Europe as a whole. The manufacturer and the farmer throughout wide areas must be able and willing to exchange their products for currencies the continuing value of which is not open to question." Marshall's astute observations on the restorative power of economic self-reliance could just as aptly be applied to the Afghan people today—demoralized and broken after decades of conflict.

4 A "Strategic" Way Forward

Hope for Afghanistan and the region lies in establishing durable economic integration fueled by shared prosperity. If any lesson should be learned from nearly two decades of war in Afghanistan, it is that this volatile and fragile region requires great powers and regional actors to cooperate on shared challenges while competing on many nuanced levels to achieve such ends. Sustainable geopolitical competition today appears more relational than ever, and isolationist strategies no longer offer a safe haven in today's profoundly interconnected world. Self-assessment regarding America's "longest war" is imperative because the missteps and challenges are not exclusive to U.S. engagement in Afghanistan. The discussions presented in this chapter, and the following chapters of this volume, can help guide U.S. strategies to navigate the modern GPC amidst shifting power structures. Ideally, lessons learned during America's "longest war" will inform policymakers at home and abroad to develop more holistic approaches, including strategies that serve all populations, even those who may not share similar value systems.

Further, this chapter has highlighted that the U.S. can still rebuild a meaningful presence in Afghanistan to facilitate regional stability through cooperative economic intervention. Establishing such a presence will require a willingness to engage in strategic cooperation with rival powers, like China, to effectively connect Afghanistan with regional and global markets. Multilateral cooperation is the first step in developing a stable and prosperous Afghanistan—free from its daunting legacy of crisis and aid dependency.

Finally, we must be willing to embrace the hard lessons learned in Afghanistan, honestly, confront what must be changed, and then be willing to act. U.S. foreign policy and geopolitical tactics must be reanalyzed to ascertain what works in this interconnected, twenty-first-century global environment. Afghanistan's unmitigated crisis of aid dependency continues to destabilize the entire region, urgently calling for the U.S. and other great powers to work in concert to unlock this landlocked nation's full potential as the "Heart of Asia."

When viewed through the geopolitical lens of great power competition, it becomes abundantly clear that Afghanistan and the wider region pose strategic challenges that cannot be effectively met with short-term tactical solutions. After all, as noted by Starr and Farhadi in *Finish the Job: Jumpstart Afghanistan's Economy*, "… if the economy stabilizes and resumes a path of sustainable growth, levels of political and security tensions will subside. Thus, economic development in Afghanistan is not something to be pursued after political stability and security have been established; rather, it is what must be achieved in order to forge political stability and communal peace" (2012, p. 6).

References

1. Ariana News (2023) Afghanistan's mining industry never achieved its potential, despite US efforts. SIGAR, Jan 10. Available: https://www.ariananews.af/afghanistans-mining-industry-never-achieved-its-potential-despite-us-efforts-sigar/
2. Biden J (2022) Remarks by President Biden on a successful counterterrorism operation in Afghanistan. The White House, Aug 1. Available at: https://www.whitehouse.gov/briefing-room/speeches-remarks/2022/08/01/remarks-by-president-biden-on-a-successful-counterterrorism-operation-in-afghanistan/
3. Blumenthal et al (2022) Chinese investment in Afghanistan's lithium sector: a long shot in the short term. Brookings Online, Aug 3. Available at: https://www.brookings.edu/blog/up-front/2022/08/03/chinese-investment-in-afghanistans-lithium-sector-a-long-shot-in-the-short-term/
4. Eastman MK (2022) Vietnam and Afghanistan: two generations, two wars, one aftermath. Fayetteville Observer, Sept 1. Available at: https://www.fayobserver.com/story/opinion/2022/09/01/vietnam-and-afghanistan-two-generations-two-wars-one-aftermath/7955378001/

5. Ellyatt H (2021) China's Xi calls Putin his "best friend" against a backdrop of souring U.S. relations. CNBC, June 5. Available at: https://www.cnbc.com/2019/06/05/putin-and-xi-meet-to-strengthen-ties-as-us-relations-sour.html
6. Fieldstadt E (2021) The last soldier to leave Afghanistan—nicknamed 'Flatliner'—was uniquely prepared for that moment. NBC News, Sept 1. Available at: https://www.nbcnews.com/news/world/last-soldier-leave-afghanistan-nicknamed-flatliner-was-uniquely-prepared-moment-n1278260
7. Gul A (2022) Russia says 'inclusive' Afghan Government key to recognizing Taliban. VOA News, June 15. https://www.voanews.com/a/russia-says-inclusive-afghan-government-key-to-recognizing-taliban-/6618812.html
8. Hanif S (2021) Minerals, drugs, and China: how the Taliban might finance their new Afghan government. The Conversation. Available at: https://theconversation.com/minerals-drugs-and-china-how-the-taliban-might-finance-their-new-afghan-government-167169
9. Hirose K, Imai K, Lyall J (2017) Can civilian attitudes predict insurgent violence? Ideology and insurgent tactical choice in civil war. J Peace Res 54(1):47–63. https://doi.org/10.1177/0022343316675909
10. Lyrichkova A, Stolyarov G (2020) Russia has $1.5 billion plan to dent China's rare earth dominance. Reuters. Available at: https://www.reuters.com/article/russia-rareearths/russia-has-15-billion-plan-to-dent-chinas-rare-earth-dominance-idUSL8N2F73F4
11. McNamara RS, VanDeMark B (1995) In retrospect: the tragedy and lessons of Vietnam. Random House, NY
12. Ott M (2021) Afghanistan: echoes of Vietnam? Wilson Center Research Institute, July 13. Available at: https://www.wilsoncenter.org/blog-post/afghanistan-echoes-vietnam
13. Price G (2021) Why Afghan nation-building was always destined to fail. Chatham House, Sept 10. Available at: https://www.chathamhouse.org/2021/09/why-afghan-nation-building-was-always-destined-fail
14. Scheyder E (2019) Exclusive: pentagon races to track U.S. rare earths output amid China trade disputes. Reuters. Available at: https://www.reuters.com/article/us-usa-rareearths-pentagon-exclusive/exclusive-pentagon-races-to-track-us-rare-earths-output-amid-china-trade-dispute-idUSKCN1U727N
15. Shah S (2023) Chinese firm signs $540 million oil-and-gas deal in Afghanistan; Beijing is expanding its presence in Central Asia after the U.S.'s exit from Afghanistan, with an eye on mineral riches. Wall Street Journal, Jan 5, New York, NY
16. Smith E (2019) Russia does not want to exploit US-China trade war, finance minister says. CNBC, Oct 19. Available at: https://www.cnbc.com/2019/10/19/siluanov-russia-does-not-want-to-exploit-us-china-tensions.html
17. Special Inspector General for Afghanistan Reconstruction (2023) Afghanistan's extractives industry: U.S. programs did not achieve their goals and Afghanistan did not realize widespread economic benefits from its mineral resources. SIGAR 23-10 Audit Report
18. Special Inspector General for Afghanistan Reconstruction (2022) Collapse of the Afghan National Defense and Security Forces: an assessment of the factors that led to its demise. Interim Report
19. Special Inspector General for Afghanistan Reconstruction (2022) Why the Afghan government collapsed. SIGAR 23-05-IP, Evaluation Report
20. Starr FS (2022) Central Asians seize the initiative. Tashkent Conference, July 26

Adib Farhadi is an Assistant Professor and Faculty Director of Executive Education at the University of South Florida. His research, which is at the intersection of religion, politics, economics, and conflict with a particular focus on the "Silk Road" Central and South Asia (CASA) Region, is the subject of his recent book, Countering Violent Extremism by Winning Hearts and Minds. He is Editor and Chief for The Great Power Competition book series and the

Director of the Great Power Competition Initiative. Formerly, Dr. Farhadi served in senior positions for Afghanistan and has extensively advised the United States government and various other international organizations. Dr. Farhadi earned his Ph.D. in Economics from the University of Canberra, M.A. from New York University, and B.S. from East Carolina University.

A Problem We Fueled: Learning Lessons from Corruption in Afghanistan

Catherine Bateman

Abstract Although U.S. officials came to see corruption as undermining the mission in Afghanistan, they failed to sufficiently appreciate or act on a crucial dimension of the problem: U.S. policies and practices incentivized graft and greatly contributed to the growth of corruption after 2001. Layers of contracting enriched the well-connected; weak oversight led to fraud and abuse; the volume of injected money was too much for Afghanistan's small economy to absorb; and U.S. partnerships with corrupt, abusive actors perpetuated a culture of impunity that alienated the population. The U.S. strategy in Afghanistan for years relied on the survival of the Kabul government; Afghan leaders knew this, and that blunted whatever leverage Washington seemingly had over the political elite in Kabul. The United States repeatedly prioritized stability over justice, and in the end, got neither. U.S. policymakers should take several lessons from Afghanistan, to apply in similar contexts: strengthen U.S. oversight and accountability for civilian and military spending; calibrate levels of U.S. spending to not exceed a country's absorptive capacity; invest in understanding the nature and scope of corruption; hold host country leaders accountable particularly for high-level corruption; prioritize anticorruption controls when developing the capacity of partner security forces; and invest long-term in civil society actors who are building social cohesion, transparency, and accountability within their political system.

Keywords Corruption · Anticorruption · Justice · Foreign aid · Military contracting

Among those involved in the U.S.-led intervention in Afghanistan from 2001 to 2021, it is now cliché to cite endemic corruption as one of the main factors in the failure of the Afghan republic and its security forces—and by extension, as a factor in the Taliban's victory. Why examine the role of corruption in those failures? The United States may be unlikely to invest twenty years of blood and treasure in another country in the near future. But the history of U.S. attempts at state-building abroad that

C. Bateman (✉)
United States Institute of Peace, Washington, DC 20037, USA
e-mail: cbateman@usip.org

A. Farhadi and A. Masys (eds.), *The Great Power Competition Volume 4*,
https://doi.org/10.1007/978-3-031-22934-3_2

entailed major deployments of U.S. military forces—from Cuba, Panama, Nicaragua, and Haiti in the early 1900s, to post-WWII Germany and Japan, to South Vietnam, Cambodia, Afghanistan, and Iraq—suggest otherwise [24]. Political promises and poor track records notwithstanding, the United States will probably attempt similar efforts again. Afghanistan demonstrates how poorly the U.S. government handled the nuanced challenges and tradeoffs around the abuse of power for private gain. It is vital that we learn from these mistakes. Discussions on reconstruction of a post-conflict Ukraine are already underway [7]. Moreover, smaller-scale efforts, like U.S. operations in Iraq, Syria, Yemen, Somalia, and the Philippines, should be informed by lessons from Afghanistan.

This chapter argues that although US officials—civilian and military—came to see rampant corruption as undermining the mission in Afghanistan, they failed to sufficiently appreciate or act on a crucial dimension of the problem: U.S. policies, practices, and relationships greatly contributed to the growth of corruption, incentivizing graft and making the problem worse. Layers of contracting enriched the well-connected and led to shoddy work; weak oversight of projects resulted in waste, fraud, and abuse; the sheer volume of money injected into a small, poor economy, without predictability in how long funding might last, incentivized Afghans to profit while they could; and U.S. partnerships with corrupt, abusive figures perpetuated a culture of impunity that alienated parts of the population from their government and the international presence.

"The paradox is that we saw ourselves as solvers of the problem, when we were part of the problem," political scientist Dipali Mukhopadhyay observed.[1] The American narrative about corruption in Afghanistan often assumes the tone of a self-assured and patronizing doctor diagnosing his patient and writing a prescription—or rushing the patient into emergency surgery. U.S. officials spoke of corruption as "part of Afghan culture," usually without acknowledging how foreign military spending and aid fueled corrupt behavior.[2] Understandably, this was offensive to Afghans; it suggests that Afghans have inherently lower moral standards than Westerners. (The chapter will return to this misperception to discuss how difficult it is for societies to transition from high to low levels of corruption.) Moreover, a simplistic view of corruption as "cultural" conveniently absolves internationals from understanding and taking responsibility for their role in perpetuating it.

This chapter first explores insights from the broader literature on corruption and anticorruption, touching on the Afghanistan case and laying out common dilemmas for fragile states and outside actors operating in them. It then tells the story of corruption and anticorruption in Afghanistan from 2001 to 2021: how U.S. practices contributed to a sharp rise in corruption, how corrupt behavior undermined U.S. goals, and what the United States tried to do about it. The chapter also considers

[1] Author's conversation with Dipali Mukhopadhyay, Washington, DC, July 12, 2022.

[2] For example, during a November 2021 U.S. Senate Foreign Relations Committee hearing on lessons from Afghanistan, Ranking Member Senator James Risch (ID) remarked, "…the culture in Afghanistan was so different than the culture that we are used to dealing with... The corruption issue is a huge issue as you try to stand up a nation and move forward. If you cannot get a handle on that, if it is endemic in the culture, it is a problem" [35].

Afghan perceptions, and reasons behind former President Ashraf Ghani's failure to reduce corruption, despite his ambitious plans. This account lays the foundation to ask, could the United States have done something differently to achieve a different outcome? The chapter closes with lessons for U.S. policymakers and practitioners.[3]

1 Understanding Corruption and Why It Persists

Within the international aid community, the focus on corruption risks in fragile states dates back to only the mid-1990s. Social science research on corruption gained momentum in the 2000s, and since then, more empirical studies about what works to control corruption have been conducted. Our understanding about corruption is still evolving. A standard definition is "the abuse of entrusted power for private gain" [34]. The scholar Alina Mungiu-Pippidi puts forward a compelling framework for corruption. She posits that societies that manage to control corruption distribute public goods and services "on the principle of ethical universalism (in which everyone is treated equally and fairly)." Societies that have not achieved this are characterized by "particularism (treating individuals differently according to their status)." Particularism, in turn, often leads to the use of public office for private profit. Mungiu-Pippidi writes:

> Public corruption thrives on power inequality and the incapacity of the weak to prevent the strong from appropriating the state and spoiling public resources. Particularism encompasses a variety of … transaction types, such as clientelism, bribery, patronage, nepotism, and other favoritisms. [21, p. 23]

Corruption exists in every society. The norms of a given country fall somewhere along a continuum between particularism and universalism. About two-thirds of countries in the world today are more corrupt than noncorrupt. Controlling corruption is something that, in general, only wealthy countries have succeeded in doing, over decades of building institutions and norms around integrity. This change often occurred alongside the emergence of market capitalism and modernity [21, pp. 23, 24]. Controlling corruption costs money—for instance, to pay civil servants a living wage, and to provide public services such that people do not need to bribe bureaucrats in order to get access to scarce resources [9, p. 61].

To understand why it is difficult for a society to move from high to low levels of corruption, it is useful to think of corruption in an equilibrium framework. People's behavior often depends on their expectations about what other people will do. In

[3] This chapter draws on research conducted in mid-2022, using primary and secondary sources, including interviews with Afghans and Americans who participated in or are familiar with anticorruption efforts in the 2001–21 period. The analysis presented here is also informed by the author's firsthand experience as a diplomat in Afghanistan in 2010–12; her work for the Office of the Special Inspector General for Afghanistan Reconstruction, a U.S. independent oversight agency; and her research at the Center for a New American Security and the United States Institute of Peace. More broadly, the author has sought to contextualize her analysis of the Afghanistan case within the academic literature on corruption and anticorruption.

high-corruption societies, individuals are likely to face higher costs for resisting or exposing corrupt behavior (loss of income or access to services, physical danger, etc.), than for engaging in it. Thus the high-corruption equilibrium sustains itself, even if the general population disproves of corruption and would rather live in a less-corrupt society. In a low-corruption equilibrium, the opposite holds: an individual usually faces higher costs for engaging in corruption than for resisting it. To move from a high- to low-corruption equilibrium, a critical mass is needed to significantly change social norms [9, pp. 6–15].

1.1 Foreign Interventions

Corruption is acknowledged as one of the greatest challenges in post-conflict and conflict-affected states. Corrupt activities undermine economic and social development, the fair distribution of public resources, the morale and effectiveness of security forces, and the legitimacy and functioning of the state. However, after more than two decades of efforts by wealthy countries and donor agencies to counter corruption in developing countries, results have been disappointing. There is little consensus on what works [6, 21, 27].

Discussions on corruption tend to take place in the aid community, with rare cross-pollination with military and national security circles. Yet in looking at post-2001 Afghanistan, corruption must be understood as inseparable from the war itself. How power and resources were distributed—and the fairness or justice of that distribution, by both domestic and foreign actors—had enormous consequences for both state-building and the war effort. The anticorruption literature increasingly grapples with the vicious feedback loop between corruption and conflict: Corruption fuels grievances against the state, and those who benefit from war and illicit economies face incentives to perpetuate violence; in turn, war and insecurity create space for illicit activities that tie corrupt officials to drug trafficking and even insurgent or terrorist networks contesting the state's authority—all of which characterized post-2001 Afghanistan [6, 12, 18].

In Afghanistan, policymakers at times debated what types of corrupt activities were most destructive to the Afghan state, its legitimacy and functioning, and public trust in state institutions. A meta-analysis of the literature, also drawing on evidence from the field, found that "in many fragile contexts, the most damaging forms of corruption are not petty or bureaucratic in nature; rather endemic political and grand corruption often underpin an institutionalised system of graft in which illicit rents are passed upwards through the chain of command" [15, p. 18].

Crucially, the literature also highlights an uncomfortable truth: donors share the blame for *worsening* corruption in fragile states. Where aid money flows for reconstruction, political pressure on donors to spend and demonstrate quick progress leads to the injection of too much money, too fast, into an economy that cannot effectively absorb it. As we will see in the Afghanistan case, reconstruction assistance paled in comparison to foreign military spending. Combined with weak oversight by donors and military forces, this massive infusion of money has the perverse effect of exacerbating corruption. Further, democratization and economic liberalization may also expand opportunities for corrupt behavior, through vote-buying and the privatization of state assets [10, 23].

A devastating effect of outside actors unintentionally fueling corruption is that the domestic population sees this, resents the vast disparity between the international community's promises and its practices on the ground, and becomes distrustful of and alienated from the foreign-backed government, the donor community, and international military forces.

All this seems to suggest that anticorruption efforts should be rigorously pursued. But here perhaps the most troubling dilemma presents itself in the context of political fragility. As Rose-Ackerman observes, "In a state with very weak institutions, corruption may be a short-term way to hold the system together and prevent violent disintegration. Political leaders buy off powerful private actors with patronage," and vice versa. The flip side is that "although it may be risky and difficult to counter corruption in post-conflict peacebuilding, if the problem is allowed to fester, it can undermine other efforts to create a stable, well-functioning state with popular legitimacy" (Rose-Ackerman 2008, p. 328). Put more simply, there is a tension between peace, or political stability, and justice. Maintaining stability or cohesion in the short term may require appeasing corrupt actors who have the power to contest state authority. Yet accommodating such actors also profoundly undermines the legitimacy of the state, and perpetuates injustice and impunity.[4]

For those seeking to maintain a fragile peace but also build sound governance in the longer term, the literature offers no easy answers. Anticorruption reforms can have destabilizing effects. Reform initiatives can be co-opted by corrupt leaders and regimes themselves, and turned against their critics. If citizens' hopes for change meet with disaster, they can become disillusioned with the government and reform leaders [16].

This is not, however, a reason to do nothing. A range of efforts have shown promise: protecting civic space for bottom-up accountability mechanisms to develop, and providing the tools needed by civil society actors to hold their government accountable (e.g., support for independent media, access to information laws, and legal aid); peace agreements incorporating anticorruption measures; international actors helping to arrange for corrupt leaders to go into exile; tailoring anticorruption interventions to the local context based on sound political economy analyses; and pursuing reforms incrementally, but also being ready to scale up when windows of opportunity open (Rose-Ackerman 2008, pp. 339–340) [15]. Donors should also place greater emphasis on the principle of "do no harm" in their approach to anticorruption [16].

[4] This analysis draws from a conversation with a former State Department official, June 29, 2022.

2 Corruption and Anticorruption in Afghanistan, 2001–21

Journalists, researchers, aid organizations, U.S. officials and military veterans who served in Afghanistan, and U.S. congressional staff and oversight agencies have written exhaustively about corruption in Afghanistan, and, to a lesser extent, what was done to combat it. Drawing on this range of sources, as well as author interviews and unpublished materials, this section addresses how and why corruption expanded in Afghanistan, the role of outside actors in fueling it, and efforts to rein in corruption. An important thread of this story is that corruption is a security issue: not only does it undermine the legitimacy and effectiveness of security forces with whom the United States partners, but U.S. policies and operations can unintentionally incentivize corruption.

2.1 On Afghan Society and Politics

A fair treatment of the richness of Afghan society, culture, and politics is outside the scope of this chapter, but some context is called for.[5] Historically, Afghan society has been highly informal and kinship-based. The central government has exerted only limited power outside the cities, extending patronage to tribal leaders in exchange for their cooperation. This was partly a function of geography, as the country's notoriously rugged and inaccessible terrain meant it was difficult to consolidate large landholdings and the political structures to administer them. As a means of distributing power and resources, patronage and nepotism (or forms of particularism, in Mungiu-Pippidi's framing) are part of how Afghan politics and government have operated—as is true today in most countries.

It is worth addressing the common misperception that Afghan society was not "ready" for democracy. This matters to discussions around controlling corruption, in that truly democratic political structures are accountable to the citizenry and afford citizens a say in how resources are distributed. Free and fair elections give voters an opportunity to vote out corrupt and abusive leaders. In his book on Afghanistan's political transition and the 2004–2005 elections, Scott Smith observes:

> Afghan politics, because they are tribal, are both highly participatory and intensely local. … Non-institutionalized systems like Afghanistan's are often far more difficult to decode, since the loci of consensus are buried in tribal lore, family relations, financial connections, long-standing feuds and other considerations … Authority is instead earned or lost on strength of arms, sophistication of alliances, and use of money.… State structures have always [*sic*] superficial, neither embedded nor autonomous. (2010, p. 11)

Thus Afghan politics have been understandably opaque to many foreign observers. Nevertheless, if participation is at the core of democratic politics, this concept was not at all alien to Afghanistan. The country had experienced some degree of political

[5] For excellent treatment of this subject, see Barfield [1] and Mukhopadhyay [20].

modernization in the twentieth century, with a peaceful constitutional monarchy under King Zahir Shah, an elected parliament, and a new constitution adopted in 1964. As Smith argues, "democracy in Afghanistan was not as far-fetched a concept as it appeared to many Western analysts" (Smith 2010, p. 12; see also Larson [17], pp. 14, 15).

2.2 *Influx of Money Amid Weak Controls*

Afghanistan under the Taliban regime of 1996–2001 had already suffered more than two decades of war, which brought not only unspeakable violence and the destruction of infrastructure, but also a devastated social fabric, with millions of refugees having fled the country, and the killing of tens of thousands of tribal and community leaders and educated elites. After 2001 and peaking in roughly 2010–2012, the vast influx of foreign money associated with both the war and the reconstruction effort distorted the Afghan economy and patterns of political patronage [3, 29, pp. 8, 9]. The scale of injected resources massively expanded the opportunity for bribery, collusion, extortion, fraud, and outright theft. "The major difference between before and after [2001] was the scope of corruption," said Fara Abbas, who served in multiple Afghan ministries and eventually in the Afghan National Security Council. "There was corruption before, but that was largely nepotism, promoting people from your tribe or ethnicity into positions of power. But after the intervention, the scope significantly magnified. Due to the unprecedented flow of international aid, corruption largely and exclusively centered around money. Money became a currency of power. It almost didn't matter who you were, but if you had money, you could get things done."[6]

U.S. government funding was by far the largest source of resources. According to the Special Inspector General for Afghanistan Reconstruction,[7] the United States alone spent $145 billion on reconstruction assistance and $837 billion on warfighting over the 20-year intervention [31, 32]. Military contracting and sub-contracting went to security, transportation, construction, and logistical services. At its peak in 2010, international aid—for economic growth, governance, social services, civil society, as well as support to Afghan security forces—was roughly $16 billion, a figure equal to the country's entire Gross Domestic Product that year [3].[8] It is important to note that not all this money stayed in Afghanistan; some unknown but significant portion went to overhead and administrative costs, or was invested outside the country. A

---

[6] Conversation with Fara Abbas, former Afghan government official, June 28, 2022.

[7] The Special Inspector General for Afghanistan Reconstruction (SIGAR) is a congressionally-mandated, independent U.S. government oversight agency tasked with preventing waste, fraud, and abuse in the U.S. reconstruction effort in Afghanistan. See www.sigar.mil for all SIGAR publications.

[8] The $16 billion in aid in 2010 was about 64 times the annual amount of humanitarian assistance to Afghanistan during the 1996–2001 Taliban regime [3].

U.S. Treasury official once remarked that no adequate accounting has ever been done of how much money was spent *on* versus *in* Afghanistan.[9]

Nevertheless, the injection of funds was enormous and had profound, distortionary effects. As an unnamed State Department official told SIGAR, U.S. officials were "so desperate to have the alcoholics to the table, we kept pouring drinks, not knowing [or] considering we were killing them" [36]. In one of the poorest countries of the world, where one breadwinner might be responsible for 20 family members, there was a strong incentive to exploit the resource flows while they lasted. According to a SIGAR lessons learned report on corruption,[10] in 2011 Afghan business leaders told the U.S. embassy that "bribery and corruption were 'pervasive, accepted, and arguably even encouraged' at that time, whereas in previous years, greater shame had been attached to these behaviors" [29, p. 6]. Several financial flows, other than aid and military spending, also grew after 2001 and were vulnerable to corruption: Afghan domestic budgetary revenues, revenues from the opium trade, Kabul Bank deposits (of which nearly $1 billion were stolen in a massive case of bank fraud), illicit trade in natural resources like minerals and timber, the takeover of public lands, and illegal land development [3].

To make matters worse, the United States and other foreign actors failed to put in place adequate oversight measures. Within U.S. agencies, rates of budget execution became proxies for measuring effectiveness, placing great pressure on agencies to spend quickly—and by extension, creating disincentives for agencies to spend time and resources on strong oversight. *Washington Post* reporter Rajiv Chandrasekaran wrote, "In Kabul, USAID officials assiduously tracked their 'burn rate.' In mid-2010, they were thrilled when it reached $340 million per month. The figure became a point of pride and was mentioned... as a sign of progress" ([4, p. 207]; also SIGAR [32], p. 13). Agencies were also hamstrung by a chronic shortage of employees who were trained in basic contract oversight. According to a SIGAR report, "at one point in 2011, USAID would have had to send nearly its entire overseas workforce to work only in Afghanistan in order to meet the U.S. government's average ratio of dollars to contracting officers" [32, p. ix].

Weak contracting and procurement controls enabled embezzlement, waste, fraud, and collusion by both Afghans and internationals. For example, an investigative report by journalist Aram Roston spurred a congressional investigation in 2010 into protection rackets along U.S. convoy routes. The resulting report, *Warlord, Inc.*, detailed how local warlords extracted tens of millions of dollars a year from U.S. contractors and sub-contractors, in exchange for trucking convoys' safe passage. Allegedly, local warlords used some of this money to pay off the Taliban and other groups to not attack the convoys [33]. In an examination of this and other high-profile cases of contract corruption, scholar Jennifer Murtazashvili laid out how Afghan entrepreneurs, politically connected to the Americans or to Afghan elites, gained access to massive contracts and sub-contracts; they then inflated costs and

[9] Conversation with U.S. Treasury Department official, Washington, DC, 2016.

[10] The author was the lead writer for this SIGAR report.

likely channeled substantial sums outside the country. A new class of Afghan elite, based not on tribal or popular support but on financial wealth, was created [13, 22].

Another dimension of poor oversight was the lack of strong monitoring and evaluation of contracts, which, done well, seeks to ensure that a contract keeps to the terms of the award and produces the intended outcome. Over the course of the intervention, the Defense and State Departments and USAID eventually strengthened their monitoring and evaluation systems in Afghanistan. But too often, these systems were too weak to detect when, for example, a road project used inappropriate standards and was poorly constructed, or when a governance program was ineffective and should have been adjusted or ended. Better monitoring and evaluation could have identified where reconstruction contracts, grants, and contractual agreements were resulting in waste, fraud, and abuse [32, pp. 9, 16–21].

An even more opaque inflow of money—payments by the Central Intelligence Agency to Afghan government officials—illustrates the contradiction between the counterterrorism mission and fighting corruption. A 2013 *New York Times* article detailed how the CIA had delivered "bags of cash" to the offices of the Afghan president for years, totaling tens of millions of dollars, to secure access, information, and collaboration in the war on terror. The money reportedly went to "paying off warlords and politicians, many of whom [had] ties to the drug trade and, in some cases, the Taliban" [25]. Perversely, however, the payments sometimes went to the very same individuals whom the State Department and law enforcement agencies were investigating for corruption and seeking to hold accountable [29, p. 43]. Barnett Rubin, an expert on Afghanistan who served as a senior advisor to Special Representative for Afghanistan and Pakistan Richard Holbrooke, pointed out that under Afghan law, the payments amounted to illegal corruption. "One part of U.S. policy corrupted Afghan officials while other parts tried to investigate and root out corruption," he wrote in the *New Yorker*. "Given the interest that defined the mission, concerns about corruption did not trump those of covert action" [26]. Rosenberg reported that in the words of one American official, "The biggest source of corruption in Afghanistan was the United States" [25].

2.3 *Corruption and Violence*

Investigative journalist Anand Gopal argues that corruption amid the Afghan war has been linked to extraordinary levels of violence. Echoing the cases detailed in *Warlord, Inc.*, Gopal recounted how U.S. forces hired a taxi driver to protect hundreds of trucks traveling between two cities. The driver hired armed men and began detaining, harassing, and killing people in villages along the route; the people turned to the Taliban for protection. This incited more attacks on the route—the driver's original goal. The driver could then point to the increased Taliban attacks, and demand more money from U.S. forces to protect the route. He could also pay off the Taliban to not attack the route. In this way, Gopal said:

> Corruption in Afghanistan was extreme forms of violence. Not just bribes; that was the least of Afghans' problems. It was, if you don't pay off someone, they will break into your house and shoot you or take away your loved ones, or call in an airstrike. If we are looking at the war from an Afghan perspective, there's an extraordinary cost. … A lot of violence in the last few years was meted out by the Afghan government, as an extreme form of corruption. Extortion rackets, killing people for not paying protection money. This was a daily occurrence in the countryside. [13]

Such violence took an enormous toll on the Afghan population. For the United States, the more direct and devastating consequence was that corruption helped drive people to the insurgency. By 2009, U.S. military and civilian leaders acknowledged that predation by corrupt Afghan officials, and Afghans' general perception of corrupt and criminal activity by their government, was a key factor in Taliban recruitment [8, pp. 847, 848, 29, pp. 35–37].

2.4 *Elections: Missed opportunities for legitimacy and accountability*

Over the 20-year intervention, elections comprised a central piece of efforts to establish a democratic system in Afghanistan. The 2001 Bonn Agreement set out a "road map" for the transition to a new political order, culminating in general elections in 2004 for a new president and parliament. Elections were meant to underpin the legitimacy of the new government. They would be followed by presidential and parliamentary elections in 2009/10, 2014, and 2018/19.

The Afghan public initially demonstrated enthusiasm for the electoral process. Registered voter turnout in 2004 was 84% [31, p. 24]. In the experience of the 2004 and 2005 elections, however, shortcuts taken were the seeds of later failures. A vetting system found that more than a thousand potential candidates had links to illegal militias, and hundreds of complaints were made alleging human rights abuses by candidates, but few were disqualified.[11] The participation of so many candidates alleged to have committed serious human rights violations set a dangerous precedent. It undermined Afghan confidence in parliamentary elections and fed a culture of impunity. Voter turnout in the 2005 elections was 49%, not low, but a significant drop from 84% the year before [31, p. 24].

Smith argues that after the 2004 election, "the institutional considerations were increasingly forsaken for short-term political concerns. This was the result of the pressure of time, the deteriorating security situation, an uncoordinated international community, and an Afghan political class whose members consistently and grossly

[11] The 2004 Afghan constitution barred anyone convicted of a crime against humanity from running for public office—a reflection of Afghan desire for justice and accountability for atrocity crimes. But no one had ever faced such charges, and there was no system to vet candidates based on allegations of human rights abuses. The alternative was to vet them for connections to illegal armed groups [14, p. 5].

put personal gain above the national interest" [28, p. 3]. In subsequent election cycles, voter turnout dramatically declined, dropping to 19% in 2019.[12] Each election was further marred by widespread claims of electoral fraud. Lower voter participation likely reflected concerns about electoral fraud and corruption, general disillusionment with the government and political process, and voters' fear for their personal safety, amid Taliban campaigns to intimidate voters through violence [31, pp. 23–25].

On top of this was a pervasive sense of foreign interference, as diplomats, particularly U.S. officials, stepped in to resolve political crises around election disputes. U.S. officials seem to have judged these steps necessary to avert political disaster, but over time, the pattern of interference contributed to Afghans' sense that their country's political process was not their own [17, pp. 14, 15, 28]. Thus a critical democratic process, which might have grounded Afghans' participation and confidence in their political system and strengthened the legitimacy of the state, was subverted by a combination of political expediency, short-sightedness, and the war itself. A byproduct was the erosion of Afghans' confidence in U.S. intentions and presence in their country, as well.

3 U.S. and Afghan Anticorruption Efforts

3.1 The Surge Years

By the latter part of the Bush administration and as President Barack Obama entered office in 2009, according to SIGAR, "a consensus began to emerge among [the Department of Defense], State, and USAID that corruption was undermining core U.S. goals by materially fueling the insurgency and turning the population against the Afghan government" [29, p. 35]. This mattered because U.S. officials increasingly recognized there may be no military solution to the war; the American exit strategy was to strengthen the Afghan state and its security forces so that they could take responsibility for keeping the Taliban at bay, protecting the population, and delivering basic services. Pervasive corruption was a core obstacle to those goals: extortion by bureaucrats and police angered and alienated citizens; corruption on a grand scale by politically-connected elites drained the state of resources, often channeling it outside the country; and corruption within the Afghan army and police—ammunition and medicine sold on the black market, "ghost soldiers" padding the payrolls, salaries pocketed by commanders, promotions based on connections not merit—eroded morale and combat readiness.

In his history of the war, Carter Malkasian writes, "The argument went that the government could never stand on its own if corruption, weak delivery of goods and services, and unfair policies drove people to the Taliban". Department of Defense (DOD) buy-in for this view propelled a broader response to corruption, as DOD and

[12] Voter turnout was about 39% in the 2009 and 2014 presidential elections [31, p. 24].

International Security Assistance Force (ISAF) brought resources to the table that civilian agencies had not previously been able to marshal. Bolstering governance and fighting corruption became key parts of the U.S. counterinsurgency strategy.

A surge in civilian personnel and assistance accompanied the 2009–2011 U.S. troop surge. U.S. agencies and ISAF stood up a flurry of anticorruption-related bodies and initiatives. These focused on preventing U.S. funds from going to the insurgency, improving oversight of U.S. contracting and procurement, strengthening the rule of law and accountability within Afghan government institutions, and supporting civil society and the media.[13] For instance, DOD and ISAF task forces assembled expertise in forensic auditing, criminal investigations, and contracting oversight, to build a more detailed picture of where and to whom contracting dollars went. The Department of Justice deployed more federal prosecutors and law enforcement agents to train and mentor specialized Afghan units for corruption-related investigations, prosecutions, and drug enforcement. The Treasury Department, Drug Enforcement Administration, and other agencies collaborated to help Afghanistan's central bank stand up a financial intelligence unit [29, pp. 40–42].

On the Afghan government side, anticorruption institutions had been created in the early 2000s and shown minimal results. Now, with heightened U.S. attention, donors ramped up assistance to and pressure on the Afghan government to strengthen them and create new ones. The story of the Major Crimes Task Force (MCTF), created in 2009 and mentored by U.S. law enforcement agents, typifies the trajectory of many of these anticorruption bodies. Initially, the MCTF developed capacity to investigate high-level corruption, organized crime, and kidnapping, and to build cases for prosecution by the Attorney General's Office. Then in 2010, U.S. law enforcement and intelligence personnel worked closely with the MCTF to investigate and arrest Muhammad Zia Salehi, an aide to President Karzai and the chief of administration of the Afghan National Security Council. It was a test case for prosecuting high-level corruption. But within hours of Salehi's arrest, Karzai ordered his release. The United States backed off the case and it was eventually dropped. It emerged that Salehi was allegedly being paid by the CIA and distributed the palace slush fund.

Karzai subsequently moved to restrict international mentors' access to the MCTF and other anticorruption units. One of the two MCTF prosecutors on the Salehi case, who were demoted and reassigned, was a woman. She told Sarah Chayes, who has chronicled corruption in Afghanistan, "When we really needed help... no one came. We were treated as if we were dispensable" [5, pp. 142, 143]. The Afghanistan Analysts Network reported, "Almost the entire original leadership of the unit emigrated to the US following the shutdown of the Salehi investigation and [*sic*] has experienced instability in its leadership ever since" [2].

The U.S. decision to back away from the Salehi case was perceived as acquiescence to such impunity. That episode, followed by the political failure to hold Afghan leaders accountable for nearly $1 billion stolen in the Kabul Bank scandal,

---

[13] See SIGAR's 2016 report 'Corruption in conflict' for a thorough accounting of these U.S. and ISAF initiatives. In the interest of full disclosure, the author was senior analyst and leader writer for this report.

further entrenched a pattern of impunity for corrupt Afghan officials at the highest level [29, pp. 43, 44, 53–55, 5, pp. 140–144]. With regard to the kind of corruption that was likely most destructive to the Afghan state and state-society relations, the international community did not find an effective way to counter it. The U.S. strategy in Afghanistan ultimately relied on the survival of the Kabul government; Afghan leaders knew this, and that blunted whatever leverage Washington seemingly had over the political elite in Kabul.

The irony of anticorruption efforts during the surge years was that precisely when the United States was more focused on the threat of corruption, it was also pouring more financial resources than ever into Afghanistan—faster than U.S. agencies could properly oversee those resources, despite their best efforts. If one basic ingredient in the scourge of corruption was cash, resourcing a 150,000-troop warfighting effort was somewhat incompatible with efforts to control corruption—and another way in which security and counterterrorism goals came into conflict with governance goals.[14]

3.2 *Declining Attention and the Ghani Years*

In the remaining years of the intervention, U.S. efforts on anticorruption emphasized the technical, but continued to fail to exert political pressure on the Afghan government. The U.S. military presence, and along with it domestic political attention on Afghanistan, dramatically declined. American and international diplomats were consumed by various crises such as the dispute over the 2014 presidential election, the Taliban's temporary military takeover of Kunduz city in 2015, the rise of an Islamic State affiliate in Afghanistan, and steadily deteriorating security overall.

In the meantime, anticorruption benchmarks remained constant elements of foreign aid compacts between donors and the Afghan government; the latter would commit to specific steps, and aid was loosely conditioned on those steps. SIGAR produced three congressionally mandated assessments of the government's anticorruption efforts, in part tracking these benchmarks. There were some bright spots, such as the relatively effective financial intelligence unit of the Central Bank. But SIGAR highlighted that a "lack of resources and security has been a continued detriment for detectives, investigators, prosecutors, and judges in Afghanistan," and some were dependent on donors for reliable funding ([30, p. 31]; also Bjelica [2]). The conditions outlined in compacts and aid agreements often turned into paper-pushing exercises. "The government started to have an approach of deceiving the donor community," observed a young Afghan professional who worked for donor agencies

---

[14] At their peak in Afghanistan, U.S. military forces numbered roughly 100,000, with an additional 50,000 NATO and coalition forces in country.

and Afghan civil society organizations. "The government would report progress that wasn't there."[15]

A troubling question is how President Ashraf Ghani, who came into office championing the fight against corruption, completely failed to dismantle the networks he condemned. A full accounting of that failure is beyond the scope of this chapter, but it is possible to draw initial theses. Ghani recruited young technocrats, many of them educated abroad, into his administration, seemingly betting that they would remake the institutions where they landed. Though this cadre doubtless included many officials who served with integrity, many others became enmeshed in corrupt circles. This dynamic is reflected in studies on corruption (and in the equilibrium framework discussed earlier in this chapter), which find that it thrives within systems. Without profound changes to the structure of the system itself, individuals have little hope of overturning the prevailing norms. Another key factor was that Ghani never built a strong political coalition around him; he himself proved too weak politically to empower his government's law enforcement and judicial institutions to hold accountable corrupt senior officials. He lacked the political capital to upend the existing system.

Further, in a tragic echo of exactly what plagued U.S. anticorruption efforts, during Ghani's tenure the Taliban made steady gains on the battlefield and the war demanded increasing policy attention and resources—crowding out efforts to improve governance. In addition, as the United States signaled its desperation to leave Afghanistan, the incentives for Afghan officials to extract as much rent as possible remained strong. Aref Dostyar, a scholar at the Kroc Institute, noted, "after the major [troop] withdrawal in 2014, many people came to doubt whether the state would last. If they don't expect it to last, people look to gain as much as they can. And if it doesn't last, they are not going to be held accountable."[16]

4 Afghan Perceptions

Both the American war and reconstruction effort in Afghanistan were prosecuted with a large dose of hubris, to which the outcome attests. At the core of the U.S. failure to help establish a legitimate, functioning state was a failure to approach the country with humility. We were not eager to question our assumptions and policies, nor determined to hear and learn from Afghans themselves. It is valuable, therefore,

[15] Conversation with a former Afghan staff member for donor agencies and civil society organizations, by Zoom, June 3, 2022; also conversation with Aref Dostyar, Scholar in Residence at the Kroc Institute for International Peace Studies, by Zoom, July 7, 2022.

[16] Conversation with Aref Dostyar, Scholar in Residence at the Kroc Institute for International Peace Studies, by Zoom, July 7, 2022. The foregoing analysis in the text also draws on the author's conversations with multiple former Afghan government officials in June and July 2022, as well as a series of unpublished papers that the U.S. Institute of Peace and the American Institute of Afghanistan Studies commissioned from young Afghan scholars, former officials, journalists, and activists, on the failure to reach a political settlement of the conflict.

to consider how Afghans experienced and perceived corruption during the U.S.-led intervention.

Initially welcoming foreign intervention after 22 years of war and devastation, Afghans were promised democracy and economic development. They then witnessed some of the most abusive warlords, who had committed atrocities in the 1990s civil war, elevated into senior government positions. Afghans were increasingly subject to the indignities of petty corruption for almost any public service they needed. As Sarah Chayes puts it, the flow of patronage was not down but upward, "in the form of gifts, kickbacks, levies paid to superiors, and the purchase of positions" [5, p. 59]. With massive increases in the foreign troop presence and reconstruction effort, Afghans then saw enormous contracts make politically-connected compatriots fabulously wealthy. They learned of senior Afghan officials and bank shareholders embezzling exorbitant amounts, with hardly any negative consequences imposed by international donors. Most of that stolen wealth was not reinvested in the country, but rather laundered in the Dubai real estate market [11, pp. 85–93]. Afghans risked their lives to vote, only to learn that the purported winner engaged in widespread fraud, and to watch foreigners intervene to determine the outcome.

For many Afghans, it was inconceivable that a global superpower did not intend for this outcome. As Chayes recounted one Afghan friend telling her, "We just assume America wants the corruption. We have no other way of explaining... how America has behaved itself" [5]. Or as a young Afghan told this author, "blame is put more on the U.S. because—it's like the good side and bad side. Nobody expects much out of the bad side. The United States is supposed to be the good side. They should have known; they should have done better."[17]

Meanwhile, U.S. officials spoke about corruption as "endemic" and "part of Afghan culture," rarely acknowledging the U.S. role in enabling it. The not-so-subtle implication was that Afghans were morally deficient, inherently inclined to corrupt behavior. That suggestion was not only hypocritical but also likely offensive, and may have played a role in the deterioration of the U.S. relationship with Karzai. This does not excuse, however, those members of the political class who put gross self-enrichment above the public interest.

Phil Raveling, who served twice as CIA base chief in Kandahar Province, saw corruption as a factor in the U.S. failure in Afghanistan. Speaking about corruption and the dependencies that U.S. policies created, he put it this way: "We built the house. We poured fuel on it, and at the end, lit it on fire. And then blamed the occupants for everything that happened. The result speaks for itself. We're certainly not better off, nor are the people of Afghanistan."[18]

Was Afghans' loss of faith in their government and the international presence any wonder?

[17] Conversation with a former Afghan staff member for donor agencies and civil society organizations, by Zoom, June 3, 2022.

[18] Conversation with Phil Raveling, former CIA senior intelligence officer, by Zoom, July 19, 2022.

5 Conclusion and Lessons

Could the United States have done something differently to achieve a different outcome?

Though corruption was a major factor in the failure of the state-building effort and in the Taliban takeover, it should not become a bumper-sticker explanation. The U.S. government made many strategic missteps: the shortsighted decision to deny the Taliban movement any space in the post-2001 political order; the failure to prioritize a comprehensive political settlement; the excesses and intelligence failures of the U.S. counterterrorism effort; a costly counterinsurgency campaign that was premised on unrealistic expectations and timelines; the failure to adequately protect civilians from violence; the assumption that Pakistan could be persuaded to deny safe haven to the Taliban; and personnel tours of one year or less, which short-circuited institutional knowledge about the country, the war, and the U.S. effort itself. All these factors and more contributed to the tragic outcome, from which the Afghan people suffer more than anyone.

Nevertheless, different U.S. choices and policies vis-à-vis the challenge of corruption might have made a difference in the final outcome. The United States repeatedly prioritized stability over justice, and in the end, got neither. The following takeaways do not offer a precise road map for countering corruption in fragile states. Yet the Afghanistan case suggests seven important lessons that should inform U.S. policymakers and practitioners:

Recognize when and how U.S. influence, resource flows, and relationships can contribute to corruption in a host country; adopt a "do no harm" approach.

The United States' first task in situations with corruption risks is to not make the problem worse. This requires a shift in perspective. Too often, we see corruption as a problem indigenous to the foreign country, which blinds us to the myriad ways in which U.S. practices can facilitate corruption. For the U.S. government to marshal the will and resources to change those practices, more policymakers and practitioners must recognize how U.S. actions do harm. U.S. agencies should foster an institutional culture that is attuned to such potential harm, and that rewards those who speak up about the risks of doing harm.

Gather information on and assess the nature and scope of corruption in a host country. Apply this to sustained efforts to counter corruption, particularly political and grand corruption that benefits the few at the expense of the many.

The U.S. government should invest greater resources, including intelligence assets, in understanding corrupt networks, their interests and incentives, and how U.S. practices may risk strengthening them. We can better train U.S. civilian and military personnel to recognize corruption when they see it, and know how to respond or report it. U.S. officials should then apply this improved understanding to maintain pressure on political and economic elites to take meaningful anticorruption actions. U.S. efforts should prioritize accountability for grand corruption, which can prove especially destructive in fragile contexts.

Carefully calibrate amounts of U.S. assistance and other financial resources flowing into the country; do not overwhelm a small economy with more money than it can absorb.

The billions of dollars in aid and military spending that flooded Afghanistan fueled corruption and waste on an unprecedented scale. Economists and development practitioners posit that around 15–45% of GDP is the amount of aid that a state can reasonably absorb. SIGAR reported that "for most of the 2002–2015 period, appropriated U.S. reconstruction assistance to Afghanistan surpassed 45% of Afghanistan's GDP, reaching a high of 105% in 2010" [29, p. 52]. Those figures do not even account for much larger amounts of military contracting for security, construction, transportation, and logistics. In future interventions, the U.S. government must holistically look at the level of resources it is injecting into another economy, and strive to maintain levels that do not exceed a nation's theoretical absorptive capacity. That right-sizing effort should engage economists who work on the country in question.

Establish and invest in transparency and accountability for U.S. civilian and military spending across all sectors, including strong monitoring and evaluation systems.

The United States should lead by example, demonstrating and championing norms of integrity. Mungiu-Pippidi, based on an exhaustive study of anticorruption tools, highlighted that donors "should publicize what they fund and how they structure the process of aid allocation itself." Another idea she puts forth is involving pro-reform civil society groups in the actual "planning and audits of aid projects," which "would also empower these groups and set an example for how local stakeholders should monitor public spending" [21, p. 33]. Though political expediency will often chafe against the demands of accountability mechanisms, policymakers must take the long view and have the discipline to prioritize good practice.

Understand corruption as a threat to U.S. security objectives, and place anticorruption at the center of Department of Defense efforts to build partner capacity.

DOD was slow to come to the realization that both high-level corruption among Afghan security leaders and lower-level extortion by the police were fundamentally undermining the U.S. mission in Afghanistan. Corruption undercut morale and combat readiness of Afghan forces, but also served as a driver of the insurgency. DOD should ensure it can properly oversee its contracting and know where resources are going. DOD should balance the benefits of outsourcing costly logistical needs, against the risks that this practice fuels massive waste, fraud, and abuse. Further, DOD and intelligence agencies should have strategies to extract themselves from temporarily expedient partnerships with abusive actors, rather than fall into patterns of reliance on those actors who prey on local populations.

Acknowledge that U.S. engagement in a fragile state is unlikely to develop host country institutions capable of controlling corruption, within the U.S. timeframe of involvement. Assume that the challenge of corruption will persist for the foreseeable future, and invest long-term in civil society actors developing ways to hold their political system to account.

American political timelines do not align well with the decades- and generations-long efforts that are needed to control corruption. Moreover, workable ways to build accountability and integrity in governance will more likely emerge from forces within a given state, than from outside actors. Therefore U.S. strategies and policies should not rely on host nations developing robust anticorruption capabilities. This echoes the testimony of Laurel Miller, former acting Special Representative for Afghanistan and Pakistan, to the U.S. Senate Foreign Relations Committee in 2021: "The strategic error was not in failing to recognize [disunity and endemic corruption] existed but, rather, in expecting that they could be sufficiently ameliorated fast enough to deprive the insurgency of fuel and to align with any plausible duration of American political willingness to prop up the Kabul government" [19, p. 4]. Importantly, U.S. agencies should support civil society actors—such as independent media, community groups that oversee service delivery, and legal aid providers—who are building social cohesion and pushing transparency and accountability [15, pp. 21–25].

Public and private communications about corruption in another country should consistently push norms of integrity and hold leaders to account, but should avoid language that patronizes and implies moral deficiency on the part of host country actors.

U.S. officials should deliver consistent and sustained messages around anticorruption. But those messages should be balanced with open discussion of what steps the U.S. government is taking to ensure its own policies and practices do not enable or fuel corrupt behavior—acknowledging that no entity or nationality is immune to such risks. This framing may help reinforce the notion that controlling corruption is something donors, recipient governments, and civil society must tackle collectively and over the long term.

References

1. Barfield T (2010) Afghanistan: a cultural and political history. Princeton University Press, Princeton
2. Bjelica J (2019) Afghanistan's anti-corruption institutions: too many, and with too few results. Afghanistan Analysts Network, 20 May. https://www.afghanistan-analysts.org/en/reports/economy-development-environment/afghanistan-anti-corruption-institutions-too-many-and-with-too-few-results/. Accessed on 18 July 2022
3. Byrd W (2013) Changing financial flows during Afghanistan's transition: The political economy fallout. United States Institute of Peace Peacebrief 157, 11 Sept, 2013. https://www.usip.org/sites/default/files/PB157-Changing%20Financial%20Flows%20During%20Afghanistan%E2%80%99s%20Transition.pdf. Accessed 12 July 2022
4. Chandrasekaran R (2013) Little America: the war within the war for Afghanistan, 1st edn. Vintage Books, New York
5. Chayes S (2015) Thieves of state: why corruption threatens global security. W. W. Norton & Company Inc., New York
6. Cheng CS, Zaum D (2008) Introduction—key themes in peacebuilding and corruption. Int Peacekeeping 15(3):301–309. https://doi.org/10.1080/13533310802058752. Accessed 7 July 2022

7. European Commission (2002) Ukraine: Commission presents plans for the Union's immediate response to address Ukraine's financing gap and the longer-term reconstruction, press release, 18 May. https://ec.europa.eu/commission/presscorner/detail/en/ip_22_3121. Accessed 30 June 2022
8. Farrell T, Giustozzi A (2013) The Taliban at war: inside the Helmand insurgency, 2004–2012. Int Aff 89(4):845–871. https://doi.org/10.1111/1468-2346.12048. Accessed 24 July 2022
9. Fisman R, Golden M (2017) Corruption: What everyone needs to know. Oxford University Press, New York
10. Galtung F (no date) Introduction: the corruption dimension of post-war reconstruction, workshop paper from the International Anti-Corruption Conference in May 2003 in Seoul. http://www.anti-corruption.org/wp-content/uploads/2016/11/The-corruption-dimension-of-post-war-reconstruction-TIRI-Galtung.pdf. Accessed 1 June 2022
11. George B (2020) The Kabul to Dubai pipeline: lessons learned from the Kabul Bank scandal. In Page M, Vittori J (eds) Dubai's role in facilitating corruption and global illicit financial flows. Carnegie Endowment for International Peace, Washington, DC, pp 85–93. https://carnegieendowment.org/2020/07/07/dubai-s-role-in-facilitating-corruption-and-global-illicit-financial-flows-pub-82180. Accessed 21 July 2022
12. Goodhand J 2008 Corrupting or consolidating the peace? The drugs economy and post-conflict peacebuilding in Afghanistan. Int Peacekeeping 15(3):405–423. https://doi.org/10.1080/13533310802058984. Accessed 1 July 2022
13. Gopal A (2021) 'Anand Gopal and Richard Ojeda on Afghanistan' on 'Deconstructed'. The Intercept, podcast, 21 Aug. https://theintercept.com/2021/08/21/deconstructed-afghanistan-richard-ojeda-anand-gopal/. Accessed 3 July 2022
14. Gossman P (2013) Documentation and transitional justice in Afghanistan. United States Institute of Peace Special Report No. 337, September. https://www.usip.org/publications/2013/09/documentation-and-transitional-justice-afghanistan. Accessed 20 Apr 2022
15. Jenkins M, Kukutschka R, Zúñiga N (2020) Anti-corruption in fragile settings: a review of the evidence. Deutsche Gesellschaft für Internationale Zusammenarbeit (GIZ) GmbH. https://www.giz.de/de/downloads/giz2020_en_anti-corruption_in_fragile_states.pdf. Accessed 27 July 2022
16. Johnston M, Johnsøn J (2014) Doing the wrong things for the right reasons? "Do no harm" as a principle of reform. U4 Brief December No 13. https://www.u4.no/publications/doing-the-wrong-things-for-the-right-reasons-do-no-harm-as-a-principle-of-reform.pdf. Accessed 1 July 2022
17. Larson A (2021) Democracy in Afghanistan: Amid and beyond conflict. United States Institute of Peace Special Report No. 497, July 2021. https://www.usip.org/publications/2021/07/democracy-afghanistan-amid-and-beyond-conflict. Accessed 12 July 2022
18. Le Billon P (2003) Buying peace or fueling war: the role of corruption in armed conflicts. J Int Develop 15(4):413–426. https://doi.org/10.1002/jid.993. Accessed 20 July 2022
19. Miller L (2021) Written statement, hearing before the United States Senate Committee on Foreign Relations "Afghanistan 2001–2021: U.S. Policy Lessons Learned", 17 Nov. https://www.foreign.senate.gov/imo/media/doc/111721_Miller_Testimony.pdf. Accessed 1 Feb 2022
20. Mukhopadhyay D (2014) Warlords, strongman governors, and the state in Afghanistan. Cambridge University Press, New York
21. Mungiu-Pippidi A (2018) Seven steps to control of corruption: the road map. Daedalus 147(3):20–34. https://direct.mit.edu/daed/article/147/3/20/27197/Seven-Steps-to-Control-of-Corruption-The-Road-Map. Accessed 3 July 2022
22. Murtazashvili J (2015) Gaming the state: consequences of contracting out state building in Afghanistan. Central Asian Survey 34(1):78–92. https://doi.org/10.1080/02634937.2015.1014162. Accessed 12 July 2022
23. O'Donnell M (2008) Corruption: A rule of law agenda? In Hurwitz A, Huang R (eds) Civil war and the rule of law. Lynne Rienner Publisher, Boulder, pp 225–259
24. Pei M, Kasper S (2003) Lessons from the past: the American record on nation building. Carnegie endowment for international peace, policy brief, 24 May. https://carnegieendowment.org/files/Policybrief24.pdf. Accessed 27 Oct 2021

25. Rosenberg M (2013) With bags of cash, C.I.A. seeks influence in Afghanistan. New York Times, 28 April. https://www.nytimes.com/2013/04/29/world/asia/cia-delivers-cash-to-afghan-leaders-office.html. Accessed 24 July 2022
26. Rubin B (2015) What I saw in Afghanistan. New Yorker, 1 July. https://www.newyorker.com/news/news-desk/what-have-we-been-doing-in-afghanistan. Accessed 24 July 2022
27. Savedoff W (2016) Anti-corruption strategies in foreign aid: from controls to results. Center for Global Development, Policy Paper 076, March. https://www.cgdev.org/sites/default/files/CGD-policy-paper-Savedoff-anticorruption-agenda.pdf. Accessed 2 June 2022
28. Smith SS (2010) Afghanistan's troubled transition: politics, peacekeeping and the 2004 presidential election. Lynne Rienner Publishers, Boulder
29. Special Inspector General for Afghanistan Reconstruction (2016) Corruption in conflict: lessons from the U.S. experience in Afghanistan. SIGAR 16–58-LL, September. https://www.sigar.mil/pdf/lessonslearned/sigar-16-58-ll.pdf. Accessed 17 June 2022
30. Special Inspector General for Afghanistan Reconstruction (2018) Afghanistan's anti-corruption efforts: the Afghan government has begun to implement an anti-corruption strategy, but significant problems must be addressed, SIGAR 18–51-AR. https://www.sigar.mil/pdf/audits/SIGAR-18-51-AR.pdf. Accessed 19 July 2022
31. Special Inspector General for Afghanistan Reconstruction (2021) Elections: lessons from the U.S. experience in Afghanistan, SIGAR 21–16-LLP, February. https://www.sigar.mil/pdf/lessonslearned/SIGAR-21-16-LL.pdf. Accessed 12 July 2022
32. Special Inspector General for Afghanistan Reconstruction (2021) The risk of doing the wrong thing perfectly: monitoring and evaluation of reconstruction contracting in Afghanistan, SIGAR 21–41-LLP, July. https://www.sigar.mil/pdf/lessonslearned/SIGAR-21-41-LL.pdf. Accessed 9 July 2022
33. Tierney JF (2010) Warlord, Inc.: extortion and corruption along the U.S. supply chain in Afghanistan. United States Congress, House Committee on Oversight and Government Reform, Subcommittee on National Security and Foreign Affairs. Washington, DC: U.S. House of Representatives. https://www.worldcat.org/title/warlord-inc-extortion-and-corruption-along-the-us-supply-chain-in-afghanistan/oclc/642929052. Accessed 20 July 2022
34. Transparency International (no date) What is corruption? https://www.transparency.org/en/what-is-corruption. Accessed 6 June 2022
35. United States Senate, 'Hearing before the Committee on Foreign Relations, Afghanistan 2001 to 2021: U.S. policies lessons learned', hearing transcript, 17 Nov 2021. https://www.foreign.senate.gov/imo/media/doc/11%2017%2021%20Afghanistan%202001-2021%20-%20U.S.%20Policy%20Lessons%20Learned1.pdf. Accessed 4 July 2022
36. Whitlock C (2019) 'Consumed by corruption' in 'The Afghanistan papers: a secret history of the war', Washington Post, 9 Dec. https://www.washingtonpost.com/graphics/2019/investigations/afghanistan-papers/afghanistan-war-corruption-government/. Accessed 10 Oct 2021

Catherine Bateman is a researcher and analyst with extensive prior experience in the U.S. government. Her research focuses on Afghanistan, peacebuilding, stabilization, anticorruption and U.S. foreign policy. She is currently a senior expert on Afghanistan for the U.S. Institute of Peace. Previously, Bateman was a project lead in the Lessons Learned Program at the Special Inspector General for Afghanistan Reconstruction (SIGAR), where she led reports on anticorruption, counternarcotics, reintegration of ex-combatants and gender equality. From 2016–2017, as a Council on Foreign Relations International Affairs Fellow at the Center for a New American Security (CNAS), Bateman researched and wrote on corruption as a national security issue. She has also served in intelligence and policy positions at the State Department in Washington, Afghanistan, and Sri Lanka, and was a legislative aide on Capitol Hill.

Bateman has a master's from the Johns Hopkins University School of Advanced International Studies and a bachelor's from Middlebury College. In addition to Bateman's published work at

USIP, SIGAR and CNAS, her analysis has appeared in Foreign Affairs, Lawfare, The National Interest, The Hill and Proceedings.

How America's Great Power Competitors Privatized the GPC

Ian Conway and Kathleen Cassedy

Abstract Since roughly the conclusion of Operation Desert Storm in 1991, the US military has either been in post-Cold War drawdown or fighting the Global War on Terror (GWOT), using the find, fix, finish, exploit, analyze, disseminate (F3EAD) methodology. The result being that the US now presides over a military machine with the unprecedented and unrivaled ability to overwhelm conventional foes on the battlefield with superior technology, or to hunt and kill individuals globally, but it is woefully underprepared for the type of warfare Russia and China now wage.

Keywords Great power competition · War on terror · China · Russia · United States

In the ensuing three decades since the first Gulf War, our primary great power competitors chose two different paths that arguably led them to the same place, albeit with different approaches. Russian intelligence brought their domestic organized crime groups to heel, then co-opted them, and then ultimately assimilated them. The result is that modern Russia now functions more like a nuclear-capable mafia family than a modern government. China watched American stealth aircraft fire missiles through the windows of Baghdad in a display of precision firepower and technology that they assumed was the stuff of science fiction, and then set about stealing it, reverse engineering it, and mass-producing knock offs. Further, China initiated a program to establish dual use civilian-military infrastructure and forward operating bases globally.

Both countries accurately assessed that US government agencies and industrial powerhouses had nearly zero competition across the diplomatic, information, military, and economic (DIME) spectrum globally. They also took careful note during the early years of the GWOT of how effectively a globally distributed network of loosely related entities (e.g., al-Qaeda and its successors) could counter the military

I. Conway (✉) · K. Cassedy
Helios Global, Arlington, VA 22202, USA
e-mail: iconway@heliosglobalinc.com

K. Cassedy
e-mail: kcassedy@heliosglobalinc.com

A. Farhadi and A. Masys (eds.), *The Great Power Competition Volume 4*,
https://doi.org/10.1007/978-3-031-22934-3_3

might and technological prowess of the US. So, these adversaries set about to reshape the great game in a manner that exploited, degraded, subverted, and disadvantaged America's instruments of national power.

As the competition evolved—and arguably well before the US realized that it was being attacked below the level of kinetic action—both countries became aware that American power is significantly diminished without partners and allies. That industrializing countries do not want to—or simply cannot—abide by idealistic western demands: carbon emissions standards, anti-corruption regulations, human rights standards, or democratic reforms. Both countries understood that Pax Americana was reliant on the survival of the post-WW2 order that had brought prosperity to so many while the Federal Reserve printed the global reserve currency, and the US Navy kept the seas free and open for trade.[1]

But upstream of this, the apex predator of the power and influence food chain was the American message: the shining beacon of liberty, land of the free, home of the brave. In the Cold War, America sold *itself*. Midwestern grocery stores housed an embarrassment of riches—teeming with fresh fruit and vegetables while Soviet citizens stood in bread lines. Suburban America had a car in every driveway while PRC citizens were lucky to own a bicycle.

Eventually, Russia and China figured this out too. Using different methods, both nations attacked the very *idea* of America on the global stage, but especially in the global south. And now America is losing the information war, turned inward against itself as preexisting political and socio-cultural differences were wedged and levered further apart, first by Russian political warfare operations, and then by US political operatives who had gained a feral understanding of this tradecraft's effectiveness while they were spectators to it, and then became practitioners. In the words of the scholar Matt Armstrong,[2] American foreign policy today amounts to yelling *"STOP IT!"* America has no message to sell. America's domestic strife has caused foreign consulates to recommend against travel to the US out of safety concerns. America's athletes kneel during the national anthem at the Olympic Games.

But if America's instruments of national power so far outclass our adversaries and competitors, if our adversaries have determined they must avoid a military confrontation to succeed, the US must understand what Russia and China are doing that the US is not. How can the world's greatest economy—a $25 trillion gross domestic product (GDP) global powerhouse whose nearest peer is fully 25% behind[3]—be so prone to hostile subversion operations? The US does not appear to be able to effectively compete in the great power competition (GPC) outside of implemented sanctions and military force.

The authors believe the answer to this riddle lies in the private sector. That the commercial space is the battleground of GPC. That the global marketplace and its financial infrastructure is the much talked about gray zone. And that while this may

[1] Zeihan [1].

[2] The authors wish to thank Matt Armstrong (https://mountainrunner.us/bio/) for generously lending his time and expertise to our research for this chapter, as well as coining the "STOP IT!" analogy.

[3] International Monetary Fund [2].

indeed be competition below armed conflict, it is also a hot (albeit mostly non-kinetic) war. And finally, that the quote widely attributed to Vladimir Lenin—"A western capitalist will sell you the rope with which to hang him"—may regrettably may have been accurate.

The authors do not believe the solution necessarily lies in more defense spending but rather in incentivizing the private sector to get in the fight. And ultimately, to incentivize the populace to behave like the America the world admired for so long.

A word on structure. This chapter is organized as follows:

- A brief examination of each of the DIME spectrum elements as practiced by the US;
- A brief review of how Russia and China blend, integrate, and verticalize DIME elements;
- A review of how Russia and China have verticalized DIME elements to both target commercial entities and operate as commercial entities; and
- Some conclusions, and some recommendations on how America can compete.

Although the authors strongly argue that GPC also involves regional powers and rising great powers,[4] this chapter focuses on Russia and China as America's greatest competitors. Regarding conclusions and recommendations, the authors advocate for a significant rethinking of America's GPC strategy. If in fact the United States of America actually has a GPC strategy.

On the one hand, American national security and defense strategy documents make it clear America has no intention of surrendering any squares on the chess board of great power competition. On the other hand, competing in GPC using American blood and treasure appears to all but guarantee political campaign losses. The US appears to lack the political will either to surrender or compete using conventional—or maybe more accurately—traditional instruments of national power. US foreign policy amounts to scolding when either Russia or China does something America does not like. The US is the only GPC competitor that still plays by the post-WWII rules that kept the globe relatively stable until the turn of this century. The US must adjust its rules of engagement to align with twenty-first century realities.

1 Who is in Charge of the US DIME Elements?

In the US, DIME is (mostly) hierarchically organized, with the President and Commander in Chief directing the Secretaries of the lead Federal Agencies to execute the administration's policies, with appropriate checks and balances from the Legislative and Judicial branches of government.

[4] Ryan et al. [3].

1.1 Diplomacy

In the United States, Diplomacy is the responsibility of the US Department of State,[5] conducted through the Secretary of State and Ambassadors appointed by the President.

1.2 Information

In the US, responsibility for information is de facto held in separate channels from diplomacy, but here things get a little muddier.

The Congressional Research Service asks: "Who Is Responsible for the "I" in DIME?".[6]

> Within the USG, much of the current information operations doctrine and capability resides with the military. Many consider DOD to be relatively well-funded, leading some to posit that the epicenter for all IW [Information Warfare] activities should be the Pentagon. Some fear that military leadership of the IW sphere represents the militarization of cyberspace, or the weaponization of information. In addition, the military may not possess the best tools to successfully lead information efforts across the USG. Title 10 U.S.C. 2241 prohibits DOD from domestic "publicity or propaganda," although the terms are undefined. It is unclear how IW/IO [Information Operations] relate to this so-called military propaganda ban. P.L. 115-232 tasked the State Department's Global Engagement Center (GEC) to "direct, lead, synchronize, integrate, and coordinate efforts of the Federal Government to recognize, understand, expose, and counter foreign state and foreign non-state propaganda and disinformation efforts...." P.L. 116-92 created a Principal Information Operations Advisor within DOD to coordinate and deconflict its operations with the GEC, who is the lead.[7]

Commander Mike Dahm, writing for the US Naval Institute's *Proceedings*, pulls even fewer punches:

> The Department of Defense (DoD) currently does not have an established definition for information warfare, nor does it have operational theories for information warfare that extend beyond narrowly defined concepts, like those for cyber or influence operations. More importantly, information warfare is an orphan in the Pentagon; no one is in charge.[8]

He is correct. The authors learned that as of May 4, 2022:

- Information Warfare has been deleted from Joint Publication 1-02, DoD Dictionary of Military and Associated Terms.
- Influence Operations is not a term in Joint Publication 1-02, DoD Dictionary of Military and Associated Terms.

---

[5] 2019 US Code, Title 22—Foreign Relations and Intercourse, Chapter 38—Department of State, Sec. 2732—"Public diplomacy responsibilities of the Department of State." https://law.justia.com/codes/us/2019/title-22/chapter-38/sec-2732/.

[6] Theohary [4].

[7] Ibid.

[8] Dahm [5].

- Psychological Operations is not a term in Joint Publication 1-02, DoD Dictionary of Military and Associated Terms.
- Information Operations is a term, defined "as the integrated employment, during military operations, of IRCs [information-related capabilities] in concert with other lines of operation to influence, disrupt, corrupt, or usurp the decision making of adversaries and potential adversaries while protecting our own."
- Military information support operations (MISO) is a term defined as: "planned operations to convey selected information and indicators to foreign audiences to influence their emotions, motives, objective reasoning, and ultimately the behavior of foreign governments, organizations, groups, and individuals in a manner favorable to the originator's objectives."

It seems the United States of America doesn't do strategic information operations or political warfare very well. The authors conducted several interviews with subject matter experts in this space during research for this chapter. The conclusions drawn from those conversations were concerning. It seems that where the US does conduct information operations, it is mostly at the tactical level. Strategic information operations on a par with what Russia and China execute are not seriously considered. MISO practitioners shared with the authors that it can take ten days or more to obtain written approval to post on Twitter. In social media time, that renders such messaging useless. What is trending right now will be irrelevant by tomorrow morning. Perhaps America is nationally uncomfortable with state-conducted strategic narrative campaigns and manipulation. Regardless, US competitors have no such qualms, and their respective use of such strategy is effective and dangerous to America and its allies and partners.

1.3 Military

The M in DIME is for military. In the US, the military means the Department of Defense (DoD). The DoD's "…mission is to provide the military forces needed to deter war and ensure our nation's security."[9] This does not include law enforcement activities, which fall under the Department of Justice and Department of Homeland Security. This creates an exploitable gap for our GPC adversaries, where the DoD's Title 10 responsibilities leave off, but US law enforcement authorities do not extend.

1.4 Economic

> A strong US economy with free access to global markets and resources is a fundamental engine of the general welfare, the enabler of a strong national defense. In the international arena, the Department of the Treasury works with other USG agencies, the governments of other nations, and the international financial institutions to encourage economic growth, raise

[9] U.S. Department of Defense website, in the "About" section [6].

> standards of living, and predict and prevent, to the extent possible, economic and financial crises.[10]

By this definition, it's interesting to observe that while a strong economy is deemed the enabler of a strong national defense, responsibility for said economy falls to the Department of the Treasury. The military fights the wars. Treasury protects the economy. As will be seen in the next section, Russia and China don't separate those functions so cleanly. In fact, it is their blending, synchronizing, and muddying of the instruments of national power that has enabled them to run circles around the US in GPC for the last decade or two.

2 Russian and Chinese Use of DIME

While America's great power competitors certainly have these equivalent departments and agencies to support the performance of statecraft, it is the creative blending of DIME elements, both in and out of government, that may make them so effective in great power competition.

Take for example Russia's unique blend of diplomacy with information warfare, often used in creative combinations with other elements of national power. A useful example can be found in Russian Foreign Minister Sergey Lavrov's April 2019 speech at his annual meeting with students and professors at the Foreign Ministry's Diplomatic Academy. He said:

> The Western liberal model of development, which particularly stipulates a partial loss of national sovereignty - this is what our Western colleagues aimed at when they invented what they called globalization - is losing its attractiveness and is no more viewed as a perfect model for all. Moreover, many people in the very western countries are skeptical about it. [...] In order to preserve their dominance and recover their indisputable authority, they use blackmail and pressure. They don't hesitate to blatantly interfere in the affairs of sovereign states.[11]

In just four deftly constructed sentences, Lavrov inventively defies nearly every rule of critical thinking through (at least) the conflation of terms, use of conspiracy, introduction of doubt, misattribution, third party credibility, gaslighting, projection, agitation propaganda, cognitive bias, logical fallacy, and reflexive control. But perhaps most effectively he passively reframes Russia's expansionist, imperial aggression as defensive in nature.

Russia is not the only adversary/competitor that blends the D and I in DIME:

The Chinese government is extremely active on western-facing social media, and a recent Twitter post by a Chinese government official is illustrative (Fig. 1).

This is what China is selling. An alternative to the terms and conditions of partnership with America and her allies. Not only is the Chinese government blending

---

[10] Understanding the Instruments of National Power [7].

[11] Tass Russian News Agency [8].

Fig. 1 https://twitter.com/MFA_China/status/1529797157632372736

diplomacy and information warfare, but there is also no shortage of military and economic salesmanship occurring here.

Before recoiling or disputing this message though, if we make a lone exception for the NSC-68 driven intolerance for communism, how different is this message from the American message between 1945 and 1991? While it is certainly true the American government hoped "they'll become more like us" (allegedly the Bush administration's motivation to allow China entry into the WTO),[12] outside of the military alliance element cited above, this was close to US policy.

Again, with the communist exception, ideological, humanitarian, democratic factors did not significantly factor into decision calculous on suitability for American ally or partner criteria until after the collapse of the Soviet Union. In fact, America has historically been exceptional at everything the Chinese government claims to offer here and has rarely had difficulty overlooking atrocious human rights abuses

[12] Delingpole [9].

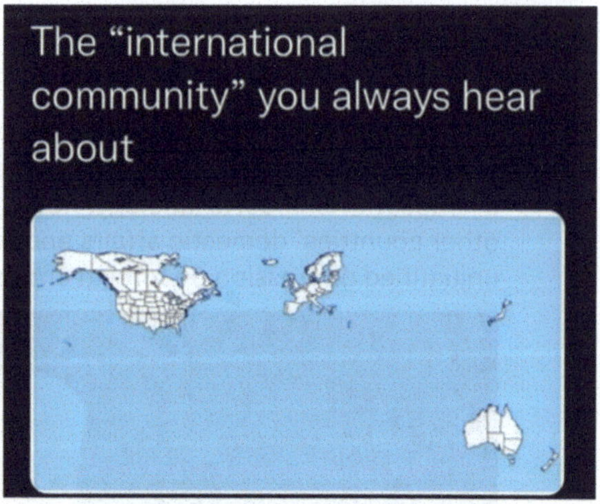

Fig. 2 Here's another made-for-Twitter meme attributed to China

and tyrannical dictators, so long as they were not perpetrated by or presided over by communists.

US history notwithstanding, what is not advertised in China's message is the terms on what they offer. Where the World Bank or the International Monetary Fund might issue loans as low as 0.25%, China is going to offer commercial rates—probably north of 6%. They will insist on collateralizing the debt with the very infrastructure the Chinese company is proposing to build for the host nation, using imported Chinese labor. And when the infrastructure project turns out to be unprofitable, China essentially drops a margin call on what is tantamount to a hard money loan and ends up owning a 99-year lease on a harbor, an airport, or a pipeline.[13]

While this messaging is highly aligned with Lavrov's statements at the start of this section, China is offering an alternative partnership. Lavrov in contrast seems to be purely aimed at deconstructing the message that sold America as an idea, since WWII (Fig. 2).

In the above, we see (allegedly) the Chinese re-framing the West as a tiny segment of nations on the planet, and they're doing so in English. So, this is clearly not intended for a domestic Chinese audience. While most Western analysts would probably start assessing the correlated GDPs or military strength of the nations presented here and dismiss this meme as infantile, it begs the question as to who is the target audience.

British researcher Carl Miller may have found the answer. In a March 18, 2022, Twitter post,[14] Miller reveals that semantic modeling of Russian messaging surrounding the Ukraine invasion exposes the target audiences: India, South Africa, Nigeria, Pakistan, Iran, and the global south generally. While in western-facing narratives Russian Foreign Minister Lavrov uses language the way an arsonist uses fire,

[13] Thorne and Spevack [10].

[14] Miller [11].

in Africa and Asia and the global south, both the Russians and the Chinese appear to be currying sympathy and seeking partnerships. But is it working?

The authors recognize that planning and executing influence operations is one thing but measuring the effectiveness and performance of these operations is something else entirely. While it is unclear how much of this is a chicken and egg scenario, advocacy for the March 2022 UN resolution condemning Russia's invasion of Ukraine and demanding immediate Russian withdrawal is telling.[15] Which nations—by region—either abstained from voting or sided with Russia?[16]

Russia and her near abroad (7): Armenia, Belarus, Russian Federation, Kazakhstan, Kyrgyzstan, Mongolia, Tajikistan

Africa (18): Algeria, Angola, Eritrea, Burundi, Central African Republic, Congo, Equatorial Guinea, Madagascar, Mali, Mozambique, Namibia, Senegal, South Africa, South Sudan, Sudan, Uganda, United Republic of Tanzania, Zimbabwe

Middle East (3): Iran, Iraq, Syrian Arab Republic

South Asia (4): Bangladesh, India, Pakistan, Sri Lanka

East Asia (4): China, Democratic People's Republic of Korea, Lao People's Democratic Republic, Vietnam

Western Hemisphere (4): Bolivia, Cuba, El Salvador, Nicaragua

It seems clear from this metric that by creatively blending diplomacy with information operations, Russia—and possibly China—are gaining favor with the countries above.

3 Case Studies

The integration of DIME elements becomes even more interesting when we examine how the Russian and Chinese governments not only verticalize instruments of national power but integrate public and private sectors as *operational* fronts. In the authors' experiences, nothing conveys such information better than case studies of real-world actions.

The authors will illustrate a case study for Russia—The Wagner Group. This shows how a media invented term is actually a network of profitable entities operating outside of any nation-state construct, in possession of all DIME elements of national power, and works to benefit Russia while enriching their shareholders. While Wagner are often referred to as mercenaries, this is likely due to fact there are not modern terms or language to describe what they are more accurately. There is no Western equivalent.

[15] Permanent Representative of Ukraine to the United Nations [12].

[16] Lawal [13].

This chapter will also look at China's Frontier Services Group (FSG), who on the other hand, operates more like a US government contractor (but with more state-backed authority due to the civil-military fusion concept), but nonetheless functions—like Wagner—a profitable entity operating nominally outside of the nation-state construct, in possession of all DIME elements of national power, and works to benefit China while enriching their shareholders.

Both would likely say they are actually copying the US. But this is different. This is effective. And again, there is no US equivalent.

The case studies will conclude with an examination of China's United Front Work Department (UFWD), a CCP-organized and administered global network of influence and espionage activities.

Given the sheer volume of African countries above who both (1) abstained from voting against Russia in the recent UN resolution, and (2) play host to Russia's Wagner Group, the first case study will focus on Wagner and their operations in Sudan specifically.

3.1 The Wagner Group (Russia)

If you hadn't heard of them prior to the outbreak of war in Ukraine, you might view the Wagner Group as a new arrival on the private military company (PMC) scene. For those who follow the PMC world, however, Wagner is a familiar and growing nemesis. Current estimates put the number of people employed by Wagner at about 10,000 people.[17] What most public reporting—and even academic writings on the subject—does not convey well is what Wagner *is not*.

Wagner is not a single incorporated PMC. It is a network of loosely affiliated and mutually supporting companies, all pointing back to a single Russian businessman, although there is no formally documented ownership since there is no monolithic company. This businessman is Yevgeny Prigozhin. Prigozhin either owns or controls (via other Russian individuals) a growing empire of international companies spread across multiple sectors but concentrated in information, security, and natural resource extraction. He holds no government appointments, although his businesses include a number of Russian government contracts. PMCs are illegal under Russian law, so Prigozhin's approach is clever and obscures direct ownership. He has multiple Russian-domiciled logistics, supplies, and administrative business entities, but when he moves into a new geography appropriate for Wagner's services, he incorporates (through one of his people) a security operations company in that location. This keeps him on the right side of Russian law since it is not a Russian company. It also frequently makes those entities difficult to monitor, because they tend to be in countries where corporate registrations are difficult to access from outside the country—and sometimes from inside the country as well.

[17] Beardsley [14].

Most reporting on Wagner says that the group began functioning as a business enterprise in 2014. A detailed report from an independent Russian journalist[18] argues the concept was first floated in Russian government and military circles in 2010, at the St. Petersburg International Economic Forum. One of the forum's invited speakers was Eben Barlow (founder of Executive Outcomes[19]), who reportedly also held a closed-door session with select members of the Russian General staff to discuss models for creating PMCs in Russia. The idea had supporters but was tabled for a few years, in part due to the Russian law prohibiting PMCs. Three years later, what would become Wagner began forming under Prigozhin's leadership, reportedly because he was seen as someone favored by Putin but with no government role, who could be trusted but also disavowed. In 2014, rumors of military retirees brought back for paid work in Crimea and Ukraine began circulating. Fast-forward nearly a decade, and Wagner's active presence has been confirmed not only in Crimea and Ukraine, but also in Syria, Libya, Venezuela, Mali, Sudan, the Democratic Republic of Congo (DRC), Madagascar, Angola, Guinea, Guinea-Bissau, Mozambique, Zimbabwe, and the Central African Republic (CAR).

The genius role of Wagner in great power competition is that it is an entirely private enterprise, with no central office/structure. It is part of a network of enterprises, purpose-built to avoid scrutiny. And it is used largely in locations where there are natural resources to be protected and/or extracted, such as oil and gas, minerals, gold, and rare elements. As these locations tend to be led by individuals with more autocratic than democratic tendencies, Wagner's approach is to offer security advisors and training to the ruling power, along with digital influence operations against the political opposition, in exchanging for mining and extraction rights on valuable natural resources. Said resources are then exfiltrated either directly to Russia or to third party nations who will convert the goods to a currency of Russia's choice. Thus, the host nation benefits from Wagner employees' military training and background, Prigozhin benefits from the mining proceeds, and Russia benefits by having a proxy presence in key countries, as well as deniability of same.

In Sudan, Wagner was invited into the country by the former President, Omar al-Bashir, in 2018, and remained after the 2019 coup that overthrew al-Bashir. Wagner reportedly came to Sudan to provide training and technical assistance to Sudanese security personnel,[20] particularly the Rapid Support Forces (RSF), which answer not to Sudan's military but to General Mohamed Hamdan Dagalo (nicknamed "Hemedti") and his family personally. The extended Dagalo family controls the most lucrative of the gold mining operations in Sudan,[21] via shell corporations and proxies, with muscle provided by the RSF. Some estimate that 70% of Sudan's mined gold product is smuggled abroad.[22]

---

[18] Malkova and Baev [15].

[19] SourceWatch [16].

[20] Meyer et al. [17].

[21] Collins [18].

[22] Ibid.

Fig. 3 Overhead image of Meroe Gold office in Khartoum, Sudan (*Source* Google Maps)

Although Dagalo and other Sudanese officials have stated as recently as the spring of 2022 that no Wagner personnel remain in Sudan, rumors and sightings of men, equipment, and vehicles continue to emerge. A first-hand account report to the authors confirms that as recently as January 2022, at least one individual in military style clothing bearing a Wagner-affiliated insignia was seen in a hotel in Khartoum. That hotel was located near the physical address for Meroe Gold (seen in overhead image below). Meroe Gold is a US-sanctioned entity affiliated with the Prigozhin network. As of winter 2022, Meroe Gold was a Sudan-registered company, although a recent report on social media[23] suggests that Meroe Gold has been dissolved, and possibly reformed as Alsolag Mining Co Ltd. (Fig. 3).

Using publicly available import–export data, the authors found that Meroe Gold in January of 2021 received shipment (using the address in Khartoum) of a helicopter. The shipper was M-Invest LLC.[24] M-Invest LLC (also US-sanctioned) is one of Prigozhin's logistics and equipment entities based in St. Petersburg, Russia (Fig. 4).

In the entity map above, we can see a portion of Prigozhin's global business holdings. Prigozhin himself is in the red circle at the top of the map. The yellow dots on the map represent people—either shareholders or directors, or others with organizational control. The blue dots represent corporate entities. As of April 2022, Prigozhin's

23 All Eyes on Wagner [19].

24 ImportGenius [20].

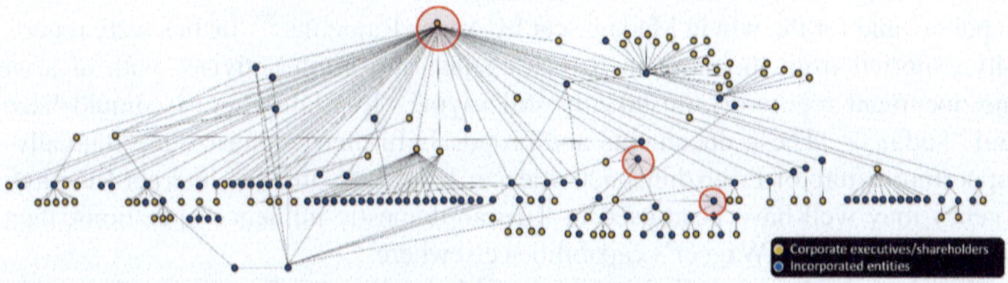

Fig. 4 Partial map of Yevgeny Prigozhin's business holdings[25]

network (only to three levels of depth) included at least 152 people/entities that he owns or controls. Moving diagonally down to the right in the entity map, the next red circle highlights M-Invest LLC. And the final red circle, near the bottom right, is Meroe Gold Co. LTD. Reporting in June 2022 strongly suggests that Meroe Gold is a vital source of undeclared gold being shipped from Sudan to Russia or Russia-controlled locations, to prop up Russia's besieged economy and skirt US and allies' sanctions.[26] Russia reportedly has $130 billion in gold secured in its central bank but needs to continue growing that reserve, to hedge against long-term damage to the Russian economy and to finance current military operations in Ukraine… and possibly elsewhere. Alternatively, Russia may be selling Sudanese gold extracted by Meroe to intermediaries in exchange for hard currency. It seems most likely that both are true.

Security training and assistance is not the only skill on offer by Wagner in these mostly African countries with exploitable natural resources. Wagner has increasingly been tied back to in-country influence operations, merging online information operations with physical threats and intimidation by Wagner personnel. Part of Wagner's tasking in Sudan when they arrived in 2018 was to help quell anti-al-Bashir protests. In documents attributed to M-Invest LLC (see paragraph above), the company:

> …proposes spreading claims that protesters were attacking mosques and hospitals. It also suggested creating an image of demonstrators as "enemies of Islam and traditional values" by planting LGBT flags among them. And it proposed a social media campaign claiming that "Israel supports the protesters." […]
>
> But most of its focus was on the protests. It recommended fabricating evidence "of arson by protesters against mosques, hospitals and nurseries, [and] stealing grain from the public store."
>
> It also suggested blaming the West for the protests and using "extensive media coverage of the interrogation of detainees, where they admit they arrived to organize civil war in Sudan." And it even proposed "public executions of looters and other spectacular events to distract the protest-minded audience."[27]

It seems that Wagner's influence operations were not limited to Sudan. They have claimed responsibility for losses by opposition parties in the Central African

---

[25] Helios Global Inc [http://heliosglobalinc.com/] proprietary, Apr 2022.

[26] Walsh [21].

[27] Lister et al. [22].

Republic, and for the win in Madagascar by Andry Rajoelina.[28] Tactics were reportedly exported from their use in Russia against anti-Putin activists, with at least one document reportedly mistakenly still saying "Russia" where it should have said "Sudan."[29] These documents and proposals further emphasize the mutually-supporting nature of Prigozhin's activities, as it suggests that his Internet Research Agency may well have been used in Russian domestic influence operations, then exported to help sell Wagner's capabilities elsewhere.

How does this benefit both Prigozhin and Mother Russia? To start with, Prigozhin is almost certainly getting richer. At the same time, Prigozhin's activities are benefiting both Russia itself and the governments where they operate, by creating closer, mutually attractive ties between the countries without the more intense scrutiny and oversight that actual Russian military activity would bring. Are Prigozhin's operations costing the Russian taxpayer anything? It doesn't look that way, as the enterprise appears entirely self-supporting. Perhaps most critically though, we should ask, how is Prigozhin's empire competing with elements of US instruments of national power, and how are Prigozhin's activities undermining the national security strategy of the United States? Are they?

While it may not be the place of the US—a sovereign state—to compete with a corporate empire, what tools are available to the US government in this circumstance outside of sanctions? Many of Prigozhin's entities mentioned here are already sanctioned by the US Treasury's Office of Foreign Assets Control, yet they remain operational. The real question may be *if* the US elects to respond, *how* will the government implement policy here? What other instruments of national power are legally available to decision makers?

3.2 Frontier Services Group (China)

While China has no readily apparent equivalent to Wagner, the closest comparison may be Frontier Services Group (FSG). In contrast to Russia, China has taken the approach of building private security contractors (PSCs), rather than PMCs. Their focus is on protecting Chinese Belt & Road (BRI) investments and projects across the globe, and providing supplementary services domestically, where military or law enforcement may be inappropriate or undesirable. Although some Chinese PSCs have grown from scratch, the largest and most known in the West is Fronter Services Group. The story of how FSG came to be a Chinese state-owned enterprise (SOE) is in many ways a prime example of China's GPC tactics.

FSG began as an American-owned company. It was founded in 2014 by Erik Prince, who also founded (and later sold) the controversial American PMC Blackwater. FSG is incorporated in Bermuda, publicly traded on the Hong Kong Stock Exchange, and it is managed from its Global Security Center in Dubai. Prince owned

[28] Harding and Burke [23].

[29] Ibid.

Fig. 5 Photo of signing ceremony between FSG executives and local government officials for construction of training center in Xinjiang[31]

and ran FSG on the public side as mainly a logistics firm for several years, adding Africa-based mobility platform capabilities, sometimes in return for African natural resource extraction deals. Prince stated from the beginning that he was interested in opening the Chinese security services market, and FSG was his vehicle for doing so. In 2017, FSG acquired the International Security and Defense College (ISDC) in Beijing. The ISDC is an end-to-end security training platform for FSG staff and for clients going abroad.

In 2018, FSG announced it had accepted investments by CITIC (a State Owned Enterprise (SOE) and previous investor, who increased their shareholding to 28.4%) and China Taiping (7.48%). With these investments came FSG Board of Directors seats. FSG was established with an eye towards logistics and security services in China and Africa, so it was a natural target for Chinese state investment. In 2019, news broke[30] that a business unit of FSG (FSG Security) had signed a deal with the management committee of an industrial park in Xinjiang, to build a "training center" (Fig. 5).

Human rights advocates and world media immediately tied this training center project to the detention, suppression, use of forced labor, and "re-education" of the minority Muslim Uighur population in the region. Although Prince hotly and consistently denied knowledge of this agreement—and the announcement was pulled from FSG's website swiftly—Prince and FSG were permanently associated in the public

[30] Zhen [24].

[31] Ibid.

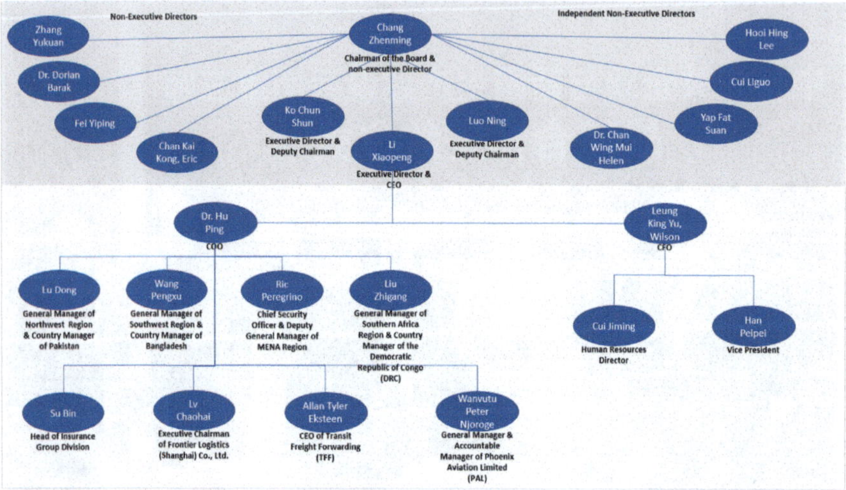

Fig. 6 FSG organizational chart[33]

eye with the Chinese government's repression of the Uighurs. Twelve days after the swiftly withdrawn announcement, a CITIC executive was named the Chairman of FSG's board of directors.[32] By 2021, CITIC and other Chinese investors had bought out Prince entirely, and he resigned from the board (Fig. 6).

Thus, in less than seven years, a company founded and run by an American was completely taken over by a Chinese SOE. The intelligence oversight ramifications of an American-led company engaged in this sort of activity appear to constitute a critical US national security weakness. A weakness almost certainly noticed by the PRC. FSG's stated purpose remains to expand along China's Belt and Road initiative locations, providing security services to Chinese-owned assets and projects in those places. The "training facility" in Xinjiang remains under contract with FSG, although the company downplays or denies it. Although the company itself states it only had 1837 employees as of the end of 2021,[34] according to an American former FSG executive, the company as of 2021 had over 35,000 personnel. (This significant discrepancy may be due to FSG counting only employees of the parent holding company.)

It is useful to recall at this point that under the CCP and President Xi, every commercial enterprise and initiative is meant to be capable of dual purposes. The Military-Civil Fusion (MCF) is a national strategy of the CCP, and its goal is to make the PLA the most advanced military in the world. According to the US Department of State, "a key part of MCF is the elimination of barriers between China's civilian

[32] Ibid.

[33] Helios Global Inc., using data from http://fsgroup.com/en/team.html.

[34] Frontier Services Group Limited [25].

research and commercial sectors, and its military and defense industrial sectors."[35] Although some prominent China scholars argue that MCF is still more aspirational than in current implementation,[36] FSG shows clear signs of the influence of MCF in guiding FSG's expansion in offerings and geolocations. They have operations throughout China, and abroad in Cambodia, Pakistan, Laos, Myanmar, Mozambique, Tanzania, the DRC, Kenya, South Africa, and more. Their 2021 Annual Report[37] states that FSG is executing a contract to build the headquarters office building for the Air Force of the Lao People's Democratic Republic (Laos). Late in 2021, FSG announced that it was acquiring Hong Kong's DeWe Security, giving FSG a larger Hong Kong presence and raising concerns about the possible use of FSG assets by China as proxy forces to suppress anti-China sentiment and protests by pro-democracy activists. DeWe is a major Chinese PSC, with projects in more than 40 countries protecting Chinese-funded enterprises and engineering projects with a value of over $20 billion USD. DeWe's founder, Li Xiaopeng, is now the CEO of FSG.[38]

Some specific applications of FSG align with the MCF strategy. FSG first made its intentions to move into Myanmar public in its 2018 annual report, in a section discussing growth plans. In March of 2019, reporting came out that FSG had entered a joint venture (JV) in Myanmar called FSG (Myanmar) Security Services Co Ltd and started recruiting for security personnel.[39] The JV had two Myanmar locals and two mainland China directors. Local rumors at the time suggested that FSG was going to provide security for a controversial hydro-electric dam project in the north of the country. Other reporting[40] drew more definitive ties between FSG's entrée into Myanmar and the Kyaukphyu deep water port in Rakhine. It turns out that CITIC Group's Myanmar entity is the main investor in that port, in a $1.3 billion project that was finalized in November 2018. Rakhine is one of Myanmar's poorest states, with ethnic unrest common. So, it seems clear that CITIC, an SOE, wanted to use FSG (now another SOE) to protect the massive government investment in the deep water port. The port itself could then be subject to MCF appropriation. Thus, China used its SOEs to commercially acquire, develop, and protect a potential forward-basing facility and port of call for PLA and PLA-N, should the need arise (Fig. 7).

FSG activities in Cambodia are also illustrative of the Chinese strategy. Under its current leadership, Cambodia has arguably become, in recent years, a de facto vassal state to China. With massive Chinese investment first in gambling and entertainment destinations for Chinese nationals, and then in Cambodia's deep water ports, Chinese influence has increased rapidly over the last decade. But despite the cozy relationship between Cambodia's government and the PLA-N,[41] the relationship between the

[35] U.S. Department of State [26].

[36] Kania and Laskai [27].

[37] Frontier Services Group Limited [25].

[38] Frontier Services Group [28].

[39] Lintner [29].

[40] AFP [30].

[41] Rising and Cheang [32].

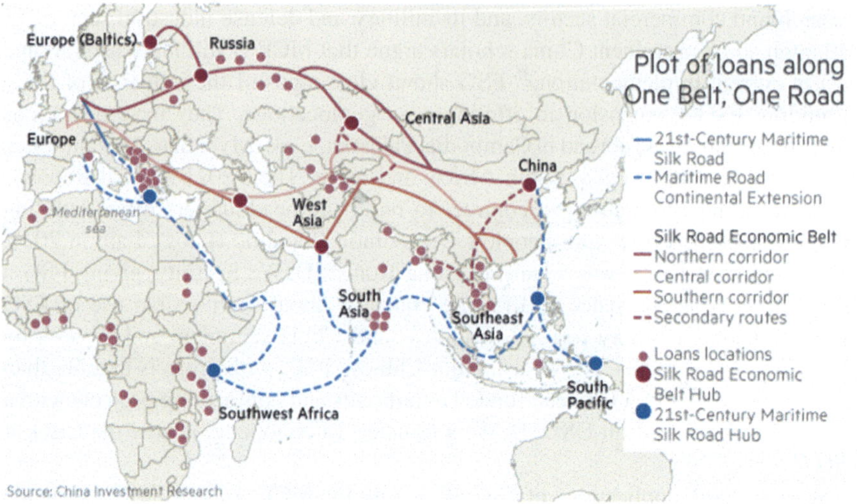

Fig. 7 2015 belt and road map[42]

general Cambodian population and Chinese nationals residing in Cambodia is not entirely friendly. The Cambodians are increasingly squeezed out of the housing and job markets in the locations where Chinese have invested heavily, creating friction and resentment. While the Chinese government wants its foreign investments protected, it is not yet prepared to deploy military forces to secure those investments. That's where FSG comes in. As of late 2020, FSG Cambodia Security Service Co ("FSG Cambodia") began providing comprehensive security services to the Bank of China Phnom Penh Branch. According to FSG's press release at the time they initiated these services,[43] FSG Cambodia is the only Chinese joint venture to hold a cash escort license there, and one of only four in the country. They also say they have over 30 security projects in country, including at Angkor International Airport. Further, they have created security training bases in Cambodia.[44]

What we can see here in Myanmar and Cambodia are not just Chinese projections of power by establishing a commercial security presence to protect commercial investment and infrastructure projects along the Belt and Road. These expansions are also chess pieces in the MCF strategy, whereby any Chinese-owned civilian resource or capability can and should be capable of being dual-purpose for military applications. Accordingly, CITIC builds a deep water port—protected by FSG—which facilitates Chinese commerce today but can also be flipped to use by the PLA-N on short notice. And in Cambodia, where government-government relations are already cozy (and Cambodian leadership is well compensated for its friendliness to China), FSG is used to protect the playground for Chinese elite, ensure the continued

---

42 China Investment Research [31].

43 Frontier Services Group [33].

44 Ibid.

flow of cash to and from the China mainland and Cambodia, and to ensure continued air access, should the country suddenly become unstable.

3.3 United Front Work Department (China)

The final case study will examine the activities of a more directly CCP-controlled entity, albeit an entity that appears to mount government operations against the private sector, target aspirational politicians, and exert influence on Chinese diaspora communities. It is used as a global tool for protecting China's interests abroad. It is also used for spotting, assessing, engaging, and retaining non-Chinese nationals to act locally on behalf of China, where the Chinese government acting directly might not be welcome, effective, or legal.

China's United Front Work Department (UFWD) is essentially the PRC's political warfare machine. The United States has no equivalent organization, or even any meaningful counter-capabilities. The authors assess this is due to the fact that the US government does not understand, appreciate, or value political warfare. The US-China Economic and Security Review Commission described the UFWD as follows:

> China uses what it calls "United Front" work to co-opt and neutralize sources of potential opposition to the policies and authority of its ruling Chinese Communist Party (CCP). The CCP's United Front Work Department (UFWD)—the agency responsible for coordinating these kinds of influence operations—mostly focuses on the management of potential opposition groups inside China, but it also has an important foreign influence mission. To carry out its influence activities abroad, the UFWD directs "overseas Chinese work," which seeks to co-opt ethnic Chinese individuals and communities living outside China, while a number of other key affiliated organizations guided by China's broader United Front strategy conduct influence operations targeting foreign actors and states.[45]

While the US-China Economic and Security Review Commission details a variety of location-specific operational thrusts and efforts, the mission of UWFD invariably distills down to classic political warfare. The major operational foci in Australia and New Zealand are cited as: Political Influence, Controlling the Media, and Subverting Unfavorable Narratives. Notably, with respect to political influence, the Commission cites the following:

> According to Clive Hamilton…United Front organizations groom members to participate in politics in target countries, both as candidates for election and as staff in important positions. The Australian Security Intelligence Organization has reportedly estimated at least ten recent Australian state and local government political candidates are connected to Chinese intelligence agencies.[46]

While the Chinese government downplays the importance of the United Front Work Department (UFWD) and regularly claims the West conflates state-sponsored

---

[45] Bowe [34].

[46] Ibid.

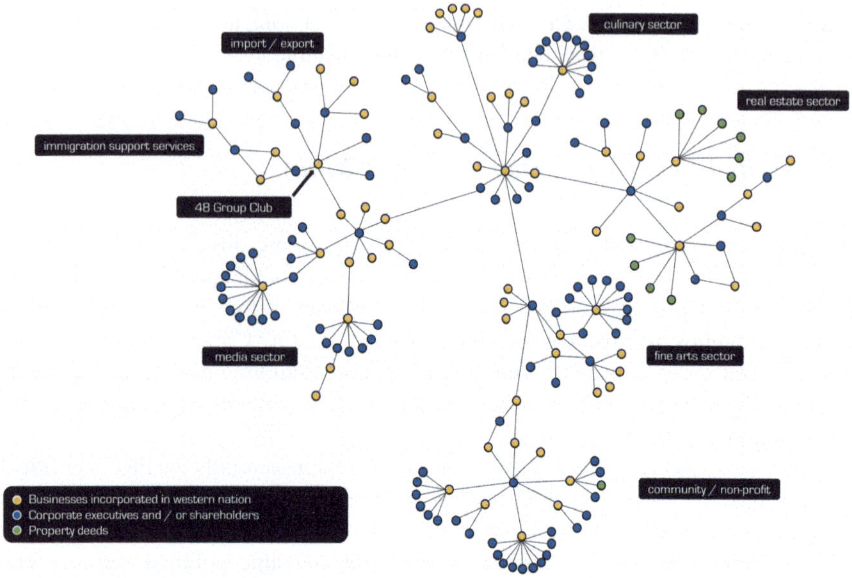

Fig. 8 Entity map of a United Front work department project[49]

influence operations with normal diplomacy and steady-state business, Georgetown University researcher Ryan Fedasiuk writes the following for the Jamestown Foundation:

> …there is a universal truth known to government bureaucrats in every country: budgets speak louder than words. […] the scale and scope of funding for the united front system belie the Chinese government's claims about its importance and function. This article synthesizes information from more than 160 budget and expense reports from national and regional PRC government and Communist Party entities. It finds that organizations central to China's national and regional united front systems spent more than $2.6 billion in 2019, exceeding funding for China's Ministry of Foreign Affairs (MFA, 2020). Nearly $600 million (23 percent) was set aside for offices designed to influence foreigners and overseas Chinese communities.[47]

The United Kingdom's 48 Group Club, described extensively and characterized as a UFWD entity in Clive Hamilton's Hidden Hand[48] was analyzed further by the authors using commercial FINTECH tools and British corporate registration data to perform network analysis and determine involved parties and related entities. The following is a depiction of the results of this effort (Fig. 8).

Notably, this analysis revealed an almost equal number of Sino and Anglo surnames listed as beneficial owners or company officers of these firms. This—on

---

[47] Fedasiuk [35].

[48] Hamilton and Ohlberg [36].

[49] Helios Global Inc [http://heliosglobalinc.com/] proprietary, Mar 2022.

the surface—appears to support Hamilton's argument that the British high-net-worth circuit is being groomed to support the CCP's agenda.

Next, profit and loss (P&L) statements and other financial reporting strongly suggests circuitous transactions between and amongst several of these entities. For example, the firm's Annual Report and Unaudited Financial Statements for the Period Ended 31 March 2020 states:

> The Chairman of the 48 Group Club, Mr. Stephen Perry is the Chairman and owner of 100 per cent of the shares of Icebreakers (UK) Limited which has been organizing the annual Icebreakers Chinese New Year Dinner for many years.
>
> During the period to 31 March 2020, Icebreakers (UK) Limited made a donation of £25,000 (2019 - £25000 to the Club from the net surplus achieved from the 2019 Icebreakers Chinese New Year dinner.[50]

Finally, note in the graphic the evidence of highly disciplined cellular structures organized around industry sectors, or verticals. This does not look anything like what analysts would expect to see in the US or Gulf State business environments. This strongly suggests top-down command and control, but notably, it is not a hierarchical structure.

Critically, none of these enterprise sectors remotely approach any space with security controls or oversight by government security officials. There is no nexus to technology, defense, infrastructure, supply chains, etc. This specific structure appears to be concerned with the acquisition of real estate, the profitable import–export trade, supporting PRC nationals relocating to the United Kingdom, developing and proliferating pro-PRC messaging, and as Hamilton asserts in Hidden Hand, currying favor and influence with high-net-worth British nationals.

While this seems perhaps benign, an examination of UFWD operations in Australia paints a remarkably more alarming story.[51] This example illustrates how organizations like 48 Group Club are used to spot, assess, and develop commercial assets that are eventually used to support the acquisition of infrastructure as part of the Belt and Road Initiative.

The company in question here is Landbridge Infrastructure, a subsidiary to Landbridge Port Group in Shandong, China. The Northern Territories government in Australia decided to privatize and lease out the Port of Darwin in an effort to bring in foreign investment and fund necessary upgrades to the port's infrastructure. Landbridge was one of 33 investors who were interested. They significantly outbid any competitors, ultimately winning with a bid of AUD $506 million (approximately 25 times the port's profit over the preceding two years), while promising to invest an additional AUD$200 million over the next 25 years.[52]

Had Landbridge been a Chinese SOE, the deal would have likely been declined on national security grounds. However, since Landbridge was viewed as a private business, it was able to bypass much government scrutiny in something of a legal

---

[50] The 48 Group Club [37].

[51] The authors wish to thank Mr. Gerald Brown [@GeraldC_Brown] for his insights and research support on this section.

[52] Garrick [38].

loophole, being subject to a significantly reduced review process.[53] It wasn't until after the deal had been secured that many entities of the government became aware of the deal and condemned it. There is speculation that local corruption may have played a factor in this acquisition as well, though it's difficult to corroborate definitively with available open source information.

Local media outlets reported that Landbridge had been cultivating local Australian politicians connected to the deal.[54] Further, the former Australian trade minister connected to the deal, Andrew Robb, conveniently left government shortly after the deal finalized, accepting a position with Landbridge as a consultant with a hefty annual salary of AUD $880,000. He also now serves on an advisory board to help promote BRI.[55]

Currently, Landbridge owns an 80% share of several key areas of the port while the Northern Territories government owns the remaining 20%. Landbridge has taken operational responsibility for many key areas of the port, including the critical military-civilian joint use-piers, which are the only facilities that can facilitate large ships.

But if Landbridge is a private company, why does this matter? First, the term "private company" has a significantly different connotations in China than it does in many Western countries. While they are not directly controlled by the Communist Party, the Party still maintains a presence within these companies, and these companies can often gain advantages through close ties with the party. While the CCP does not necessarily manage their day-to-day operations, it has significant influence with key decisions, and the lines between the state and private enterprises are increasingly blurred. This is particularly true under General Secretary Xi Jinping, who has overseen a massive strengthening of state influence over private businesses in China, increasingly adding party elements within ostensibly private companies and having party members attend executive level meetings to weigh in on key decisions. Xi further began requiring businesses to write in an expansive role for the party in their guidelines and cooperate with the party in things such as intelligence efforts as the party sees fit. Reports from the Communist Parties Central Organization Department claimed 68% of China's private companies had Communist Party bodies included by 2016, with the goal of continuing to increase this number.[56]

This is readily apparent within Landbridge. Until 2018, the owner of Landbridge, Mr. Ye Cheng, was a member of the Chinese People's Political Consultative Conference, a prominent body of the United Front Work Department.[57,58] Mr. Cheng openly proclaimed that the Darwin acquisition was his contribution to China's Belt and Road Initiative, and General Secretary Xi Jinping himself came to the opening of Darwin Port with Mr. Ye Cheng. Further, Landbridge maintains substantial ties to

53 Anderson [39].

54 Tanaka and Shimbun [40].

55 McKenzie and Massola [41].

56 McGregor [42].

57 McKenzie and Massola [41].

58 Angliviel de la Beaumelle [43].

the People's Liberation Army, even establishing a People's Armed Militia branch within the company. Taken from a statement by Landbridge in 2014:

> Comrade Ye Cheng was cited by the Provincial CPC Committee and the Provincial Government as one of the 10 outstanding individuals of Shandong province who have concerned themselves with national defence construction. The establishment of the Landbridge Group's Armed Militia manifests the true melding of a strong enterprise and an active armed militia under the Landbridge Group.[59]

Revisiting the potential strategic implications of this deal, many components of BRI appear to be about energy and resource security, widening China's circle of friends and political influence by fostering economic dependence on China, building supply route redundancy, and the ability to support naval assets abroad, all of which are supported by the Darwin acquisition. Additionally, the geopolitical location is of great significance, and it's the only location in Northern Australia that's suitable for major naval use. Critical joint-use wharfs are now operated under Landbridge's control, for a port that hosts US and allied warships, and provides basing for both US and Australian Marines.

As evident from above, the PRC cleverly blends elements of national power in a manner that can fall outside of the US', US partners' and allies' ability to identify and counter given current constraints. Arguably (and assuming for the moment that media allegations are indeed accurate) in the Landbridge/Darwin case, this operation involved:

- Use of a private sector entity to cultivate local politicians critical to the business deal
- Bypassing the central government and working with the provincial government
- Conveying information that a potentially government-controlled firm was in fact privately held
- Outbidding the competition for the deal, possibly using national treasure
- The acquisition of a civilian-military dual use transportation hub to further project power and influence across the DIME spectrum.

4 Conclusion

The following provides a summary of the implementation of instruments of national power as currently practiced by the primary great power competitors, and accordant report cards.

US: The US government generally separates responsibilities and authorities for executing policies across the DIME spectrum. Exceptions include Joint Interagency Task Forces (JIATFs), bust most often these are transient, and issue

[59] Barnes et al. [44].

specific. The US invariably separates government functions from commercial activities and maintains sharp lines between military and intelligence authorities. US policies under Executive Order 12333 prohibit collections on US Persons (who—in the authors' experience—nearly always have presence in foreign commercial networks and ecosystems).

Diplomacy: D. Can vacillate wildly from administration to administration.

Information: F. Abjectly incapable at the strategic level, mired in outdated bureaucracy, leaderless and visionless.

Military: A+. Far and away the most capable in the world.

Economic: B+. While exceedingly powerful, democratic principles and norms limit the state's ability to utilize as an instrument of power. USD as a global reserve currency is critical.

GPA: 2.25

Russia: Often labeled a kleptocracy, as discussed in the introduction, Russia functions more like a nuclear-capable mafia family than a modern government. Policy is implemented with a wink and a nod to commander's intent, designed to muddy attribution. The foundation of policy is deception. Information is the weapon of choice, and information dominance is the goal.

Diplomacy: C. Passably competent, but rife with deception.

Information: A. Outside of the strategic nuclear stockpile, the crown jewels of the Russian Federation.

Military: C. Conventional forces are technically inferior, outdated, outmoded, and poorly led. However, nuclear forces offset conventional inferiority to a significant degree, strategically.

Economic: D+. Outside of natural resources—specifically energy—this is not a great story. However, by directing/controlling private sector activities, Russia has stayed in the game.

GPA: 2.375

China: China's energy, food, resource, and intellectual property deficiencies—combined with their de facto workshop of the world status—makes China's foreign policy primarily about acquisition, and often theft. China has integrated, synchronized, and centralized all DIME elements (much easier in an authoritarian state), and nearly fully integrated the public and private sectors.

Diplomacy: B. Integrated with other elements of power, China has gained preferred partner status in many nations previously preferential to the US.

Information: B. China has a narrative to sell, and it appears to be working.

Military: B. Arguably the second most powerful.

Economic: B+. Currently the world's industrial base but lacking in resources. China's predatory mercantilism and economic agenda has created a new frontier in geopolitical leverage.

Note: The authors are of two minds regarding China. While the authors concur with Peter Zeihan's geopolitical and demographic assessment that indicates very dark days lay ahead for the CCP,[60] *the authors also find the anonymous blogger "DeepThroatIPO" to have a very compelling take on China's ability to weaponize the spoils of years of trade surplus and currency manipulation.*[61] *If both are true, China may be very destructive when they eventually again decline.*

GPA: 3.125

As any student has learned, Ds and Fs destroy your GPA. If the United States could be consistent in its diplomatic actions and was even passably competent at using Information as an instrument of national power, the US would be a clear GPC valedictorian. Worse, the latent informational capabilities of Silicon Valley exceed the combined capabilities of the rest of the planet by many orders of magnitude. There is just no political will to use American industry for US geopolitical advantage. Further, the current domestic mood leans more toward regulating and/or breaking up Silicon Valley's information behemoths, rather than harnessing them.

It is impossible to dominate the information environment when social media rules of engagement (ROE) for Military Deception (MILDEC)/Psychological Operations (PSYOP) personnel require ten days for approval before they can post or respond. In social media time, ten days is an eternity. The war is over by then. Outsourcing to the private sector is currently the only reasonable strategy.

However, the real threat from our great power competitors is that they have verticalized and commercialized the DIME construct and blended it with privateering. The West has nothing to compete with this but "STOP IT!" Can any reasonable analyst envision a scenario in which the United States brings the full weight of her national power to bear on Sudan, to eradicate Prigozhin's assets, support Degalo's military regime's transition to a palatable allied state, and bring in US companies to mine gold under diplomatically guaranteed mining concessions? We can't either.

On the other hand, what if the US Congress passes into law a bill to facilitate some variant of privateering, with some indemnification against legislative and regulatory constraints, thus facilitating American enterprise to go to Sudan and compete with Prigozhin for profit?

---

[60] Zeihan [45].

[61] DeepThroatIPO [46].

5 Summary of the Problem and the Solution

THE PROBLEM: The US may still be a great power, perhaps even *the* great power, but largely due to two primary factors, the US stands on the sidelines when it comes to "great power competition." These factors are (1) an inability to use information as an instrument of national power, and (2) the inability to use the capabilities of the private sector as an instrument of national power (or, arguably, to effectively protect the intellectual property of American industry).

Russia and China spent the last two decades building their capabilities and launching their activities to undermine American influence while elevating their own, all in the realm of strategic competition below the threshold of kinetic action. They took all that is core to the American identity—free speech, individualism, capitalism—and figured out how to turn those strengths against the US, to compete with the US for global dominance. And they did it in a manner that bypasses and/or sidelines two of America's greatest strengths: the US DoD and the Intelligence Community.

SOLUTION 1: The priority of action for the United States should be to determine exactly what message the United States is selling to the rest of the world and setting about aligning domestic policies to support that message. President Eisenhower understood this in the 1950s when the Soviets weaponized information regarding American civil rights and set about correcting it. If one can easily be made out as a hypocrite, messaging will fail.

SOLUTION 2: Next, the United States should actively market the message to the rest of the world. Critically, the execution *must* rely on private sector capabilities, speed, efficiencies, and infrastructure.

SOLUTION 3: Partnership with the United States should not be handcuffed by requirements to adhere to American democratic ideals. Neither Russia nor China sets such expectations, and many countries that could help the US counter Russian and Chinese aggression are unwilling or unable to comply with such demands from the US. Basing strategic relationships on US values—including respect for other cultures—rather than ideological concepts will yield better results.

SOLUTION 4: The United States should contemplate amending, updating, or creating exceptions to its economic and financial laws and regulations to incentivize its private sector to compete and engage with Russia and China in great power competition, on an equal footing.

If the United States government alters its thinking and expectations in alignment with the above proposed solutions, it is still possible to check the forward momentum of Russia and China. If these changes or similar are not implemented soon, America's place as the leader of the free world may already be in the rear-view mirror.

References

1. Zeihan P (2020) Disunited nations: the scramble for power in an ungoverned world. Harper Business, 03 Mar 2020
2. International Monetary Fund (2022) World economic outlook database, April 2022. https://www.imf.org/en/Publications/WEO/weo-database/2022/April/weo-report?
3. Ryan S, Conway I, Cassedy K (2022) Who are the great powers in 2021? In: Farhadi A, Masys A (eds) The great power competition, vol 2. Springer Nature, Switzerland AG. https://doi.org/10.1007/978-3-030-90390-9_3
4. Theohary C (2021) Defense primer: information operations. In: Congressional research service 01 Dec 2021. https://crsreports.congress.gov/product/pdf/IF/IF10771
5. Dahm M (2021) The reality of war should define information warfare. Proceedings, vol 147/3/1417 U.S. Naval Institute Mar 2021. https://www.usni.org/magazines/proceedings/2021/march/reality-war-should-define-information-warfare
6. U.S. Department of Defense website, in the "About" section. https://www.defense.gov/About/. Accessed 28 Jun 2022
7. Understanding the Instruments of National Power (undated) JFODS5–1: The joint forces operations & doctrine SMARTbook, 5th Ed. w/change 1 (Guide to Joint, Multinational & Interorganizational Operations), The Lightning Press. https://www.thelightningpress.com/understanding-instruments-national-power/. Accessed 28 Jun 2022
8. Tass Russian News Agency (2019) Western liberal model is losing attractiveness, Lavrov believes, 12 Apr 2019. https://tass.com/world/1053401. Accessed 01 Jul 2022
9. Delingpole J (2019) Letting China join the WTO was the worst decision the West ever made. The Spectator, 21 Sep 2019. https://www.spectator.co.uk/article/letting-china-join-the-wto-was-the-worst-decision-the-west-ever-made
10. Thorne D, Spevack B (2018) Harbored ambitions: how China's port investments are strategically reshaping the Indo-Pacific. C4ADS, 17 Apr 2018. https://static1.squarespace.com/static/566ef8b4d8af107232d5358a/t/5ad5e20ef950b777a94b55c3/1523966489456/Harbored+Ambitions.pdf
11. Miller C [@carljackmiller] (2022) Russian information operations modeling in non-Western countries. Twitter, 18 Mar 2022. https://twitter.com/carljackmiller/status/1504896238826700800
12. Permanent Representative of Ukraine to the United Nations (2022) Letter dated 28 Feb 2014, to the United Nations addressed to the President of the Security Council. https://www.documentcloud.org/documents/21314169-unga-resolution. Accessed 28 Jun 2022
13. Lawal N (2022) UN resolution: list of countries that voted against Russia's invasion of Ukraine and those that supported it, Legit, 10 Mar 2022. https://www.legit.ng/world/europe/1458290-un-resolution-full-list-countries-voted-against-support-russia-ukraine/. Accessed 01 Jul 2022
14. Beardsley E (2022) An ex-member of one of the world's most dangerous mercenary groups has gone public. All Things Considered, NPR 06 Jun 2022. https://www.npr.org/2022/06/06/1102603897/wagner-group-mercenary-russia-ukraine-war? Accessed 07 Jun 2022
15. Malkova I, Baev A [translation by Maier T] (2019) Private army for the president: The tale of Evegeny Prigozhin's most delicate mission. The Bell 31 Jan 2019. https://thebell.io/en/a-private-army-for-the-president-the-tale-of-evgeny-prigozhin-s-most-delicate-mission/. Accessed 01 Jul 2022
16. SourceWatch (2006) Executive outcomes. In: The center for media and democracy. https://www.sourcewatch.org/index.php/Executive_Outcomes. Accessed 21 Jun 2022
17. Meyer H, Arkhipov I, Rahagalala A (2018) Putin's notorious 'chef' is now meddling across Africa BNN Bloomberg 20 Nov 2018. https://www.bnnbloomberg.ca/putin-s-chef-is-now-meddling-across-africa-1.1170783. Accessed 21 Jun 2022
18. Collins T (2019) Sudan's gold: Hemedti's untold power, African Business 08 Jul 2019. https://african.business/2019/07/economy/sudanese-gold-hemedtis-untold-power/. Accessed 21 Jun 2022

19. All Eyes on Wagner [@alleyesonwagner] (2022) Meroe Gold Co Ltd employee changes. Twitter 09 Jun 2022. https://twitter.com/alleyesonwagner/status/1534899136616611841. Accessed 10 Jun 2022
20. ImportGenius (2021) Meroe Gold Co., Ltd. Government Customs Records Notifications. 28 Jan 2021. https://www.importgenius.com/russia/buyers/meroe-gold-co-ltd. Accessed 21 Jun 2022
21. Walsh D (2022) 'From Russia With Love': a Putin Ally Mines gold and plays favorites in Sudan. New York Times 05 Jun 2022. https://www.nytimes.com/2022/06/05/world/africa/wagner-russia-sudan-gold-putin.html. Accessed 06 Jun 2022
22. Lister T, Shukla S, Elbagir N (2019) Fake news and public executions: documents show a Russian Company's plan for quelling protests in Sudan, CNN. 25 Apr 2019. https://www.cnn.com/2019/04/25/africa/russia-sudan-minvest-plan-to-quell-protests-intl/index.html. Accessed 23 Jun 2022
23. Harding L, Burke J (2019) Leaked documents reveal Russian effort to exert influence in Africa. The Guardian 11 Jun 2022. https://www.theguardian.com/world/2019/jun/11/leaked-documents-reveal-russian-effort-to-exert-influence-in-africa. Accessed 23 Jun 2022
24. Zhen L (2019) Frontier services group founder Erik Prince denies knowledge of Xinjiang training base deal, South China Morning Post, 01 Feb 2019. https://www.scmp.com/news/china/politics/article/2184734/frontier-services-group-founder-erik-prince-denies-knowledge. Accessed 23 Jun 2022
25. Frontier Services Group Limited (2021) Annual Report. 30 Mar 2022. https://doc.irasia.com/listco/hk/frontier/annual/2021/ar2021.pdf. Accessed 23 Jun 2022
26. U.S. Department of State (undated) The Chinese Communist Party's military-civil fusion policy. Archived material from 2017–2021. https://2017-2021.state.gov/military-civil-fusion/index.html. Accessed 23 Jun 2022
27. Kania E, Laskai L (2021) Myths and realities of China's military-civil fusion strategy, Center for a New American Security, 28 Jan 2021. https://www.cnas.org/publications/reports/myths-and-realities-of-chinas-military-civil-fusion-strategy. Accessed 23 Jun 2022
28. Frontier Services Group (2022) Our team: board of directors and management team. Company website. http://fsgroup.com/en/team.html. Accessed 23 Jun 2022
29. Lintner B (2019) Controversial security firm moving into Myanmar. Asia Times 19 Mar 2019. https://asiatimes.com/2019/03/controversial-security-firm-moving-into-myanmar/. Accessed 23 Jun 2022
30. AFP (2019) Former Blackwater contractor in Myanmar security venture. Frontier Myanmar 21 Mar 2019. https://www.frontiermyanmar.net/en/former-blackwater-contractor-in-myanmar-security-venture/. Accessed 23 Jun 2022
31. China Investment Research (2015) Graphic of plot of loans along one belt, one road, Jun 2015. http://www.chinainvestmentresearch.org/wp-content/uploads/2015/06/cir-belt-road.png. Accessed 23 Jun 2022
32. Rising D, Cheang S (2022) China, Cambodia break ground on port, dismiss US concerns. ABC News, 08 Jun 2022. https://abcnews.go.com/International/wireStory/china-cambodia-break-ground-port-dismiss-us-concerns-85254620. Accessed 23 Jun 2022
33. Frontier Services Group (2020) FSG Cambodia providing comprehensive security services for bank of China, company press release, 27 Nov 2020. http://www.fsgroup.com/en/news/show-586.html. Accessed 24 Jun 2022
34. Bowe A (2018) China's overseas united front work: background and implications for the United States. U.S.-China economic and security review commission, Staff Research Report, 24 Aug 2018. https://www.uscc.gov/sites/default/files/Research/China%27s%20Overseas%20United%20Front%20Work%20-%20Background%20and%20Implications%20for%20US_final_0.pdf. Accessed 24 Jun 2022
35. Fedasiuk R (2020) Putting money in the party's mouth: How China mobilizes funding for United Front Work, China Brief, vol 20, issue 16, Jamestown Foundation, 16 Sep 2022. https://jamestown.org/program/putting-money-in-the-partys-mouth-how-china-mobilizes-funding-for-united-front-work/. Accessed 01 Jul 2022

36. Hamilton C, Ohlberg M (2020) Hidden hand: exposing how the Chinese communist party is reshaping the world, Oneworld Publications 2020
37. The 48 Group Club (2021) Annual report and unaudited financial statements for the period ended 31 March 2020, Companies House GOV.UK, 16 Jun 2021. https://find-and-update.company-information.service.gov.uk/company/09417061/filing-history. Accessed 01 Jul 2022
38. Garrick J (2018) Darwin port's sale is a blueprint for China's future economic expansion, The Conversation, 06 Dec 2018. https://theconversation.com/darwin-ports-sale-is-a-blueprint-for-chinas-future-economic-expansion-108254. Accessed 01 Jul 2022
39. Anderson F (2016) It's on Defence's head: FIRB chairman on Darwin Port, Australian Financial Review, 11 Mar 2016. https://www.afr.com/politics/its-on-defences-head-firb-chairman-on-darwin-port-20160310-gnfrfe. Accessed 01 Jul 2022
40. Tanaka Y, Shimbun S (2019) Concerns raised as Australian port rolls out red carpet for Chinese investment. Japan Forward, 02 Feb 2019. https://japan-forward.com/concerns-raised-as-australian-port-rolls-out-red-carpet-for-chinese-investment/. Accessed 01 Jul 2022
41. McKenzie N, Massola J (2017) Andrew Robb's secret China contract: money for nothing. Sydney Morning Herald, 06 Dec 2017. https://www.smh.com.au/politics/federal/andrew-robbs-secret-china-contract-money-for-nothing-20171205-gzzaq5.html. Accessed 01 Jul 2022
42. McGregor R (2019) How the state runs business in China, The Guardian, 25 Jul 2019. https://www.theguardian.com/world/2019/jul/25/china-business-xi-jinping-communist-party-state-private-enterprise-huawei. Accessed 01 Jul 2022
43. Angliviel de la Beaumelle M (2017) The United Front Work Department: "Magic Weapon" at Home and Abroad, China Brief, vol 17, issue 9, Jamestown Foundation, 06 Jul 2017. https://jamestown.org/program/united-front-work-department-magic-weapon-home-abroad/. Accessed 01 Jul 2022
44. Barnes P, Bateman S, Behm A, Benich P, Bergin A, Cronin P, James N, Jennings P, Wade G, Zhang F (2015) Chinese investment in the Port of Darwin: a strategic risk for Australia? Strategic Insights, Australian Strategic Policy Institute, Dec 2015. https://s3-ap-southeast-2.amazonaws.com/ad-aspi/2017-07/SI101_Chinese_investment_Darwin_v2.pdf?Ld8eqchCvcCXtq2bfBaeNmWWr21Wrkuw. Accessed 01 Jul
45. Zeihan P (2022) The end of the world is just the beginning: mapping the collapse of globalization, Harper Business, 14 Jun 2022
46. DeepThroatIPO (2020) The really, really great rebalancing, blog post 15 Dec 2020. http://www.deepthroatipo.com/the-really-really-great-rebalancing-sort-of-i-guess/. Accessed 01 Jul 2022

Ian Conway manages Helios Global, Inc., a risk analysis consultancy that specializes in applied research and analysis of asymmetric threats. He currently specializes in analysis of political warfare operations and economic subversion. He has supported DoD and homeland security programs focused on counterterrorism, counterproliferation of WMD, hard and deeply buried targets, and critical infrastructure protection.

Kathleen Cassedy is an independent contractor and open source specialist with two decades of experience. She spent the last several years identifying, researching, and analyzing modern Russian and Chinese political and economic warfare tactics; the role of foreign influence operations in gray zone problem sets; and the global geopolitical influence of multi-national entities, non-state actors, and super-empowered individuals.

Organizing US: Regime Theory for Integrated Statecraft

David C. Ellis and Charles N. Black

Abstract Complex adaptive challenges, like the intervention in Afghanistan, defy traditional government-centric, control-oriented, sector-based strategies. Rather, complexity theory advocates for deep sensitivity to local conditions, systemic interactions, and non-linear effects. Yet, the national security and foreign policy systems are not structured to adequately appreciate these aspects nor promote intergovernmental (local, state, federal) cooperation. Consequently, a situation similar to "international anarchy"—no authority above the level of the state—exists across the interagency and with respect especially to intergovernmental, multilateral, and civilian partners. Integrated statecraft in the context of strategic competition requires achieving political effects through the actions of multiple executive entities, with allies and—when appropriate—commercial partners all acting towards a common purpose. In recognizing the condition of "anarchy," insights from regime theory in the International Relations literature provides a viable approach for overcoming collective action challenges in the US national security bureaucracy. New regimes of behavior built on shared appreciation and trust can lead to ever increasing horizontal inter-governmental integration at all levels of action without change to statutory role, resourcing, or authorities. Professional Military Education (PME) and civilian universities are put forth as viable regime structures for complex challenges due to their multidisciplinary specializations, existing infrastructure, and intersections between the policy and civilian sectors.

Keywords Regime theory · Statecraft · Complexity theory · International anarchy · Afghanistan · United Sates

D. C. Ellis (✉) · C. N. Black
The Joint Special Operations University, Tampa, FL 33621, USA
e-mail: david.ellis@ellisanalyticsinc.com

C. N. Black
e-mail: cnblack66@gmail.com

A. Farhadi and A. Masys (eds.), *The Great Power Competition Volume 4*,
https://doi.org/10.1007/978-3-031-22934-3_4

1 Introduction

Underlying the myriad mistakes that compounded into abject failure of U.S. foreign policy in Afghanistan in August 2021 was a central mental model: state governments are sovereign, hierarchical social structures that control their populations and territory through a monopoly on the use of force and earn legitimacy through the distribution of material services.[1] Whether in Afghanistan, Iraq, Somalia, Libya, or Syria, the architecture of U.S. intervention since 9/11 has followed this basic premise, and in each case the result has been expensive stalemate at best or policy failure at worst. But the U.S. is not alone—this basic premise appears across the statebuilding literature and underpins much of the effort by international organizations dedicated to institutional development in post-conflict environments, such as the United Nations or the Organizations for Economic Cooperation and Development [3, 8, 47, 59, 65, 67]. Although no single "cause" ever accounts for the success or failure of a policy, evidence is mounting that the central premise of supporting partners and proxies with a central government, control-oriented mentality is insufficient for achieving sustainable political objectives. Coming to terms with the contradictions inherent to the government-centric mental model is even more urgent when placed in the context of strategic competition. The U.S. can no longer afford expensive, personnel-intensive interventions that yield little to no strategic gain. Considering that strategic competition manifests in a variety of ways across the developing world, it becomes essential to gather wisdom from the experiences of the past twenty years to consider if the architecture of intervention can be adapted to each unique, local context.

"Organizing US: Regime Theory for Integrated Statecraft" uses the double entendre of us/U.S. to connote the need for reflexivity; that is, "it's not them, it's us" and how "we" think about "their" circumstances.[2] This chapter asserts that the fundamental lesson from Afghanistan is that a complexity-based—rather than a government control-oriented—intervention architecture is necessary moving forward. Changing this one mental model alters all other elements of the intervention architecture, including the U.S. approach to politics and development at the local-to-national levels and, just as significantly, the character of population engagement strategies in each location. If the complexity frame is accurate, then the U.S. must

[1] Stephen D. Krasner describes these assumptions as a combination of international legal, Westphalian, interdependence, and domestic sovereignty. Yet, the most critical—domestic sovereignty—eludes US policy design thereby undermining the others. See Krasner [41], 11–12 and 220–238. On the U.S. Government concepts and joint doctrine related to the consolidation of partner state sovereignty, see: Joint Publication 3-24: Counterinsurgency [35], Joint Publication 3-25: Countering Threat Networks [36], Joint Publication 3-26: Counterterrorism [37], Joint Publication 3-07: Joint Stabilization Activities [34]. The *United States Strategy to Prevent Conflict and Promote Stability* [72], jointly produced by the Departments of State, Defense, and Treasury, and USAID offers a more nuanced perspective.

[2] On the need for perpetual organizational reflexivity, see Argyris and Schön [5]. In particular, Argyris and Schön recognize the importance of double-loop learning (also called deuterolearning) through which individuals and organizations "learn how to learn" by questioning their own assumptions, values and norms, behaviors, and models of practice.

develop a structure for addressing complex socio-political challenges in incredibly diverse situations without increasing the funding and personnel burden on the United States Government (USG). How might this be possible?

Rather than creating a new bureaucracy or seeking stronger hierarchy in the national security apparatus, the complexity theory frame suggests that *decentralization*, systems thinking, and a fluid, contextually relevant amalgamation of participating organizations is necessary for meeting complex adaptive challenges. The chapter suggests regime theory from the International Relations discipline offers the best structural form for complex adaptive challenges. The chapter proceeds in four sections. First, the chapter briefly explores the emerging literature on complexity, legitimacy, development, and evolutionary governance as an alternative frame for interventions. Second, it makes the case that something akin to the condition of "anarchy" as defined in International Relations exists across the USG interagency process. Third, the chapter demonstrates how regime theory could help to overcome the condition of anarchy across the interagency. Finally, it offers that the university system, especially the universities associated with Professional Military Education (PME), could serve as the structural foundation for complex adaptive responses given their inherently diverse fields of expertise, organic networks across regions and topics, and existing research infrastructure.

2 Complexity Science and Interventions

Although the concept of "stability" permeates U.S. foreign policy across departments and agencies,[3] stability in social systems is a myth [18]. There is a tendency to confuse the propensity for social systems to establish patterns of "regularity" [79] with the assumption that regular patterns *are* or *should be* "stable."[4] Stability evokes notions of resistance to change, equilibrium, balance, and endurance such that proper engineering of the social variables can lead to preferred outcomes in socio-political relations.[5] Indeed, this idea has long been a staple assumption in the social sciences, but it led Horst Rittel and Melvin Webber in 1973 to introduce the difference between "tame problems"—for which environmental conditions can

---

[3] Office of Stability and Humanitarian Affairs [13, 14, 57].

[4] Space prevents a thorough treatment of the structuration processes that yield cultural, social, and institutional change over time. Suffice it to say that there is a difference in desiring predictable institutional and policy environments to promote, for example, investment and peaceful political participation and the idea that stability is a function of the government being perceived as legitimate through the material distribution of resources and social services. On structuration, see Stones [69]. For an empirical example of populations seeking governance support from a range of informal and formal, local to national providers, see Ariotti and Fridy [6].

[5] The social engineering perspective falls currently in the realm of behavioral science and has been popularized with the concept of "nudge theory;" see Thaler and Sunstein [70]. The equilibrium foundation of the social systems assumption, however, dates back to Vilfredo Pareto's equilibrium-oriented economic analysis; see Buckley, Schwandt, and Goldstein [11].

be controlled, the variables known, and the scientific method directly applied—and "wicked problems"—for which conditions cannot be controlled, the variables not completely known, and the scientific method only problematically approximated [62]. Social systems are inherently open systems and therefore fall into the "wicked problems" category of challenges. In other words, concepts and doctrine that emanate from social engineering, linear planning, and government control-oriented precepts inherently violate the basic realities of social systems.

To be fair, complexity science was in its infancy when the U.S. intervention in Afghanistan first began in late 2001, but it has matured broadly across disciplines since then. If the purpose of this edited volume is to track lessons learned, then complexity science can contribute much to framing why the U.S. intervention architecture failed. Three complexity lenses are of particular importance in this regard: explaining the illusion of stability in social systems, describing the topology of social systems as a function of local decision making, and framing the trajectory of political and economic institutional change as a result socio-cultural evolution.

2.1 "Stability" Through a Complexity Lens

As complexity theory consolidated over the 1970s–1990s and expanded across a broader range of disciplines, a few key tenets emerged. It recognized that, even in the absence of organized structures at some theoretical starting point, extremely simple rules could lead over time to "self-organization" through the development of "strange attractors," or interactions that provide mutual benefit to two or more variables. That is to say, simple rules creating relationships among variables could lead to temporarily repeatable—or "stable"—interactions, or what some have described as "patterned disorder." The "stability" of the interactions results from "positive feedback loops" that increase the gains from relationships.[6] Once the positive feedback loops lock in relationships, the logic of "path dependence"—the costs of transitioning to new relationships and patterns outweigh the benefits, such as with the QWERTY keyboard—maintains the patterns as "regularities" [56]. All the while, there are "dissipative structures," or "negative feedback loops" (in this context insurgents, narcotraffickers, social movements, etc.), competing against the dominant regularities or interfering with the relationship patterns. This competition opens the "stable" regularities to innovation and potentially new, novel, and possibly more beneficial relationships. Should they prove more advantageous, the innovations generate their own positive feedback loops thereby displacing the existing ones and altering the path dependencies to some degree [15, 24, 53, 73, 77]. This phenomenon is called "emergence" and creates what complexity theory describes as "the edge of chaos"—the

[6] The term "positive" in this context is not normative; thoroughly odious political systems can create feedback loops that outcompete local rivals or relegate populations to economic patterns well below their full potential.

non-linear, unpredictable consequences of human decisions that defy full institutional control over individuals and society over the long term.[7]

Through the complexity lens, "stability" in political systems assumes some combination of (a) utility among existing relationships, (b) institutional path dependency, (c) socio-cultural congruity with the predominating institutions, and (d) environmental fit of the institutions along with their capacity to adapt to new, emergent relationships, opportunities, and stressors. Crucially, complexity theory also recognizes that the patterns of behavior locked in positive feedback loops exist at all levels, from the local to national to the international, which means that *governance* of human relationships exists across a range of competing informal and formal relationships, networks, and systems, not just at the level of formal, internationally recognized, sovereign government. Rather, as Colander and Kupers [15, p. 165] write,

> Complexity policy does not involve control by government; it involves affecting the economy's evolution by changing the ecostructure to better allow people to solve their problems. Since many of those problems are collective problems, collective institutions are necessary to deal with them, but these collective institutions don't have to be the state as we currently know it. Policy includes changing the nature of the state—making it so that it can deal with the problems. Alternatively, policy could include the development of new 'shared space' institutions that allow people to organize together from the bottom up in alternative collective organizations that complement or even replace the state in their subareas.

In other words, governance from the complexity lens might be, but does not have to be, synonymous with government and indeed often is not. "Stable" regularities are only likely to be disrupted or displaced if (a) new positive feedback loops prove more useful to a population or (b) if significant resources and effort are applied to support new patterns if they bring less perceived utility to the population. But the phenomenon of emergence guarantees innovation across the social, cultural, political, and economic spheres of life, which renders the idea of "stability" problematic at best, especially if the central government in weak, post-conflict environments is expected to meet myriad changing needs and expectations.

2.2 *The Topology of the System of Local Conditions*

Interventions from the complexity perspective must be cognizant of population "flows;" that is, natural social patterns, relationships, and cultural habitus that follow courses molded for them by the friction of human history over time, but, through their own mass, velocity, and force can shape their future channels and paths [73]. Interventions that metaphorically "flow" with populations encounter less friction and, therefore, mitigate the costs associated with applying the resources needed to "dam up" the flows, control them, and maintain them against the weight of socio-cultural expectations. In short, if an intervention requires extensive resourcing even among

[7] [53, p. 104] describes the "edge of chaos" as a threshold, "the point where systems are in an unstable, agitated state."

passively supportive populations, then it should be assumed that the challenge lies in the mental models underpinning the intervention strategy, not a problem with the implementing partner's capacity. If the USG presumes the latter, then the strategy will not flow with the population and institutions as they are, friction will be guaranteed, and gaps in governance will allow competitors to innovate with potentially more useful positive feedback loops (such as Frumin recounts in this volume with regard to the Taliban's provision of justice).

Population flows, as complex adaptive systems, are a function of systems dynamics: individuals and organizations engage in a wide variety of interactions, which in turn rely on a variety of interactions about which they are largely unaware, which again rely on other interactions, so on and so forth. Wicked problems and nonlinear dynamics stem from the unknowable number of relationships and interactions operating "behind the scenes" (meaning, outside the scope of one's understanding or capacity to measure) and emergence. There is no such thing as "*the* system;" rather, all systems are (a) a function of interrelationships between and among nodes, (b) dependent on multiple perspectives, and (c) bounded by the limited awareness of the perceiver [78]. Since all systems are a function of a choice by or a predisposition of the observer,[8] a systems "understanding" is always incomplete and best approximated through purposeful, transdisciplinary or multi-stakeholder exploration, especially when the explicit interest is to design and intentionally bring change into the world [55].

One highly useful lens through which complex population flows can be evaluated is through network topology, the form and character of the patterns of interaction between nodes [63]. All flows begin at the most discrete, local level and aggregate up to higher levels of abstraction. Network topology recognizes that subnational (micro-level) systems can operate concurrently with but independent of regional (meso-level) and national (macro-level) systems; there is always space between what governments mandate and that over which they can effectively extend their authority. For this reason, self-organization and emergence at the lowest level can over time spread across a range of networks, generate system-wide effects, and lead to a "phase transition" (a fundamentally different pattern of interactions or ordering principles) in the behavior of the system as a whole [53, 63]. Hilton Root [63, p. 42] explains, "While a phase transition can transform the system's topology, it is that same topology, created of its components' interactions, which will determine whether those component networks survive the perturbation, or fail. The resilience to survive a shock depends as much on the system's topology as on the impact of the shock." Resilience, in this context, does not mean the strength of the government's ability to withstand political challenges or armed insurgency as such, but "the ability of the macro system to absorb subsystem disturbances while retaining its own system of rules and procedures" [63, p. 87] or "the capacity of individuals, a community, a system, or a population to survive and thrive in the face of threats and challenges…"

[8] Here the concept of paradigms [44] is instructive, whether professional, disciplinary, cultural, and/or ideological.

[66, pp. 80, 81]. Resilient networks, in turn, require a multiplicity of redundant, positive aims relationships so that malign actors cannot effectively impede interaction among the population through "cut points," or targeted breaks in the networks [66]. In practical terms, the complexity lens demands that the USG consider the following attributes of its intervention strategy.

Does the strategy focus mainly on the Western institutional-functional form of a state, or does it align popular flows with the provision of socio-culturally recognizable governance patterns?

Does the strategy seek to expand network connections, relationships, and trust for redundancy at both the institutional and civil society levels, or is it principally government-centric and, therefore, creating identifiable cut points?

Does the strategy adopt a bureaucratic specialization approach to government, or does it intentionally intersect departments and disciplines to build social resilience and the potential for emergent, positive feedback loops?

The network topology lens is essential for any intervention strategy precisely because resilience—either for maintaining existing social regularities or promoting phase transitions—depends upon the organic linkages at the local level. Failing to flow with populations, linking them together where there are gaps, and acting in ways that improve upon their existing patterns decreases the likelihood that an intervention will be sustainable once external support is terminated. Systemic interventions, then, require integrated approaches with diverse stakeholder perspectives, not just departmental specialization and functional bureaucracy.

2.3 *Evolutionary Governance, Development, and Legitimacy*

If complexity theory is right, and phase transitions can occur over time through a series of localized positive feedback loops connecting together through new and emergent relationships, then sustainable interventions require extraordinary sensitivity to local conditions and the ability to analyze, assess, and intervene in a variety of natural and social systems. Already there is a body of literature forming that recognizes the necessity of beginning with local logics of interaction. For example, in *The Complexity of Self Government: Politics from the Bottom Up*, Lane [46] explores the capacity for political self-organization—a foundational complexity tenet—leading to socio-political emergence based on basic, often cultural, locally relevant rules. In this way informal governance as a social ordering construct impacts the behavior of formal government over time. In the economic development arena, analyses of fitness landscapes, network connections, and production knowledge bases, among many other non-linear systems-based factors, are now entering the discourse even though linear management intervention strategies continue to dominate the institutional culture.[9] Similarly, Judith E. Innes and David E. Booher offer a framework of emergent, "collaborative governance" for social resilience. They write, "The complexity and rapid change in contemporary society have created increasing awareness among policy leaders of the limits to hierarchical control by government agencies and to formal expertise in solving problems" [31, pp. 188, 189]. They position collaborative governance in contrast to traditional, control-oriented government mental models, stating:

Collaboratively rational processes are about engaging with other members of a community to jointly learn and work out how to get better together in the face of conflict, complex changing conditions, and multiple conflicting sources of information. Such processes are not only about finding new ways to move forward, but they are ultimately about building community and governance capacity to be resilient in the face of the inevitable new challenges [31, p. 191, 192].[10]

Yet as noted previously, USG concepts, and certainly most military joint doctrine, continue with government-based, control-oriented mental models.

Within the complexity theory framing, certain lessons learned become unavoidable:

Moving complex adaptive social systems requires forging new connections and pathways of interaction; that is, strategically changing the topology of social interactions through "nurture network" [17] activities across the local-to-national levels;

Sensitivity to local conditions is paramount in interventions since populations determine value based on often hidden or obscured logics of rationality;

Absent an emic or authentic immersion in the culture [64] and a deep appreciation of local path dependencies, cultural narratives, and identity palettes, such local conditions cannot be planned for;

Co-creation with—instead of planning for or upon—the local population is a predicate for achieving new advances in perceived utility; and

All of these factors run contrary to the government-based, control-oriented mental model.

Perhaps the most fruitful way of synthesizing these lessons and moving forward conceptually is by adopting the framework of Evolutionary Governance Theory, which synthesizes complexity theory, the social construction of reality, structuration, multi-level governance, and narrative communication as they apply to socio-political development [9]. Here it is useful to reflect on Colin Wight's observation that "state actions should not be attributed to the state as an originating subject, but rather should be understood as the emergent, often unintended and complex result of what rival agents within states have done and are doing on a complex strategic terrain" [76,

[9] For an in-depth critique of the development aid industry through a complexity theory lens, see Ramalingam [60]. For an exploration of the importance of network connections and human development see, for example, Hartmann [25] and Hausmann and Hidalgo [27]. For a wider ranging discussion of complexity, economics, and development, see Arthur, Beinhocker, and Stanger [7].

[10] Of further note, Innes and Booher specifically align collaborative governance with Argyis and Schön's concept of deuterolearning (see footnote 2).

p. 225]. In other words, a key lesson of Afghanistan from the complexity theory lens is that the effort should have been conceived less as statebuilding (i.e., government building) and more as statecraft—meaning "the crafting of the state" as an imagined community [4] and series of resilient and intersecting governance networks.

3 Anarchy and Regime Theory for Statecraft

Given the complexity theory framing, how might the U.S. structure itself for interventions built around complexity dynamics when hierarchy and control are not possible? One highly useful concept, the *Joint Concept for Integrated Campaigning* [33], issued by the Department of Defense's (DOD) Joint Chiefs of Staff in 2018, advises the Joint Force to integrate activities among the military and align them across interagency, intergovernmental, multilateral, and civilian (IIM-C) partners. The necessity for the JCIC stems from the realization that sustainable strategic effects in the current and future operating environment can only be achieved through shaping the environment with a persistent, active campaigning orientation. Moreover, the JCIC is imbued with the presumption of complexity, non-linear effects, the political significance of perception, and the importance of collaboration in a transdisciplinary way. Since this must occur below the level of armed conflict, even the DOD recognizes it will often need to play a supporting role to the IIM-C.

The challenge presented by the JCIC for the DOD derives from the reality that only within the military is there a vertical hierarchy or chain of command. Working across the IIM-C, on the other hand, relies on partners' decisions to voluntarily contribute personnel, resources, and materiel based on shared interests or objectives. The requirement for this cooperative behavior can be observed across intergovernmental, multilateral, and civilian partners. However, many in the USG presume there to be hierarchy and authority across interagency partners since they are elements of the Executive Branch of government with a statutory chain of command flowing from the president.

In practice, the interagency chain of command concept falls apart between the strategic level of the National Security Council (NSC) and the tactical level of the embassy team. As a result, a new theory of operational level interagency campaigning is necessary to make the ideas in the JCIC possible. If no meaningful authority exists across the IIM-C, then voluntary collective action becomes a necessity, and it is this very condition that underlies the notion of "anarchy" among states. It is proposed here that regime theory from the International Relations literature provides the mechanism for implementing the JCIC and that Professional Military Education (PME) institutions and civilian universities provide the existing structures necessary to make it effective.

It is an axiom in the discipline of International Relations that the fundamental condition among independent, sovereign states is that of "anarchy." Anarchy does not mean chaos; rather, it is defined as the absence of authority above the level of the state [51, 74]. The effect of anarchy is that there is no formal political hierarchy among the actors in the system, meaning that any and all unified action among them is completely voluntary. While anarchy might make cooperation difficult, it is still certainly possible, if not commonplace [12, 52, 74]. The fact that politics is an "infinite game"—that is, one without final outcomes—means that there are incentives to balance long-term interests against the demands of near-term ones. While the impulse is to prioritize the latter at the expense of the former to maximize power, even states find ways to overcome anarchy to achieve more important policy objectives.

Within the U.S. government there is some degree of hierarchy and authority, particularly within the Executive Branch. In theory, the president of the United States is a singular authority that delegates his prerogative to designated officials who, in turn, execute his decisions. Although Executive functions are distributed across numerous departments and agencies, in theory unitary, hierarchical government should be possible. This is especially true with respect to the military for which the chain of command and responsiveness to direct orders are fundamental values. Additionally, the NSC, consisting of the leadership of the Executive's departments and agencies, provides a framework for coordinated action. However, there is a significant divergence from the theory and practice of interagency activities at the center of policy in Washington, as compared to U.S. embassies, military geographic combatant commands, and the various USG entities carrying out policy on the ground. In other words, there is a "missing middle" in coordination and collaboration among interagency partners.

Of course, with intergovernmental, multilateral, and civilian actors, the condition of anarchy is indisputable; action only occurs due to voluntary commitment since no hierarchy or direct control exists or is highly contingent where it does. With respect to the interagency, the effect of anarchy exists due to a variety of tensions built into the USG system. In short, the tensions elucidated below create the conditions where cooperation across Executive departments and agencies remains voluntary in essence, often despite presidential or NSC decisions or in the ambiguity in how those decisions are written and conveyed.

3.1 Reasons for the Condition

3.1.1 Principal-Agent Dilemma

Within the Executive Branch, the distribution of functional specialties across numerous departments and agencies results in the well-known Principal-Agent (P-A) Dilemma. The P-A Dilemma reflects the potential for divergence between the interests, objectives, and decisions of the principal (main decision maker) and the agent (implementer) to whom the task is delegated. In theory, the USG bureaucracy works at the direction of the president and implements his decisions in a timely and orderly fashion. In reality, there is tremendous opportunity in the USG for "agency slack," "autonomy," and "slippage." Each of these options results in a decision by subordinates somewhere in the chain of implementation to diverge from the principal's intent, whether through bureaucratic delay, reinterpretation or misinterpretation of the intent, genuine confusion about the intent, a conflict with other Executive decisions, or a simple disagreement with the policy choice [29]. The P-A Dilemma is diminished in the military through adherence to the chain of command, but even it is subject to the phenomenon where the bureaucratic process creates the opportunity for agency slack, autonomy, and slippage. But, where hierarchy, authority, or accountability do not effectively exist, the P-A Dilemma becomes a significant challenge. Below the NSC level and above the embassy team, the P-A Dilemma is most pronounced since the interagency system is furthest removed from immediate accountability to the president himself or the ambassador who serves as the president's direct representative [28].

3.1.2 Congressional Mandates and Earmarked Funding

Due to the structure of the USG, generating coordination and cooperation across departments and agencies reduces down to a series of bargaining games based on perceptions of organizational self-interest. Indeed, the coherence of the "national interest" in U.S. foreign policy has long been a contested frame among scholars, especially those studying the USG [40].This is due to multiple factors. First, the USG has often been considered a relatively "weak state" as a consequence of the distribution of influence across Congress and the numerous departments and agencies within the Executive Branch. "Strong states," on the other hand, are generally characterized as having direct, accountable, and less diffuse executive influence over policy. In the US system, there are numerous entry points for interest groups to influence legislation, the missions and policies of the bureaucracies in Executive departments and agencies, and fluctuations in influence between Congress and the presidency over time [30, 45, 61, 71]. Second, the specialization of modern government into separate, issue-oriented bureaucracies has resulted in paradigmatically-constrained domain expertise upon which decision-makers must rely for policy alternatives [20, 21]. Although expertise creates a degree of efficiency within the particular domain, it often obstructs bureaucrats' ability and need to take into consideration other paradigms and interests. Third, where Congress reinforces these structural divisions through earmarked funding streams or political support, the disincentives for coordination and cooperation strengthen "preference heterogeneity" or, more colloquially, "siloes of excellence" [40, 48]. The problem of preference heterogeneity impedes integrative behavior because bureaucrats' organizational goals and preferences diverge despite sharing interest across a higher set of national, collective outcomes [29]. And fourth, it must be remembered that leaders within the Executive bureaucracy are themselves typically politicians. While party politics might keep them generally within the policy lanes established by the president, there is still ample room for political appointees to carve out space for their own agendas.

3.1.3 Bureaucratic Politics and Careerism

The fundamental structural problem for integrated statecraft is that "…the 'decision-maker' of national policy is obviously not one calculating individual but is rather a conglomerate of large organizations and political actors" [2, p. 3]. The interests of these dynamic conglomerates diffuse across the USG resulting in professional bureaucracies for which organizational prerogatives, constituencies, and personnel career interests generally diverge. While the P-A Dilemma results from the structure of the USG, there are also P-A Dilemmas within the departments and agencies since they, too, are conglomerates of individuals who have their own perspectives and interests. For a career civil servant, advancement requires fidelity to the organization's mandate, not overall USG effectiveness in realizing the "national interest." Moreover, discreet organizational habitus (mutually reinforcing processes, identities, routines, and cultures), legislative mandates, planning cycles, and promotion criteria make aligning interests across the USG exceedingly difficult as a general business practice [2, 10, 26]. When combined with increased specialization, Peter Haas notes, "These trends contributed to the emergence of what Dorothy Nelkin has called 'the policy role of the knowledge elite.' The proliferation of new agencies and the practice of staffing them with professionals also contributed to the erosion of centralized control over public bureaucracies, which has occurred despite widespread efforts since World War II to curb the discretion of bureaucratic administrators" [23, p. 10]. In other words, the P-A Dilemma at the structural level of the USG is further complicated by the P-A Dilemma within the bureaucracies themselves, which must accept the agency slack, autonomy, and slippage that arise from reliance on specialized offices and personnel.

3.1.4 Collective Action

As a result of these realities in the USG interagency system, there is an unmistakable and persistent collective action problem. Collective action problems arise when a range of actors have shared interests and objectives but lack the trust and cost sharing mechanisms that make it rational to work together and subordinate near-term organizational interests to longer-term collective aims. In their absence, both bureaucracies and individuals are incentivized to operate on the basis of self-interest with all achieving worse results than if they joined forces [58, 74]. For the reasons described above, the USG system actually disincentivizes the norms, values, and behaviors necessary for interagency collective action. For all intents and purposes, the USG foreign policy system faces the same condition as what International Relations scholars describe as anarchy. Nevertheless, it is still possible to achieve national level objectives when mutual interests coincide [40, 50].

4 Why Regimes?

In the context of the interagency, formal hierarchy and authority exist, but the functional behaviors enabling cooperation at the operational-to-strategic level do not. With respect to international and civilian partners, neither the hierarchy nor functional behaviors exist. How then, can anarchy in integrated statecraft be remedied?

To overcome the condition of anarchy and encourage cooperation, states created a social structure called a "regime." Stephen D. Krasner defines regimes as "sets of implicit or explicit principles, norms, rules, and decision-making procedures around which actors' expectations converge in a given area…" [42, p. 2]. Regimes can exist as informal, decentralized agreements that regulate behavior and require mutual reporting on activities, or they can be formalized organizations with offices and bureaucracies. In the former case, the regime manifests in something like the General Agreement on Tariffs and Trade, which can potentially over time become the latter, such as the World Trade Organization.

Regime Theory provides eight functional reasons why collective action problems can be redressed through regimes, even informal ones.

Permanency: Regimes tend to be created for persistent issues and therefore generate norms and Standard Operating Procedures (SOPs) for enduring engagements. Since they are not temporary, regimes constitute a "form of cooperation that is more than following short-run self-interest" [32, p. 173]. Although discreet topics might periodically change, the sector or issue might persist as an enduring need for cooperative action. Regimes overcome problems of "credible commitment" when long-term interests would otherwise be sacrificed for short-term needs. The time-inconsistency tradeoff usually results in suboptimal outcomes, but persistent positive interactions through regimes create reputation effects that make each instance of cooperation more probable [29].

Strategic Appreciation: Enduring regimes, especially those with professional bureaucracies, have the capacity for long-term institutional expertise and historical memory that span across administrations. Although policies, priorities, and rule and decision procedures might change with administrations, the regimes retain the knowledge of how to navigate across them and leverage more enduring regulations and capabilities of participating actors [1, 29].

Trust-Building: Persistent engagement over time by personnel creates relationships, patterns of cooperation, mutual expectations, and customs of behavior that result in trust when each participant or organization regularly meets its obligations. In this way even informal regimes can generate order without formal structure because they become "joint-maximizers rather than self-maximizers" [68, p. 139]. Collective action problems typically derive from distrust about one another's commitment to an agreement absent hierarchy or an adjudicating authority. Trust building through regular, repeated norms, values, and SOPs encourages leaps of faith that enable participating entities to achieve greater gains, which consequently encourage future cooperative behavior and discourage defection from the regime's expectations. Regimes tend to be more flexible and adaptive with less bureaucracy and can therefore respond more rapidly to emerging requirements. As bureaucracy grows or becomes more entrenched, which often helps with regularizing the patterns of interaction, the opportunity cost of rigidity gets introduced. Resolving this tension for integrated statecraft will be a key strategic design factor.

Collective Wins: Opportunities for identifying mutually beneficial outcomes, often through issue-linkage, improve as trust-building increases. Each participating entity brings a different set of interests, priorities, and objectives that absent cooperation are hard to achieve [22]. Cooperative action, on the other hand, often outweighs "egoist self-interests" and "incentivize[s] cooperation" among departments and agencies [42, p. 2]. In this way unity of purpose and action are possible even in the absence of hierarchy [49].

Epistemic Community: A key effect of regimes is the promotion of "shared awareness" about the sector or issue for which it is established. Consistent with the notion of wicked problems, each actor brings to the regime a different experience and perspective, which then produces broader sensitivity to and knowledge about contributing systemic factors. This epistemic (or knowledge) community can bridge formal organizational and cognitive boundaries that would otherwise impede trust-building and the identification of collective wins. With persistent engagement and high levels of trust, the regime can serve as a mechanism for interorganizational learning and generate concepts for unity of action even within a condition of anarchy. The epistemic nature of regimes improves the propensity for cooperation and trust-building once personnel formulate a shared awareness and interpretation of the systemic factors [19, 23]. Tactical-level Interagency Task Forces (IATF) are often effective precisely because they are micro-epistemic communities as co-located groups ingesting, digesting, and acting upon common information while being bound by reputational effects of small groups in collective action.[11] Above the IATF level in the USG, however, the epistemic community dissipates, and the organizational politics listed above dominate to produce the anarchy-like context.[12]

Systemic Sensemaking Process: Because regimes are designed to generate shared awareness, participating entities become more aware of their own organizational paradigms and operational blinders. As individual organizations they are more likely to act on the basis of incomplete information resulting in judgments and decisions that fail to align with social complexity. Regimes, therefore, serve as systemic sensemaking platforms that improve knowledge of the multitude of variables impacting the system while fomenting the linkages with entities designed to address them.[13]

Synthesis: With shared awareness, trust, and collective wins comes the opportunity for achieving synthesis of strategy instead of mere deconfliction or synchronization [1]. There is a significant distinction here. Synthesis of strategy across entities denotes "co-creation" or the intentional, reinforcing utilization of one another's

[11] These factors reflect the benefits of small group interactions; see Olson [58], pp. 33–43 and 53–57.

[12] For similar insights among states, see Haas [23], p. 20.

[13] On the necessity of sensemaking across new situations and issues, see Weick [75].

authorities, resources, and capabilities to achieve a collective objective. Deconfliction and synchronization, on the other hand, imply separate initiatives undertaken according to idiosyncratic organizational paradigms that are simply arrayed out so as not to overtly undermine one another's intended effects. Metaphorically the difference is the satisfaction one would receive from the flavor derived from a cookie recipe as opposed to tasting the individual ingredients sequentially.

Habitus: Through persistent engagement in a structured environment, a culture of interaction develops. When the culture becomes entrenched, it creates what is called "habitus," or a fusion of norms, values, and SOPs aligned with the structure's operating principles [10]. Highly effective regimes generate a role identity for participants that, in turn, improves expectations about objectives, trust, and collective action. Regimes with a strong habitus, regardless of the degree of formalization, can generate extensive cooperative behavior so long as they stay within established authorities. They can act effectively without having to seek extra permission from leadership.

4.1 Why Integrated Statecraft Regimes for Strategic Competition?

Although collective action across the USG has been regularly attempted, the problems of "the interagency" are well known. Yet, the complexity of national security and national defense in the context of strategic competition makes it an absolute necessity. Contemporary strategic competition has a multi-dimensional character with increasing numbers of interdependent actors requiring integrated approaches to shape the environment and address strategic challenges. The range of issues, such as Illegal, Unregulated, and Unreported Fishing (IUUF), the Arctic, terrorism, and the Chinese Communist Party's Belt and Road Initiative, exceed the USG tradition of giving one or two departments or agencies "coordinating authority" for a given policy. Rather than deconfliction or synchronization, strategic competition demands the artful, integrated application of societal, cultural, and state power—statecraft—in pursuit of national interests. More directly, in strategic competition, departments and agencies must reform the independent, specialization-oriented decision-making culture and seek synthesis. The USG needs more effective structural ways to resolve *dilemmas of common interests* and *dilemmas of common aversion* [68]. Dilemmas of common interests, characterized by the desire to *ensure* a common outcome, require collaboration while dilemmas of common aversion, characterized by the desire to *avoid* certain outcomes, require coordination [68].

Of the two dilemmas, achieving common interests is the harder case because it often requires designing, resourcing, and implementing new interventions in the system. In other words, the costs of collaboration and co-creation are often perceived as high and as impediments to normal operations. Organizations' natural tendencies are to act as self-maximizers, free-ride on the initiatives of other departments and agencies, and consequently undermine long-term cooperation [29]. Dilemmas of common aversion are thought to be less burdensome because the requirement is more akin to coordination or deconfliction of behaviors that would otherwise result in undesired outcomes. Shared awareness and mutual appreciation of the challenges generally suffice to promote collective action [29]. States have learned through delegating such issues to regimes that they can "…reap gains from specialization, as well as capture policy externalities, facilitate collective decision-making, resolve disputes, enhance credibility, and lock-in policy biases. The larger these benefits, the greater the likelihood that states will choose to delegate to an [international organization]" [29, p. 23].

While there is often an impulse in the USG is to create or adapt a formal bureaucracy to deal with such challenges, the International Relations literature on regimes in the condition of anarchy is clear: the change in research focus from formal international organizations to less formal regimes occurred because the process of collaboration and coordination among principals continued to generate collective action wins irrespective of the degree of formality. In other words, formal bureaucracy does not necessarily result in collective action in anarchy, but regimes can promote the conditions for voluntary agreement nevertheless [43]. Research on epistemic communities suggests that even relatively small groups—as few as thirty-five well-positioned people—can significantly impact policy development, especially during times of crises and novel, emergent conditions [1]. The benefit of these small, transdisciplinary teams is that they can often derive cooperation options with their specialty knowledge that higher echelons of policymakers cannot discern due to their lack of familiarity with the range of authorities and capabilities available to them [29]. If the JCIC's main message to the Joint Force is to compete daily by supporting IIM-C partners, then synthesis-oriented, integrated statecraft is a necessity in strategic competition. Overcoming the inherent anarchy across the IIM-C is, therefore, a prerequisite.

5 Universities as Regimes for Wicked Problems

Integrated statecraft as imagined in the JCIC will face constant barriers to cooperative action due to the lack of a structure between the policy and tactical levels. Rather than create a formal, hierarchical bureaucracy, the anarchical condition of IIM-C activities is more appropriately conceived as requiring a loose, adaptive regime with very limited formal structure. No matter the benign intentions of a formal bureaucracy, the fact that IIM-C activities require voluntary participation and collective wins to achieve persistent collective action suggests that an alternative, regime structure is necessary.[14]

It is proposed here that universities, especially those found in the PME system, could serve as the regime structures for the USG for numerous reasons. With more

[14] On this point, the value of alliances in international politics is instructive. See, for example, Morgenthau [54], p. 197–202.

than a dozen undergraduate and graduate level PME institutions, staffed with scholars and professors of practice across disciplines, there is ample opportunity for the DOD to provide a baseline of regime support to engage in strategic competition while meeting the Joint Chiefs of Staff's vision for PME faculty to generate more force-relevant scholarship [16]. Moreover, partner, ally, and interagency personnel often attend PME programs, which could further enhance IIM-C initiatives. Such a regime structure would have many benefits for dealing with complex adaptive challenges:

Swarming for Wicked Problems: Integrated statecraft will occur in complex, emergent, and rapidly changing environments and require wide ranging expertise drawn from multiple disciplines to adequately explore the systemic factors at play and the local contexts in which they manifest. In other words, integrated statecraft will require "swarming" transdisciplinary epistemic communities together as needs arise. Since individuals, departments, and agencies are designed for specialization, each intrinsically operates on the basis of bounded rationality and cannot possibly sense the systemic interactions lurking in other sectors. Without divergent perspectives, they are individually bound to satisfice with suboptimal outcomes [38]. Only universities have the inherent diversity of perspectives necessary to swarm effectively.

Strategic Appreciation: Faculty at universities tend to stay for long periods of time and could naturally provide a level of transdisciplinary strategic appreciation difficult—if not impossible—to replicate in other kinds of USG departments and agencies. Similarly, a central deficiency across the USG is sensitivity to local worldviews, politics, and economic dynamics. Universities could serve as hubs for accruing relevant expertise of translating policy objectives into meaningful local engagement through subject matter expertise [39]. The larger "knowledge system" could include think tanks, professional associations, non-governmental organizations, and even intergovernmental organizations.

Existing Structures: PME institutions and even civilian universities provide already existing, funded, and staffed structures suitable for regime-oriented activities. Aside from some minor event organizing costs, there would be negligible new investment—perhaps, no new investment—needed to fund the structures. At best, a reprioritization of existing resources could be undertaken without the cost of creating a new USG bureaucracy that would likely become rigid and culturally self-regarding. Since the costs of coordination are often seen as a barrier, collective action problems persist [39]. But so long as the benefits of meeting and co-creating solutions through existing structures outweigh the costs, such barriers can be overcome. This is particularly true when the participating organizations have a high degree of interdependence and requirement for enduring interaction [39].

Predictable Engagement: PME institutions and civilian universities have predictable patterns of operations around which or in accordance with regime activities could be planned. As such, they could become predictable platforms for trust-building and collective action since time, place, and experience would result in a form of habitus. Additionally, repeated experience generates mutual expectations about how to share information, accomplish collective objectives, and thereby lower the transaction costs of cooperation [39]. As already-existing structures, universities could serve as regimes to achieve economies of scale on high issue density topics (repetition of need and linked issues) for which ad hoc solutions would be more expensive or prohibit collective action [39].

Education and Research: It is feasible that PME institutions and civilian universities could become sector or issue specialists for strategic competition and provide the opportunity for designated action officers and executives to learn deeply about the focus area prior to assuming mission responsibility. Additionally, they could be sites for interagency preparation so that joint and interagency personnel could generate shared awareness while solidifying relationships prior to working together operationally. As thought leaders, the universities could engage in agenda-setting, issue popularization, brokering, and innovating policy ideas [29, 80]. Through such education and research functions, they could assist in identifying win–win outcomes that make collective action a rational use of limited personnel, time, and money.

Although this chapter has been framed in the context of the statebuilding class of wicked problems, universities have the requisite attributes to serve as regimes for a range of governance and governmental challenges. Applied to Afghanistan, a regime structure such as the one imagined here would have greatly helped the intervention assess the socio-political condition as it was, the potential network topology that could help it evolve into the future, and the cultural habitus in accordance with which new relationships, connections, and institutions would need to be co-created. Universities as regimes of systemic exploration could help "craft the state," in other words, by serving as "basins of attraction" to overcome the inherent barriers to collective action that persistently plague USG and broader IIM-C activities.

6 Conclusion

Statebuilding is among the most complex, dynamic foreign policy undertakings precisely because it requires both the foreign policy application of statecraft and the "crafting of the state" application. From the complexity lens, statebuilding is about bottom-up resilience, which requires multiple systems lenses, empirical research, and non-kinetic engagement congruent with local cultures and needs. The main lesson from Afghanistan should be that wicked problems demand reflexivity, humility, longer time scales, and a commitment to appreciating systems dynamics. The USG's traditional functionally divided, government-centric, control-oriented mental model fails in such situations.

Despite this structural reality, it is still possible to generate coherent action across the USG even in the absence of hierarchy. When viewed from the perspective of structural anarchy in International Relations, the regime theory literature provides a viable means for overcoming collective action problems and for generating meaningful co-created action. The regime theory approach is all the more pertinent in the context of USG engagement with intergovernmental, multilateral, and civilian partners with whom all action must by nature be voluntary and mutually beneficial. Rather than creating new structures, universities provide the ideal, already existing infrastructure required to serve as regime platforms for complex, collective action challenges. While Afghanistan constitutes a tragic failure for the U.S., it could serve as the spark for systemic reform for strategic competition and twenty-first century governance more broadly.

References

1. Adler E, Haas PM (1992) Conclusion: epistemic communities, world order, and the creation of a reflective research program. Int Organ 46(1):367–390
2. Allison G, Zelikow P (1999) Essence of decision: explaining the cuban missile crisis, 2nd edn. Longman, New York
3. Aman S (2016) State building in post cold war period: a critique of 'responsibility to protect' and 'humanitarianism.' Rev Hum Rights 2(1):87–103
4. Anderson B (1995) Imagined communities: reflections on the origin and spread of nationalism, Revised. Verso, New York
5. Argyris C, Schön DA (1996) Organizational learning II: theory, method, and practice. Addison-Wesley Publishing Co, Reading, MA
6. Ariotti MH, Fridy KS (2020) Informal governance as a force multiplier in counterterrorism: evidence for Burkina Faso. JSOU Press, Tampa
7. Arthur WB, Beinhocker ED, Stanger A (2020) Complexity economics: dialogues of the applied complexity network. The Santa Fe Institute Press, Santa Fe
8. Belloni R, Costantini I (2019) From liberal statebuilding to counterinsurgency and stabilization: the international intervention in Iraq. Ethnopolitics 18(5):509–525
9. Beunen R, Van Assche R, Duineveld M (eds) (2015) Evolutionary governance theory: theory and applications. Springer, New York
10. Bourdieu P (2020) Habitus and field: general sociology, Volume 2 Lectures at the Collège de France 1982–83 (trans: Collier P). Polity Press, Medford
11. Buckley W, Schwandt D, Goldstein JA (2008) Society as a complex adaptive system. E:CO 10(3):86–112
12. Bull H (1977) The anarchical society: a study of order in world politics, 2nd edn. Columbia University Press, New York
13. Bureau for Conflict Prevention and Stabilization (2022). U.S. Agency for International Development. https://www.usaid.gov/who-we-are/organization/bureaus/bureau-conflict-prevention-and-stabilization
14. Bureau of Conflict and Stabilization Operations (2022). U.S. Department of State. https://www.state.gov/bureaus-offices/under-secretary-for-civilian-security-democracy-and-human-rights/bureau-of-conflict-and-stabilization-operations/
15. Colander D, Kupers R (2014) Complexity and the art of public policy: solving society's problems from the bottom up. Princeton University Press, Princeton
16. Developing Today's Joint Officers for Tomorrow's Ways of War: The Joint Chiefs of Staff Vision and Guidance for Professional Military Education & Talent Management (2020). Joint Chiefs of Staff. https://www.jcs.mil/Portals/36/Documents/Doctrine/education/jcs_pme_tm_vision.pdf?ver=2020-05-15-102429-817
17. Ellis DC (ed) (2020) The network illusion: how a network-centric special operations culture impedes strategic effect. JSOU Press, Tampa
18. Ellis DC, Black CN (2018) Complexity organizational blinders, and the SOCOM design way. JSOU Press, Tampa
19. Finnemore M (1993) International organizations as teachers of norms: the united nations' educational, scientific, and cultural organization and science policy. Int Organ 47(4):565–597
20. Frieden J (1988) Sectoral conflict and foreign economic policy, 1914–1940. Int Organ 42(1):59–90

21. Gourevitch PA (1984) Breaking with orthodoxy: the politics of economic policy responses to the depression of the 1930s. Int Organ 38(1):95–129
22. Haas E (1980) Why collaborate? Issue-linkage and international regimes. World Politics 33:370–375
23. Haas PM (1992) Introduction: epistemic communities and international policy coordination. Int Organ 46(1):1–35
24. Harrison NR (ed) (2006) Complexity in world politics: concepts and methods of a new paradigm. State University of New York Press, Albany
25. Hartmann D (2019) Economic complexity and human development: how economic diversification and social networks affect human agency and welfare. Routledge, New York
26. Hastedt GP (2003) American foreign policy: past, present, future, 5th edn. Prentice Hall, Upper Saddle River
27. Hausmann R, Hidalgo CA (2014) The atlas of economic complexity: mapping paths to prosperity. The MIT Press, Cambridge
28. Hawkins DG, Jacoby W (2006) How agents matter. In: Hawkins DG, Lake DA, Nielson DL, Tierney M (eds) Delegation and agency in international organizations. Cambridge University Press, Cambridge, pp 199–228
29. Hawkins DG, Lake DA, Nielson DL, Tierney M (2006) Delegation under anarchy: states, international organizations, and principal-agent theory. In: Hawkins DG, Lake DA, Nielson DL, Tierney M (eds) Delegation and agency in international organizations. Cambridge University Press, Cambridge, pp 3–38
30. Ikenberry GJ, Lake DA, Mastanduno M (1988) Introduction: approaches to explaining American foreign economic policy. Int Organ 42(1):1–14
31. Innes JE, Booher DE (2018) Planning with complexity: an introduction to collaborative rationality for public policy, 2nd edn. Routledge, New York
32. Jervis R (1983) Security regimes. In: Krasner SD (ed) International regimes. Cornell University Press, Ithaca, pp 173–194
33. Joint Concept for Integrated Campaigning (2018). Joint Chiefs of Staff. https://www.jcs.mil/Portals/36/Documents/Doctrine/concepts/joint_concept_integrated_campaign.pdf?ver=2018-03-28-102833-257
34. Joint Publication 3-07: Joint Stabilization Activities (2022). Joint Chiefs of Staff. https://pksoi.armywarcollege.edu/wp-content/uploads/2022/03/jp3_07-2022.pdf
35. Joint Publication 3-24: Counterinsurgency (2018). Joint Chiefs of Staff. https://www.jcs.mil/Portals/36/Documents/Doctrine/pubs/jp3_24.pdf
36. Joint Publication 3-25: Countering Threat Networks (2016). Joint Chiefs of Staff. https://www.jcs.mil/Portals/36/Documents/Doctrine/pubs/jp3_25.pdf
37. Joint Publication 3-26: Counterterrorism (2014). Joint Chiefs of Staff. https://www.hsdl.org/?view&did=759133
38. Keohane RO (1984) After hegemony: cooperation and discord in the world economy. Princeton University Press, Princeton
39. Keohane RO (1983) The demand for international regimes. In: Krasner SD (ed) International regimes. Cornell University Press, Ithaca, pp 141–171
40. Krasner SD (1978) Defending the national interest: raw materials investments and U.S. foreign policy. Princeton University Press, Princeton
41. Krasner SD (1999) Sovereignty: organized hypocrisy. Princeton University Press, Princeton
42. Krasner SD (1983) Structural causes and regime consequences: regimes as intervening variables. In: Krasner SD (ed) International regimes. Cornell University Press, Ithaca, pp 1–21
43. Kratochwil F, Ruggie JG (1986) International organization: a state of the art on an art of the state. Int Organ 40(4):753–775
44. Kuhn TS (1996) The structure of scientific revolutions, 3rd edn. The Chicago University Press, Chicago
45. Lake DA (1988) The state and American trade strategy in the pre-hegemonic era. Int Organ 42(1):33–58

46. Lane R (2017) The complexity of self government: politics from the bottom up. Cambridge University Press, New York
47. Lotz C (2010) International norms in statebuilding: finding a pragmatic approach. Glob Gov 16(2):291
48. Lyne MM, Nielson DL, Tierney MJ (2006) Who delegates? Alternative models of principals. In: Hawkins DG, Lake DA, Nielson DL, Tierney M (eds) Delegation and agency in international organizations. Cambridge University Press, Cambridge, pp 41–76
49. Martin LL (1992) Interests, power, and multilateralism. Int Organ 46(4):765–792
50. Mastanduno M (1988) Trade as a strategic weapon: American and alliance export control policy in the early postwar period. Int Organ 42(1):121–150
51. Mearsheimer JJ (2001) The tragedy of great power politics. W.W. Norton & Company, New York
52. Milner H (1991) The assumption of anarchy in international relations theory: a critique. Rev Int Stud 17(1):67–85
53. Morçöl G (2014) A complexity theory for public policy. Routledge, New York
54. Morgenthau HJ (1993) Politics among nations: the struggle for power and peace, Brief Edition (revised: Thompson KW). Boston, McGraw-Hill
55. Nelson HG, Stolterman E (2014) The design way: intentional change in an unpredictable world. The MIT Press, Cambridge
56. North DC (1990) Institutions, institutional change and economic performance. Cambridge University Press, New York
57. Office of Stability and Humanitarian Affairs (2022) U.S. Department of Defense—Undersecretary of Defense for Policy. https://policy.defense.gov/OUSDP-Offices/ASD-for-Special-Operations-Low-Intensity-Conflict/Stability-and-Humanitarian-Affairs/
58. Olson M (1998) The logic of collection action: public goods and the theory of groups, 17th printing. Harvard University Press, Cambridge
59. Principles for Good International Engagement in Fragile States & Situations (2007) Organization for economic cooperation and development. https://www.oecd.org/dac/conflict-fragility-resilience/docs/38368714.pdf
60. Ramalingam B (2015) Aid on the edge of Chaos: rethinking international cooperation in a complex world. Oxford University Press, Oxford
61. Risse-Kappen T (1991) Public opinion, domestic structure, and foreign policy in liberal democracies. World Politics 43(4):479–512
62. Rittel HWJ, Webber MM (1973) Dilemmas in a general theory of planning. Policy Sci 4:155–169
63. Root HL (2020) Network origins of the global economy: east vs west in a complex systems perspective. Cambridge University Press, New York
64. Greene-Sands RR, Arakelian D (2018) Advancing SOF cultural engagement: the Malinowski model for a qualitative approach. JSOU Press, Tampa
65. S/2021/854—Concept Note for the Security Council High-Level Open Debate on Peacebuilding and Sustainable Peace, on the Theme "Diversity, State-Building and the Search for Peace", to be Held On 12 October 2021. United Nations Security Council. https://www.securitycouncilreport.org/atf/cf/%7B65BFCF9B-6D27-4E9C-8CD3-CF6E4FF96FF9%7D/S_2021_854.pdf
66. Slaughter AM (2017) The chessboard and the web: strategies of connection in a networked world. Yale University Press, New Haven
67. States of Fragility 2016: Understanding Violence (2016) Organization for economic cooperation and development 102. https://read.oecd-ilibrary.org/development/states-of-fragility-2016_9789264267213-en#page1
68. Stein AA (1983) Coordination and collaboration regimes in an anarchic world. In: Krasner SD (ed) International regimes. Cornell University Press, Ithaca, pp 115–140
69. Stones R (2005) Structuration theory. Palgrave Macmillan, New York
70. Thaler RH, Sunstein CR (2021) Nudge: The Final Edition. Yale University Press, New Haven

71. Trubowitz P (1992) Sectionalism and American foreign policy: the political geography of consensus and conflict. Int Stud Quart 36(2):173–190
72. United States Strategy to Prevent Conflict and Promote Stability (2020) U.S. Department of State, Bureau of Conflict and Stabilization Operations. https://www.state.gov/wp-content/uploads/2021/01/2020-US-Strategy-to-Prevent-Conflict-and-Promote-Stabilit-508c-508.pdf
73. Urry J (2003) Global complexity. Polity Press, Malden
74. Waltz KA (1979) Theory of international politics. McGraw-Hill Inc, New York
75. Weick KE (1995) Sensemaking in organizations. SAGE Publications, Thousand Oaks
76. Wight C (2006) Agents, structures and international relations: politics as ontology. Cambridge University Press, New York
77. Williams A (2020) Political hegemony and social complexity: mechanisms of power after gramsci. Palgrave Macmillan, Cham
78. Williams B, Hummelbrunner R (2009) Systems concepts in action: a practitioner's toolkit. Stanford University Press, Stanford
79. Winch P (1995) The idea of a social science and its relation to philosophy, 2nd edn. Routledge, London
80. Young OR (1991) Political leadership and regime formation: on the development of institutions in international society. Int Organ 45(3):281–308

David C. Ellis is President of Ellis Analytics, Inc., a firm dedicated to helping organizations navigate complexity and diverse socio-cultural landscapes. He served as a Resident Senior Fellow at the Joint Special Operations University from 2016–2022 where he co-developed the Applied Research Inquiry (ARI) interagency and civil society co-creation format. He has facilitated dozens of ARIs across a range of themes, including strategic competition, counter violent extremism, and organizational improvement. He holds a doctorate in international relations and comparative politics from the University of Florida and a master's degree in international development from The George Washington University. He served as an intelligence analyst in the USSOCOM J2, deployed to Afghanistan in support of Special Operations Forces from 2010–2011, and worked closely with interagency personnel. His current research focuses on the intersection of complexity, organizational learning within the special operations enterprise, and integrated campaigning and statecraft.

Charles Black is the Co-Founder and a Managing Partner at Xundis Global, LLC, which specializes in helping clients navigate complexity. He is a retired Marine Corps Officer with diverse experiences that span over three decades of service on four continents. Post military service his endeavors span from the board room, C-Suite to start-ups. He has served as a Non-Resident Senior Fellow and Senior Researcher at the Joint Special Operations University and holds a Senior Professional position at John Hopkins University Applied Physics Lab. He is working on his dissertation related to special operations and special operations forces.

A Cyber Solarium Commission Approach to Cyber Security Cooperation

Bryce T. Englin

Abstract In 2019 the United States Government (USG) established the bipartisan Cyberspace Solarium Commission (CSC) as part of the 2019 John McCain National Defense Authorization Act (NDAA). The CSC provided an opportunity for the United States (US) Congress to direct the USG towards a goal that develops a consensus strategic approach to protect the USG in cyberspace against cyber-attacks of significant consequences. The CSC legislation directs the USG to establish policy and focus on that ensures adequate protection are in place to deter attacks in the cyberspace environment. The CSC provides a good first step towards laying out a clear direction and recommendations on how the US can develop policy that prevents attacks in the Cyber Environment. The CSC furthermore advocates this new approach to cyber strategic using layered approach towards cyber deterrence. One area in this new strategic approach worth looking at is the concept of cooperation between the US and the multiple allies and partners we work with around the globe. This CSC does not dictate policy, but provides more than 75 recommendations that help make this approach a reality. Within the scope of these recommendations and layered approach includes clear language that directs the US to work with allies and partners and emphasizes this specifically as element within part CSCs Layer 1 objective. This book chapter looks at the CSCs layer 1 objectives and discusses their application within the area of cyber security cooperation and covers ideas including security cooperation policy and application within the cyber environment, cyber security cooperation leadership considerations, and specific security cooperation (SC) safeguards and risks considerations that can make cyber security cooperation inititiaves a reality.

Keywords Cyber solarium · Cyber environment · Cyberspace · Cyber-Attacks · Cyber deterrence

The United States (US) Department of Defense (DoD) clearly understands the intent of the directions, recommendations, and objectives put forth by congress as part

B. T. Englin (✉)
Defense Security Cooperation University, Washington, DC 20301-2800, USA
e-mail: bryce.t.englin.civ@mail.mil

A. Farhadi and A. Masys (eds.), *The Great Power Competition Volume 4*,
https://doi.org/10.1007/978-3-031-22934-3_5

of the Cyber Solarium Commission (CSC). In fact it is not hard to argue that key elements within the official March 2020 CSC report clearly incorporate DoDs defend forward cyber strategy. Through this strategy DoD aims towards placing a "broad focus on disrupting and halting malicious cyber activity at its source, including activity that falls below the level of armed conflict."[1] Defending forward as a strategy in the context of cyber is a great idea and theory but, there are still many limitations. One of my biggest personal observations over my 25 plus years across my active and civilian public service and 12 years of working in and around Security Cooperation (SC) is a concern in getting leaders to visualize the makeup and culture of the cyber environment and empahsis in understanding how the vitality of the cyber environment within the scope of SC. United States Cyber Commands (USCYBERCOM) April 2018 command vision describes the Cyber environment as a "fluid environment of constant contact and shifting terrain. New vulnerabilities and opportunities continually arise as new terrain emerges. No target remains static; no offensive or defensive capability remains indefinitely effective, and no advantage is permanent."[2] This idea was further emphasized during an April 29, 2021 virtual Lunch and Learn event I attended conducted by the Office of Secretary of Defense (OSD) Policy highlighting a recent cyber brief to legislators on Capital Hill. During this meeting, OSD cyber policy highlighted that "our understanding of cyber across the USG is still evolving."[3] In the same meeting, OSD cyber policy officials conveyed that "elected officials on both sides of the aisle in Washington DC carry a high level of support towards getting the USG to focus on objectives outlined in the CSC."[4] These two statements alone highlight just how important officials within the USG legislative and executive branches understand the importance of protecting the cyber environment. These statements also highlight some of the gaps that exist when it comes to the USGs understanding of the cyber environment.

The CSC through its layered approach provides a good starting point and recommendations that put the USG in a direction where we are equipped to strategically defend ourselves in the cyber environment. The CSC does not dictate cyber policy creation and instead emphasizes an approach that policy creation is the responsibility of each USG executive department and organizational entity. DoD has culturally and historically tended to take a very progressive approach when focusing on defensive cyber-strategy creation and implementation. The USG emphasis regarding current cyber policy and strategy began in the 1990s. In July 1996, President Clinton signed Executive Order 13010, which highlighted cyber specifically as a vital area of critical infrastructure and one of many areas requiring Critical Infrastructure Protection (CIP).[5] The emphasis on protecting cyber across the USG as part of CIP was

---

[1] https://media.defense.gov/2018/Sep/18/2002041658/-1/-1/1/CYBER_STRATEGY_SUMMARY_FINAL.PDF.

[2] https://www.cybercom.mil/About/Mission-and-Vision/#:~:text=Vision%20Document%20as%20of%20April%202018.

[3] OSD Cyber 101 MS Teams Meeting April 29, 2021.

[4] OSD Cyber 101 MS Teams Meeting April 29, 2021.

[5] https://mccrary.auburn.edu/work/insights/reflections-on-executive-order-13010/.

further codified in policy with the issuing of Presidential Decision Directive 63 in May 1998.[6] Protection of cyberinfrastructure as a part of CIP is vitally important as the infrastructure containing cyberspace houses the cyber domain, cyber environment, and cyber operating areas where operations are taking place. This itself is critically evident as DoDs reliance on cyberspace has significantly grown over the last 25 years. A case could even be made that the cyber environment and cyber operating areas within cyberspace should be considered entities within specifically what is considered the cyber domain.

DoD Joint Publication 3–12 highlights the idea of cyberspace as part of the cyber domain specifically defining cyberspace as "A global domain within the information environment consisting of the interdependent networks of information technology infrastructures and resident data, including the Internet, telecommunications networks, computer systems, and embedded processors and controllers."[7] The idea of a "global domain" as expressed within the Cyberspace JP 3–12 definition clearly makes a case that DoD considers Cyber and to a greater extent cyber environments and cyber operating areas as an operational domain on par with air, land, sea, and space.

Merriam-Webster's dictionary specifically defines a domain as "a sphere of knowledge, influence or activity."[8] Look anywhere over the last 25 years and it is not hard to envision how much of our daily lives, work, operations, and multiple activities are accomplished within what is very much a sphere where knowledge, influence and activities are conducted. Examine further and we find that it not hard to make a case that USG and DoD conduct these same types of operations within a cyber sphere that has emerged to form its own domain, environment, and operational areas. The biggest challenge still today is understanding what the cyber domain is and defining it in a way that individuals can normalize it. The biggest difference between the cyber domain and other identifiable warfighting domains such as land, sea, air, and space are the physical aspects and layers associated within these respective domains. Land, sea, air and space have tangible aspects we can feel or visualize. Cyber is different and does not come with the same level tangibility that allows an individual to visualize or feel the environment. Retired Israeli Major General Dan Efrony provides an exemplary comparable view regarding the differences between cyber and land, sea, air and space domains by highlighting that importance of cyber viewing it as "a new creation; a fifth domain, being added to the well-known four, namely, land, sea, air and lastly, space which became accessible to humanity in the middle of the twentieth century. While these four domains are the components of the cosmos, the fifth is a human creation—a pure ongoing technological development with continuous and significant impacts on the whole universe."[9] Efronys view clearly highlights that a discernable difference exists between the specifics of cyber and land, sea, air, and space domains.

---

[6] https://irp.fas.org/offdocs/pdd/pdd-63.htm.

[7] https://www.jcs.mil/Portals/36/Documents/Doctrine/pubs/jp3_12.pdf?ver=2018-07-16-134954-150.

[8] https://www.merriam-webster.com

[9] https://csrcl.huji.ac.il/sites/default/files/csrcl/files/dan_efrony.pdf.

More importantly, his explanation provides a visualization of the cyber domain as a boundless domain without physical or geographical borders.[10] Efrony's perspective matches up well with US Cyber Commands 2018 Command vision highlighted earlier that the cyber environment is a "fluid environment of constant contact and shifting terrain."[11]

Putting Efrony and US Cyber Commands vision together presents an opportunity to visualize that the cyber domain and operational environment are a place that is ever changing, shifting, and never fluid. These visuals in turn present us with an understanding that allows for the creation of philosophical viewpoint leaders can use to develop policies focusing on operations within the cyber domain and operating environment. As we start thinking about policy development across cyber domain and operating areas, it is possible to incorporate other areas with an intent of developing robust cyber strategy and posture. When thinking about cyber domain and operational policy development, one area where DoD could consider the application of the Defend Forward strategy identified earlier and[12] incorporate recommendations within the layered approach emphasized within the CSC as a way to furthur develop and refine cyber strategy and focus on operational efforts across cyber domain and operating environment. A specific area of these efforts identified within Layer 1 of the March 2020 CSC focuses on the idea of Shaping Behavior.[13] This recommendation presents itself in the area focusing on Strengthening Norms and Non-military Tools.[14] This area of the CSC provides a specific recommendation that USG and DoD look at using "A system of norms, built through international engagement and cooperation, promotes responsible behavior and, over time, dissuades adversaries from using cyber operations to undermine any nation's interests."[15] The specifics of this idea especially in "using a system built through international engagement and cooperation" highlights an opportunity to think about the idea of how the USG and DoD might use cyber cooperation and engagements with international allies and partners we work with around the globe.

The recommendation regarding the area of "international engagement and cooperation" put forth by the CSC falls into an area of what DoD refers to as Security Cooperation (SC). Using SC as a tool to implement this specific Layer 1 recommendation demonstrates one direction DoD could take towards attaining a key recommendations outlined in the CSC.[16] The idea of using SC to meet CSC objectives present some interesting opportunities while also leading to additional questions.For instance, one must ask what does a normal cyber end-state look like when modifying

---

[10] Ibid.

[11] https://www.cybercom.mil/About/Mission-and-Vision/#:~:text=Vision%20Document%20as%20of%20April%202018.

[12] https://media.defense.gov/2018/Sep/18/2002041658/-1/-1/1/CYBER_STRATEGY_SUMMARY_FINAL.PDF.

[13] Ibid.

[14] https://drive.google.com/file/d/1ryMCIL_dZ30QyjFqFkkf10MxIXJGT4yv/view.

[15] Ibid.

[16] https://drive.google.com/file/d/1ryMCIL_dZ30QyjFqFkkf10MxIXJGT4yv/view.

behavior for partners and allies? Perhaps another greater question worth asking is what types of cyber international engagement and cooperation could DoD conduct with partners and allies? These two questions present an opportunity to look and discuss how DoD might use SC towards modifying behaviors within the scope of international cooperation and engagement as outlined in Layer 1 of the CSC.

The specific ideas and questions associated within the scope of this Layer 1 area of the CSC highlight the need for a deeperlevel of understanding in how SC as a tool can support specifically identified Layer 1 CSC objectives. My goal over the course of this chapter is to dive deeper into four areas that highlights how DoD could use SC to help meet the specifically identified recommendation in the area of working with allies and partners to meet CSC Layer 1 objectives. The first part of this chapter will introduce SC. Here we will spend time explaining what SC is, provide historical background, and discuss how SC could fits within cyber-related missions. The second area of the chapter will further discuss SC programs and explain how DoD and other USG executive branch entities, including Department of State (DOS), Department of Homeland Security (DHS), and Department of Justice (DOJ) and their specific role in SC. This part of chapter will highlight different SC and Security Assistance (SA) programs and explain how they can support Layer 1 CSC. In the third part of chapter our discussion will move towards making SC a reality. The specifics here will cover SC and SA program processes and explain how the USG and our allies and partners, benefit from utilizing SC and SA programs to support cyber initiatives. The fourth and final part of the chapter will touch on a variety of risks and concerns associated within the areas ofculture and cyberinfrastructure. The goal across this final part of the chapter is highlight risks and challenges cyber community members may encounter when dealing with allies and partners, explain how SC policy can mitigate specific risks, and understand other specific considerations in dealing with new and emerging cyber SC areas to include cyber infrastructure. The ideas covered in this chapter are meant to provide a cursory introduction and enhance the readers understanding across the area of Cyber SC.

1 SC Explained

The USG and DoD find value through interactions and cooperation with allies and partners. A key element across these areas is the foundation they create within the scope of Security Cooperation (SC) that the USG and DoD utilize as a goal of achieving US National Security objectives. Joint Publication 3–20 defines SC as: "all DoD interactions, programs, and activities with foreign security forces (FSF) and their institutions to build relationships that help promote US interests; enable PNs to provide the US access to territory, infrastructure, information, and resources; or to build and apply their capacity and capabilities consistent with US defense objectives."[17] SC in essence is interactions with foreign partners and allies that

[17] https://www.jcs.mil/Portals/36/Documents/Doctrine/pubs/jp3_20_20172305.pdf.

contribute towards building relationships, promoting US interests, gaining access, and helping ally and partner objectives align with USG national security objectives.

Fully understanding SC policy starts by briefly looking at the history of SCO policy to understand how SC has turned into a tool that helps the US achieve national security objectives. The concept of what is known as SC in terms of specific policy begins in 1961 with the passing of the Foreign Assistance Act (FAA)[18] authorization. A 2011 Congressional Research Service (CRS) report highlights that the FAA was implemented by the Kennedy administration "to organize and implement US foreign assistance programs with a commitment to long-range economic assistance to the developing world."[19] The FAA allowed the US to provide foreign assistance and sell military equipment to allies and partners, however, the policy lacked the element of congressional oversight needed to provide the SC system with the needed legislative checks and balances. Fifteen years later, congress would finally implement the additional set of checks and balances as part of the SC policy with the passing of the Arms Export Control Act (AECA) in 1976.[20] A review of the 2020 CRS report highlights that the AECA updated policy which clarified the executive branch role and: "provided the President with the statutory authority to control the export of defense articles and services, the statutory authority for the Foreign Military Sales (FMS) program, and specific criteria for Direct Commercial Sales (DCS)."[21] The AECA of 1976 also clarified the legislative branches role in the SC process by establishing a "requirement for congressional consideration of certain foreign defense sales proposed by the President. This procedure includes considering proposals to sell major defense equipment and services or retransfer such military items to other countries."[22] Since 1976, many provisions associated with AECA have evolved into many different SC programs. The importance of SC Policy, begins with the FAA and AECA serving as the cornerstones that define the process, rules, an path the USG utilizes when executing SC.

Policy defines the path forward to execute the functions and process of SC. SC itself is a group of programs used by the USG and DoD to cooperate and work with a broad range of allies and partners worldwide. DoD views SC as a national security tool used to help allies and partners build capacity to respond to challenges and work alongside the USG. At a much broader level identified SC initiatives allow the USG to build relationships around common defense objectives and interests with different branches and departments across the USG Executive Branch. Many of these shared interests come about through foreign defense and security establishment engagements, disaster or emergency response, Security Assistance (SA) programs,

[18] https://www.govinfo.gov/content/pkg/USCODE-2017-title22/html/USCODE-2017-title22-chap32.htm.

[19] https://sgp.fas.org/crs/row/R40089.pdf.

[20] https://www.govinfo.gov/content/pkg/USCODE-2018-title22/html/USCODE-2018-title22-chap39.htm.

[21] https://sgp.fas.org/crs/natsec/R41916.pdf.

[22] Ibid.

and others. SC is about developing partnerships and encouraging allies and partners to support USG led and inspired foreign assistance objectives that provide stability. Across the executive branch the DoD serves as the lead executive agency for executing SC actions in the areas supporting defense policies and national security interests.[23] However, in these areas the DOS defines and sets policy guidelines for SC and SA programs. SC and SA programs and initiatives are considered tools of US national security that contain foreign policy implications, and therefore fall into the scope of USC Title 22. Simply put, DOS establishes policy[24] and relies on DoD to execute the different programs due to the sizable workforce of defense-related SMEs who can support multiple SC and SA program areas.

The overarching goal of using SC and SA programs is about attaining results based on identified initiatives, including objectives, end-states, laws, and policies. One area of interest in the scope of SC and SA programs worth looking at are initiatives focusing on cyber cooperation. The reality is that goals and outcomes should be focused on these same factors. However, due to the differences in how cyber and physical environments are constructed, cyber cooperation initiatives outcomes are going to be different. If we look back at the intent of CSC Layer 1 outcome [25]along with US Cyber Command lead forward approach, and Major General Efrony's view of the cyber environment we can highlight an idea that outcomes assocaited with each cyber situation and environment are not the same. A perfect example that highlights these differences can be found within a December 2020 virtual training cyber agreement signed between the United States and Australia.[26] This first of its type of agreement "enables U.S. Cyber Command to incorporate Australian Defence Force feedback into USCYBERCOM's simulated training domain, the Persistent Cyber Training Environment."[27] This specific platform,referred to as the PCTE and according to the US Army Program Executive Office (PEO) STRI is "a training platform supporting standardized Joint Cyberspace Operations Forces individual sustainment training, team certification, mission rehearsal and provides the foundation for collective training exercises."[28] PEO STRI highlights that this agreement leverages the PCTE in a way that leverages existing connectivity to facilitate the sharing of resourcesand provides additional cyber "maneuver space."[29] This specific cyber agreement is the first of its kind that provides a key USG ally with access to a cyber domain training environment that provides an opportunity to train in a"fluid environment of constant contact and shifting terrain.[30] The benefits of working within

---

[23] https://www.defense.gov/About/.

[24] https://samm.dsca.mil/chapter/chapter-1#C1.1.

[25] https://drive.google.com/file/d/1ryMCIL_dZ30QyjFqFkkf10MxIXJGT4yv/view.

[26] https://www.cybercom.mil/Media/News/Article/2434919/us-and-australia-sign-first-ever-cyber-agreement-to-develop-virtual-training-ra/.

[27] Ibid.

[28] https://peostri.army.mil/persistent-cyber-training-environment-pcte.

[29] Ibid.

[30] https://www.cybercom.mil/About/Mission-and-Vision/#:~:text=Vision%20Document%20as%20of%20April%202018.

the operational and configurable cyber operational environments within the PCTE is that both the US and Austrailia have an opportunity to cooperate on cyber activities that fulfill the identification of cyber domain and operating area highlighted by Major General Efrony.,US Cyber Command, and JP 3-12 earlier in this chapter.

By signing this first of its kind agreement and bringing Australia into the PCTE environment the US is fulfilling key objectives of what we can refer to as Cyber SC initiative with a key US all who is one part of the many partners and allies the USG works with. A key policy objective of this Cyber SC initiative within the scope of the CSC is that. The PCTE environment provides an opportunity for the US and Australia to work together towards developing meeting a Layer 1 recommendation of modifying behavior as outlined earlier in the chapter. Figuratively, the concept of working towards modifying behavior comes back to having a specifically identified outcomes with an included set of countries or other entities, and a specific cyber aim. In theory, this agreement allows the US and Australia to utilize the PCTE to set up and develop cyber engagements and initiatives focused on key cyber outcomes. Such outcomes could perhaps focus on training that deals with cyber behavior or tactics employed by common adversaries or other entities considered proxies serving on behalf of these adversaries who are striving to conduct harmful cyber domain and operational environment operations.

The benefits of the agreement signed bytween the US and Australia regarding use of the PCTE highlight the possibilities and potential role SC can play towards helping the USG and DoD work to meet a CSC layer 1 recommendation of working with allies and partners to modify behavior. We see that DoD and DoS involvement in employing cyber cooperation initiatives with allies and partners. The idea conveyed here leads to a another question which should allow us to ask if there are other SC programs out there that could be used to support Cyber cooperation initiatives.

2 Using SC and SA Programs to Support Cyber Initiatives

The previous section of this chapter highlighted the definition of SC in policy according to JP 3–20. The previous portion explained that SC consists of "all DOD interactions, programs, and activities with foreign security forces (FSF)."[31] The USG accomplishes SC under USC Title 10 or DoD authorities,[32] while SA falls under USC Title 22 or DOS authorities.[33] The specific execution of an identified Cyber SC and SA objectives begins with the USG and an ally or partner identifies a potential cyber concern. Utilizing an SC or SA program to support a cyber initiative requires that the concern line up with policy objectives that help the US achieves national security objectives.[34] Even if an SC or SA initiative meets national security objectives, there

[31] https://www.jcs.mil/Portals/36/Documents/Doctrine/pubs/jp3_20_20172305.pdf.

[32] Ibid.

[33] Ibid.

[34] https://samm.dsca.mil/chapter/chapter-1#C1.1.

are considerations the USG must takebefore committing to support a cyber initiatives. Two such consideration including highlighting how the initiative aligns with DoDs defending forward strategy and broad focus towards "disrupting and halting malicious cyber activity at its source”[35] and the CSC Layer 1 recommendation of strengthening norms and changing behaviors[36] and highlighting if this use “international engagement and cooperation to promote responsible behavior and over time dissuade adversaries from using cyber operations to undermine nations interests”[37] meets the overall aim and intent. Multiple possibilities exist where SC or SA cyber initiatives can provide mutual beneficial, however, the end goal is about ensuring the intent of such an initiative apply the appropriate level of cyber deterrence and promote responsible behavior in a way that line up with USG National Security Objectives. If the USG determines that a cyber SC or SA initiative meets these intents, the focus then shifts towards finding an appropriate program or initiative to support and meet the intent of the cyber objective.

Finding an SC or SA initiative starts with identifying the respective SC or SA program authority that could support the cyber initiative. One of the most useful tools that can assist the USG or ally and partner nations is the Defense Security Cooperation University (DSCU) School of Security Cooperation Studies (SSCS) SC and SA programs guide.[38] This guide is updated annually by DSCU faculty and provides a comprehensive overview of all available SC and SA programs. When the USG and allies or partners review this guide they must be aware that not every identified program can support a Cyber initiatives. A review of the most recent guide highlights 11 various SC and SA programs[39] categories that contain a total of 105 SC and SA programs.[40] If we look at these 105 SC and SA programs, there are approximately 64 or 61% of them that could support cyber initiatives. The key to authorizing SC or SA programs in support of a cyber initiatives is understanding the outlined purposes, authorizations, guidance, and eligibility of the program and how its applies towards USG cyber and cooperation policy.[41]

Using existing SC and SA programs to support cyber initiatives with allies and partners can be beneficial. An important distinction regarding to the use of SC or SA programs to support cyber initiatives is understanding how to maximize the effectiveness of different programs. There may be a level of hesitancy on behalf of those engaging in cyber cooperation with allies and partners for the first time. An important thing to be aware of with regard to SC and SA programs is that the USG, DoD and DOS have a set of effectivepolicies in place that ensure maximum benefits

[35] Ibid.

[36] https://drive.google.com/file/d/1ryMCIL_dZ30QyjFqFkkf10MxIXJGT4yv/view.

[37] https://drive.google.com/file/d/1ryMCIL_dZ30QyjFqFkkf10MxIXJGT4yv/view.

[38] https://www.dscu.mil/documents/publications/security_cooperation_programs_handbook/security_cooperation_programs_handbook.pdf.

[39] Ibid.

[40] https://www.dscu.mil/documents/publications/security_cooperation_programs_handbook/security_cooperation_programs_handbook.pdf.

[41] Ibid.

and protections for the USG and our allies and partners. When looking at the specifics associated with the different protections policies associated with these SC and SA programs across DSCA policy within the Security Assistance Management Manual (SAMM) and appropriate US Military Department SC Manuals it is easy to see that SC policy can easily apply to cyber and protections ensure appropriate precautions are taken to protect and limit the release of certain types of technology. Understanding specific ramification of policy and the protections are key towards making cyber SC and SA initiatives when focusing on the cyber domain and cyber operational environment a reality.

3 Making Cyber Initiatives a Reality

The concept of using SC or SA programs to turn cyber initiatives within the cyber domain and cyber operational environment into reality may come across as a somewhat new idea. The reality is that these types of cooperation are already happening. Think back to earlier in the chapter and the highlighting of the bilateral agreement signed between the USG and Australia granting Australia access to the PCTE environment. Thinking about the agreement and potential outcomes that could come from the US and Australia working together in the PCTE a case could be made that work in the area of an SC and SA Program category known as International Armaments Cooperation (IAC) could be happening. IAC as an SC program strives "to share in the development, funding, and production of mutually required weapons systems or components."[42] A takeaway in understanding the application and execution of Cyber initiatives in the scope of SC and SA programs are understanding the differences when compared with implementing these same types of programs within traditional and tangible platforms. A specific difference for instance is tangibility. The USG along with our allies and partners can touch an F-16 or like platform. Stakeholders have a clear vision of how to use an F-16 within the scope of land, air, and maritime domain operations. The use of SC and SA programs focusing on the cyber domain and cyber operational environment differ from a traditional platform due to the cyber domain make-up. The catalyst for this perspective requires us to think back to retired Israeli Major General Dan Efrony's thoughts regarding the "characteristics"[43]of the cyber domain and his view regarding cyber as a location or space that is "boundlessness and manmade."[44]If applying US Cyber Commands 2018 Command vision describing cyber as a "fluid environment of constant contact and shifting terrain"[45] we find that that the outcomes striving to be attained and SC

[42] https://www.dscu.mil/documents/publications/security_cooperation_programs_handbook/security_cooperation_programs_handbook.pdf?id=1.

[43] https://csrcl.huji.ac.il/sites/default/files/csrcl/files/dan_efrony.pdf.

[44] https://csrcl.huji.ac.il/sites/default/files/csrcl/files/dan_efrony.pdf.

[45] https://www.cybercom.mil/About/Mission-and-Vision/#:~:text=Vision%20Document%20as%20of%20April%202018.

and SA programs being used to achieve those outcomes have additional complexities that must be navigated.

The specifics regarding these types of complexities begin with understanding the variables, factors, and connections required to attain the end-state goal. Once the path to achieve a goal have been clarified, the next step is research and designing a way that allows leaders to understand the visuals of the domain, operating environment, and objectives to reach the goal. A visual representation and perspective provides leaders with little awareness and understanding of cyber with a presentation that allows them to digest the initiative and understand the objectives the USG is striving to accomplish with our allies and partners. Development and implementation of the actual cyber initiatives requires an understanding of allies and partners existing cyber operations, operating environments, planned capability employment, and cyber ambitions. Certain aspects should be available based on existing working relationships, however understanding an ally or partners existing cyber operations, cyber institutional capacity, and defense cyberculture are pivotal in moving an initiative forward. The more willingess displayed by an ally or parnter when developing initiatives the easier it becomes to identify challenges and gaps that could hinder the initiatives and end goal. As an example, in working with an ally or partner the USG finds a gap or issues within an element of the partners cyber institutional capacity. The identified gap is small but could have a significant impact on the cyber initiative if not mitigated. As a solution the USG implements an additional part into the cyber initiative providing a solution and mitigating the gap.

Identifying gaps is vital as specific gaps can have a significant impact on initiative implementation. A key critical area where gaps may exist and could be detrimental to initiative implementation is within an allies' and partners' existing cyber institutional capacity. The lack of specific institutional capabilities can be detrimental towards success of an initiative. The United States Agency for International Development (USAID) 2009 Human and Institutional Capacity Development (HICD) Policy Paper and 2008 Fast Track Initiative Guidelines for Capacity Development in the Education Sector defines institutional capacity: "as the quality of leadership, incentives, systems, resources, and personnel that produces results based on the mission, goals, and objectives of the institution. It refers to both the organizational capacity as reflected by the institution's public mandate, legitimacy, resources, and systems and the human capacity reflected by the motivation, status, technical, and managerial skills of its leaders and staff."[46] From this policy we should be able to understand that cyber institutional capacity consisting of existing cyber elements and institutions that allies and partners have in place to support cyber-related initiatives. Specifics associated within the area of cyber institutional capacity have reached a level of importance that USAID has incorporated the technology perspective as part of the 2020–2024 USAID Digital Strategy.[47] The strategy strives to "position the Agency to advance our mission—to end the need for foreign assistance—through digitally supported programming that fosters the Journey to Self-Reliance in our

[46] https://www.edu-links.org/sites/default/files/media/file/ICA_Tool_FINAL_Sept2014.pdf.

[47] https://www.usaid.gov/sites/default/files/documents/USAID_Digital_Strategy.pdf.pdf.

partner countries and maximizes the benefits while managing the risks that digital technology introduces into the lives of the communities we serve."[48] Self-sufficiency amongst allies and partners is a viable and achievable end-state towards working for. This type of end-state cannot be achieved without implementing new and augmenting existing cyber institutions at strategic and operational levels. The intent of refining cyber institutions at these levels set the opportunity to fill in gaps and help allies and partners utilize initiatives to realize an appropriate end state. The greater amount of investment placed committed towards developing and refining existing institutions by allies and partners makes it easier to fill gaps, achieve an applicable end state, and attain self-sufficiency.

When looking at existing institutions and institutional capabilities we must understand that they come in tangible and intangible forms. Understanding specifics regarding tangible and intangible areas can help the USG build initiatives that help develop institutions that align with CSC Layer 1 outcomes that "promotes responsible behavior and dissuades adversaries from using cyber operations to undermine any nation's interests."[49] As institutions are developed and refined over time it is possible to help an ally or partner develop responsible cyber behaviors that dissuade themselves or in other cases specifically identified adversaries from conducting unwanted cyber activities. Multiple ways exist that allow the US to help allies and partners build cyber institutional capacity, focus on eliminating gaps, and changing behaviors. A potential way to highlight the benefits of cyber cooperation is possible by looking at the following hypothetical situation. The USG and Australia were recently made by Philippine Ministry of Defense (MOD) s of recent hacking activity undertaken by groups acting as a proxy on behalf of a terrorist organization operating in the Philippines. The US and Australia conclude through conducting a forensic cyber analysis that these hacking attempts are similar to the critical infrastructure attack that took the Colonial Pipeline down and off-line for a lengthy period of time in May 2021.[50] The Philippine MOD requests assistance through United States Indo Pacific Command asking for assistance in having the US and Australian military cyber forces work together, conduct scans, and identify cyber gaps within the Philippines' critical fuel storage infrastructure areas. The Phillippine government allows the US and Australia to securely access and start scanning network areas of the critical fule storage infrastructure. During the scan the US and Austrailia observe gaps and find openings where the hackers could gain access and provide this information to the Philippines. The Philippine government takes the information and requests additional assistance from the US and Australia to help close these gaps. The US and Australia make a decision to utilize the PCTE and construct a cyber domain operational environment design using the information provided by the Philippine military. Over the course of a few days US and Australia cyber forces work together within the PCTE and develop a critical cyber fix plan. The US and Australia conduct a virtual meeting with the Philippines and provide the recommended solution

[48] Ibid.

[49] https://drive.google.com/file/d/1ryMCIL_dZ30QyjFqFkkf10MxIXJGT4yv/view.

[50] https://www.energy.gov/ceser/colonial-pipeline-cyber-incident.

to fix the most critical vulnerabilities. The Philippine MOD requests further assistance from the US and Australia and grants both countries with the appropriate level of access to apply fixes to the critical cyber infrastructure.

In this hypothetical situation the US and Australia worked together to help a strategic partner to put critical cyber fixes in place and fix a critical cyber gap within the country's critical infrastructure. Both the US and Australia had a national security and SC related interest in supporting this initiative as the pipelines with critical cyber gaps have been routinely used to refuel US and Australian navy ships during port calls in the Philippines. This situation also presented US with an opportunity to identify some institutional cyber capacity gaps and was able to help the Philippine MOD put together a couple of events and engagementsfocusing on critical cyber infrastructure protection development. The US and Australia's focus here provided assistance in helping the Philippines achieve a CSC Layer 1 objective of changing existing instituational behavior to focus on identifying gaps within the cyber areas of critical infrastructure and apply appropriate deterrence objectives to ensure the proxy of an identified terrorist group was unable to hack and destroy critical infrastructure.

This situation is meant to highlight the importance of institutional capacity and other factors vital to making an SC or SA initiative a reality. The concerns in this scenario are a small example of what could be different cyber concerns raised by any of the allies and partners the USG works with. In this specific scenario the USG had signed an agreement with Australia, both countries had a common national security interest, and realized benefits of using the PCTE to develop a solution. The specific national security interest in this scenario was the criticality of the fuel infrastructure used to refuel American and Australian ships stopping in the Philippines for port visits. The situation also allowed the US to identified gaps and provided recommendations that helped the Philippines further develop institutional capacity in the area of critical infrastructure cyber protection development. Cyber institutional capacity is one of many areas within the scope of cyber security cooperation. Objectively, institutional capacity is important in that it contributes to the ally and partners ability to develop, successfully implement a cyber initiative, and reach the desired end state. The example here highlights a focus on cyberinfrastructure and shares what a Cyber SC or SA initiative might look like in reality. The point here leads to an important and final discussion area within this chapter focusing on both cultural and technological risks as part of the cyber SC and SA initiatives. Cultural and technological risks are an essential consideration within the big picture the USG must think about when working with allies and partners. The specificty of certain risks require critical thought espeicially when dealing with allies and partners who are engaging with China, Russia, North Korea, Iran, proxies, and potentially other rogue states[51] in an ever-growing Cyber Great Power Competition (GPC) and Cyber Strategic Competition.

[51] https://drive.google.com/file/d/1ryMCIL_dZ30QyjFqFkkf10MxIXJGT4yv/view.

4 Understanding the Risks

When decisions are made to implement cyber capabilities and initiatives using SC and SA authorities there are a set of cultural and technological risk considerations the USG must think about prior to moving forward with implementating and executing acyber cooperation initiative. The concerns with cultural and technological implications are vital to helping USleaders ensure success in working with allies and partners on cyber initiatives. The importance of risk considerations that come with the deployment of cyber initiatives across infrastructure cannot be understated. Extra levels of consideration must also be given to infrastructure developed and provided by other partners or adversaries. In most instances USG led cyber cooperation initiatives limit the access allies and partners have to USG cyber networks. If an initiatives requires access to USG networks or cyberinfrastructure operating lanes, these requests are always vetted by appropriate DoD and DOS agencies and will have appropriate mitigation factors applied and fall into the scope of a technology transfer or access decision associated with the project.[52] Most leaders should be aware that technology can only be released when an authorized technology transfer decision has been made and the amount of release is specific to the level an ally or partner has a need to know. An important perspective to understand is that process regarding these types of decisions are made by appropriately identified USG cyber stakeholders and are traditionally very limited in scope.

It is not uncommon to see cyber leaders deal with cultural issues when decisions regarding access and transfer of cyber technology come up. Up front cyber leaders must understand that they cannot use the same approach and considerations with allies and partners that they may use when focusiong on USG specific cyber operations. Cyber leaders interfacing with allies and partners must understand that there are limits to the amount of control and responsibility they have when dealing with an initiative. Individuals must undertand that their shifts to one of advising and consulting when working with allies and partners. Potential interests of trying to take control or dictate cyber operations will find these actions as potentially detrimental to the existing relationships. Most cyber leaders have no issue making this type of shift, however, there are some who may find making this shift challenging and must understand these types of cultural traits and attitudes are potentially detrimental to the overall big picture working relationship with allies and partners.

A cyber leaders or Subject Matter Expert (SME) encountering an SC and SA cyber project or initiative for the first time must understand that allies and partners may not conduct cyber operations in a manner they are use to. Certain instances may produce the same end result, but the steps used to get there may be completely different. Cyber leaders or SME may find differences in approachchallenging and may have the challenges in understanding differnt approaches and might feel interactions and developing of cyber relationships are counter. Individuals with a limited understanding of dealing with cultural issues, exchanges, and differing ally and partner attitudes focusing on cyber must be cognizant of the differences and be aware of

[52] https://samm.dsca.mil/chapter/chapter-3#C3.1.

missteps that ould affect an existing USG relationship. A great first step cyber leaders and SMEs can take when dealing an emerging cyber ally or partner is spending time inquiring and learning how certain cyber operations are conducted and why they might be done a certain way. Taking initiative to spend extra time learning why and how an ally or partner is conducting cyber operations in a specific way can help eliminate anxiety and provide a better understanding and appreciation for an ally and partners cyber institution and operations. As concerns arise it is important to discuss these issues with SC stakeholders. There may be a feeling to immediately stop or halt things, in these situation, immediate SC stakeholder involvement is a must. Working with fellow SC stakeholders allows a consensus approach and relay concerns through appropriate diplomatic channels. Decision regarding specifics of SC and SA initiatives are approached through a flat approach focused towards gaining consensus. This approach is helpful in mitigating risks that could hamper an existing ally or partner relationship. Building relationships is a key foundation of SC and SA interactions with allies and partners.

Crucial aspects of an SC and SA relationships are importance when executing cyber-related SC or SA initiatives and engagements. One aspect within the scope of a relationships is concern that an ally or partner utilizes the cyber technology, capability, or initiative as intended. This is a risk consideration, however, allies and partners receiving cyber capabilities to include training must understand the US does implement safeguards and requirements to ensure adequate protections are in place. One such protection consideration are end-use requirements implemented within the scope of Foreign Military Sales (FMS) terms and conditions,[53] FAA section 505 agreements,[54] and policy outlined within the DSCA SAMM.[55] Specifically identified requirements outlined in the previous identified policies apply to all technologies, including cyber and they ensure cooperation initiatives are utilized as intended. DSCA SAMM C8.6.1. highlights this in policy highlighting "potential end-use violations as unauthorized use of defense articles, defense services, or technical data, including information provided during CONUS or in-country training."[56] Appropriate and inappropriate use of Cyber technology and training is no different and there are considerations that must be taken to try and minimize risk of cyber technologies being used to support potential end use violation.

A previous real-world example highlighting a potential cyber end use violations could be the Egyptian government decision to cut off internet within the country during the Arab Spring in January 2011.[57] Building on the real world scenario, land from a hypothetcial situation, let's say the USG had previously provided Egypt's Ministry of Defense (MOD) with a cyber package and training

[53] https://samm.dsca.mil/figure/figure-c5f4.

[54] https://www.govinfo.gov/content/pkg/USCODE-2017-title22/html/USCODE-2017-title22-chap32-subchapII-partII-sec2314.htm.

[55] https://samm.dsca.mil/.

[56] https://samm.dsca.mil/chapter/chapter-8.

[57] https://www.theguardian.com/technology/2011/jan/28/egypt-cuts-off-internet-access.

that included network configuration and internet controls. As part of the hypotheticals, perhaps the partner experiences unrest and protesting similar to the real-world situation of the Arab Spring.Egypt makes a decision to have the MOD use the technology and training provided to shut off all cell phone and internet access across the country. The lack of accesshighlighted by the Guardian "caused concern among observers who believe that internet access—which the Egyptian government limited earlier this week by cutting off social networks—is essential to ensure that government acts responsibly towards its citizens."[58] Furthering this concern is a perspective from Andree Toonk from BGPmon, a monitoring site that checks the connectivity of countries and networks highlighted that "88% of the 'Egyptian internet' has fallen off the internet and that all of the major ISPs seem to be almost completely offline. With the government taking a shotgun approach by ordering ISPs to stop routing all networks."[59] In this situation it is worth wondering how the USG might be viewed if direction to take down the internet had come from members of Egypt's military and the activities to shut down the internet used the cyber equipment and network training received from the US. The hypothetical added to the real world situation presents an extreme use of cyber to suppress internet and cell phone access, however it is a consideration that must be thought through. It is highly unlikely the USG would ever providing cyber capabilities and training at this level to ally or partner with a questionable history of abd motives. The goal here is meant to highlighting the associated technological risks that come about when utilizing with allies and partners.

Technological risks associated within cyber security cooperation and specifically cyber SC and SA initiatives are an area of concern. The technological concerns that arise are the risks that must be thought through. One risk worth a higher level of discussion are risk considerations that come with employment of technology across cyberinfrastructure. The US must maintain a sense of awareness when deciding to employ technologies across cyberinfrastructure put in place by in other countries by countries the USG considers adversaries in the scope of the Great Power Competition (GPC) and emerging strategic competition. Since 2015 the USG has been locked in a GPC with China and Russia. The previous US Chairman of the Joint Chiefs of Staff Marine Corps General Joseph Dunford explained the GPC in 2019 as a place where "Russia and China are competitors to the United States and both nations are looking to overturn the current rules-based international order."[60] Gen Dunford follows this up explaining that "Military buildups are only part of the picture, though, the chairman said. All aspects of international relations—economic, diplomatic, political, and even cultural—come into play in great power competitions"[61] Cyber and more specifically technologies such as cyberinfrastructure are aspects that have not really been envisioned as in the scope of the GPC when thinking

[58] Ibid.

[59] https://www.theguardian.com/technology/2011/jan/28/egypt-cuts-off-internet-access.

[60] https://www.defense.gov/News/News-Stories/Article/Article/1791811/dunford-describes-us-great-power-competition-with-russia-china/.

[61] Ibid.

about competition. In many instances, however, emerging allies and partners military have come to rely on China to develop and build cyberinfrastructure within their respective countries. The Council on Foreign Relations (CFR) has assessed China's Digital Silk Road (DSR) initiative[62] and has a presentation that highlights 26 countries activily participating as part of China's DSR.[63] The 26 countries on the list are located within each US Combatant Command and include quite a few long-term and emerging SC and SA partners and allies[64] such as the United Kingdom, South Korea, Saudi Arabia, and Egypt.[65] In the scope of existing cooperation relationships both the United Kingdom[66] and South Korea[67] have a great deal of involvement with the US on IAC initiatives, whileSaudi Arabia and Egypt have benefitted long-term cooperation relationships with the US that have been a stabilizing force across the Middle East region.

An interesting aspect of the CFR presentation is the DSR Memorandum of Understanding (MOU) signed between South Korea and China.[68] The CFR explains that the MOU may not be fully binding[69] however, there are multiple concern that arise as the US and South Korea work together researching and developing future cyber technologies, including "6G, quantum technology, and biotechnology."[70] A concern that does factor into this agreement is the effect of Chinese Intellectual Property theft which could be interpreted as being open in the scope of the agreement. The damange from Chinese Intellectual property theft in an economic sense is estimated to cost the US between $225 billion and $600 billion per year.[71] The bigger concern outside the scope of economics and lost revenue is the theft of intellectual property based on technology research and potential for the Chinese military to use this technology to gain a decisive advantage. Co-development and cooperation of cyber initiatives can provide the US and our allies and partners with benefits that clearly meet the SC IAC objectives to "share in the development, funding, and production of mutually required weapons systems or components."[72] The considerations the US must take when decision to embark on co-development, research, and production, require deliberate thought regarding the risks and levels of protection undertaken to prevent economic and technololgical losses in the area of intellectual property

[62] https://www.cfr.org/china-digital-silk-road/.

[63] Ibid.

[64] Ibid.

[65] Ibid.

[66] https://obamawhitehouse.archives.gov/the-press-office/2012/03/14/joint-fact-sheet-us-and-uk-defense-cooperation.

[67] https://www.whitehouse.gov/briefing-room/statements-releases/2021/05/21/fact-sheet-united-states-republic-of-korea-partnership/.

[68] https://www.cfr.org/china-digital-silk-road/.

[69] Ibid.

[70] https://www.whitehouse.gov/briefing-room/statements-releases/2021/05/21/fact-sheet-united-states-republic-of-korea-partnership/.

[71] https://foreignpolicy.com/2019/10/16/china-intellectual-property-theft-progress/.

[72] https://www.dscu.mil/documents/publications/security_cooperation_programs_handbook/security_cooperation_programs_handbook.pdf?id=1.

theft. Another concern are the associated risks surrounding actual installation of physical cyberinfrastructure.

For the US, associated risks within the scope of physical cyberinfrastructure require specific consideration when dealing with items installed by China and the companies chosen to develop and deploy infrastructure and technology. If we look at the DSR in action in Africa, we clearly see Chinese state owned technology companies conducting digital engagements with 47 of 54 countries on the continent.[73] In addition to digital engagements, China has also installed and deployed cyberinfrastructure backed by companies either state owned or having a state-own stake in 33 countries on the continent of Africa where the US has SC or SA involvement.[74] Another concerns associated with Chinese backed cyberinfrastructure builds and deployments are the requirements dictated by Chinese law. The most concerning issues are found within Chinas 2017 National Intelligence,[75] 2016 Cybersecurity,[76] and 2014 Counter-Espionage[77] laws. The intent of and provisions associated within the laws are vague and raise concerns for the implementation of Cyber SC and SA initiatives. Article 37 of China's 2016 Cybersecurity Law requires that participating countries store all information on servers in mainland China.[78] The idea of storing information on servers in mainland China may provide a sense of relief and allow for redundancies. The concern with this law is whether the country or China own the knowledge and information due to storage location. Article 7 of China's 2017 National Intelligence Law and China's 2014 Counter Espionage law states that companies must turn over data to China when requested.[79] The concern with these laws are the ownership rules and authority China could utilize to request that a state-owned company turn data turned over to the government whenever requested. The risks associated when these rules are put together paint a troubling picture and increase the risk of the US and ally or partner defense-related knowledge being stolen or inappropriately received.

---

[73] https://www.uscc.gov/sites/default/files/Feldstein_Testimony.pdf.

[74] https://chinatechmap.aspi.org.au/#/map/f1-Angola,f1-Burkina%20Faso,f1-Burundi,f1-Ben in,f1-Bouvet%20Island,f1-Botswana,f1-Congo%20%5BDRC%5D,f1-Central%20African%20R epublic,f1-Congo%20%5BRepublic%5D,f1-C%C3%B4te%20d'Ivoire,f1-Cameroon,f1-Cabo% 20Verde,f1-Djibouti,f1-Algeria,f1-Egypt,f1-Western%20Sahara,f1-Eritrea,f1-Ethiopia,f1-Gab on,f1-Ghana,f1-The%20Gambia,f1-Guinea,f1-Equatorial%20Guinea,f1-Guinea-Bissau,f1-Ken ya,f1-Comoros,f1-Liberia,f1-Lesotho,f1-Libya,f1-Morocco,f1-Madagascar,f1-Mali,f1-Mauritani a,f1-Mauritius,f1-Malawi,f1-Mozambique,f1-Namibia,f1-Niger,f1-Nigeria,f1-R%C3%A9unio n,f1-Rwanda,f1-Seychelles,f1-Sudan,f1-Saint%20Helena,f1-Sierra%20Leone,f1-Senegal,f1-Som alia,f1-S%C3%A3o%20Tom%C3%A9%20and%20Pr%C3%ADncipe,f1-Eswatini,f1-Chad,f1-Togo,f1-Tunisia,f1-Tanzania,f1-Uganda,f1-Mayotte,f1-South%20Africa,f1-Zambia,f1-Zimbab we,f1-South%20Sudan,f4-State%20controlling%20stake,f4-State-owned%20enterprise.

[75] https://cs.brown.edu/courses/csci1800/sources/2017_PRC_NationalIntelligenceLaw.pdf.

[76] https://www.chinalawtranslate.com/en/2016-cybersecurity-law/.

[77] https://www.chinalawtranslate.com/en/anti-espionage/.

[78] https://thediplomat.com/2017/06/chinas-cybersecurity-law-what-you-need-to-know/.

[79] https://www.cnbc.com/2019/03/05/huawei-would-have-to-give-data-to-china-government-if-asked-experts.html.

Other concerns that come up when looking at these laws include requirements placed by the Chinese government places on state-owned companies tor companies with a state stake of ownership. The requirements in these laws explain why the USG and the Federal Communication Commission have banned imports from Huawei and ZTE equipment into the US.[80] Allowing the importation of these companies' items subjectsUSG cyber borders and networks to greater risk of Chinese hacking. The US has banned the use of these companies cyberinfrastructure other countries have not. This presents another concern and risk the US must think about when deploying SC and SA cyber initiatives. If Huawei, ZTE, or another Chinese-based company installs physical cyberinfrastructure for specific ally or partners, how does the US ensure cyber technology and training provided for defense purposes are not compromised? If we think about the perspective of Chinese law discussed earlier the quesiton must be asked how does Chinese law affect US knowledge, information, cyber technology and training that flow across Chinese developed cyberinfrastructure? How does the USG approach this type of situation and ensures adequate protection of items being used on different types of installed physical cyberinfrastructure? The CSC does highlight that USG physical infrastructure protection is more in the areaof critical infrastructure protection,[81] however, a case could be made that infrastructure does have an element of concern associated with the CSC Layer 1 focus of "influencing behaviors and outcomes," When thinking about China's installation of infrastructure and measuring it against deploying SC and SA cyber initiatives across specific cyberinfrastructure, the US must be aware of how China will interpret their laws and the intent towards achieving a specific outcome, desire, or advantage through an existing or potential gap in our cyber relationship with an ally or partners.

The CSC highlights the multiple adversaries such as China, Russia, Iran, North Korea, terrorists, and state-based actors and identifies lengths these adversaries will go to in order to gain access to critical USG infrastructure, knowledge, and information. Each adversaries has different motives and aims and is working to achieve specific outcomes. The CSC highlights that adversaries have made multiple attempts over a period of time with outcomes and aims focused towards using vulnerabilities and backdoors to access US cyber systems, critical infrastructure, knowledge, and information. Regardless adversaries will continue to utilize aggressive behaviors to gaining access, causing havoc, or creating issues in the Cyber Domain and operating environment. The Public cyber sphere and domain will continue to be an emerging and growing area for the USG as we continue moving through the decade of the 2020s. The USG does have mechanisms in place with different SC and SA programs that can enable cyber cooperation initiatives. The CSC lays out recommendations and provides a roadmap to help the USG increase partner involvement in the area of using Cyber initiatives and engagements to achieve a successful end state. The success of using cyber SC and SA initiatives requires the US to think about appropriately mitigating and balancing risks. Care must be used when moving

---

80 https://www.reuters.com/technology/us-fcc-votes-launch-further-crackdown-huawei-zte-equipment-2021-06-17/.

81 https://drive.google.com/file/d/1ryMCIL_dZ30QyjFqFkkf10MxIXJGT4yv/view.

forward to cyber cooperation with partners and allies and in the scope of the GPC and emerging Strategic Competition, greater involvement is a must.

5 Conclusion

The overall course of the chapter focused on the idea of using cyber cooperation to support Layer 1 objectives as outlined in the 2019 CSC report focused on by "shaping behavior by strengthening norms and over time using through international engagement and cooperation to promote responsible behavior and dissuades adversaries from using cyber operations to undermine any nation's interests."[82] The discussion and topics in the chapther have gone deeper and discussed many different issues focusing on the area of cyber security cooperation. We started the chapter by providing background on the CSC, identified limitations associated with Layer 1 of the CSC. This section highlighted a couple of questions based on CSC layer one limitations and highlighted four discussion areas looking at how SC and SA initiatives in the scope of cyber security cooperation could help meet CSC layer one objectives.

The first area discussed SC and highlighted the differences between SC and SA in the policy. In this part of the chapter the history of SC through the scope of policy to including the establishment of the FAA and AECA. The second area explored how SC and SA programs might support cyber initiatives. Here we discussed the cyber environment, analyzed SC and SA program policy, and determined how a program could support initiatives within the cyber domain and operating environment. The third area discussed the process of making cyber SC and SA initiatives a reality. The focus here looked at how the USG develops SC and SA initiatives, partner and ally considerations when developing initiatives and engagement, and appropriate protections and concerns USG stakeholders may have when implementing cyber-related SC and SA initiatives. Our fourth area focused on cultural and technological risks associated with SC and SA initiatives. The discussion here looked at focused on cultural considerations and challenges when dealing with allies and partners and risks and concerns associated with deployment of cyber SC and SA initiatives across physical cyberinfrastructure not put in place by the US. Discussions in the area of risks related to physical cyberinfrastructure tended to highlight China because of associated concerns with their involvement and initiatives in multiple areas around the world.

The overarching takeaway from this chapter is that the cyber domain and operating area are broad. The 2019 CSC and 2020 CSC report are an appropriate first step in the right direction to help the USG lay public sector cyber policy groundwork. Policy in the scope of the CSC is limited in many areas Layer one of the CSC report provides an opportunity to discuss how cyber cooperation could be used to work with allies and partners and allies in meet cyber objectives. The path towards using SC

[82] https://drive.google.com/file/d/1ryMCIL_dZ30QyjFqFkkf10MxIXJGT4yv/view.

and SA programs to support cyber initiatives is still very much in its infancy and will require furthur refinement in the area of policy interpretation when incorporating SC and SA programs into a cyber initiatives. The existing US SC and SA enterprises and system already contain appropriate precaution and protection levels that are able to mitigate concerns. The existing limiting factor is actual execution of incorporating SC and SA programs and allowing them to meet layer one objectives within the CSC. The USG must assess and think through the different cyber risks, however, there is a need to be flexible and agile as moving cyber SC and SA initiatives forward. China's involvement in the DSR and grand cyber ambitions make the country a cyber player within the scope of the GPC and emerging strategic competition. As things go deeper, the US must be cognizant of the cyber roles that Russia, Iran, North Korea, and other nation-states will play as they continue looking for ways to destabilize the US and our ally and partner cyber activities, thereby, making establishment of cyber in sphere of the public cyber sector all the more critical.

References

1. Arthur C (2011) Egypt cuts off internet access. The Guardian. Guardian News and Media, 28 Jan 2011. https://www.theguardian.com/technology/2011/jan/28/egypt-cuts-off-internet-access
2. Australia Strategic Policy Institute (ASPI) International Cyber Policy Centre (2022) Mapping China's Tech Giants. http://chinatechmap.aspi.org.au/#/map/f1-Angola,f1-Burkina%20Faso,f1-Burundi,f1-Benin,f1-Bouvet%20Island,f1-Botswana,f1-Congo%20%5BDRC%5D,f1-Central%20African%20Republic,f1-Congo%20%5BRepublic%5D,f1-C%C3%B4te%20d'Ivoire,f1-Cameroon,f1-Cabo%20Verde,f1-Djibouti,f1-Algeria,f1-Egypt,f1-Western%20Sahara,f1-Eritrea,f1-Ethiopia,f1-Gabon,f1-Ghana,f1-The%20Gambia,f1-Guinea,f1-Equatorial%20Guinea,f1-Guinea-Bissau,f1-Kenya,f1-Comoros,f1-Liberia,f1-Lesotho,f1-Libya,f1-Morocco,f1-Madagascar,f1-Mali,f1-Mauritania,f1-Mauritius,f1-Malawi,f1-Mozambique,f1-Namibia,f1-Niger,f1-Nigeria,f1-R%C3%A9union,f1-Rwanda,f1-Seychelles,f1-Sudan,f1-Saint%20Helena,f1-Sierra%20Leone,f1-Senegal,f1-Somalia,f1-S%C3%A3o%20Tom%C3%A9%20and%20Pr%C3%ADncipe,f1-Eswatini,f1-Chad,f1-Togo,f1-Tunisia,f1-Tanzania,f1-Uganda,f1-Mayotte,f1-South%20Africa,f1-Zambia,f1-Zimbabwe,f1-South%20Sudan,f4-State%20controlling%20stake,f4-State-owned%20enterprise
3. Chapter 1: Defense security cooperation agency. Security assistance management manual. Defense Security Cooperation Agency, 2021. https://samm.dsca.mil/chapter/chapter-1
4. Chapter 2: Defense security cooperation agency. Security Assistance Management Manual. Defense Security Cooperation Agency, 2021. https://samm.dsca.mil/chapter/chapter-2
5. Chapter 3: Security assistance management manual. Defense Security Cooperation Agency, 2021. https://samm.dsca.mil/chapter/chapter-3
6. Chapter 4: Defense security cooperation agency. Security Assistance Management Manual. Defense Security Cooperation Agency, 2021. https://samm.dsca.mil/chapter/chapter-4#C4.3
7. Chapter 32: Foreign assistance. USC title 22—foreign relations and Intercourse. United States Government, 2017. https://www.govinfo.gov/content/pkg/USCODE-2017-title22/html/USCODE-2017-title22-chap32.htm
8. Chapter 39: Arms export control act. USC title 22—foreign relations and Intercourse. United States Government, 2018. https://www.govinfo.gov/content/pkg/USCODE-2018-title22/html/USCODE-2018-title22-chap39.htm
9. China Law Translate. 2014 Counter-Espionage Law of the People's Republic of China. China Law Translate. China Law Translate, 22 Aug 2019. https://www.chinalawtranslate.com/en/anti-espionage/

10. Cilluffo F et al (2021) Reflections on executive order 13010. McCrary Institute, Auburn University. https://mccrary.auburn.edu/work/insights/reflections-on-executive-order-13010/
11. Cyber Solarium Commission Final Report. CSC Final Report. Cyber Solarium Commission, March 2020. https://drive.google.com/file/d/1ryMCIL_dZ30QyjFqFkkf10MxIXJGT4yv/view
12. Definition of Cyber Space. DOD dictionary of military and associated terms. Joint Chiefs of Staff, August 2021. https://www.jcs.mil/Portals/36/Documents/Doctrine/pubs/dictionary.pdf
13. Department of Defense (2018) Department of Defense Cyber Strategy 2018. Defense.Gov. http://media.defense.gov/2018/Sep/18/2002041658/-1/-1/1/CYBER_STRATEGY_SUMMARY_FINAL.PDF
14. Domain. The Merriam-Webster.Com Dictionary. www.merriam-webster.com/dictionary/domain. Accessed 9 Feb 2022
15. Efron D (2018) The cyber domain, cyber security, and what about the international law? The federmann cyber security research center—cyber law program. https://csrcl.huji.ac.il/sites/default/files/csrcl/files/dan_efrony.pdf
16. Fact Sheet: United States—Republic of Korea Partnership. The White House. The United States Government, 22 May 2021. https://www.whitehouse.gov/briefing-room/statements-releases/2021/05/21/fact-sheet-united-states-republic-of-korea-partnership/
17. Federation of American Scientists: Intelligence Resource Program." Critical Infrastructure Protection (PDD 63). Federation of American Scientists: Intelligence Resource Program, 22 May 1998. http://irp.fas.org/offdocs/pdd/pdd-63.htm
18. Figure C3.F1.: Defense Security Cooperation Agency. Security Assistance Management Manual. Defense Security Cooperation Agency, 2021. https://samm.dsca.mil/figure/figure-c3f1
19. Figure C5.F4.: Defense Security Cooperation Agency. Security Assistance Management Manual. Defense Security Cooperation Agency, 2021. https://samm.dsca.mil/figure/figure-c5f4.
20. Foreign Assistance Act of 1961: Authorizations and Corresponding Appropriations. Federation of American Scientists: Project on Government Secrecy. Congressional Research Service, 29 July 2011. https://sgp.fas.org/crs/row/R40089.pdf
21. Huang Y, Smith J (2019) China's record on intellectual property theft is getting better and better. Foreign Policy, 17 Oct 2019. http://foreignpolicy.com/2019/10/16/china-intellectual-property-theft-progress
22. Institutional capacity assessment tool: G2G education toolkit. Education Links. United States Agency for International Development, Jan 2013. https://www.edu-links.org/sites/default/files/media/file/ICA_Tool_FINAL_Sept2014.pdf
23. Joint Fact Sheet: US and UK Defense Cooperation. Obama White House Archives. National Archives and Records Administration, 14 March 2012. https://obamawhitehouse.archives.gov/the-press-office/2012/03/14/joint-fact-sheet-us-and-uk-defense-cooperation
24. Joint Publication 3-12 Cyberspace Operations. Joint Chiefs of Staff, 8 June 2018. www.jcs.mil/Portals/36/Documents/Doctrine/pubs/jp3_12.pdf?ver=2018-07-16-134954-150
25. Kharpal A (2019) Huawei says it would never hand data to China's government. Experts Say It Wouldn't Have a Choice. CNBC. CNBC, March 5, 2019. https://www.cnbc.com/2019/03/05/huawei-would-have-to-give-data-to-china-government-if-asked-experts.html
26. Kurlantzick J et al (2021) Chinas Digital Aid: The Risks and Rewards. Council on Foreign Relations. Council on Foreign Relations. https://www.cfr.org/china-digital-silk-road/
27. National Intelligence Law of the People's Republic of China. Cyber Security and International Relations. Brown University, 27 June 2017. https://cs.brown.edu/courses/csci1800/sources/2017_PRC_NationalIntelligenceLaw.pdf
28. NCI Information Systems, Inc. Persistent Cyber Training Environment (PCTE)—PEO STRI. U.S. Army Program Executive Office STRI. http://peostri.army.mil/persistent-cyber-training-environment-pcte. Accessed 9 Feb 2022
29. Office of cybersecurity, energy security, and emergency response. Colonial Pipeline Cyber Incident. Energy.Gov 2021. www.energy.gov/ceser/colonial-pipeline-cyber-incident

30. Person and Shepardson D (2021) US FCC votes to advance proposed ban on Huawei, ZTE gear. Reuters. Thomson Reuters, 18 June 2021. https://www.reuters.com/technology/us-fcc-votes-launch-further-crackdown-huawei-zte-equipment-2021-06-17/
31. Security Cooperation Programs (2021) DSCU All Publications Website. Defense Security Cooperation University. https://www.dscu.mil/documents/publications/security_cooperation_programs_handbook/security_cooperation_programs_handbook.pdf?id=1
32. "Section 2314: Furnishing of Defense Articles or Related Training or Other Defense Service on Grant Basis. USC title 22—foreign relations and Intercourse. United States Government, 2018. https://www.govinfo.gov/content/pkg/USCODE-2017-title22/html/USCODE-2017-title22-chap32-subchapII-partII-sec2314.htm
33. Testimony before the U.S.-China Economic and Security Review Commission Hearing on China's Strategic Aims in Africa. Steven Feldman Testimony. US-China Economic and Security Review Commission, 8 May 2020. https://www.uscc.gov/sites/default/files/Feldstein_Testimony.pdf
34. "The Management of Security Cooperation (Green Book)." DSCU Green Book. Defense Security Cooperation University, May 2021. https://www.dscu.mil/documents/publications/greenbook/01_Chapter.pdf?id=1
35. The US export control system and the export control reform initiative. Federation of American Scientists: Project on Government Secrecy. Congressional Research Service, 28 Jan 2020. https://sgp.fas.org/crs/natsec/R41916.pdf
36. United States Agency for International Development (USAID). Digital Strategy 2020–2024. United States Agency for International Development (USAID) 2020. www.usaid.gov/sites/default/files/documents/USAID_Digital_Strategy.pdf
37. U.S. Cyber Command. Achieve and maintain cyberspace superiority: command vision for US cyber command. United States Cyber Command 2018. www.cybercom.mil/Portals/56/Documents/USCYBERCOM%20Vision%20April%202018.pdf?ver=2018-06-14-152556-010
38. U.S. Cyber Command. US and Australia Sign First-Ever Cyber Agreement to Develop Virtual Training Range. U.S. Cyber Command, 4 Dec 2020. www.cybercom.mil/Media/News/Article/2434919/us-and-australia-sign-first-ever-cyber-agreement-to-develop-virtual-training-ra
39. U.S. Department of Defense. Dunford Describes U.S. Great Power Competition with Russia, China. U.S. Department of Defense, 21 Mar 2019. www.defense.gov/News/News-Stories/Article/Article/1791811/dunford-describes-us-great-power-competition-with-russia-china
40. U.S. Department of Defense. Our Story. U.S. Department of Defense, www.defense.gov/About. Accessed 9 Feb 2022
41. Wagner J (2017) China's cybersecurity law: what you need to know. The Diplomat. The Diplomat, 1 June 2017. https://thediplomat.com/2017/06/chinas-cybersecurity-law-what-you-need-to-know/
42. Xie M, Deng (2019) 2016 China cybersecurity law. China Law Translate. China Law Translate, 27 Nov 2019. https://www.chinalawtranslate.com/en/2016-cybersecurity-law/

Bryce T. Englin is an Instructor of Security Cooperation Management at the Defense Security Cooperation University (DSCU) School of Security Cooperation Studies (SSCS). He has been Serving as an instructor at DSCU SSCS since April 2019. His Duties include instructing across the Security Cooperation Office (SCO) Depament with focus on instructing Security Cooperation Office (SCO) office personnel heading out to work abroad in SCO Office Operations, SCO International Training Management (ITM) and SCO Foreign Military Sales (FMS) Case Life Cycle Management.

Bryce is a retired U.S. Air Force Senior NCO (Jun 1996–Aug 2016) with 14 years of overseas assignments. Post US Air Force and prior to joining DSCU SSCS include duties at in the USINDOPACOM/J6 and the Air Force Culture and Langauge Center at Maxwell AFB in Alabama. His assignments & duties consists of approximately 12 years supporting SC and IA. SC related assignments include Office of Military Cooperaition Egypt, U.S. Embassy, Egypt (2004–2007).

Plans and Policy Directorate, J5, Headquarters, United States Forces Japan (2007–2010). Office of Security Cooperation, Djibouti, U.S. Embassy, Djibouti (2013–2015). AFRICOM, CENTCOM, EUCOM FAO Program Analyst, Air Force Culture and Language Center, Maxwell AFB, AL (2017–2019).

His, education includes an MBA Grantham University, Lenexa KS (2017), B.S. Business Administration/Logistics, Touro University International, Cypress CA (2008), AAS Information Management, Community College of the Air Force, Maxwell AFB, AL (2006). Bryce's SC research areas focused on Cyber and Cyberspace Security Cooperation within the Cyber Operational and Tactical Domain.

Putting the Afghan Puzzle Together: Afghan and U.S. Perspectives on Legitimacy

Amy Frumin

Abstract Over the last two decades, the United States tried to rebuild the puzzle of Afghanistan with little appreciation for the Afghan population's own reference image of the state in the past, present, or future. The puzzle pieces with which the U.S. wanted to work were incongruent with the reality of the Afghan puzzle. The Taliban, on the other hand, had lived through the creation of that puzzle and defined some of the shapes of those pieces. The Taliban offered a picture of the complete puzzle that looked more familiar to the Afghans than that which the United States offered. Because the United States subordinated the Afghan perspective to the U.S. perspective, the United States and U.S.-backed Afghan Government lost the competition to win over the Afghan population to the Taliban. The competition was about establishing legitimacy. Below is a brief view of theories of legitimacy. Using the case study of justice and the rule of law sector, the chapter illustrates how the United States chose to prioritize the establishment of legal-rational legitimacy, in contrast to the traditional forms of legitimacy the Taliban pursued. An overview of the Afghan perspectives explains why they found the traditional form of legitimacy more compelling than the legal-rational form. A small but important point on predictability explains why justice was central to the construction of legitimacy for the Taliban. A section on the U.S. perspective highlights that while U.S. policy acknowledged the importance of the local perspective in building legitimacy, in practice, many of the U.S.-funded projects reflected the American perspective more so than the Afghan.

Keywords Justice · Rule of law · Predictability · Afghanistan · Counterinsurgency · Taliban · Legitimacy

This exploration of perspectives requires a few disclaimers. The Afghan and American perspectives are not monolithic. There are various subgroups within each actor, each with their own perspectives. This paper does not specifically look at all the subgroups, but rather attempts to appreciate the aggregated perspective of the Afghan people and the U.S. policy approach to Afghanistan. This paper also does not delve

A. Frumin (✉)
Washington, District of Columbia (DC) 20001, USA
e-mail: abfrumin@gmail.com

A. Farhadi and A. Masys (eds.), *The Great Power Competition Volume 4*,
https://doi.org/10.1007/978-3-031-22934-3_6

into how various other state actors influenced the relationship between Afghanistan and the United States. This is not because actors such as Pakistan or Russia are not important to consider, but because the Afghan and American perspectives are primary. Additionally, hindsight itself is a perspective. The exercise of appreciating perspectives from hindsight will not necessarily forestall similar missteps in the future, but it may inform U.S. policymakers' approach to engagements in the future. At the very least, the aggregate U.S. policy community must appreciate its own biases and the perspective of the local population of the community it seeks to support.

1 Theories of Legitimacy

The meaning and source of legitimacy have been debated for centuries. It is crucial to understand theories regarding how legitimacy is constructed before reviewing the disconnect between the Afghan and U.S. perceptions of legitimacy. This section provides a brief overview of some of the key concepts of legitimacy.

Max Weber offered three sources of political legitimacy: rational-legal, charismatic, and traditional. Rational-legal legitimacy rests on the belief in formal rules and institutionalized procedures. The government establishes and enforces law and order, and therefore gains public trust and is considered legitimate. This form of legitimacy requires a set of institutions. Charismatic legitimacy is conferred on charismatic leaders who are able to convince a population to follow the regime/leader. Traditional legitimacy is founded in social customs and norms based on history.

Dr. Muthiah Alagappa[1] identified four separate but related sources or elements of legitimacy: normative, procedural, performance, and consent. Normative legitimacy is based on shared beliefs—a system of norms and values. These norms and values can be based on nationalism, religion, ethnicity, etc. In procedural legitimacy, the leader or regime is considered legitimate if they adhere to established rules. Performance legitimacy is based on the acceptable and effective use of state power. Consent of the governed comprises the last form of legitimacy. This refers to allegiance to the government, not just passive acceptance.

Florian Weigand distills the analysis into two forms of legitimacy: instrumental or substantive. Governments earn instrumental legitimacy by responding to the needs of the governed, based on the population's own conceptualization of those needs. In substantive legitimacy, a population with shared values judges that a leader has the right to exercise control [18].

Among these theorists there is one common thread—legitimacy is in the eye of the beholder, or the governed. All of the theorists reference shared norms and values as a component of legitimacy. Other components of legitimacy relate to the degree to which the norms are codified into laws and systems of governance, how the leadership adheres to those laws, and how a leader uses its power to carry out

[1] See Alagappa (editor), (1995), *Political Legitimacy in Southeast Asia: The Quest for Moral Authority*, Stanford University Press.

the will of the people. Not all elements of legitimacy must be present to confer legitimacy on a leader. However, in constructing legitimacy, all of these facets should be considered. Most importantly, legitimacy must be considered from the perspective of the governed.

2 Justice as a Case Study, and the Small but Important Missing Piece: Predictability

Justice and the rule of law offer an effective case study to understand the different perceptions of legitimacy in Afghanistan because they were the chosen mechanisms through which the Taliban staked their claim on legitimacy. This case study illuminates the incongruities between the American and Afghan approach to rule of law and justice. It also highlights how the Taliban managed to use justice as the key corner piece in the puzzle.

Notably, many of the theories of legitimacy fail to recognize the unique circumstances of war-torn countries—namely, the instability and lack of predictability in a population's daily lives. Justice and rule of law underpin predictability in a society. All people, and markets for that matter, seek predictability. Predictability is the ability to anticipate what might come. It is what allows us to plan for the future. Predictability in a society is guaranteed through: (1) the understanding of the rules or norms of a society, and (2) the application and enforcement of those rules or norms. Both are components of legitimacy according to the theorists. For example, traffic lights offer predictability in traffic patterns to avoid collisions and improve traffic flows. The drivers must know the meaning of the various colors and follow the rules. When the rules are broken, the driver is fined. In war zone, predictability takes on amplified importance. When the very basic requirements of life are threatened (i.e., food, water, shelter), then a leader or regime's ability to provide a level of predictability to obtain these basics becomes exponentially more important and can contribute to their legitimacy.

Predictability's links to legitimacy are through the application and enforcement of those common norms, or the rule of law. In this chapter, rule of law refers to the system (formal or informal) that includes the creation or adoption of the norms or laws of a society, adjudication when norms or laws are violated, and subsequent remediation. Justice is a subcomponent of this system. In the example above, it is the police officer that enforces the predictability of traffic patterns to avoid accidents. All drivers submit to and understand the rules of the road and the police are thus considered legitimate to enforce those rules. In many ways, the provision of predictability is a precursor to Weber's traditional or rational-legal sources of legitimacy (assuming the rules and norms are codified), and of the sources of legitimacy suggested by Alagappa and Weigand.

3 Bias in Building Legitimacy

State-building activities, such as the ones the United States carried out in Afghanistan, must translate these concepts and sources of legitimacy into policy for action in order to successfully create a governance system the state's population deems legitimate. The concept of state-building itself relies on an assumption that state institutions are necessary—if not sufficient—to confer legitimacy on a government. Further, Western states are more likely to fall prey to a confirmation bias that the Western-style governments and institutions are inherently legitimate. This bias often leads to mirror imaging, where the goal in state-building is to create a Western-style government that performs well, with the assumption that the newly built government will therefore achieve legitimacy. That legitimacy is based on the rational-legal approach of Weber, which focuses on institutions. Further, if that leader or regime conforms to the established rules, they will have procedural legitimacy and if they perform well, they will have performance legitimacy. All of this will lead to consent legitimacy as described in Alagappa's theories.

This Western bias relegates non-Western elements of governance to incidental. Western practitioners often overlook the contribution of local governance mechanisms to legitimacy, even if the population confers normative legitimacy upon them and they embody Weigand's instrumental and substantive legitimacy. State-building policies should seek to strengthen that which the governed population considers legitimate and not overemphasize sources of legitimacy that do not resonate with the population.

4 The Afghan Perspective: The Puzzle Pieces

The Afghans' perspective on legitimacy is grounded in the Afghan experience. When the United States first send sent troops into Panjshir Valley following the attacks of September 11 in 2001, Afghanistan was ranked 89th of 90 developing countries on the Human Poverty Index[2] according to the United Nations Development Report [18]. Afghanistan had already endured 23 years of continuous conflict. Violence, instability, and a lack of predictability characterized Afghans' daily lives. Ninety-nine percent of the population of approximately 31 million is Muslim; Islam was a constant in Afghan culture throughout the various conflicts.

The period of intense conflict began in 1973, when a coup overthrew the Afghan monarch Mohammed Zahir Shah and ended 40 years of relative stability. Just five years later, the assassination of Gen. Mohammed Daoud Khan, the first President of Afghanistan, resulted in the creation of a Soviet-aligned Communist state based on Islamic principles. Internal rivalries within the Afghan Communist Party led to

[2] The Human Poverty Index measures income deprivation, literacy deprivation, social services deprivation, and employment deprivation.

infighting among the leadership, while in the countryside, conservative Islamists fought against some of the reforms that came with Daoud Khan's regime.

This Islamist movement developed into the Mujahedeen, who fought the Communists. A central theme in the conflict between the Mujahadeen and the Communists was justice:

> Law and justice became a stake in the struggle between the Communist government and the insurrection, but also in the competition between elites (tribal notables, university graduates, and religious clerics) over who was to regulate social relations. While the Afghan state during the twentieth century had gradually secularized the judicial system, the victory of the Mujahadeen by the end of the 1980s gave the religious scholars (IT) a monopoly on the legitimate exercise of law [3].

To prop up the flailing Afghan Communist regime, the Soviet Union invaded in 1979, triggering a bloody war that lasted a decade. By 1982, 6.1 million Afghans had fled to Pakistan and Iran [9]. During this period, the United States supported the Mujahadeen in their conflict with the Soviets as part of a Cold War proxy battle. When the Soviets withdrew in 1989, war lords and regional leaders fought a series of civil wars throughout the country. The various fighting factions all claimed to uphold Islamic law. Kabul became a battle ground and fell to the Taliban—which the Pakistani Taliban movement supported—in 1996. "Most Afghans, exhausted by years of drought, famine and war, approve of the Taliban for upholding traditional Islamic values" [9]. From 1996 until 2001, Afghans toiled under the harsh Taliban rule, under which men and women were publicly executed, beaten, and had body parts amputated in soccer stadiums for violations of the ethics and morality codes of the Taliban.

After the Americans removed the Taliban from power in 2001, Hamid Karzai became the interim President of the newly founded Afghan Government, which U.S. policy officially recognized as the legitimate ruling body in Afghanistan. With Pakistan's assistance, the Taliban responded by diligently setting out to reestablish their control over the countryside. The Taliban's involvement in the justice sector gave them insights into local disputes, which they often exploited to their own political ends. Additionally, their dispute resolution capabilities afforded them elements of legitimacy, as discussed below. One tactic they used to institutionalize their system was to set up courts in the villages, mostly in Pashtun areas. In some cases, the Taliban literally and figuratively decapitated the previous justice system by killing the local elders and replacing the community's access to justice with their own mechanisms. "Some communities have been enticed to accept 'Taliban justice' to restore order and avoid violence and intimidation" [14].

Despite the constant need to stay on the move to avoid Coalition Forces, the Taliban managed to establish a bureaucratic system based on clerics—often recruited from Deobandi Madrassas—that provided a relatively predictable judicial system. "First, the movement integrated judges into a centralized organization that was inspired by the institutional architecture of the Taliban regime of the 1990s. A rotation and monitoring system was designed to ensure judges' good practices and their impartiality on local issues" [3]. Thus, the Taliban's system embodied both procedural and performative elements of legitimacy—as the governing power, they established and adhered to clear rules, and successfully imposed these rules upon the population. Further, this system was based on the tenets of Islam familiar to the Afghans, providing a normative element of legitimacy. The lure of predictability was so strong that communities preferred to conform to the rules of the Taliban's imposed system, rather than fighting back against the Taliban's violence and risk living in a society without the relative justice and security provided by the Taliban.

In contrast to the Taliban's system of justice, the United States and the Afghan Government they supported pragmatically used patronage networks to establish the interim Afghan Government [10]. Though Karzai was Pashtun, an attribute the Americans sought to appease the large Pashtun population, he lacked his own militia. Upon arriving in Kabul in December 2001 to take over as interim Afghan President, he was met by Tajik warlord Muhammad Fahim who asked, "Where are all your men?" Karzai's brave response was, "You are my men" [11]. Without his own militia, Karzai had to rely on coopting the support of strong men with militias, which in turn relied on a patronage system. The American military took a similar approach to warlords after the initial invasion. The decisions to install Karzai and support warlords at the local level, thereby creating the need for this patronage system, has had ripple effects throughout Afghanistan's recent history [13]. Corruption in Afghanistan grew as donor funding flowed in, outstripping the absorption capacity[3] in Afghanistan. The international aid community in Afghanistan and the Afghan Government itself were simply not able to oversee the hundreds of millions of dollars in aid assistance. USAID workers were pressured to keep "burn rates"[4] high and performance indicators for programs were often based on how much money was spent. With little oversight and a lot of opportunism, the patronage networks took advantage of the situation [4]. The resulting endemic corruption often undermined American efforts, as well as the Afghan Government's legitimacy, and security [12]. Corruption became its own puzzle piece, making it more difficult for the interim Afghan Government to reflect the picture of legitimacy the Americans supported.

5 Implications for Legitimacy: How Those Puzzle Pieces Fit Together

As described above, Afghanistan "has witnessed numerous inter- and intrastate conflicts as well as an entire spectrum of competing forms of political organizations, from monarchy to socialism to communism, not to mention attempts at theocracy" since 1978 [6]. The Afghan population's lack of faith in the successive failed governments reinforced the decentralized systems of governance based mostly on normative (Alagappa) and substantive (Weigand) sources of legitimacy carried out by strong men or warlords to achieve traditional (Weber) legitimacy. Successive Afghan

[3] See SIGAR report p. 26 for more on absorption capacity. https://www.sigar.mil/pdf/lessonslearned/SIGAR-21-46-LL.pdf.

[4] In USAID terms "Burn rates" indicate if allocated money was spent on a project.

governments demonstrated a lack of performance legitimacy (Alagappa) and instrumental legitimacy (Weigand). Importantly, the Afghan people grew to expect very few services from the central government. The reference picture the Afghans had of their puzzle did not include a centralized government system.

With these diminished expectations and[5] considering the number of local disputes that grew out of the wider conflicts, Afghans most fundamentally sought predictability. The centrality of predictability is often overlooked in literature on legitimacy. However, the value of predictability in countries in perpetual conflict cannot be overstated. Just as traffic laws help to avoid collisions, predictability can help maintain civility in society. Predictability can only function if the population understands the norms or rules of that society. Predictability feeds substantive (Weigand) and normative (Alagappa) sources of legitimacy leading to traditional legitimacy (Weber). If codified in rules and institutions, predictability can become rational-legal in Weber's terms. Ideas that are foreign or based on principles that do not conform with the population's weltanschauung[6] will lack substantive and normative legitimacy (Alagappa) and traditional legitimacy (Weber).

When a society's governing norms are violated, the civility of that society is undermined. If a suitable leader exists, the aggrieved person may turn to them to adjudicate the violation of norms. Notably, this adjudicator must have the capability to enforce their ruling, or their performance legitimacy is undermined. If executed within the bounds of what the population considers acceptable, the man (in this Afghan case) can be considered legitimate based on the Weberian idea of traditional legitimacy and Alagappa's normative and performance conceptualizations of legitimacy. Additionally, this man will have instrumental and substantive legitimacy based on Weigand's analysis. Notably, the population views the leader as legitimate despite the lack of government institutions that would be required to achieve Weber's rational-legal form of legitimacy.

We can use a concrete example to better understand the Afghans' perspective on legitimacy: the average Afghan farmer sought the predictability to know that if he took his harvest to market, he would run into one person who might demand some money to pass on the road, but the farmer could still sell his products for a profit. The farmer knows the trip to market would be worthwhile. While many Americans would consider this corruption, Afghans would disagree. Rather, Afghans see this as price of doing business, similar to a toll on an American road. However, when predictability is undermined and that farmer cannot know how many people he might meet on the road and how much it would cost him to get his product to market, then he is unable to make an informed decision about whether it is worthwhile to take his product to market. His life has become unpredictable. It is predictability, rather than creation/existence of formal rule of law institutions, that he expects a legitimate governance system to provide.

---

[5] Robert Lamb [7] does discuss predictability as a dimension of legitimacy. He says that legitimacy is a necessary but not sufficient dimension and it includes transparency and credibility.

[6] A particular philosophy or view of life; the worldview of an individual or group.

If his life is predictable, then if asked for a bribe by someone whom he feels has no right to a "toll," the Afghan farmer knows the local governance system has a mechanism to address this injustice. He can bring his grievance to a local leader who is both trusted to resolve such conflicts and has the capabilities to enforce the ruling. These mechanisms ensure that Afghan farmer trusts the informal governance system to deliver de facto justice. Conflicts in Afghanistan are and have been overwhelmingly resolved through these types of informal mechanisms through the shura (a consultative council at the local level) and tribal systems. In Pashtun areas, an estimated 99 percent of disputes are resolved through informal mechanisms [, p. 8]. Elders sit in judgement of the violations of those rules that govern a village. Their associated militias often enforce their rulings.8

"Complex insurgencies are powered by injustice. The ability to adjudicate and resolve disputes in a way that the population believes to be fair is a major step towards establishing the legitimacy an insurgent requires for success" [, p. 6]. The justice sector is perhaps the first point of contact for many with the governance apparatus, especially in environments where government services are neither present nor expected, like Afghanistan. A government that can provide predictability through the justice sector will gain elements of legitimacy. The Taliban understood this when it declared in 1996 that their governance priorities were security and justice; security is a crucial part of civility in society and is closely linked to justice [, p. 6]. "While the Taliban gained power through arms, they also won the war through the establishment of judicial system that allowed them to embody the state in the eyes of many Afghans." [3, 8, 8]

With credentials from a madrassa and the enforcement capabilities of the Taliban, these judges were able to credibly adjudicate and enforce rulings, thereby creating legitimacy in the eyes of the rural Afghans. The Taliban dispensed of institutionalizing the judicial system through books or courthouses, intentionally avoiding Weber's concept of rational-legal legitimacy. Partly due to the Taliban insurgency's need to be on the move to avoid capture, embedding the justice system into the broader Taliban administration also lent legitimacy to the other elements of the Taliban governance structures across political, military, and judicial functions [3]. It must be noted that while the Taliban did have many elements of legitimacy, the case can also be argued that the Taliban lacked the consent legitimacy discussed by Alagappa and often ruled by brutal coercion.

The legitimacy sought and conferred upon the Taliban through the justice sector and the provision of predictability was based in the one constant Afghans did have throughout the previous decades of war: Islam.

6 America's Reference Image for the Puzzle and Perspective on Legitimacy

After enduring over two decades of war and violence, the Afghans were initially willing to give the Americans and the U.S.-endorsed Afghan leadership a chance to improve their situation. However, the rent-seeking reality of the Afghan Government quickly diminished the Afghan population's hope for a less violent future. The failure of the U.S.-backed Afghan Government to convince the population that it provided rational-legal, procedural, or performative elements of legitimacy was compounded by America's meandering Afghan strategy. The American approach failed to appreciate the most basic requirement of the Afghan people—the need for predictability, and that predictability's relationship with legitimacy through the rule of law and justice sector. The Taliban's success in providing that predictability, especially vis-à-vis the American-backed alternative, was a key to their achievement of legitimacy.

Many critiques of the American intervention in Afghanistan begin by lamenting the lack of coherent and consistent policy towards Afghanistan [2]. Immediately following the 2001 invasion, the American goal was to defeat Al Qaeda and depose the Taliban Government to ensure that Afghanistan was no longer a safe haven for terrorists who could threaten the American homeland or American allies. While the U.S. military quickly defeated Al Qaeda and removed the Taliban Government, they were unable to prevent the Taliban from dissolving back into the population. Denied formal governance mechanisms, the Taliban waged a protracted insurgency for the next two decades. During that period, the United States recognized that unless they planned to remain in Afghanistan in perpetuity, the Afghans themselves needed the capability to keep Al Qaeda from resurfacing. The American goals of the mission broadened from counterterrorism into counterinsurgency (COIN) and nation building.

The COIN approach, which the American military eventually adopted, considered the Afghan population to be the center of gravity. The United States' COIN doctrine, detailed in the Department of Defense's (DoD's) Joint Publication 3–24, says, "Legitimacy is the Main Objective… Without the host nation achieving legitimacy, COIN cannot succeed" (COIN 2009, pp. III-11). The updated COIN doctrine considers that "Legitimacy is achieved by the HN [host nation] government through being perceived as effective and credible and by providing an environment for the population to maintain predictable and acceptable living conditions" (COIN 2018, pp. 1–8). The COIN doctrine offers a list of four factors in legitimacy: mandate (the perceived legitimacy of the means through which the regime/leader's authority is established), manner (perceived legitimacy of the way a regime/leaders acts to meet expectations of the population), consent (extend of compliance by the population to the regime/leader), and expectations (whether a leader meets expectations of the population and international community, which could be based on the quality or quality of services delivered by the regime/leader) (COIN 2018, pp. 1–8). These factors are reminiscent of the theories put forward by Weber, Alagappa, and Weigand. Notably, the doctrine boldly states (literally—the text is bolded) that the population decides legitimacy, with the caveat that the government's ability to meet the expectations of the international community is also a factor of legitimacy.

DoD COIN doctrine also emphasizes the need to understand the operational environment, and not being too threat-centric in that understanding. In other words, understanding the local perspective is critical to a successful COIN mission. The Americans recognized they needed more fidelity on the root causes of the conflict in Afghanistan to be able to address the population's core grievances and create a system that enabled the formal Afghan Government structures to establish their legitimacy. The U.S. Government deployed several efforts like the Provincial Reconstruction Teams, District Stability Teams, Cultural Support Teams, and the controversial Human Terrain System to try to better understand the dynamics on the ground and address grievances. It was a dubious endeavor to task Americans, most of whom had little cultural understanding or linguistic capability, to gather the ethnographic fidelity necessary to appreciate the complex dynamics of a country like Afghanistan during an active war. "As a member of General Stanley McCrystal's assessment team observed, implementing an effective counterinsurgency campaign requires 'a level of local knowledge that I don't have about my own hometown'" [, p. 76]. [12]

While the COIN doctrine does touch on some key points with regards to legitimacy and the local perspective, translating these concepts to action on the ground fell short, despite the $145 billion spent on reconstruction projects on the ground.[7] Many of the projects represented a mirror imaging of what Americans themselves would want, or thought Afghans needed. They did not necessarily address the Afghans' actual or perceived needs. Roads, healthcare, education, and a myriad of other services were provided through massive projects that were seldom co-created with the Afghans. "The West has placed too much emphasis on economic development, despite the fact that successful state-building will depend much more on establishing security and functioning, responsive judicial system" [, p. 19]. Jabar Naimee, a former governor in four provinces of Afghanistan, told the Special Inspector General for Afghanistan Reconstruction (SIGAR) [5] [8] that "in the majority of districts, we never even heard the real problems of the people. We made assumptions, conducted military operations, brought in government staff, and assumed it would lead to security and stability" (SIGAR 2021, p. 76). Without the appropriate local context, U.S.-funded projects were manipulated by local leaders towards their own ends, exacerbating local rivalries and/or undermining the legitimacy of the Afghan Government.

For example, the United States spent an estimated three billion dollars on roads projects over two decades in Afghanistan. With the required security for construction workers, some estimate that roads in Afghanistan cost a half a million dollars per kilometer.[9] The roads did serve a tactical purpose: creating what the military call lines

[7] Note that not all elements of the U.S. Government followed COIN doctrine. COIN Doctrine drove military actions, often supported by civilian counterparts in the U.S. Government.

[8] The SIGAR was charged with providing independent oversight of American reconstruction efforts in Afghanistan.

[9] During the author's time in Afghanistan from 2006–2007 that was the average number used to estimate road projects.

of communication between villages, towns, and cities. What the Americans neglected to understand was that most Afghans did not have cars, nor did the often travel far from their villages. Most Afghans are farmers, with over 70% of the population living in rural areas. Lines of communication were not a critical need; to the Afghans, the roads only impeded the flow of water to their fields. Culverts (which allow water to pass under the road) were part of the road design. Engineers dictated the location of culverts, not the local population. In the Panjshir Valley, a steep mountain valley about 90 minutes from Kabul, the farmers came out at night with pickaxes to dig a channel across a newly built road to allow water to flow to their fields. When finally confronted as to why they were destroying the very expensive road America gifted to the Panjshiris, a farmer explained that the engineer-designed culvert allowed the water to flow exclusively to the farm upstream. The upstream farmer began charging the downstream neighbor for access to the water. This exemplifies how the Americans made assumptions about Afghan needs and how Afghans would perceive the services provided. In this case, the road project was not a road project to the farmer with no car. Rather, it was a poorly executed water project. The project did not provide the Afghan Government procedural, performative, or instrumental legitimacy, nor did it help to win "hearts and minds" in return for the American investment. In some cases, these types of interventions instead exacerbated local tensions.[10]

The COIN doctrine did look beyond development projects to try to meet the perceived needs of the Afghans and thereby contribute to the legitimacy of the Afghan Government, and the Americans also attempted to address the justice sector through a variety of programming. The justice sector is a linchpin for predictability in Afghanistan, and as argued above, contributed to the perceived legitimacy of the Taliban. If, as Brigadier Macay put it in 2007, 'the population is the price', it is won partly through providing justice" [, p. 6]. Between 2003 and 2015 America spent more than a billion dollars on programs related to rule of law. The first U.S. Government Integrated Civil—Military (Civ-Mil) Campaign Plan for Support to Afghanistan, published in 2009, pushed to create a formal justice system as a viable alternative to the local dispute resolution system [, p. 24]. The U.S. Government considered justice as one of the 11 key counterinsurgency transformative effects. A 2011 update to the Civ-Mil Campaign Plan identified the lack of justice as a key driver of conflict and the number one governance campaign objective for 2014. "Improving the government's capacity to render justice equitably and incarcerate those in violation of laws will improve the legitimacy of the state [8, 14, 15]."

Much of the U.S. effort was spent on creating a formal justice system—books of Afghan laws were printed and sent to the provinces for reference by judges; courthouses were built. However, as in the example of the roads projects, the new institutions often failed to meet the Afghans' own needs for justice. The courthouse in the Panjshir province was rarely found occupied. Upon asking the building caretaker where the judge was, he answered, quite simply, that the judge went home for lunch because there was no kitchen in the new courthouse building. Since the walk was

[10] My life in Panjshir, 2006.

so long between his home and the courthouse, he rarely came back after lunch.[11] These rule of law programs often resulted in structures spotting the countryside of Afghanistan—monuments to the American fundamental lack of appreciation of how Afghans prefer to adjudicate their grievances. Americans tried to create new puzzle pieces that fit together to create a formal justice system, but it was one that was unrecognizable to the Afghans. Not only were people often unable to read the laws due to the low literacy rates, but the principles of the laws themselves were foreign to the Afghans. The imported system was based on the retributive justice of the American system, whereas the shura justice system employed restorative justice based on tenets of Islam. Afghans had been solving their problems through restorative justice for centuries.

7 Implications for the American Conception of Legitimacy

While the American policy that COIN doctrine informed appreciated the importance of justice in establishing legitimacy (even informal forms of justice), their programs favored rational-legal (Weber) and procedural (Alagappa) elements of legitimacy over traditional legitimacy. With this narrow programmatic interpretation, the Americans sought to improve the performance legitimacy of the legal system to establish a legal-rational point of reference for the Afghan Government's legitimacy in the rule of law sector. The updated 2011 Civ-Mil Campaign Plan proffered a solution that linked the traditional forms of justice to the formal system. This was problematic for several reasons.

By 2011, corruption and inefficiencies plagued the formal system. One Afghan succinctly captured the perspective of much of the population: "If you are in the wrong but rich, go to the court; if you are in the right but poor seek community arbitration" [, p. 7]. The Civ-Mil Campaign Plan highlighted the lack of access to the formal Afghan judicial system as a primary concern, with additional concerns about quality and speed. The Americans tried to address these concerns through projects such as the Rule of Law Field Force-Afghanistan, support to the Department of the Huqhuq, [1][12] and USAID's Afghanistan Rule of Law Stabilization Program. However, these programs all aimed to strengthen the formal justice system either on its own or through linking it with the traditional system, as proscribed in the COIN doctrine: "To enhance HN [host nation] legitimacy, justice reform should build upon the existing legal frameworks in the HN. This may include common law, civil law,

[11] My life in Panjshir, 2006.

[12] Within the Afghan Government itself there was a system that linked the informal and formal justice systems. The Afghan Department of the Huquq, established in 1921, in the Afghan Ministry of Justice links the traditional system to the tools of the Afghan Government's judicial and legislate framework by registering and tracking petitions, helping petitioners understand the paperwork, forwarding cases to the court system and working with the police to ensure compliance with judgements. The military did support the huqhuq. (General Directorate of Huquq—Ministry of Justice (moj.gov.af)).

criminal codes, traditional or religious law, and international law" (COIN 2018, p. A-13). The Americans were unwilling to let the traditional systems stand on their own. As Rubin Barnett said, "Trying to compete with the Taliban's successful dispute resolution would have meant allowing Sharia, and that's not something we could do politically" [, p. 75]. [12]

The reality was that various legal systems were in competition to be the legal system of choice for the Afghan people, and therefore earn the prize of some conferred legitimacy onto the winner's governance type. The result was radical legal uncertainty for the Afghans.

The proliferation of competing authorities results in conflicting norms and verdicts, which profoundly destabilize social structures and aggravate disputes around certain issues, particularly property relations and ethnic, religious, class, gender, and age hierarchies. Law, which always is an essential realm in the structuring of the state and society, therefore plays an even more decisive role as it becomes a key stake in the civil war [3].

The formal legal system lost the competition for various reasons. To begin with, it was made up of an amalgamation of laws from previous Afghan Constitutions, often with input from foreign advisors. Additionally, the proclivities of different donors and advisors impacted the Frankenstein set of laws that constituted the "formal system." For example, in drafting the financial laws, advisors trampled upon some of the Afghan local traditions to include the tenets of Sharia law [, p. 74]. These laws were unfamiliar to the Afghans and therefore could not confer the forms of legitimacy that were based on a society's norms to include Weber's traditional, Alagappa's normative, and Wegand's substantive form of legitimacy. [12]

Further undermining the legitimacy of the formal legal system was the rampant corruption of judges and the police. The system did not demonstrate performance legitimacy (Alagappa) or instrumental legitimacy (Weigand) because it did not respond to the needs of the majority of the Afghan population. Again, the American efforts to support the formal legal system thus undermined the overall legitimacy of the Afghan Government it supported. Moreover, the Taliban pointed to the corruption in the formal system as yet another reason to reject the Americans and their "puppet government in Kabul," amplifying Afghans' historical resentment of the imposition of foreigners on their land.

American attempts to establish a formal system only increased uncertainty and reduced predictability. The Taliban were able to exploit the weaknesses of the American-supported governance services and formal legal system to support their own claims to legitimacy, appearing by comparison to be the only entity that could offer predictability back into the lives of many of the average Afghans.

8 Conclusion: Putting the Puzzle Together

Afghanistan has puzzled the United States for over two decades. The American approach to rebuilding that puzzle was premised on the need to establish a legitimate Afghan central government that provided services, to include a formal system to administer the rule of law. However, despite the COIN doctrine's words about the criticality of understanding the local perspective which ostensibly drove operations in Afghanistan, the Americans tended to mirror image the programming on the ground to provide governance services they would themselves value, rather than sponsor projects that met the Afghans actual needs based on the Afghans' own perspective. The justice sector case study, not unlike other sectors in Afghanistan, highlights that the Americans had the wrong reference image for the Afghan puzzle; they discounted the perspective of the Afghans, and were working with puzzle pieces that did not exist in Afghanistan and would likely never fit into the puzzle. The Americans introduced an entirely new justice system that was based on an entirely different legal approach—retributive, rather than the restorative approach of the shuras more familiar to the Afghans. The Taliban, on the other hand, developed a model for legitimacy based on the reference picture of Afghanistan that was much more familiar to its population. Starting with the two key puzzle pieces of security and justice, the Taliban was able to ensure a level of predictability in Afghans' lives through the application and enforcement of justice based on the familiar Islamic tenets; tenets which were arguably the only constant in Afghan society throughout decades of conflicts. Building off these puzzle pieces, the Taliban embodied elements of legitimacy that led to their eventual control over the entire country.

The lesson for America from this case study is that in future competitions, America must appreciate the reference image the governed have of their own history and their vision for the future. It is their perspective that is paramount. A more thorough appreciation of the Afghan perspective, and cognizance of American confirmation bias, would have illuminated the modest initial requirements to begin building legitimacy for the Afghan Government and to form a realistic counter to the Taliban's offering of legitimacy. Additionally, America must appreciate that it will likely never have the ethnographic fidelity required to figure out how the puzzle pieces are put together. Nor can we introduce puzzle pieces that do not resonate with the population, such as the formal justice system or centralized government in Afghanistan. Ultimately, America must co-create with the population towards the population's vision of the future. Only then will the puzzle pieces fit together.

References

1. The American Institute of Afghanistan Studies (2010) Rule of law in Afghanistan Conference Report. https://afghan-institute.org/rule-of-law-in-afghanistan/. Accessed 10 June 2022
2. Azizian N (2021) Easier to get into war than to get out: the case of Afghanistan. Harvard Belfer Center. https://www.belfercenter.org/publication/easier-get-war-get-out-case-afghanistan. Accessed 13 May 2022

3. Baczko A (2021) How the Taliban justice system contributed to their victory in Afghanistan. Social Science Research Center. https://items.ssrc.org/insights/how-the-taliban-justice-system-contributed-to-their-victory-in-afghanistan/. Accessed 29 May 2022
4. Brown Z (2021) Aiding Afghan local governance: what went wrong. Carnegie endowment for international peace. https://carnegieendowment.org/2021/11/08/aiding-afghan-local-governance-what-went-wrong-pub-85719. Accessed 25 May 2022
5. Dorronsoro G (2009) The Taliban's winning strategy in Afghanistan. The carnegie endowment for international peace. https://carnegieendowment.org/files/taliban_winning_strategy.pdf. Accessed 15 May 2022
6. Khan H (2015) Islamic law, customary law, and afghan informal justice. United States Institute of Peace, Special Report 363. https://www.usip.org/publications/2015/03/islamic-law-customary-law-and-afghan-informal-justice#:~:text=As%20Afghanistan%27s%20nascent%20democracy%20works,settling%20of%20disputes%20outside%20Kabul. Accessed 1 June 2022
7. Lamb R (2014) Rethinking Legitimacy and Illegitimacy: A New Approach to Assessing Support and Opposition across Disciplines (https://www.csis-website-prod.s3.amazonaws.com)
8. Ledwidge F (2009) Justice and counter-insurgency in Afghanistan: a missing link. RUSI J 154(1)
9. PBS (2011) A Historical timeline of Afghanistan. https://www.pbs.org/newshour/politics/asia-jan-june11-timeline-afghanistan. Accessed 10 May 2022
10. Rahimi H (2021) Lessons from Afghanistan's History: How Not to Fix a Failed State. The Diplomat. https://thediplomat.com/2021/10/lessons-from-afghanistans-history-how-not-to-fix-a-failed-state/. Accessed 15 May 2022
11. Rubin E (2009) Karzai in his labyrinth. New York Times, 4 Aug 2009. https://www.nytimes.com/2009/08/09/magazine/09Karzai-t.html. Accessed 17 June 2022
12. SIGAR (2021) What we need to learn: lessons from twenty years of Afghanistan reconstruction. https://www.sigar.mil/pdf/lessonslearned/SIGAR-21-46-LL.pdf. Accessed 10 June 2022
13. Torabi Y (2020) Afghanistan after America—the corruption epidemic. https://www.afghanistanafteramerica.com/. Accessed 17 May 2022
14. United States Government (2011) United states government integrated civilian-military campaign plan for Afghanistan, Revision 1. https://nsarchive2.gwu.edu/NSAEBB/NSAEBB370/docs/Document%209.pdf. Accessed 10 May 2022
15. United States Government (2009) United States government integrated civilian-military campaign plan for Afghanistan. https://www.comw.org/qdr/fulltext/0908eikenberryandmcchrystal.pdf. Accessed 10 May 2022
16. United States Joint Chiefs of Staff (2018) FM 3–24 Counterinsurgency. https://www.jcs.mil/Portals/36/Documents/Doctrine/pubs/jp3_24.pdf. Accessed 5 May 2022
17. UNDP (2001) Human development report 2001: making new technologies work for human development. UNDP. https://hdr.undp.org/system/files/documents//hdr2001enpdf.pdf. Accessed 6 June 2022
18. Weigand F (2015) Investigating the role of political order of conflict-torn spaces. http://eprints.lse.ac.uk/62691/1/Investigating-Legitimacy-in-the-Political-Order-of-Conflict-torn-spaces.pdf. Accessed 12 June 2022

Amy Frumin is a consultant currently leading a project for the Joint Staff, Joint Force Development, in the Office of Irregular Warfare and Competition. Previously, she supported U.S. Central Command in the Operations and Intelligence Directorates, and worked as a strategist at U.S. Special Operations Command Central. Ms. Frumin was also a faculty member in the Department of Graduate Studies, Joint Special Operations University, U.S. Special Operations Command. Ms. Frumin has also worked as a trainer preparing military and civilian personnel to deploy to Afghanistan. During her International Affairs Fellowship at the Council on Foreign Relations, Ms. Frumin wrote about the efficacy of U.S. foreign assistance in stability operations.

In 2006 and 2007, Ms. Frumin served as a U.S. Agency for International Development (USAID) officer at the Provincial Reconstruction Team in Panjshir, Afghanistan. Before deploying to Afghanistan, Ms. Frumin worked for USAID's Office of Transition Initiatives. She has also worked for the United Nations in the Department of Peacekeeping Operations in Kosovo and for UNICEF's Office of Emergency Operations. Additionally, Ms. Frumin has written for several think tanks and political journals, including Foreign Affairs, PRISM, the Center for Strategic and International Studies' Post-Conflict Reconstruction Project, the United Nations Association for the USA, the World Policy Institute, and the French publication Politique Americaine. Amy earned a BA in Political Science with a minor in International Development at McGill University and an MS in Development Studies from the London School of Economics and Political Science.

The Taliban's Exploitation of Afghan Information and Information Systems

Mark Grzegorzewski, Michael Spencer, and Ken Brown

Abstract In this chapter, we discuss an often-overlooked aspect of the Taliban's insurgency, their use of modern communications infrastructure to influence the information environment. Prior to U.S. involvement in Afghanistan, the country had little modern communications infrastructure, and those devices that could connect to the Internet were banned under Taliban rule. During this period, the Taliban had virtually monopolistic control over Afghanistan's information space. As the United States and its partners brought modern communications infrastructure to Afghanistan, information security was overlooked by the Coalition. The Taliban were able to exploit information systems to access the information space to enable their insurgency, by appealing both to internal and external audiences, to promote a different version of the Taliban. We detail that story, and in doing so argue the Taliban's exploitation of the Afghan information and information systems was one factor in the failure of the U.S.-backed Afghan government. The lesson learned: security should never be an afterthought when promoting an open information environment. Once information and information systems vulnerabilities are discovered, the promise of an open information environment will be exploited by malign actors.

Keywords Taliban · Afghanistan · Information · Communication · Influence · United States

1 Introduction

Following the attacks of September 11, 2001, the United States singular political focus become Al-Qaeda in Afghanistan. With CIA operators already on the ground in Afghanistan, U.S. Special Operations Forces were introduced in early-October

M. Grzegorzewski (✉) · M. Spencer · K. Brown
Joint Special Operations University, Tampa, FL 33621, USA
e-mail: Mark.Grzegorzewski.ctr@socom.mil

M. Spencer
e-mail: mgspencer@usf.edu

A. Farhadi and A. Masys (eds.), *The Great Power Competition Volume 4*,
https://doi.org/10.1007/978-3-031-22934-3_7

2001 where they linked-up with insurgency forces pre-vetted by the CIA, popularly known as the Northern Alliance [1]. It took Special Forces and Agency operatives, supported by conventional forces, almost 2 months to sweep the Taliban from rule in Afghanistan. Initially, the purpose of "Enduring Freedom" was to "prevent Afghanistan from becoming a safe haven or sanctuary for al-Qaeda, and to make sure al-Qaeda is not there in Afghanistan, and, therefore, a destabilizing force in the region" [2]. Soon, the Afghanistan counterterrorism mission expanded into a state-building effort over multiple administrations [2].

What is remarkable is both the expansion of the mission by the Department of Defense, and the U.S. government's hubris that it could create a functioning government in a country where there was not even a common national identity (if there was an incipient national identity, it did not correspond with the borders of the Afghan state [3]). The U.S. and Coalition forces attempted to forge Afghans together in a common national identity so the Afghan-state could develop and survive. As observed by Warren Chin, since World War II the U.S. has "continued to see its security through the prism of technology advance [4], and Afghanistan was no different in that it was assumed modern information systems would play a largely positive role in creating a stable Afghan state [5]. In many respects these information systems were beneficial for the government. They were also beneficial for the Taliban.

It is this aspect, the use of information systems to exploit the information environment, that we proffer as one factor in why the Afghan government failed. The irony in the Taliban exploitation of the Afghan communications infrastructure is twofold. First is the curious notion of an Islamic fundamentalist movement leveraging modern communication infrastructure to promote itself to the modern world. Second is that much of the communications infrastructure brought to Afghanistan came from Western funding so the Afghan state could consolidate its hold over its territorial space. While Afghan information systems brought modern conveniences to the country, security concerns were not adequately addressed, as demonstrated both by Taliban attacks on the infrastructure and then the Taliban's use of that same infrastructure to exploit the information environment. To further demonstrate that security was not a primary concern, Estonia's National Cyber Security Index ranked Afghanistan 134 out of 160 on its ranking of state's cybersecurity practices, ranging from "protection of digital services" to "cyber crisis management" [6]. Ultimately, the open architecture coupled with poor cybersecurity practices allowed the insurgency to coordinate, message, and influence the information environment. This should have been anticipated. This tale, convenience trumps security, is a tale as old as time in the cybersecurity community. It was a tale repeated in Afghanistan.

In what follows, we discuss the importance of state-facilitated communications in creating a nation. Next, we discuss social movement theory and the impact of communications infrastructure to better understand the insurgency against the Afghan government. This includes a discussion on the quantitative findings in the collective action literature which demonstrate a strong linkage between the arrival of communications technology and the increased ability of an insurgency to organize. In the next section we detail how the Taliban exploited vulnerabilities in Afghanistan's communication infrastructure to influence the information environment. Last, we offer lessons learned in Afghanistan, which can be applied more broadly, concerning introducing communication infrastructure into a new environment.

2 Communications Technology: Opportunities for Collective Action

As the U.S. and Coalition partners set-up the new Afghan government, an Afghan national identity had to be fashioned. The tribal-based country consisted of about 31 million people with 14 recognized ethnic groups, and no one group with an absolute majority [7]. In the national identity literature, some understand the nation as an "imagined community," meaning that people who will never meet each other think of themselves as belonging to the same community [8]. It is through this prism of national identity creation that we understand the Afghan nation-state project. Within this context, the U.S. and newly formed Afghan government established its Internet, Communications, and Technology (ICT) policy, "to improve public services delivery, accelerate economic development and social change for the benefit of every Afghan" [9]. In essence, ICTs were used to link Afghans together with each other and the state. Open Communications architecture and access to the information environment would be key to knitting an Afghan identity together. The hope was that suddenly events happening in the Afghan metropolis and then later discovered in the countryside would no longer be a world away. It would be happening in the same country, to the same people, Afghans. However, due to "poor regulatory and legal frameworks, and weak monitoring and oversight mechanisms" [10], there was frequent exploitation of Afghan open communications architecture.

Technology is agnostic. It can be used by anyone for purposes beyond the innovators original vision. In the case of Afghanistan, it was used by an enemy of the state, the Taliban insurgency. In Afghanistan's insecure environment, the U.S. government continued to provide communications support and the Taliban were able to leverage open communications infrastructure and mobilize their movement through the information space. To be clear, the information environment is understood as "comprised of and aggregates numerous social, cultural, cognitive, technical, and physical attributes that act upon and impact knowledge, understanding, beliefs, world views, and, ultimately, actions of an individual, group, system, community, or organization. The IE also includes technical systems and their use of data. The IE directly affects and transcends all OE."[1] Complementary to the information environment is information systems which are defined as "A discrete set of information resources organized for the collection, processing, maintenance, use, sharing, dissemination,

[1] Joint Chiefs of Staff. (2018). Joint concept for operating in the information Environment (JCOIE). https://www.jcs.mil/Portals/36/Documents/Doctrine/concepts/joints_concepts_jcoie.pdf

or disposition of information."[2] These two concepts interrelate in that information systems "in the physical domain create and direct the flow of information in the information domain which, in turn, affects human perceptions, attitudes, and ultimately decision-making in the cognitive domain."[3] Therefore, physical attack, or threat of attack, can shape the information environment, and when the information system is non-discriminatory, it can be leveraged by malign actors. Once human perceptions are shaped in the information environment (or domain), it can have real world physical effects over who controls the information systems.

As with any other social movement, the Taliban's reformation depended on its pre-existing organization in Afghan society and the strength of its beliefs to influence the people [11]. The Taliban collectively organized with specific aims and goals [12]; operated outside normal channels; and operated with a sense of organization. These factors allowed the Taliban to organize in pursuit of political opportunity. The Taliban's goal was to return to power and build an Islamic Emirate based upon Sharia throughout Afghanistan [13]. To achieve this aim, the Taliban needed to alter its image within Afghanistan and the world. Reduced information costs facilitated information, communication, and coordination [14, 15] in pursuit of collective action. The Taliban ultimately came to understand the power of information and exploited Afghan information systems. This should have been anticipated.

Research indicates that insurgencies can leverage the same communication infrastructure used by civilians to report on an insurgency [16]. Additional research indicates that the dual use nature of technologies, used by both the government and private sector, have a strong causal link between the introduction of mobile communication networks into unregulated environments and provides the insurgency an ability to coordinate collective action, thereby increasing violence. On the other hand, communications infrastructure can also allow government security forces to tap into insurgent's communications, thereby allowing them to be monitored, and leading to a decrease in violence. Due to its dual use, communications infrastructure can both enable and disempower an insurgency. It all depends on how effectively the technology is utilized by different actors [17].

Other research considering the relationship between mobile communications and opportunities for collective action focus on causal motivations. In research focusing on the African continent to test a connection between phone availability and collective action, the results were inconclusive [18]. However, researchers still surmised that depending on a group's population, its ruralness, and population density, "mobile phones increase the likelihood of violent collective action by decreasing specific types of barriers to organization" [19]. Further research on political violence in Africa found "cell phones as a communication technology allows political groups

[2] Office of the DoD chief information officer. (2020). DOD INSTRUCTION 5000.82 ACQUISITION OF INFORMATION TECHNOLOGY (IT). https://www.esd.whs.mil/Portals/54/Documents/DD/issuances/dodi/500082p.pdf

[3] Cordray III, R. & Romanych, M. J. (2005) Mapping the information environment. IO Sphere, 7–10

to overcome collective action problems more easily and improve in-group cooperation, and coordination" [19]. As the authors note, despite potential nefarious uses, increasing communications infrastructure in developing countries is largely a net-gain. This outcome must be weighed against the ability of malign actors to use the same infrastructure to overcome communications costs and challenge the state.

Finally, research has also tested the association between ICTs, to include mass media and social media, and its pacifying effects on African populations and its ability to generate increases in collective action [20]. Findings demonstrate a strong relationship exists between the state's ability to leverage ICTs to reinforce nationalism, and with ICTs and the growing promotion of sectarian and extremist appeals on social media, which can lead to collective violence. While demonstrating a strong linkage between ICTs and insurgency organization, the collective research reinforces a larger point: technology, if not properly secured, is a non-discriminatory good. Unless actors consider how to secure their information systems, the information environment can be exploited by malign actors and thereby undermine state efforts.

3 Information Convenience Over Information Security

To facilitate the flow of information within Afghanistan, since 2002 the Department of Defense (DOD), State Department, and USAID have invested over $2.6 billion (the DOD contributed 96% of this total) in the Afghan Information and Communications Technology sector [21]. The investments "largely supported the Afghan National Defense and Security Forces and the Afghan media." Of the $2.5 billion, the DOD contributed, the majority, over $1.8 billion, went to command-and-control equipment indicating that the DOD's primary focus in Afghanistan was to supply communications equipment and support for the Afghan National Army and Police [22]. The State Department allocated "$83 Million for ICT support of Afghan media and rule of law development" with the majority, over $42 million, going to "Afghan Media, Radio/TV Stations and General News Coverage" [22]. Finally, USAID underwrote "over $44 million, primarily on activities intended to increase the ICT capacity of Afghan ministries" with the majority, $24.4 million, going to "promote the use of mobile technologies and improve the transparency of payment systems" [22].

While never an ICT-saturated country, Afghanistan after 2001 could count 37 different ICT solution providers [22]—services that create, disseminate, store, and manage information. By 2016, the country also had 5 Mobile Network Operators [23] (also known as a wireless service provider) [24], the largest providers being Roshan and MTN. For the year in which statistics were last displayed, 2013, Afghanistan could also boast of almost 90% mobile Telephone population coverage; over 330,000 3G Broadband Subscribers; 87 TV operators; telephone penetration into 72% of the country; and over $2 billion (USD) investment in the telecom sector [22]. The state-owned Afghan Telecom owned the majority of the physical communications infrastructure, after inheriting it from the U.S. and other partners [24]. Additionally, Afghanistan developed 52 Internet Service Providers (ISP) that allowed Afghans to gain a point of presence on the Internet [22]. These numbers by themselves are impressive, especially when starting from essentially nothing. As seen in Fig. 1, this Afghan communications infrastructure was both seen as a target and a tool by the Taliban.

2001: Prior to the US invasion of Afghanistan the Internet was essentially banned by the Taliban.[33] The Internet and ICTs were virtually nonexistent in Afghanistan and in some regions would remain that way for decades.[34] Globally, cell phone use was limited to voice calls or messaging services. Internet access was regulated to computer use through DSL or phone lines.[35]
2004: The Taliban and other militant Islamists began coordinating their operational and political messaging via mobile phones. Yet, network availability remained scarce in most regions.
2005: To account for the lack of mobile phone reception, the US-backed Afghan government created an array of cell towers across the country capable of providing service to about 1 million Afghans.[36,37]
2006: The Taliban create their first website called Al-Emarah. The websites content is initially provided in three languages: Dari, Pashtu, and in English.[38]
2007: The Taliban is recognized as being technologically sophisticated and exceedingly capable of utilizing ICT and social media to enhance its propaganda and messaging.[39]
2007: The main hardware platforms available in Afghanistan are Motorola and Nokia's monochrome flip phones. These phones have the capability of sending out mass text messages to targeted customers. The Taliban use these mobile phones to provide intimidating reminders that the Taliban are still present and active in Afghanistan.[40]

[33] The Economic Times-Tech. (2022, July 1). "How the Taliban Turned Social Media into a Tool for Control." The Economic Times. https://economictimes.indiatimes.com/tech/technology/how-the-taliban-turned-social-media-into-a-tool-for-control/printarticle/85511668.cms

[34] Campbell, E. (2021, October 12). *How Digital Rights Are Key to Protecting Afghans Under the Taliban*. MEI Publications. https://www.mei.edu/publications/how-digital-rights-are-key-protecting-afghans-under-taliban

[35] Culpan, T. (2021, August 22). *How a Technology Revolution Powered the Taliban's Return*. The Business Standard. https://www.tbsnews.net/features/panorama/how-technology-revolution-powered-talibans-return-291436

[36] Teller Report. (2021, August 28). *The Taliban Uses Social Media as a Tool Of Threat and Enticement*. Teller Report. https://www.tellerreport.com/news/2021-08-28-the-taliban-uses-social-media-as-a-tool-of-threat-and-enticement.B1ZzT7Ndbt.html

[37] The Economic Times-Tech. (2022, July 1). *How the Taliban Turned Social Media into A Tool For Control*. The Economic Times. https://economictimes.indiatimes.com/tech/technology/how-the-taliban-turned-social-media-into-a-tool-for-control/printarticle/85511668.cms

[38] Taneja, K. (2021, August 17). *From 'Night Letters' to the Internet: Propaganda, the Taliban and the Afghanistan Crisis*. Observer Research Foundation. https://www.orfonline.org/research/from-night-letters-to-the-internet/

[39] Culpan, T. (2021, August 22). *How a Technology Revolution Powered the Taliban's Return*. The Business Standard. https://www.tbsnews.net/features/panorama/how-technology-revolution-powered-talibans-return-291436

[40] Culpan, T. (2021, August 22). *How a Technology Revolution Powered the Taliban's Return*. The Business Standard. https://www.tbsnews.net/features/panorama/how-technology-revolution-powered-talibans-return-291436

Fig. 1 Afghan communications from 2001–2021

2007: By the end of the year U.S. telecom companies supported by NATO forces began constructing what would be an extensive network of cell phone towers. The Taliban take advantage of this increased access to reach more audiences throughout Afghanistan, update Western media outlets of Taliban activities, answer questions, and provide on recent events and battles. For example, Taliban victories that are either not known or ignored the mainstream media outlets.[41]
2009: The Taliban establishes their first YouTube channel.[42]
2011: The Taliban begin posting on Facebook and Twitter.[43]
2012: 3G Internet services are introduced in Afghanistan beginning a dramatic increase of social media use across Afghanistan and throughout the world. ICTs became pivotal to Afghanistan's economic, political, and social life.[44]
2012: Mobile technologies and social media platforms provide increasing access to online spaces. Afghans can congregate and express their opinions and ideas on these public spaces. These spaces become very useful to journalists, educators, and vulnerable groups who need to share information across a variety of platforms.[45]
2015: By the end of the year, risks associated with service disconnections and/or Internet shutdowns increase.[46]
2015: The Taliban extend their use of social media platforms to WhatsApp and Telegram.[47]

[41] Culpan, T. (2021, August 22). *How a Technology Revolution Powered the Taliban's Return*. The Business Standard. https://www.tbsnews.net/features/panorama/how-technology-revolution-powered-talibans-return-291436
[42] Puyosa, I. (2021, October 17). *The Taliban's Social Media Warfare Operation Toward Seizing Kabul*. Toda Peace Institute. https://toda.org/global-outlook/the-talibans-social-media-warfare-operation-toward-seizing-kabul.html
[43] Puyosa, I. (2021, October 17). *The Taliban's Social Media Warfare Operation Toward Seizing Kabul*. Toda Peace Institute. https://toda.org/global-outlook/the-talibans-social-media-warfare-operation-toward-seizing-kabul.html
[44] Campbell, E. (2021, October 12). *How Digital Rights Are Key to Protecting Afghans Under the Taliban*. MEI Publications. https://www.mei.edu/publications/how-digital-rights-are-key-protecting-afghans-under-taliban
[45] Campbell, E. (2021, October 12). *How Digital Rights Are Key to Protecting Afghans Under the Taliban*. MEI Publications. https://www.mei.edu/publications/how-digital-rights-are-key-protecting-afghans-under-taliban
[46] Cowie, J. (2012, November 29). *Could It Happen in Your Country?* Renesys. https://web.archive.org/web/20121201001304/http:/www.renesys.com/blog/2012/11/could-it-happen-in-your-countr.shtml
[47] Puyosa, I. (2021, October 17). *The Taliban's Social Media Warfare Operation Toward Seizing Kabul*. Toda Peace Institute. https://toda.org/global-outlook/the-talibans-social-media-warfare-operation-toward-seizing-kabul.html

Fig. 1 (continued)

4 Mobilization

Prior to 2001, the Taliban actively banned the Internet throughout Afghanistan, destroying Afghans' phones, cameras, and televisions, and enacting harsh punishments for their use. The reasoning was to keep technology out of the hands of the populace and "control those things that are wrong, obscene, immoral, and against Islam" [40–42]. After the U.S. led invasion of Afghanistan, the Information Technology (IT) sector would become one of the few successes in the war-ravaged nation. Private sector companies, in partnership with the new Afghan government, expanded

2017: The World Bank estimates that approximately 11.4 percent of the Afghanistan population are routinely using the Internet, a sharp increase from 0 percent when the Taliban lost power to US and coalition forces after the events of 9/11.[48]
2018: Mobile phone networks provide Afghans broadband access to the Internet and social media accounts throughout most of the country.[49,50] Social media participation rises to more than 10 percent. Facebook becomes the preferred means of accessing local news and information. The Taliban takes over the city of Kandahar in the south.[51]
2019: Cell phone use escalates to over 22 million persons, approximately 70 percent of the population has at least some access to mobile services and the Internet.[52] Realizing the importance of cell phone access the Taliban begins extorting "protection taxes" from telecom providers by threatening to destroy ICT infrastructure.[53,54]
2021: Afghanistan's overall Internet use in January is estimated to reach 89 percent of the Afghani population, increasing over 13 percent from prior year. Afghanistan satellite providers increase from 6 to 40 over the past decade.[55,56] This increase allows the Taliban to contact more Afghans and spread their propaganda and recruitment incentives to more regions.[57]

[48] Thorbecke, C. (2021, August 19). *How the Taliban Uses Social Media to Seek Legitimacy in the West, Sow Chaos at Home*. ABC News. https://abcnews.go.com/Technology/taliban-social-media-seek-legitimacy-west-sow-chaos/story?id=79500632

[49] Tech Policy Press. (2021, August 17). *What the Taliban's Victory Means for Access to Technology in Afghanistan*. Tech Policy Press, Technology and Policy. https://techpolicy.press/what-the-talibans-victory-means-for-access-to-technology-in-afghanistan/

[50] Guo, E. (2018, October 21). *Afghanistan's Real Internet Lives on Its Streets*. New York Magazine, Global Tech. https://nymag.com/developing/2018/10/afghanistan-sneakernet-internet-physical-file-sharing.html

[51] Taneja, K. (2021, August 17). *From 'Night Letters' to the Internet: Propaganda, the Taliban and the Afghanistan Crisis*." Observer Research Foundation. https://www.orfonline.org/research/from-night-letters-to-the-internet/

[52] The Economic Times-Tech. (2022, July 1). "How the Taliban Turned Social Media into A Tool For Control". The Economic Times. https://economictimes.indiatimes.com/tech/technology/how-the-taliban-turned-social-media-into-a-tool-for-control/printarticle/85511668.cms

[53] Tech Policy Press. (2021, August 17). *What the Taliban's Victory Means for Access to Technology in Afghanistan*. Tech Policy Press, Technology and Policy. https://techpolicy.press/what-the-talibans-victory-means-for-access-to-technology-in-afghanistan/

[54] Dominque, G. (2016, January 21). *How the Taliban get their money*. DW, Asia. https://www.dw.com/en/how-the-taliban-get-their-money/a-18995315

[55] Campbell, E. (2021, October 12). *How Digital Rights Are Key to Protecting Afghans Under the Taliban*. MEI Publications. https://www.mei.edu/publications/how-digital-rights-are-key-protecting-afghans-under-taliban

[56] Kemp, S. (2021, February 11). *Digital 2021: Afghanistan*. Dataportal. https://datareportal.com/reports/digital-2021-afghanistan

[57] Mehran, W. (2021, August 24). *How Social Media Helped the Tech-Savvy Taliban Retake Afghanistan*. The Guardian. https://www.theguardian.com/commentisfree/2021/aug/24/tech-savvy-taliban-afghanistan-propaganda-fundraising-recruitment

Fig. 1 (continued)

mobile services throughout Afghanistan [43]. Believing the U.S. and NATO forces were collaborating with mobile phone companies to track and precision target fighters through their cell phone signals caused the Taliban to systematically destroy many of the newly constructed telecom towers throughout the southern region and prompted mobile companies to acquiesce to Taliban demands that cell tower signals be shut down for 10 hours each night. These acts demonstrated the Taliban could control Afghanistan's mobile phone networks, and the Afghan government and Coalition

2021: In the eight months prior to the Taliban offensive into Kabul, Taliban social media accounts post drastically increase to 38,000 pieces of propaganda.[58]
2021: Dramatic increases in Internet activity occur in both Afghanistan and globally. Some of the greatest increases are in in the Middle East and North Africa (MENA) countries and throughout the Central and South Asia. This growth is spurred on by the introduction of 4G coverage and massive investment in these regions which had fallen behind the rest of the world.[59,60]

[58] Winter, C., Alrhmdon, A., & Sayed, A. (2021, August 28). *The Taliban's Vast Propaganda Machine has a New Target*. Wired. https://www.wired.co.uk/article/taliban-propaganda-news-afghanistan
[59] Campbell, E. (2021, October 12). *How Digital Rights Are Key to Protecting Afghans Under the Taliban*. MEI Publications. https://www.mei.edu/publications/how-digital-rights-are-key-protecting-afghans-under-taliban
[60] Vale, L. (2021, February 1). *What Internet Penetration, Growth, and Usage Means for the Arab World*. Albawaba, Business. https://www.albawaba.com/business/what-internet-penetration-growth-and-usage-means-arab-world-1408518

Fig. 1 (continued)

forces were all but powerless to stop them. These actions also severely angered the local Afghan population, which was counterproductive to the Taliban winning over the hearts and minds of the people [44, 45].

Over time, Taliban leadership began to realize that as an insurgency, the expansion of Afghanistan's telecom towers and the country's growing access to the Internet could be an excellent mechanism for spreading propaganda, raising funds, and recruiting. By 2005, their first website, Al-Emarah, was created and used to disseminate propaganda locally in five languages. Within five years, the Taliban's interests had rapidly evolved, and they were investing in social networks like Twitter to amplify their message throughout the world, engage in diplomacy for international recognition, and reinventing their image [46]. Further, the Taliban "deployed fake accounts and bots to amplify their messaging and undermine the Afghan government" [47]. It was during this period that they launched a series of successful social media campaigns. Utilizing teams of dedicated volunteers, they focused on getting Taliban hashtags trending on Twitter and disseminating different variations of messaging on Facebook and WhatsApp. These activities coincided with the massive growth of Internet access throughout Afghanistan [48, 49]. Within a few more years the Taliban had their own multimedia production team, and their key spokespersons all had hundreds of thousands of followers on multiple platforms, and their messaging was reaching young recruits across the world [42, 43].

For the months leading up to the offensive that would put Afghanistan back in Taliban control, social media was successfully used to project the appearance of strength tempered with moderation. Afghans were led to believe in the inevitability of Taliban's return and the international community was promised a more enlightened Taliban leadership [47]. As the offensive was taking place, each abandoned Afghan stronghold or captured city was accompanied by pro-Taliban messaging on multiple social media platforms touting Taliban victories and achievements and exaggerating failures of the Afghan government. The Taliban's social media blitz was correlated with Afghan Army units leaving their posts after learning that Taliban victory was all but assured and that they should surrender or face severe consequences [50].

The Taliban's 2.0 strategy was not simply about changing local public perception via distributing propaganda and/or convincing the international community of its legitimacy, it was also about using information and information systems as a source for intelligence collection that could be used against their opponents [48, 49]. Social media accounts and government databases enabled the Taliban to track down and retaliate against Afghans who were loyal to the prior Afghan government and/or to U.S. and Coalition forces. These measures were so effective that in the latter months of 2021 Afghans were rushing to erase all traces of their digital lives out of fear of Taliban reprisals [41, 47].

5 Recruiting

As an insurgency, the Taliban were considered a mobile fast-moving military entity relying on small arms and Rocket Propelled Grenades for is combat operations. When confronted by the vastly superior Coalition forces, the Taliban quickly realized they would need to engage in a psychological campaign paired with kinetic warfare. Advances in information system platforms turned out to be the answer to this problem and provided the Taliban with new abilities to shape information, new methods to recruit new members, and a broader array of tools from which to threaten or intimidate adversaries [47].

Domestically, social media allowed the Taliban to provide unique cultural, religious, and politically based propaganda tailored specifically to different Afghan communities. It also permitted Taliban members to emphasize the shared values and culture considered important to urban populations and local tribal units alike. The goal was to target relevant issues and change the perception of the Afghan people in favor of the Taliban, framing them as fellow Afghans and explaining why they were against the US and its coalition forces [28]. Social media and direct messaging were excellent tools that provided a constant reminder to Afghans of their shared interests [51]. Daily messages and livestream events in multiple languages (Pashtu, Urdu, Dari. Arabic, and English) allowed Taliban representatives to speak directly to local communities and their leaders in their own dialects further demonstrating their similarities [30, 31]. As the Taliban grew more efficient at engaging in this type of influence it pulled ahead of the U.S. and Afghan governments in the battle of narratives and positioned itself as representing the true interests of the Afghan people [33]. While the Afghan public opinion of the Afghan government steadily declined [51] over the 20 years of U.S. involvement, it did not mean that Afghans favored the Taliban's return [52]. This made the use of technology for social mobilization even more important. The Taliban exploitation of information allowed them to quickly sell a counter narrative and spin any incident. Essentially, it allowed the Taliban to be first with the news, even if it was false, which impacted the resonance of U.S. messaging and the support the Afghan government received from the population [52].

Globally, the Taliban utilized a strategy involving a constant barrage of social media posts combined with pro-Taliban commentaries aimed at depicting themselves as peaceful and stable liberators to the international community as opposed to an oppressive and ruthless regime. The objective was to deceive the West and demonstrate their legitimacy to rule Afghanistan to the rest of the world [26, 53]. To accomplish this, the Taliban learned to speak the language of the West. They catered to the views of Western media outlets and journalists and eventually gained acceptance in many circles. The result was having their messaging repeated in a surprising number of global media outlets [33]. With the Taliban having limited financial resources and Afghanistan falling well behind the modern world in ICT and infrastructure availability, social media access and mobile/smart phones availability became a highly effective tool for disseminating propaganda. Another boon for the Taliban was how the Afghanistan ICT environment lacked reliable fact checkers and that media literacy among the populace was low, allowing propaganda to be misconstrued for fact [33, 54]. In this environment, the Taliban could simply outperform its rivals by posting and sharing more, often with more exaggerated content across their growing network of authentic and inauthentic platforms [26, 30, 39].

Prior to the 2021 offensive, the Taliban had been already successfully shaping the Afghan information environment for years, stressing their inevitable victory over the U.S. and Coalition invaders, along with the promise of a better future under its rule. These actions, over time, generated the belief among the Afghan public that the Taliban's victorious return was all but inevitable [39, 55]. The more victories and captured territory the Taliban acquired, all of which were recorded with digital cameras and cell phones, the more propaganda could be sent across multiple social media channels displaying the Taliban's strength. In turn, this meant more battlefield victories occurring without firing a shot, thereby creating a self-fulfilling prophecy [56]. After the success of the 2021 offensive, the Taliban shifted their propaganda focus away from the aforementioned tactics and a general agitating of the Afghan population against the U.S., its Coalition forces, and the Afghan government towards one of integration. The emphasis was now on displaying how the Taliban were providing good governance, security, and stability, and unifying the country [39]. Taliban spokesman began to redefine their messaging presenting the Taliban as a responsible, legitimate, and accepted ruling body worthy of international respect and assistance [39, 57].

The Taliban realized to win the hearts and minds of the greater Afghan people they would need to show the softer side of their character. Initiatives were put in place to aid the poor, establishing rebuilding efforts, and celebrating holidays. All these actions were accompanied by video, commentary, and interviews spread throughout multiple social media and messaging outlets [57]. Taliban leadership understood to succeed in their transition they would have to alter the perception of the populace to one where the country was at peace and the Taliban could provide law and order and a stable government [57, 58]. The Taliban's three main spokesman, Zabiullah

Mujahid, Dr. M. Naeem, and Suhail Shaheen, were all highly effective at spreading this new message of peace and stability. The three-spokesman combined have over 800,000 Twitter followers and used the platform as the primary means for their new propaganda campaign. Zabiullah Mujahid alone went from tweeting 2 to 3 times per day prior to the offensive to upwards of 20–50 tweets per day afterwards [30]. Taliban representatives were also busy. They doubled their efforts to spread pro-Taliban narratives throughout mainstream media outlets with few if any checks and balances or counter arguments from the press [26, 30].

6 Internal Influence

Clever messaging and directed social media influence campaigns targeted at Afghan security forces were instrumental in encouraging them to put down their weapons and not fight the inevitable Taliban takeover [30]. The Taliban utilized time-tested propaganda tools such as videos of soldiers in training or conducting successful direct-action missions to both recruit and strengthen their fighter's spirit and resolve [26]. As soon as Taliban forces achieved any degree of military success, the actions were immediately recorded and sent out in mass on social media and messaging services throughout Afghanistan [59]. Taliban soldiers were also trained in video production. These soldiers primarily used hand-held digital cameras and smart phones to record the frontline as Taliban forces swept into Afghanistan. Additionally, videos were recorded whenever Afghan soldiers surrendered without a struggle [31, 54]. These videos, combined with a message of intimidation, offers of amnesty, and portrayals of U.S. forces abandoning the Afghan government, established a belief in Afghanistan that a Taliban victory was all but certain and resistance was futile. The result was that many of the Afghan government's best defendant cities surrendered without a fight [26, 31]. The quick victory over the Afghan government and its military force was not only a shock to the US and Coalition forces but to the entire world. It not only questioned the validity and purpose of the decades long war but demonstrated how social media influence campaigns used with basic communications technology could be used effectively to mobilize, disrupt, and engage states with superior military and intelligence collection capabilities [26].

After regaining control of Afghanistan, the Taliban further expanded their use of Twitter, establishing thousands more accounts both official and anonymous to soothe the population with images of the Taliban providing stability and peace [59]. In urban areas where the more tech savvy Afghans reside, the Taliban used cell phones and social media to inform locals of what to do (and what not do), how to remain safe, and what it would be like to experience an easy transition of government. Any reports from the prior government officials or international media outlets regarding crackdowns or reprisal attacks were quickly countered and overwhelmed with upbeat messages of the Taliban providing unity and peace. Taliban journalists combed the streets of newly conquered cities and villages providing videos of the populace accepting Taliban rule and of surrendered oppositional officials endorsing the Taliban

[26, 31]. Despite Taliban assurances that Afghan residents would be safe and free from prosecution, many Afghans who had worked for the prior government or US coalition forces and considered their previous professions and social status a possible threat to Taliban rule began editing or outright deleting their social media accounts. Their fear was that Taliban enforcers would use online surveillance and/or search their social media histories to root out potential dissenters or insurgents [26]. This included individuals or groups such as Afghan or foreign journalists that may have portrayed the Taliban in a negative light or representing a threat to Afghanistan [27, 34].

To make matters worse, the US also left behind crucial command-and-control systems which the Taliban are using to improve the communication capabilities among its leadership [59, 60]. To keep these command-and-control systems in check, expand upon its social media and messaging capacity, and enhance their digital surveillance capabilities, the Taliban plans to fully control Afghanistan's ICT entities, such as the state owned Afghan Telecom [34, 61]. This could be problematic for the Taliban since while the risk of disruptions or outright destruction of Afghanistan's ICT infrastructure under the old government was considered low to moderate, it is growing as rival militant groups seek to hamper the Taliban's ability to rule [62].

7 External Influence

The US Department of State has yet to officially designate the Taliban as a terrorist organization. This has placed many social media companies in compromising positions. While many extremist groups have been banned entirely from social media platforms, the Taliban has been able to operate unobstructed on certain platforms for years [59, 63]. The main platform for Taliban propaganda dissemination has been Twitter. Twitter has restricted some Taliban activities in violation of its policies against "the glorification of violence, platform manipulation, and spam," but the Taliban have navigated around these restrictions with ease [33, 34]. Dozens of Taliban representatives routinely used Twitter as part of their disinformation campaigns, including tweeting propaganda to over a million followers worldwide. This has allowed the Taliban to post more Twitter messages than the Afghanistan Ministry of Defense. It also allowed the Taliban to contradict other more reliable media outlets and civilian advocacy groups quickly and effectively [64].

Relatively unrestricted access to Twitter caused the service to become the Taliban's main social media platform to conduct influence campaigns [55]. For example, when the Taliban wanted to increase their messaging prior to the 2021 offensive and defeat of the Afghan government, scores of new accounts started popping up on Twitter all sharing the same five videos legitimizing the Taliban's authority to rule Afghanistan. In less than 24 hours these videos combined had attracted more than half a million views [28]. In contrast, Facebook from the outset categorized the Taliban as a "dangerous organization," claiming it had been permanently banning Taliban accounts and/or any accounts that "offered praise, support, or endorsement" for Taliban activities. Facebook also dedicated a team of multilingual experts to help identify any additional attempts by Taliban representatives to gain access to their platform [59, 65]. One of Facebook's subsidiaries, WhatsApp, took a different approach claiming they could not ban the Taliban since WhatsApp is by nature an encryption service, and what cannot be read cannot be banned [34, 59]. By the time the Taliban transitioned to power, the WhatsApp encrypted text messaging service had become one of the Taliban's primary messaging platforms since it could be used to directly influence local populations without U.S. monitoring [66].

YouTube and TikTok followed Facebook placing a blanket ban on almost all Taliban content by citing Executive Order 13129 [28, 58, 59]. However, YouTube remains undecided on what to officially do with Taliban content, pointing out the confusion that exists between the U.S. State Department's list of foreign terrorist organizations, which as of yet does not include the Taliban, and the U.S. Treasury Department's list that classifies the Taliban as a Specially Designated Global Terrorist that can have its assets frozen and be blacklisted from working with U.S. firms [34, 67]. Even with Facebook and YouTube regularly banning Taliban accounts and mainstream hosting services dropping their websites, Taliban representatives have grown proficient at swiftly restoring their websites on rival hosting services and utilizing a blitz of text messages to attract new followers and/or redirect existing customers to other platforms such as Twitter [68]. The Taliban's flexibility in changing platforms, altering messaging content, varying the spelling within hashtags, using encryption or encrypted apps, and utilizing multiple spokesmen all with multiple accounts, made eliminating its presence on major Western platforms almost impossible [26, 28]. The Taliban also realized that by not accurately listing their members names and not posting content containing violence or hateful rhetoric, which violated company policies, they could still make it clear to the viewer where the content came from and what the messaging was intended for [57]. The result was Twitter, WhatsApp, Facebook, YouTube, Telegram, and Tik Tok providing the Taliban a broad array of platforms from which to share their propaganda. Even platforms where the Taliban are banned are still considered porous since they could be temporarily accessed via new accounts or unaffiliated third parties could be hired to provide pro-Taliban rhetoric and further shape public perceptions among populations they would otherwise not have access to Refs. [28, 30].

After the Taliban took back Afghanistan, social media platforms like Facebook and YouTube began allowing Taliban propaganda to be published despite their long-term bans. These companies found themselves in a difficult position between recognizing the Taliban as a legitimate government and deciding to what extent they would continue to restrict access to their platforms [57]. Posts on Facebook increased 120% and included an increase of more than 49,000 Taliban followers. In response, Facebook activated an emergency response team to assess at what level the Taliban should be allowed to access their products [57]. The Taliban is now routinely using multiple social media platforms with numerous accounts to post new messaging, images, memes, videos, etc. about the benefits of current Taliban rule. Taliban generated YouTube videos received tens of thousands of views up from only 1000 the prior year [57]. The common thread being reiterated is that the Taliban are Afghanistan's legitimate rulers and that they are here to stay [57]. Social media analysts believe the Taliban are being advised by at least one public relations company on how to create viral messaging, amplify existing messages, create cross-platform content, and identify and push key content. The goal is to enhance the Taliban brand and presented in a favorable light in a similar fashion to political and corporate media campaigns [57].

8 Lessons Learned

That the Taliban used the Internet architecture to their advantage should not be surprising, but it nonetheless offers lessons for the U.S. and our international partners. First, is the fact that any technology we build or deploy as a public good will greatly complicate our ability to dominate the information environment. With the advent of social media, the Internet, and other communications technologies, it has become increasingly difficult to win the war of words, particularly where the adversary has an inherent advantage in its cultural ties, language skills, and knowledge [58]. Rather than treating digital development as a critical, stand-alone goal therefore, it must be designed and implemented in consonance with a cohesive, consistent, and well-developed messaging campaign.

Secondly, the risks of digital development can be mitigated by how we design and implement programs. In Afghanistan, for instance, the U.S. treated the Internet as a public good. Although this seemed intuitive in view of U.S. strategic goals and long-standing policy of promoting an open Internet, that was not the only approach available. Instead, understanding the inherent risks, the U.S. and Afghan governments could have adopted a more limited model in which government and commercial users were the priority, at least while the security situation matured. This appears to be the approach the Taliban are taking, in which the Internet serves primarily government ends, with restricted public access [69]. At the same time, however, technology should not be seen as the main tool to addressing our information operations problems. No matter how many restrictions the U.S. or host nation place on the Internet, a determined and technically savvy foe will still be able to exploit the infrastructure to achieve their goals. This is evidenced in China, which, despite its much-vaunted internet monitoring and suppression program, is experiencing increasing difficulties controlling online protests [70].

Moreover, metrics for success must be tailored to our goals, achievable, and consistent with our operational design. Although this observation might appear obvious, experience demonstrates that the U.S. repeatedly learns (and unlearns) these lessons at great cost [71]. Applying this to the Afghanistan digital development effort and the Taliban's exploitation thereof, there are at least two important components to consider. First, in Afghanistan, development success was often measured by quantitative means, with limited accounting for who had access, how it was being used, and whether the communications infrastructure was supporting the coalition's strategic goals. Instead, investments were made on the basic assumption that providing technology would inherently achieve development goals and facilitate economic, social, and physical advances. Second, the associated reporting process and pressure to achieve stated goals created an incentive for good news stories over accurate assessments. As a result, over time, the picture presented to leadership grew increasingly divorced from reality, leading to additional mistakes and poor investment decisions. Digital development, therefore, must be adapted to the physical, social, cultural, and economic realities of the state in which it is occurring, and its progress must be measured appropriately.

A final important lesson is that the U.S. must take a long view in the architectures we develop. As we are witnessing with the fall of the Afghan government, the Taliban are now exploiting the information systems the U.S. installed to propagate their narrative and undercut domestic adversaries. When the U.S. builds infrastructure, it should do so with the understanding that it may very well be turned against itself and its allies at some point. Thus, any architecture installed should provide avenues for future manipulation and collection and we must be proactive in ensuring the right policies, processes, capabilities, and authorities are in place to exploit that infrastructure, when required.

9 Conclusion

In 2021, the Taliban achieved the seemingly impossible when they rapidly swept to victory despite twenty years of US and allied efforts to create a stable, secure, and unified Afghanistan. While there are multiple reasons why the Taliban succeeded, this chapter explored the underappreciated role played by modern communications technologies installed as part of the Coalition's development efforts.

To examine these effects, the chapter outlined the post-9/11 operations and their goals, then discussed the role of communications infrastructure in creating a nation. The chapter then reviewed social movement theory to examine the insurgency's impetus, discussed the evolution of the Afghan communications infrastructure, and analyzed how the Taliban turned it against the government. Finally, we provided some critical lessons learned for implementing communications infrastructure development projects in an insecure environment.

Overall, this chapter determined that the Coalition's unrealistic goals, poorly executed development programs, and use of an open communications architecture combined to create an unfavorable information environment that the Taliban exploited with great success. While it is unfortunately too late to apply these lessons to Afghanistan, if learned and instituted, they offer significant benefits for future operations.

References

1. US Army (2003) Operation Enduring Freedom: October 2001–March 2002. CMH Pub 70-83-1. https://history.army.mil/html/books/070/70-83/cmhPub_70-83.pdf
2. United States Senate Committee on Foreign Relations (2009) Exploring three strategies for Afghanistan. Senate Hearing 111–321. https://www.govinfo.gov/content/pkg/CHRG-111shrg55538/html/CHRG-111shrg55538.htm
3. Ruggiero F (2021) How to avoid another state-building failure after Afghanistan. Atlantic Council. https://www.atlanticcouncil.org/blogs/new-atlanticist/how-to-avoid-another-state-building-failure-after-afghanistan/
4. Centlivres P, Centlivres-Demont M (2000) State, national awareness and levels of identity in Afghanistan from Monarchy to Islamic State. Central Asian Surv 19(3–4):416–425
5. Chin W (2019) Technology, war and the state: past, present and future. Int Aff 95(4):765–783
6. Wafa Z (2014) National cyber security strategy of Afghanistan (NCSA). Islamic Republic of Afghanistan Ministry of Communications and IT. https://mcit.gov.af/sites/default/files/2020-08/National%20Cybersecurity%20Strategy%20of%20Afghanistan%20%28November2014%29.pdf
7. National Cyber Security Index (n.d.) "Afghanistan." National Cyber Security Index. https://ncsi.ega.ee/country/af/
8. Morning Edition (2013) Afghans confront sensitive issue of ethnicity. National Public Radio. https://www.npr.org/2013/05/08/179079930/afghans-confront-senstive-issue-of-ethnicity
9. Anderson B (2006) Imagined communities: Reflections on the origin and spread of nationalism. Verso books
10. Afghanistan Ministry of Communications and Information Technology (2017) ICT policy for Afghanistan: a digital agenda for development and social change 2018–2022 (Draft). Draft National ICT Policy—Afghanistan. https://mcit.gov.af/sites/default/files/2018-12/information%20and%20communications%20technology%20Policy%20of%20MCIT%20.pdf
11. Privacy International (2021) Afghanistan: what now after two decades of building data-intensive systems? Privacy International. https://privacyinternational.org/news-analysis/4615/afghanistan-what-now-after-two-decades-building-data-intensive-systems
12. McCarthy JD, Zald MN (1977) Resource mobilization and social movements: a partial theory. Am J Sociol 82(6):1212–1241
13. Council on Foreign Relations. (n.d.) "The Taliban." Council on Foreign Relations. https://www.cfr.org/taliban/#!/
14. Dorronsoro G (2009) The Taliban's winning strategy in Afghanistan. Carnegie Endowment for Peace. https://carnegieendowment.org/files/taliban_winning_strategy.pdf
15. Bimber B, Flanagin AJ, Stohl C (2005) Reconceptualizing collective action in the contemporary media environment. Commun Theory 15(4):365–388
16. Shirky C (2011) The political power of social media: Technology, the public sphere, and political change. Foreign affairs 28–41
17. Shapiro JN, Weidmann NB (2015) Is the phone mightier than the sword? Cellphones and insurgent violence in Iraq. Int Organ 69(2):247–274
18. Shapiro JN, Siegel DA (2015) Coordination and security: how mobile communications affect insurgency. J Peace Res 52(3):312–322
19. Bailard CS (2015) Ethnic conflict goes mobile: mobile technology's effect on the opportunities and motivations for violent collective action. J Peace Res 52(3):323–337
20. Pierskalla J, Hollenbach F (2013) Technology and collective action: the effect of cell phone coverage on political violence in Africa. Am Political Sci Rev 107(2):207–224
21. Warren T (2015) Explosive connections? Mass media, social media, and the geography of collective violence in African states. J Peace Res 52(3):297–311
22. Special Inspector General for Afghanistan Reconstruction (SIGAR) (2019). SIGAR 16-46-AR/Afghanistan's ICT Sector. SIGAR. https://www.sigar.mil/pdf/audits/SIGAR-16-46-AR.pdf

23. Islamic Republic of Afghanistan (2014) ICT Solution Providers. Afghanistan Telecom Regulatory Authority. https://web.archive.org/web/20140407111943/http://atra.gov.af/en/page/6986/8943
24. Islamic Republic of Afghanistan (2014) Telecom Statistics. Afghanistan Telecom Regulatory Authority. https://web.archive.org/web/20140407123229/http://atra.gov.af/en/page/telecom-statistics-2013
25. Islamic Republic of Afghanistan. (2014) ISPs. Afghanistan Telecom Regulatory Authority. https://web.archive.org/web/20140407112052/http://atra.gov.af/en/page/6986/6997
26. The Economic Times-Tech (2022) How the Taliban turned social media into a tool for control. The Economic Times. https://economictimes.indiatimes.com/tech/technology/how-the-taliban-turned-social-media-into-a-tool-for-control/printarticle/85511668.cms
27. Campbell E (2021) How digital rights are key to protecting Afghans under the Taliban. MEI Publications. https://www.mei.edu/publications/how-digital-rights-are-key-protecting-afghans-under-taliban
28. Culpan T (2021). How a technology revolution powered the Taliban's return. The Business Standard. https://www.tbsnews.net/features/panorama/how-technology-revolution-powered-talibans-return-291436
29. Teller Report (2021) The Taliban uses social media as a tool of threat and enticement. Teller Report. https://www.tellerreport.com/news/2021-08-28-the-taliban-uses-social-media-as-a-tool-of-threat-and-enticement.B1ZzT7Ndbt.html
30. Taneja K (2021) From 'Night Letters' to the Internet: Propaganda, the Taliban and the Afghanistan crisis. Observer Research Foundation. https://www.orfonline.org/research/from-night-letters-to-the-internet/
31. Puyosa I (2021) The Taliban's social media warfare operation toward Seizing Kabul. Toda Peace Institute. https://toda.org/global-outlook/the-talibans-social-media-warfare-operation-toward-seizing-kabul.html
32. Cowie J (2012) Could it happen in your country? Renesys. https://web.archive.org/web/20121201001304/http://www.renesys.com/blog/2012/11/could-it-happen-in-your-countr.shtml
33. Thorbecke C (2021) How the Taliban uses social media to seek legitimacy in the west, sow chaos at home. ABC News. https://abcnews.go.com/Technology/taliban-social-media-seek-legitimacy-west-sow-chaos/story?id=79500632
34. Tech Policy Press (2021) What the Taliban's victory means for access to technology in Afghanistan. Tech Policy Press, Technology and Policy. https://techpolicy.press/what-the-talibans-victory-means-for-access-to-technology-in-afghanistan/
35. Guo E (2018) Afghanistan's real internet lives on its streets. New York Magazine, Global Tech. https://nymag.com/developing/2018/10/afghanistan-sneakernet-internet-physical-file-sharing.html
36. Dominque G (2016) How the Taliban get their money. DW, Asia. https://www.dw.com/en/how-the-taliban-get-their-money/a-18995315
37. Kemp S (2021) Digital 2021: Afghanistan. Dataportal. https://datareportal.com/reports/digital-2021-afghanistan
38. Mehran W (2021) How social media helped the tech-savvy Taliban retake Afghanistan. The Guardian. https://www.theguardian.com/commentisfree/2021/aug/24/tech-savvy-taliban-afghanistan-propaganda-fundraising-recruitment
39. Winter C, Alrhmdon A, Sayed A (2021) The Taliban's vast propaganda machine has a new target. Wired. https://www.wired.co.uk/article/taliban-propaganda-news-afghanistan
40. Vale L (2021) What internet penetration, growth, and usage means for the Arab World. Albawaba, Business. https://www.albawaba.com/business/what-internet-penetration-growth-and-usage-means-arab-world-1408518
41. Xataka (2021) How the use of technology by the Taliban has changed. Tekdeeps. August 18, 2021. https://tekdeeps.com/how-the-use-of-technology-by-the-taliban-has-changed/
42. Aishani (2021) Decoding Taliban's use of social media and the anti-Taliban influencers of Afghanistan. Digital Forensic Research and Analytics Center. https://dfrac.org/en/2021/09/28/decoding-talibans-use-of-social-media-and-the-anti-taliban-influencers-of-afghanistan/

43. Iyengar R (2021) The Taliban's social media Dilemma. CNN Business. https://www.cnn.com/2021/08/29/tech/taliban-afghanistan-social-media/index.html
44. Kumar R (2021) Taliban targeting Afghanistan's crucial power, IT infrastructure. Aljazeera. https://www.aljazeera.com/news/2021/7/15/taliban-afghanistan-it-electricity-power
45. Shachtman N (2008) Taliban threatens cell towers. Wired, Security. https://www.wired.com/2008/02/in-iraq-when-th/
46. Herald D (2008) Taliban attack Afghan cell towers. The Associated Press. https://www.heraldextra.com/news/2008/mar/26/taliban-attack-afghan-cell-towers/
47. Pakzad K (2021) How the Taliban are using social media for their propaganda. Teller Report. https://www.tellerreport.com/tech/2021-09-05-how-the-taliban-are-using-social-media-for-their-propaganda.ry4s9kdGGt.html
48. Jensen B (2021) How the Taliban did it: Inside the 'operational art' of its military victory. Atlantic Council 15:2021
49. Atiq S (2021) The Taliban embrace social media: 'We too want to change perceptions'. BBC News Asia. https://www.bbc.com/news/world-asia-58466939
50. Stengel R (2021) The Taliban want you to keep your phone on. The New York Times, 23 Aug 2021. https://www.nytimes.com/2021/08/23/opinion/taliban-afghanistan-social-media.html
51. Sadr O (2018) Reviewing Taliban narratives. The Strategy Bridge. https://thestrategybridge.org/the-bridge/2018/5/7/reviewing-taliban-narratives-1
52. ABC News, BBC, and ARD (2009) Afghanistan: where things stand. ABC News, BBC, and ARD Poll. http://abcnews.go.com/images/PollingUnit/1083a1Afghanistan2009.pdf
53. Goudsouzian T (2021) The battle for narrative in Afghanistan. LeMonde Diplomatique. https://mondediplo.com/outsidein/afghanistan-narrative
54. Lamdacore (2021) How the Taliban did it: inside the 'operational art' of its military victory. Pakistani Defense, Country Watch. https://defence.pk/pdf/threads/how-the-taliban-did-it-inside-the-operational-art-of-its-military-victory.720431/
55. Bahar H (2020) Social media and disinformation in war propaganda: how Afghan government and the Taliban use Twitter. Media Asia. https://www.researchgate.net/publication/344382295_Social_media_and_disinformation_in_war_propaganda_how_Afghan_government_and_the_Taliban_use_Twitter#pfd
56. Bodetti A (2016) The Taliban's latest battlefield: Social Media. The Diplomat. https://thediplomat.com/2016/09/the-talibans-latest-battlefield-social-media/
57. Decker B, Frenkel S (2021) Taliban ramp up on social media, defying bans by the platforms. NY Times. https://www.nytimes.com/2021/08/18/technology/taliban-social-media-bans.html
58. Burgaz S (2021) Against all odds: Taliban and social media. Asia Today. https://theasiatoday.org/opinions/against-all-odds-taliban-and-social-media/
59. Sharma B (2021) How Taliban took over Afghanistan with the help of basic technology. https://www.indiatimes.com/technology/news/taliban-aghanistan-technology-skills-548057.html
60. Benveniste A (2021) Afghan journalists are 'absolutely petrified' amid the Taliban takeover. CNN Business. https://www.cnn.com/2021/08/15/media/journalists-at-risk-in-afghanistan/index.html
61. Ankersen C, Martin M (2021) The Taliban, not the West, won Afghanistan's technological war. MIT Technology Review. https://www.technologyreview.com/2021/08/23/1032459/afghanistan-taliban-war-technological-progress/
62. Sardari W (2020) Internet and Afghanistan. Int Res J Eng Technol 7(06):1158–1161
63. MaDory D (2021) What's next for the internet in Afghanistan? Kentik. https://www.kentik.com/blog/whats-next-for-the-internet-in-afghanistan/
64. Griffen A (2021) Twitter does not say it will ban Taliban posts after Afghanistan takeover—but will ensure it follows rules. Independent. https://www.independent.co.uk/tech/taliban-twitter-afghanistan-facebook-rules-b1903946.html
65. Samriddhi R (2021) Neo-Taliban turns digital: a reconquest strategy. TGP, The Geopolitics. https://thegeopolitics.com/neo-taliban-turns-digital-a-reconquest-strategy/
66. Franceschi-Bicchierai L (2021) WhatsApp can't ban the Taliban because it can't read their texts. Vice, Motherboard. https://www.vice.com/en/article/93yvy5/whatsapp-says-its-not-banning-the-taliban-because-it-cant-read-their-texts

67. The White House (1999) Blocking property and prohibiting transactions with the Taliban. The White House, Executive Order 13129. https://home.treasury.gov/system/files/126/13129.pdf
68. Cullford E, Singh K (2021) Taliban"s Afghanistan takeover presents fresh challenge for social media companies. Rueters, Technology. https://www.reuters.com/technology/talibans-afghanistan-takeover-presents-fresh-challenge-social-media-companies-2021-08-17/
69. Singer P (2001) Winning the war of words: information warfare in Afghanistan. Brookings. https://www.brookings.edu/research/winning-the-war-of-words-information-warfare-in-afghanistan/
70. Stokel-Walker C (2021) The battle for control of Afghanistan's internet. Wired. https://www.wired.co.uk/article/afghanistan-taliban-internet#:~:text=When%20the%20Taliban%20last%20ruled,having%20a%20role%20in%20society.
71. Cheng L, Diana L (2018) Protest meets party control: renegotiating social norms online in present-day China. Brookings. https://www.brookings.edu/opinions/protest-meets-party-control-renegotiating-social-norms-online-in-present-day-china/
72. Cordesman AH (2021) Learning the right lessons from the war in Iraq and Syria: archives of the key metrics from 2004–2019. CSIS. https://www.csis.org/analysis/learning-right-lessons-war-iraq-and-syria-archives-key-metrics-2004-2019

Mark Grzegorzewski serves as the Senior Academic Instructor in the Department of Strategic Intelligence and Emerging Technology at the Joint Special Operations University. He is also an Army Cyber Institute non-Resident Fellow and the Special Operations Journal books review editor. He has recently co-edited and contributed a chapter to a JSOU Press edited monograph titled "Big Data for Generals" and produced a Lawfare piece titled "Taking the Elf Off the Shelf: Why the U.S. Should Consider a Civilian Cyber Defense" and has an article with the Modern War Institute titled "In Search of Security: Understanding the Motives Behind Iran's Cyber-Enabled Influence Campaigns." Academically, Dr. Grzegorzewski interests include cyberspace and irregular warfare, cyberspace proxies, nation-state motivations in cyberspace, and developing unconventional cyberspace capabilities.

Michael Spencer is a professor at Saint Leo University where he teaches courses in Democracy, Democratic Institutions, and Historical Immigration. He is also the founder and a Board of Director for the Halcyon Institute, a new technology-based research and policy analysis think tank dedicated to the development and integration of dual use digital technologies. Prior to this endeavor he was the Director of Synergia Innovare a non-profit research company focused on identifying viable research lines for private sector and government investing. Dr. Spencer holds a Ph.D. in Government with a specialization in International Security, HMBA, and a Masters in International Management (MIM). He currently resides north of Tampa Florida with his wife and three children and a completely insane puppy affectionately named Pyscho.

Ken Brown is a former Naval Officer and Intelligence Professional with extensive experience in the Special Operations community. He currently serves with the Office of the Chief Digital and Artificial Intelligence Officer in the Department of Defense. Ken holds a Ph.D in Government with a specialization in International Relations, a Juris Doctor, and a Masters of Arts in National Security Studies. He currently lives in the metro DC area.

500 Men: An Alternate Approach to America's War in Afghanistan

John Alan Hennings

Abstract Using Major Fernando Luján's "Light Footprints: The Future of American Military Intervention" as an analytical framework, this chapter identifies the pros and cons of a counterfactual or alternate chain of events. What if America and the West had kept its post-9/11 intervention in Afghanistan small-scale, with more limited objectives? What if they had delimited the military personnel filling the "boots on the ground" to special operations forces with uniquely relevant expertise? The analysis concludes that such an approach would have avoided major pitfalls of Operation ENDURING FREEDOM, but incurred greater risk in other areas. Sustainable success would have been more possible, but a faster, far less expensive failure might have been more likely. In addition, many of the achievements of Operation ENDURING FREEDOM have effects reaching beyond the fall of Kabul. Those effects would have been lost as an opportunity cost of a smaller intervention.

Keywords Afghanistan · Light footprint · Intervention · Reconstruction · Development · Engagement · Security force assistance · Counterinsurgency · Foreign internal defense

1 Introduction

The purpose of this chapter is to consider an alternate and counterfactual chain of events regarding America's war in Afghanistan and then compare the effects of those events with what actually happened. The "What ifs" posited in this examination are twofold. First, what if America and the West had kept its post-9/11 intervention in Afghanistan small-scale, with more limited objectives? Second, what if they had delimited the military personnel filling the "boots on the ground" to those with uniquely relevant expertise?

J. A. Hennings (✉)
University of South Florida, Tampa, FL 33620, USA
e-mail: jahennings@usf.edu

A. Farhadi and A. Masys (eds.), *The Great Power Competition Volume 4*,
https://doi.org/10.1007/978-3-031-22934-3_8

2 Prologue

On 17 July 1973, General Mohammed Daoud Khan deposed his cousin, King Mohammed Zahir Shah, as the head of government in Afghanistan. Daoud Khan had previously served as Prime Minister of Afghanistan in the 1960s. At that time, he had been an enthusiastic recipient of foreign aid, including aid from the Soviet Union. As an autocratic President, he eventually distanced himself from Soviet aid and the domination that came with it. He sought closer ties with the West and with Middle Eastern powers [3, 24, 30].

In 1978, the pro-Communist People's Democratic Party of Afghanistan, or PDPA, deposed and murdered Daoud Khan. Nur Muhammad Taraki, Hafizullah Amin, and Babrak Karmal led the coup, which came to be known as the Saur Revolution. Taraki initially headed the Revolutionary Council, which the PDPA modeled on the Supreme Soviet. The PDPA attempted to remake Afghan society along Communist lines. Predictably, this led to popular dissent, then brutal repression, and finally open revolt. Fierce factionalism within the PDPA further complicated their efforts to retore order [11, 23, 26, 30].

Amin deposed Taraki in September 1979 and had him killed a few weeks later [30, 32, 34]. The senior leaders of Soviet foreign policy were alarmed by Amin's removal of their confederate. They were also fearful that Amin would soon reject their sponsorship and turn to the United States. Earlier in the year, the Taraki regime had requested Soviet intervention to restore order [7, 34]. At the time, the Soviet Union had refused the request, but they overturned that decision in light of current events and sent the 40th Army into Afghanistan on Christmas Eve, 1979. The Soviets poisoned and shot Amin and installed Babrak Karmal [4, 7, 12].

The Soviet Army now faced a burgeoning revolt. Many Afghans considered the revolt a defensive *jihad*, invoking their moral obligation to defend Islam and its people against external aggressors—aggressors who, in this case, were literal godless Communists. The U.S., Saudi Arabia, and Pakistan supported multiple militias of *mujahideen* (warriors of the jihad) to varying degrees. China and Iran also provided support. The Soviet Army attempted to depopulate the countryside. Muslims from other nations joined the mujahideen, including Osama bin Laden. Mohammad Najibullah replaced Karmal as President of Afghanistan in 1987. The Soviets and the Afghan government they installed signed peace accords with Pakistan and the U.S. in 1988. The Soviet Army left Afghanistan the following February [7, 13, 24, 30]. At this point, the Soviet Union was already beginning to fragment as republics declared independence. On Christmas Day in 1991, 13 years and a day after the Soviet invasion of Afghanistan, General Secretary Mikhail Gorbachev resigned and the Soviet Union dissolved [8]. In the meantime, the mujahideen had continued to fight the Afghan Communist government the Soviets had installed, and they toppled it in April of 1992. They then began to fight one another for control of the nation, inflicting an entirely new level of devastation on a nation already ravaged by war [7, 13, 24, 30].

With support from Pakistan's military intelligence arm, a group of religious extremists arose to quell the chaos. They called themselves the *Taliban* (literally, "seekers") [25, 40]. The Taliban took Kandahar in 1994 and Kabul in 1996. Some mujahideen militias became part of the Taliban, but others fought against them. The most effective opposition confederation was known as the Northern Alliance [13, 24, 30]. Also in 1996, Sudan expelled Osama bin Laden and his terrorist organization *al Qa'ida* ("The Base"). With the Taliban's blessing, bin Laden relocated al Qa'ida's headquarters and training camps to Afghanistan [21]. At the beginning of September 2001, the Taliban controlled approximately 90% of the territory of Afghanistan, with the Northern Alliance controlling the rest [30].

On 9 September 2001, an al Qa'ida hit squad, posing as journalists, murdered Ahmad Shah Massoud. Massoud had been the most effective military leader in the Northern Alliance and previously in the mujahideen. On 11 September 2001, another al Qa'ida unit hijacked four aircraft and used them to murder 3000 people in the United States of America. On 17 September, acting on behalf of the United States, Pakistani government officials formally asked Mullah Omar, the Taliban leader, to surrender bin Laden. Mullah Omar refused. The next day, the U.S. Congress authorized the President to use military force against those responsible for the attacks and anyone who harbored them [24]. Eight days after that (26 September), the Central Intelligence Agency inserted a team of operatives (call sign "Jawbreaker") into Afghanistan to make contact with forces opposing the Taliban [9].

3 What Really Happened: How Coalition Forces Planned and Executed the War

In mid-September, General Tommy Franks, the Commander of U.S. Central Command, directed his staff to plan for conventional American ground combat forces to deploy into Afghanistan as a follow-on to special operations forces (SOF). The final plan for Operation ENDURING FREEDOM planned for 10,000 to 12,000 American Soldiers and Marines [45].

On 7 October, the United States and the United Kingdom began bombing Al Qa'ida and Taliban targets in Afghanistan. This was the official beginning of Operation ENDURING FREEDOM, which President Bush described as the first phase in the broader war on terrorism [17, 24]. Twelve days later—less than six weeks after the 9/11 attack—the first military special operations teams crossed the border into Afghanistan. These teams consisted primarily of U.S. Army Special Forces (with embedded supporting forces such as U.S. Air Force Combat Controllers) and British Special Air Service soldiers. Once in country, the teams met with CIA officers. Those officers introduced the teams to the Afghan resistance forces with whom CIA had established (or re-established) a relationship. These included both the Northern Alliance and anti-Taliban Pashtuns in the south [2, 24, 29].

On 30 October, General Franks spoke at a press conference in Uzbekistan. He explained that the objective of the "operation in Afghanistan is the destruction of the Al-Qaeda terrorist network and the destruction of Taliban leadership which provides safe harbor for that terrorist organization." He also discussed "opening a land bridge so that we can move massive amounts of humanitarian assistance for more than seven and a half million people in Afghanistan who need it," and the as-yet-undetermined role of military forces with respect to a post-Taliban government in Afghanistan [18].

It is noteworthy that humanitarian assistance to the populace (a function of governance) and the military's relationship with the future government of Afghanistan were already under discussion. In fact, humanitarian food drops had begun even before the insertion of special operations forces, and President George W. Bush has written that it was his intent to begin stabilizing the country and helping "the Afghan people to build a free society" [15]. When the special operations teams were in place and introductions were complete, the battle for Afghanistan began in earnest. The Afghan resistance forces conducted the overwhelming bulk of the ground fighting, while the American and British special operations forces coordinated their efforts and directed air support. Under this unprecedented onslaught, Taliban rule began collapsing across Afghanistan. Bamyan, Herat, Kabul, and Jalalabad fell in November, and Kandahar, the home of the Taliban and its senior leaders, fell in early December. Many Taliban forces were still present in Afghanistan, but as insurgents and fugitives, not as a ruling power [24, 45].

During this timeframe (November–December 2001), elements of the U.S. Marine Corps, U.S. Army Rangers, U.S. Navy SEALs, and specialized U.S. counterterrorism units had all came into Afghanistan to conduct direct action missions against key objectives. For example, one mission was a raid on a compound associated with Mullah Omar; another took control of an airfield in Kandahar. These missions supported the resistance effort to retake the country from the Taliban. In some cases, they also supported larger U.S. objectives in the Global War on Terrorism. However, in addition to these strike forces, the U.S. Army's 10th Mountain Division, among others, began moving into Afghanistan and Uzbekistan [24, 45]. With this influx, control of the ground began to shift away from special operations forces [29]. They operated under the command of the Combined Forces Land Component Commander (CFLCC), a subordinate element of U.S. Central Command with its forward headquarters at Karshi-Khanabad (K2) Air Base in Uzbekistan.

Simultaneously, external powers were maneuvering to direct what happened next. In mid-November, the United Nations Security Council passed Resolution 1378. The resolution advocated a "central role" for the UN in the establishment of an interim government in Afghanistan. The plan for that interim government was written in early December—not in Afghanistan, but in Bonn, Germany, although representatives of Afghan groups were present [24]. The United Nations' Bonn Conference selected Hamid Karzai as the head of that initial post-Taliban government [45].

At the same time, many of the Taliban and their Al Qa'ida allies had ensconced themselves in caves in the Tora Bora mountains, about six miles from the Pakistan border. The Afghan militias resisting the Taliban fought a pitched battle against them.

Bin Laden is believed to have escaped the caves and entered Pakistan as the battle drew to a close in mid-December [24, 45].

A few days later (20 December), UN Security Council Resolution 1386 established the International Security Assistance Force (ISAF), to be led by the North Atlantic Treaty Organization (NATO).

The intervention took a significant escalatory step on 5 December 2001, when the United Nations (UN) talks on Afghanistan established the Bonn Agreement. Annex 1 of that agreement begins as follows: "The participants in the UN Talks on Afghanistan recognize that the responsibility for providing security and law and order throughout the country resides with the Afghans themselves." In the next sentence, with no acknowledgment of the irony, the participants "pledge their commitment to do all within their means and influence to ensure such security." The next paragraph asks the international community to help establish and train the Afghan security forces. The third paragraph asks for a UN-mandated force to establish "security for Kabul and its surrounding areas," which could be "expanded to other urban centres and other areas." The final paragraph of the annex asks the force "to assist in the rehabilitation of Afghanistan's infrastructure" [6].

On 20 December, the United Nations Security Council responded by adopting Resolution 1386. This resolution authorized the establishment "as envisaged in Annex 1 to the Bonn Agreement… for 6 months of an International Security Assistance Force [ISAF] to assist the Afghan Interim Authority in the maintenance of security in Kabul and its surrounding areas." The Bonn Agreement and UNSC Resolution 1386 also asked that other militaries withdraw from Afghanistan or submit to the authority of ISAF [6, 43].

Two days after Resolution 1386, Hamid Karzai was sworn in as head of the Afghan Interim Authority. Over time, ISAF grew into the conventional command that controlled all outside forces in the war against the Taliban, led reconstruction and development, and supported several successive national governments of Afghanistan [24]. It lasted not six months, but thirteen years under the same name, eventually being replaced by the much smaller Resolute Support Mission.

In early February, General Franks testified before Congress that the initial objectives of Operation ENDURING FREEDOM were essentially complete. He described the continuing actions of U.S. and Coalition Forces in Afghanistan as follows:

- gaining and exploiting intelligence (including detainees and sensitive sites) in order to prevent future terrorist attacks and to develop greater understanding of al Qa'ida
- locating and defeating remaining pockets and surviving leaders of Taliban, al Qa'ida fighters, and resistance to the government of Afghanistan
- investigating to confirm or deny the existence of research into or production of chemical, biological, or radiological weapons
- supporting Afghan forces as required
- conducting and supporting civil military operations.

By mid-February, CFLCC (Forward) had moved to Bagram Air Base within Afghanistan and taken the name CJTF-Mountain, solidifying conventional forces' leadership role over America's war in Afghanistan. In April 2002, President Bush announced and Congress approved a $38 billion humanitarian and reconstruction aid package, and Karzai's Interim Government became the Transitional Government of Afghanistan [24, 45].

In early March of 2002, U.S. Secretary of Defense Donald Rumsfeld publicly concurred that U.S. forces had not only completed the initial military objectives laid out in early October, but in many cases, surpassed them.

By the middle of 2002, there would be 10,000 troops in Afghanistan, an order of magnitude increase over November 2001 [1]. To both Washington leadership and General Franks—a veteran of Vietnam and DESERT STORM whose background was in artillery, infantry, and cavalry operations—12,000 personnel was still a small-scale intervention. The conventional wisdom was that one of the Soviet Union's greatest mistakes in Afghanistan was going in with a large occupying force that provoked resistance [15, 19, 20, 24, 45]. By 2003, the U.S. began implementing full-spectrum counterinsurgency, complete with governance and economic development, and they realized they were understrength for the task. At that point, the conventional wisdom shifted, and the new thinking was that it was the Soviet Army's indiscriminate attacks on the populace, not merely their force presence, that provoked the Afghans [45].

Several of the activities on General Franks' list above continued for nineteen more years. The peak force level was in mid-2011, when there were 100,000 U.S. troops (another order of magnitude) and almost 40,000 non-U.S. Coalition Forces in Afghanistan [16]. The first half of the second decade included a long, slow drawdown. Troop levels fluctuated, but after 2015, they stayed below 20,000 [35]. The first half of 2021 included a precipitous withdrawal of foreign forces and the fall of the government of Afghanistan to the Taliban.

4 The Real World Results

In the wake of 2021, which saw the defeat of Afghanistan's military, the rapid collapse of its government, and the ongoing humanitarian crisis within its borders, it is difficult to grasp and accept the scale of the good that was done in Afghanistan in the previous 20 years. This is especially true when one considers that it was all achieved during an active insurgency, when security was still an issue in many parts of the country. As seen below, the improvement across multiple indicators ranges from significant to stunning (e.g., a 185% increase in GDP).

The list of indicators below shows scores for the first and last years for which there are data, beginning in 2000. The years reflected in the data are shown in parentheses after the score. An asterisk (*) indicates that the best performance was reached prior to the last recorded year, and performance subsequently declined.

Health and sanitation		
Life expectancy at birth (years)	56 (2000)	65 (2020)

(continued)

(continued)

*Immunization, measles (% of children ages 12–23 months)	27 (2000)	66 (2020)
Maternal mortality ratio (modeled estimate, per 100,000 live births)	1450 (2000)	638 (2017)
Mortality rate, under-5 (per 1000 live births)	129 (2000)	58 (2020)
Births per 1000 women ages 15–19	154 (2000)	58 (2020)
Prevalence of stunting, height for age (% of children under 5)	59.3 (2004)	38.2 (2018)
*Prevalence of undernourishment (% pop.)	40 (2001)	26 (2019)
People using at least basic drinking water services (% pop.)	28 (2000)	75 (2020)
People using safely managed drinking water services (% pop.)	11 (2000)	28 (2020)
People using at least basic sanitation services (% pop.)	22 (2000)	50 (2020)
Economy		
*Annual GDP (constant 2015 US$)	7.23B (2002)	20.62B (2020)
*GDP per person employed (constant 2017 Purchasing Power Parity US$)	5572 (2002)	9227 (2020)
Education		
Literacy rate, adult total (% of people ages 15 and above)	31 (2011)	37 (2021)
School enrollment, primary (% gross) (Score > 100% reflects addition of out-of-age range enrollments)	21 (2000)	107 (2019)
General and pre-vocational secondary completion (% of relevant age group)	16 (2005)	58.3 (2019)
Literacy rate, youth total (% of people ages 15–24)	47 (2011)	56 (2021)
Number of girls in school (per 100 boys)	0 (2000)	64 (2020)
Energy		
Access to electricity (% pop.)	1.6 (2000)	97.7 (2020)
Access to clean fuels and technologies for cooking (%pop.)	6 (2000)	33 (2020)
Access to Information		
Individuals using the Internet (% of population)	0 (2001)	18 (2020)
*Mobile cellular subscriptions (per 100 people)	0 (2000)	58 (2020)

Even these indicators tell only some of the good news. Five million refugees returned. By the end of 2008, the U.S. Agency for International Development (USAID) had built 715 km of major highways, built or refurbished 670 health clinics, trained 10,600 health workers, constructed over 600 schools, bought more than 60 million school textbooks, and trained 65,000 teachers [24]. In 2019, 54,861 women were in college in Afghanistan. The number of students in school overall increased by a multiple of ten in the twenty years since the Taliban fell. Governance over industry and public services, physical infrastructure, energy generation, and access to health care all improved dramatically [41]. The World Bank also reported that crop and fishery production skyrocketed, as did women's participation in the economy and government.

Of course, a sizable improvement is easier to achieve when performance is so low at the start point. In 2000, Afghanistan had already experienced more than two decades of ongoing conflict, a few years of drought, and several years under the oppressive and extremely primitive rule of the Taliban. Furthermore, the accomplishments above were achieved with the support of many actors outside of Coalition Forces. These included the United Nations, national aid organizations such as the United Kingdom's Department for International Development (DFID), and international humanitarian organizations such as the International Red Cross, among many others. It also includes the diligent work of the Afghan people themselves—the most important element. Of course, one could easily argue that the Coalition Forces and their work were critical enablers to every single improvement.

Nevertheless, the positive indicators are an incomplete picture. First, the progress was almost entirely funded by external sources and consequently unsustainable. Second, as indicated by the asterisks above, many of the indicators rose to a peak and then declined. Other World Bank indicators, not shown above, stayed stagnant or even deteriorated. Participation in the labor force and percent of the labor force employed both declined. Correspondingly, the percentage of the population in poverty or experiencing food insecurity increased by some measures. As in the developing world globally, urban populations grew more quickly than cities could manage. A 2021 lessons learned report from the Special Inspector General for Afghanistan Reconstruction (SIGAR) showed that fear for personal safety actually increased over time.

The security assistance picture is a similar study of extremes. The Coalition trained, equipped and funded hundreds of thousands of troops, many of whom performed admirably in combat on multiple occasions. This author personally observed that the Afghan special operations forces (i.e., Commandos and Special Forces) were consistent high performers, as were many of the tactical police and counternarcotics forces. And they were not alone. There were heroes in every kind of unit, and the insurgency unfortunately gave them many opportunities to demonstrate their courage and ability.

But Afghan security forces were largely made in their patrons' image; they were dependent upon technology, aviation, and logistics that were not sustainable without either the Coalition itself or the eight thousand contractors it funded. Moreover, morale and loyalty plummeted in the wake of Coalition withdrawal [14]. Afghanistan is a collectivist culture. Soldiers and police were forced to calculate odds of success and consider what position was most survivable for not only themselves, but their families, clans, and tribes. Even prior to the collapse, the corruption, illiteracy, and drug abuse that were endemic to the society affected the military, also. Retention was also poor. In an environment with low employment, that meant that some of the skills sown by the Coalition would inevitably be reaped by the Taliban.

In many ways the Coalition, and especially the U.S., helped fuel its own opposition. The Taliban's ideology is antithetical to the Iranian regime, so Iran supported U.S. efforts to topple the Taliban in 2001 [19]. However, Iran was also opposed to a long term presence of U.S. forces in the region, and they eventually became a strong source of lethal aid to the Taliban insurgency [24]. In the U.S. in the early twentieth century, Prohibition exploded the market for illegal alcohol and vastly increased the resources and power of organized crime. The tremendous influx of aid money and foreign government assistance similarly poured gasoline on the fire of Afghanistan's endemic crime and corruption. The demand for near-immediate results increased the funding and decreased the oversight, with predictable effects (SIGAR 2015). That corruption severely undermined support for the new government and in many cases directly funded the insurgency [24]. While many Afghans welcomed the Coalition's emphasis on human rights, others resisted what they saw as Western attacks on their understanding of Islam and Afghan culture [42].

Most importantly, the high numbers of U.S. and other NATO forces caused the Afghans to cede leadership and responsibility for the conflict to the Coalition. In 2003, Coalition forces began a formal counterinsurgency campaign that "hinged on Coalition forces taking ownership of specific areas of operation (AOs) and conducting a mix of reconstruction, security, and information operations to win the support of the population" [45]. The Coalition actually used the phrase "area ownership" to describe each subordinate unit's responsibility for the territory, in direct contradiction to the sovereignty of the government they were there to support.

Prior to this, the relatively low troop numbers (by conventional, not light footprint standards) had not constrained the Coalition [45]. Its objectives had been more limited, and its force requirements far lower. This shift in strategy ballooned both spending and force requirements, which topped out in the eventual surge [42]. This massive investment of not only forces and funds, but national prestige, destroyed any leverage the Coalition had over the Afghans [33].

Afghanistan has a long history of exploiting outside imperial powers for their resources. They have even judged the effectiveness of their leaders by the amount of resources they can extract. Historians have observed that empires pay tribute to Afghanistan, not the other way around [5]. In this case, military and civilian Afghan officials knew that the Coalition nations had taken responsibility for outcomes in Afghanistan away from the Afghans themselves, and hence would not fail to expend forces and resources to see it succeed. At least, not for more than a decade.

To achieve these mixed results, the United States spent twenty years and $145 billion in military and non-military aid. The United States also spent $837 billion on warfighting [42]. Other nations contributed billions more. Associated Press reporter Ellen Knickmeyer created this summary of other costs:

> American service members killed in Afghanistan through April: 2448.
> U.S. contractors: 3846.
> Afghan national military and police: 66,000.
> Other allied service members, including from other NATO member states: 1144.
> Afghan civilians: 47,245.
> Taliban and other opposition fighters: 51,191.
> Aid workers: 444.
> Journalists: 72.

5 The Light Footprint Approach: Definition, Applicability, Strengths, and Limitations

In March 2013, the Center for New American Security published an article called "Light Footprints: The Future of American Intervention," by then-Major Fernando M. Luján. At the time, the author was already a U.S. Army Special Forces veteran of multiple deployments to Latin America, Iraq, and Afghanistan. He was also a foreign area specialist and a participant in the Afghanistan-Pakistan Hands program. The Hands program provided personnel an opportunity to gain expertise in the region and one or more of its languages. It then placed them in key positions working with host nation personnel (generally).

In the article, Luján points out that long-term, large-scale deployments of conventional military forces in Afghanistan and Iraq had exhausted both the force and the budget. Consequently, smaller-scale, less costly interventions had become appealing. He described these light footprint operations broadly as "a combination of air power, special operators, intelligence agents, indigenous armed groups and contractors, often leveraging relationships with allies and enabling partner militaries to take more active roles." He cites the long-term but highly cost-effective U.S. support to Colombia as a positive example. Luján makes clear, however, that the light footprint option is no panacea. It has limitations and risks, as well as strengths. I will spend the next several paragraphs summarizing his work to enable its use in this analysis.

The justification for the light footprint approach rests on two pillars. First, some security challenges cannot be resolved quickly. This is especially true of irregular conflicts against borderless, non-state actors. Second, large-scale interventions generate U.S. domestic pressure to withdraw, regardless of consequences. That pressure increases over time.

Light footprint operations as described by Luján are typically:

- Civilian-led
- Small
- Indirect
- Long-term
- Preventive

"Civilian-led" means in support of the U.S. embassy country team and under the control of the chief of mission. This is essential because in these types of conflicts, the political requirements for resolution are more critical than the military requirements. Thus, military action must be integrated with and in support of other elements of national power. A military-led response treats the symptoms with no hope of curing the disease.

"Small" means "as minimalist and non-intrusive as possible... to reduce the partner nation's dependency on U.S. resources and minimize the chances of a backlash from the local population." Luján cites examples ranging from about 50 to approximately 1000 military personnel and contractors. Even at the high end, this would be a population that could be airlifted out in a single day, reducing the chance that a partner would ever come to think of America's commitment as irrevocable.

Just as the relatively minute size of the intervention affects the partner nation, it also has the following five implications for U.S. forces:

1. The intervention force cannot mitigate risk through sheer numbers. They must be tactically adept to protect themselves and others and culturally adept to pick up on warning signs while working within a foreign environment.
2. A small force in a fluid situation far from its higher headquarters must work with a high degree of autonomy to be effective. With that autonomy comes responsibility and a requirement for maturity.
3. Necessary support elements like close air support; intelligence platforms; quick reaction or personnel recovery forces; and logistical and medical facilities must be based outside partner nation territory, yet close enough to be useful.
4. A sizable portion of the intervening force may be allies or contractors, instead of U.S. forces—perhaps even all of it.
5. Small-scale interventions have neither the personnel nor the resources to accomplish difficult tasks in multiple locations by themselves, and they cannot create new systems to do so. That leads to the next characteristic of light footprint operations.

"Indirect" means "by, with, and through" host/partner nation personnel. This relegates the American outsiders to training, mentoring, advising, and supporting the people and organizations who actually accomplish the mission. It preserves the independence and initiative of the host nation actors, which will remain important long after the intervention force has returned home. It also increases the local forces' freedom to apply solutions which are uniquely suited to their environment. Foreigners are unlikely to see these solutions, but locals can sustain them more effectively than methods or equipment imported from foreign lands. Finally (and to some, unfortunately), it significantly reduces the United States government's ability to control outcomes.

Luján caveats this point by saying that even in a light footprint intervention, there will be exceptions to the indirect approach. "If local actors are unable or unwilling to act in time," the U.S. must preserve unilateral capability to address fleeting targets and imminent threats.

As discussed previously, direct action raids and strikes to remove key adversary personnel from the battlefield were an important component of the U.S. intervention in Afghanistan. However, for light footprint operations, Luján advocates that unilateral direct action should be the exception, not the rule. Consequently, he discusses the value of these strikes and raids and their drawbacks. He again cites research literature showing that these "decapitation" missions disrupted adversary forces and reduced their attacks against U.S. forces and the host nation in the short term. However, if the adversary group weathers the storm, it becomes more resilient over time. The author further fortifies his argument that these actions should be used judiciously by citing the potential negative impact on popular support for the intervention and the supported government.

"Long-term" means accepting the fact that the problems driving the intervention are complex, and solutions are unreachable in the immediate future. The author cites research finding that conflicts of the kind that these interventions would support generally exceed ten years in length. In light footprint operations, the U.S. is generally helping a host nation develop security forces and institutions so they can address those complex problems themselves. In practice, this work requires years or decades of close working relationships. That deliberate pace prevents these four known hazards of "rushing the job":

1. An unvetted force vulnerable to infiltration, corruption, or enemy indoctrination; in particular, a rapid influx of U.S. resources greatly exacerbates the tendency toward corruption.

2. An unbalanced force full of rapidly trained infantrymen but dependent upon the U.S. for aviation, logistics, intelligence, maintenance, and medical care capabilities—all of which take longer to produce.

3. A maladapted force, "mirror imaged" after its mentors, and thus unsuited to the conditions of the environment or the circumstances of the conflict.

4. An illegitimate force lacking ethics and expertise. These qualities take multiple generations of engagement to cultivate. Without them, they cannot earn the respect of the government they serve and the populace they protect, which is vital to their operational effectiveness.

Finally, "preventive" does not mean that a disaster or crisis has not already occurred. The intervention is likely a response to such events. It means that the objectives are very limited, focused on preventing the disaster from happening again or the crisis from worsening. In fact, the long-term relational nature of the work makes it possible to create a network that is uniquely capable to "influence the situation indirectly and discreetly" for a variety of potential challenges, not just those that provoked the intervention.

For example, Luján cites a longitudinal study finding that an intervention consisting of small numbers of foreign advisors supporting an embattled government does not significantly enhance the government's chances of all-out victory over an insurgency. However, this kind of intervention substantially reduced (but did not eliminate) the chances of insurgent victory, implying that preventing defeat until a settlement can be reached is a viable goal. The author also approves of using these interventions to help contain a conflict or (relevant to Afghanistan) keep an adversary too pressured and off-balance to plan and execute attacks against the United States. Regardless, the U.S. must accept the following as it sets goals and strategies for light footprint operations.

- Changes to the host nation will be incremental, not transformational, except potentially across generations.
- The U.S. can best influence a partner through relationships and through that partner's interests, but the U.S. has very little influence over those interests. It can only work to understand them (which takes time and deliberate effort) and leverage them accordingly.

6 The Hypothetical Results

Clearly, Operation ENDURING FREEDOM began as a light footprint operation. As outlined above, in 2001, the U.S. and Allied forces in Afghanistan were largely special operations forces working closely with Afghan ground forces. The other forces conducted unilateral missions that generally fit under the exceptions with which Luján caveated his "indirect" rule. The number of U.S. forces in Afghanistan at the end of 2001 was still about 1000. But larger numbers of conventional forces began preparing to move in before the year was out. If the U.S. had planned and executed a truly light footprint approach in Afghanistan—one that matched the description Fernando Luján published in 2013—what would it have looked like? What would the results have been?

First, there would have been no ISAF. The leader of the U.S. Mission would have been an ambassador or charges d'affaires, operating as an emissary of the President with the support of the U.S. Department of State and authority over all U.S. government operations in country, including the military. The Chief of Mission would certainly have had military personnel on the country team. With an ongoing military mission in country, the Department of Defense might have augmented the Embassy staff with liaisons, planners, and intelligence personnel to coordinate with the civilian agencies and enable direct reporting. None of that approaches the size or authority of a headquarters command and staff.

Second, special operations forces would have led the military effort, as they have in many similar interventions. Luján explained that a small-scale intervention force operates with little support or security, well-removed from their military command and control, coordinating closely with the embassy country team they support. Those conditions entail personnel who can protect themselves and others and operate with a high degree of autonomy. They must also be as capable of working with a U.S. Foreign Service officer as they are a local tribal chief.

There are several special operations career fields that select and train for those conditions. Add the requirement to train and mentor foreign combat forces to counter an insurgency, and the candidates for lead force narrows to the U.S. Army Special Forces (SF), popularly known for their distinctive headgear as "Green Berets." Other forces would support the Green Berets or conduct specialized missions.

SOF who work closely with the local populace, such as SF, align regionally and train in the languages and cultures of their region. The 5th Special Forces Group (Airborne), or "5th Group", is the unit aligned with the Middle East and Central Asia. Consequently, it is the unit most likely to have personnel who speak Pashto, Dari, or Farsi—the languages which are most useful in Afghanistan. It was 5th Group Green Berets who entered Afghanistan to meet with CIA operatives and Northern Alliance forces on 19 October 2001.

Special Forces Groups are overwhelmingly male and were even more so in 2001. At the time, an SF group was composed of three operational battalions with accompanying headquarters and support units. Each battalion similarly consisted of three companies as well as headquarters and support elements, known as a Special

Forces Detachment Charlie (SFOD-C). SFOD-Cs become the core of Special Operations Task Forces when deployed. Each company consisted of approximately six Special Forces Operational Detachment-Alphas, (SFODAs, also known as ODAs or A Teams) and one Special Force Operational Detachment Bravo (SFOD-B or B Team for short). B Teams consist of approximately 3 officers and 7 non-commissioned officers providing leadership, guidance, and specialized support to the SFODAs. When deployed, B Teams function as an Advanced Operations Base.

SFODAs are the core tactical SF maneuver unit. They are 12-man teams, ideally composed of 10 non-commissioned officers (NCOS, or sergeants) trained in two or more military specialties, a warrant officer who is a former special forces NCO, and a captain in command.

Every Green Beret is a minimum three-time volunteer (Army, Airborne, and Special Forces). Most enter SF from other military career fields, adding to the versatility already designed into the team. In addition to language and culture, each man is highly trained in small unit tactics, and the team is trained and equipped to operate without additional security personnel. As independent teams, SFODAs maintain and operate advanced communications gear, provide medical care, collect and analyze intelligence, use special weapons and explosives, and run engineering projects. They have additional training in psychological operations; working with humanitarian, civil government, and civil society groups; and coordinating air support. However, they often deploy with supporting specialists in those fields to augment their capabilities and capacity.

A light footprint operation that matched then-Major Luján's description would have initially been composed of at least one battalion of 5th Group personnel, or about 3 ODBs and 18 ODAs. An additional force may have deployed into Afghanistan to perform unilateral direct action missions, or it may have based in a nearby Central Asian republic and inserted as required. With associated embedded support personnel, the number of active duty forces in country would come to (very approximately) 500 men. Additional contract support or allied forces would round out the 1000 that Luján describes as the top end.

The command and control and the rest of the support required would have been based in Pakistan or a Central Asian republic. A force like this would have been capable of working alongside Afghan forces in more than a dozen locations throughout the Texas-sized country. The Special Forces would have embedded with the Afghan forces to coordinate their operations, provide U.S. intelligence and fire support, and guide their efforts to root out Taliban and Al Qa'ida remnants. As time passed, the other battalions of 5th Group would relieve them, and perhaps another Special Forces groups would enter the rotation. The country team leadership in the U.S. Embassy would help maintain continuity of purpose across rotations. Furthermore, 5th Group personnel would return often enough to maintain their relationships with the key personnel on the Afghan side; to know the culture, the terrain, and the adversary; and to tell whether or not progress was being made. With so few forces and no overarching U.S. military headquarters in Kabul, there would have been no question that the responsibility for Afghanistan's security resided with the Afghans themselves.

What could that have gained? First, it might have avoided all the problems reviewed above that resulted from the far larger force—the support for the resistance from America's adversaries, the backlash from the portion of the populace most concerned with tradition and sovereignty, the enormous increase in corruption and criminality, and the task ownership by the Coalition forces.

Second, it would have been far, far less costly, as measured by the deployment strain on the conventional military, the number of U.S. forces killed and wounded, and the hundreds of billions spent on the war. For the real Operation ENDURING FREEDOM, simply supplying the force through some of the worst infrastructure and most forbidding terrain created an exorbitant burden [17]. It also constrained America's foreign policy options with Pakistan, Russia, and the Central Asian states to the north.

Third, the organs of government and civil society—to include security forces—would have had an opportunity to emerge organically. If they had done so, they would likely have been far more sustainable.

So, the opportunity cost of the U.S. and the Coalition's actual decisions are quite apparent in hindsight, and they are high. The opportunity cost of the light footprint option might be even worse, depending on one's priorities.

First, it might not have been acceptable to the American populace or the government. After the U.S. military took ten weeks to overthrow the Taliban (admittedly with Afghans doing most of the fighting), the American populace's confidence in that military was very high [17]. Most seemed to expect that the military could capture or destroy the remaining Taliban and Al Qa'ida resistance and make Afghanistan a stable, democratic republic that was no longer a terrorist safe haven within a reasonable time and cost. Many believed and some still believe it was our duty to do so [33].

This evokes memory of Secretary of State Colin Powell's warning before the invasion of Iraq in 2003 of "once you break it, you are going to own it," colloquially known as the Pottery Barn rule [10]. The obvious problem with that reasoning, in parlance any elementary school child would understand, is that Afghanistan was already broken when we got there. By September 2001, Afghanistan was utterly destitute, in a state of ongoing and epic humanitarian crisis, and headed into winter [24, 45]. Nevertheless, leaving them largely in that state, providing only the aid that the Afghans themselves could absorb and distribute, might well have been unacceptable to the American populace [17]. It would have required America to accept some tragedy and consciously decide not to do everything it could to prevent it.

Second, the interim government and its successors might have failed. In the past, when there was not a clear king, members of the royal family or regional warlords battled for the throne [5]. This history of internecine warfare lasted longer than the throne itself. The latest iteration was after the Communist government fell in 1992, and many of those men and their militias were part of the resistance against the Taliban. The Pashtuns are the ethnic group that has traditionally led Afghanistan. Hamid Karzai was selected partly for his legitimacy as the leader of a high status tribe of Pashtuns, but that was no guarantee [5, 24, 33, 45].

It is not clear whether U.S. backing would have been strong enough to prevent challenges without a significant presence on the ground. Another civil war would have been seen internationally and domestically as the result of American intervention, and it would have given extremists promising order an opportunity just as the last one had. The United Nations and the government of Japan supported a major Coalition effort in the early 2000s to disarm, demobilize, and reintegrate the militias in the early 2000s [24]. This may have been key to preventing such a scenario.

Even Luján made the point in his article that interventions do better when there are existing institutions to support. In Afghanistan those institutions no longer existed. The Coalition was attempting to restore order to a place that had not seen order since the Communist party took control in 1978. According to Luján, that may have been possible through a light footprint operation *if* the intervening force could have leveraged common ground between the local armed actors. Without attempting it, there is no way to know for sure if enough common ground existed. Without that common ground, Luján admits that if a crisis posed a direct threat to U.S. national security, it could justify a larger intervention to provide temporary security while institutions develop.

Third, a light footprint operation might not have prevented another major terrorist attack on the American homeland. It would be simplistic and inaccurate to credit the absence of a major terrorist attack in the two decades since 9–11 to Operation ENDURING FREEDOM alone. Increased security in the homeland, aggressive investigations by law enforcement, a global reordering of intelligence priorities, and U.S. and partner nation operations worldwide all played a part in that success. However, it is likely that investigations begun in Afghanistan and the denial of a safe haven in what had previously been key terrain for Islamist extremists played an important role. In a 2002 press briefing, U.S. Secretary of Defense Donald Rumsfeld put it simply: "A terrorist under fire in the mountains of Afghanistan is a terrorist who has bigger problems than trying to plan the next attack on the United States" [17]. A light footprint operation in Afghanistan would have attempted to put that exact same pressure on the terrorists, but indirectly, through Afghan partners, which may or may not have been enough.

Finally, just as there was an opportunity cost to a large-scale intervention, there would have been an opportunity cost to a light footprint. Many Afghans would have died at a far younger age. Others would never have been born. Those that survived would have far less education, economic opportunity, and exposure to the greater world around them.

Most of the hundreds of thousands of Afghan immigrants and refugees who have made their home in the West would likely never have left Afghanistan. The developed world, and especially the U.S., has benefited from the intelligence, drive, and patriotism of immigrants, and especially refugees. They are a source of language skill and cultural knowledge for the military, the Intelligence Community, and the Foreign Service. They are also a source of labor, entrepreneurs, and innovation for the economy. The Afghans here in America show every sign of continuing that trend. Both within Afghanistan and beyond its borders, we cannot yet estimate the impact these healthier, more educated, more empowered Afghans will have on the world.

In addition to the human opportunity cost, there is a direct cost to innovation and expertise. The cost–benefit ratio is too high for it ever to be a valid justification to start a war, but armed conflict spins off new ideas like the space program [22, 27, 36–38]. The war in Afghanistan has resulted in advances in trauma medicine, prosthetics, body armor, language translation, vehicles, weapons systems, human development, civil engineering, and tactics for the application of all these things. Many of these advances, such as those in small drones, have already had application far beyond the military. Some of these innovations would have occurred in a light footprint operation and perhaps others would have occurred which we will never see. However, it is unlikely that a smaller intervention would have motivated such a high volume of innovation.

7 Conclusion

The rapid, transformational change the U.S. sought in Afghanistan was unachievable at any cost. A much, much smaller intervention, with far more limited objectives, executed over an even longer period of time, would have avoided many of the pitfalls of a larger intervention. It might have achieved a sustainable equilibrium in Afghanistan—an equilibrium that would have prevented Afghanistan from once again becoming a safe haven for transnational terrorists.

However, there is no guarantee that opportunity would have born fruit. It would have incurred significant risks and might have failed far faster than the real Operation ENDURING FREEDOM did. The augur of small-scale intervention, Fernando Luján, specified that in a light footprint operation, failure must be maintained as an option, because only the host nation forces (and of course the adversary) can determine the final outcome. Afghanistan might have become a renewed terrorist safe haven far sooner.

Furthermore, the dire situation in Afghanistan and its potential effects on U.S. national security may have justified a larger intervention, although for a much shorter period than actually occurred. Luján's vision of a light footprint intervention tops out at a 1000 personnel—usually a mix of U.S. service members, allied personnel, and contracted civilian support. The institutions of Afghanistan's civil society, government, and economy were completely devastated in 2001, and more than a thousand personnel may have been necessary to get started.

Regardless, even if a light footprint intervention had failed, it would have done so at a tiny fraction of the cost in blood, treasure, and credibility that the failures of Operations ENDURING FREEDOM and FREEDOM'S SENTINEL incurred. And, as Luján pointed out, escalation would always have been an option.

The author thanks Carl D. Baker III, U.S. Army Special Forces Sergeant Major, Retired, for his expert review.

References

1. Al Jazeera (2021) Timeline: how September 11, 2001 led to US's longest war. https://www.aljazeera.com/news/2021/9/6/timeline-how-september-11-2001-led-to-uss-longest-war. Accessed 23 April 2022
2. American Special Ops (n.d.) Jawbreaker—CIA Special Activities Division. http://www.americanspecialops.com/cia-special-operations/jawbreaker/. Accessed 30 May 2022
3. Arnold A (1985) Afghanistan: the Soviet invasion in perspective. Hoover Press, Stanford
4. Baker P (2019) Why did Soviets invade Afghanistan? documents offer history lesson for Trump. The New York Times. https://www.nytimes.com/2019/01/29/us/politics/afghanistan-trump-soviet-union.html. Accessed 17 May 2022
5. Barfield TJ (2010) Afghanistan: a cultural and political history. Princeton University Press, Princeton
6. Bonn Conference (2001) Agreement on provisional arrangements in Afghanistan pending the reestablishment of permanent government institutions. United Nations. Published by International Committee of the Red Cross. https://ihl-databases.icrc.org/ihl-nat/a24d1cf3344e99934125673e00508142/4ef7a08878a00fe5c12571140032e471/$FILE/BONN%20AGREEMENT.pdf. Accessed 5 June 2022
7. BBC News (2009) Timeline: Soviet war in Afghanistan. http://news.bbc.co.uk/2/hi/south_asia/7883532.stm. Accessed 17 May 2022
8. BBC News (2013) Soviet Union timeline. https://www.bbc.com/news/world-europe-17858981. Accessed 17 May 2022
9. Berntsen G, Pezzullo R (2005) Jawbreaker: the attack on Bin Laden and Al-Qaeda—a personal account by the CIA's key field commander. Crown, New York
10. Betz D (2012) Counter-insurgency, Victorian style. Survival 54 no.4
11. Bradsher HS (1983) Afghanistan and the Soviet Union. Duke Press Policy Studies, Durham
12. Braithwaite R (2011) Afgantsy: the Russians in Afghanistan 1979–1989. Oxford University Press, Oxford
13. Britannica, The Editors of Encyclopaedia (2021) Afghan war. Encyclopedia Britannica. https://www.britannica.com/event/Afghan-War. Accessed 17 May 2022
14. Boot M (2021) How the Afghan Army collapsed under the Taliban's pressure. Council on Foreign Relations 16
15. Bush GW (2010) Decision points. Crown Publishing Group, New York
16. Carati A (2014) No easy way out: origins and consequences of NATO's failure in Afghanistan. Paper presented at the 13th Annual Conference of the Transatlantic Studies Association, University of Ghent, Ghent, 8 July 2014
17. Clement JG (2003) Operation Enduring Freedom as an enabling campaign in the war on terrorism. School of Advanced Military Studies, Army Command and General Staff College, Fort Leavenworth
18. Franks TR (2001) Press conference with U.S. General Tommy Franks, Commander-in-Chief of the U.S. Central Command (CENTCOM) Tashkent, Uzbekistan; October 30, 2001. The Avalon Project. https://avalon.law.yale.edu/sept11/dod_brief62.asp. Accessed 22 May 2022
19. Franks TR (2002) Testimony during Senate Hearing 107–801, Operation Enduring Freedom hearing before the Committee on Armed Services United States Senate, One Hundred Seventh Congress, Second Session. Government Printing Office, Washington
20. Franks TR, McConnell M (2004) American soldier. Regan Books, New York
21. Frontline (2002) Timeline: Al Qaeda's Global Context. https://www.pbs.org/wgbh/frontline/article/timeline-al-qaedas-global-context/. Accessed 17 May 2022
22. Gabriel RA (2013) Between flesh and steel: a history of military medicine from the middle ages to the war in Afghanistan. Potomac Books, Inc
23. Goodson LP (2011) Afghanistan's endless war. University of Washington Press, Seattle
24. Hooker RD, Collins JJ (eds) (2015) Lessons encountered: learning from the long war. National Defense University Press, Washington

25. Jones OB (2003) Pakistan: eye of the storm. Yale University Press, New Haven
26. Kaplan RD (1990) Soldiers of God: with Islamic warriors in Afghanistan and Pakistan. Knopf Doubleday Publishing Group, New York
27. Kay S (2013) America's Sputnik moments. Survival 55(2)
28. Knickmeyer E (2021) Costs of the Afghanistan war, in lives and dollars. Associated Press, 17 Aug 2021
29. Krivdo ME CJSOTF-A (Combined Joint Special Operations Task Force-Afghanistan): A Short History 2002–2014. Veritas J Army Spec Oper Hist 12(2). U.S. Army Special Operations Command
30. Lempesis J (2016) The civil war in Afghanistan: an overview. Mount Holyoke University International Relations Department. https://www.mtholyoke.edu/acad/intrel/afp/afghanoverview.htm. Accessed 23 Apr 2022
31. Luján FM (2013) Light footprints: the future of American military intervention. Center for a New American Security, Washington
32. Malik H (2016) Soviet-pakistan relations and post-soviet dynamics, 1947–92. Springer, New York
33. Miller L (2021) fghanistan 2001–2021: U.S. Policy Lessons Learned. Crisis Group, 18 Nov 2021. https://www.crisisgroup.org/asia/south-asia/afghanistan/afghanistan-2001-2021-us-policy-lessons-learned. Accessed 21 May 2022
34. National Security Archive (2019) The Soviet invasion of Afghanistan, 1979: not Trump's terrorists, nor Zbig's warm water ports. https://nsarchive.gwu.edu/briefing-book/afghanistan-russia-programs/2019-01-29/soviet-invasion-afghanistan-1979-not-trumps-terrorists-nor-zbigs-warm-water-ports. Accessed 23 Apr 2022
35. North Atlantic Treaty Organization (NATO) (2021) RSM placemats archive. https://www.nato.int/cps/en/natolive/107995.htm. Accessed 30 May 2022
36. Okur C (2013) The effect of defense R&D expenditures on military capability and technological spillover. Graduate School Of Engineering and Management, Air Force Institute of Technology, Wright-Patterson AFB
37. Ritchie HA (2016) Institutional innovation and change in value chain development: negotiating tradition, power and fragility in Afghanistan. Routledge.
38. Robbins MD (2015) Towards a political economy of military spending (Doctoral dissertation, Carleton University)
39. Rumsfeld DH (2002) Press conference with U.S. General Tommy Franks, Pentagon, Virginia, 6 Mar 2002. The Avalon Project. https://avalon.law.yale.edu/sept11/dod_brief152.asp. Accessed 8 June 2022
40. Shaffer B (2006) The limits of culture: Islam and foreign policy. MIT Press, Cambridge
41. Shah MQ (2021) What did billions in aid to Afghanistan accomplish? 5 questions answered. In The Conversation. https://theconversation.com/what-did-billions-in-aid-to-afghanistan-accomplish-5-questions-answered-166804. Accessed on 21 May 2022
42. Special Inspector General for Afghanistan Reconstruction (SIGAR) (2021) What we need to learn: lessons from twenty years of Afghanistan reconstruction. Government Printing Office, Washington
43. United Nations Security Council (UNSC) (2001) Resolution 1386. United Nations. N0170855.pdf (un.org). Accessed 5 June 2022
44. World Bank (2020) Afghanistan, Data. https://data.worldbank.org/country/afghanistan. Accessed 30 May 2022
45. Wright DP, Connors PW, Bird JR, Clay SE, Van Wey DF, Garcia LC, Farquhar SC (2009) A different kind of war: the United States Army in Operation ENDURING FREEDOM, October 2001–September 2005. Government Printing Office, Washington

John Alan Hennings is a retired U.S. Air Force lieutenant colonel with several combat deployments, including three in Afghanistan. He has worked closely with Afghan security forces and special operations forces from across the Coalition. John is an avid lifelong student. He completed a doctorate of business administration at the University of South Florida in 2020. His dissertation and other academic research have been focused on capturing the lived experiences of people working cross-culturally and how it changes them, as well as discovering processes for doing it well. He also enjoys assessing risk, building consensus, analyzing processes, identifying root causes, and collaboratively crafting solutions.

Reflections on the Fateful Collapse of the Afghan National Defense and Security Forces

Ali A. Jalali

Abstract The rapid collapse of the Afghan army in the face of the Taliban blitz in August, which contributed to the sudden fall of the Afghan government on August 15, 2021, took both Afghans, including the Taliban, and the world by surprise. A closer look at the situation, however, reveals a more complex story—for one thing, the collapse of the Afghan security forces was not abrupt, but it was a slow, and painful deterioration that began long before the fall of Kabul. According to recent assessments, neither the United States or the Afghan governments showed political commitment to deal with the challenges, including devoting the time and resources necessary to develop a professional Afghan National Defense and Security (ANDSF) which required a multi-generational effort. Further, with little attention paid to the traditional security culture of the Afghan society, international mentors and trainers helped create a force made in the image of the United States' own military that was hardly sustainable following the withdrawal of the U.S. and NATO forces. This Chapter looks at the ANDSF's collapse in the context of socio-political traditions and security culture of the Afghan society along with the shifting political context where the partner nations' changing assumptions and priorities impacted the ANDSF's mission, size, structure, capacity, and sustainability at various times.

Keywords Afghan Army · Fall of Kabul · ANDSF · US withdrawal · Afghanistan · Afghan police · Taliban · Afghan government · Force sustainability · Taliban Emirate · Tribal allegiance · Opposing narratives

1 Historical Background

Few Afghan armies have successfully monopolized the legitimate use of force. Traditionally, the army had rarely been the only military institution within a social system imbued with military pluralism. The country traditionally relied on popular uprisings to fight foreign invasions and enlisted the aid of tribal levies to beef up the regular

A. A. Jalali (✉)
Near East South Asia Center for Strategic Studies, National Defense University, Washington, DC 20319, USA
e-mail: aajalali@hotmail.com

A. Farhadi and A. Masys (eds.), *The Great Power Competition Volume 4*,
https://doi.org/10.1007/978-3-031-22934-3_9

army to crush domestic rebellions. The situation reflects the evolving nature of state-society relations since the emergence of Afghanistan as a modern state at the end of the nineteenth century [1]. It was then a loose conglomerate of tribes and ethnic communities over which the central government had varying degrees of control at various times. Until the middle of the twentieth century, the central government in Afghanistan was not strong enough to integrate the nation through a wide network of political and economic institutions. Society remained segmented and immobilized.

The lack of integration made the communities, particularly in tribal areas, semi-independent, mostly relying on their own resources and their own traditional institutions. This included local military forces that were mobilized during inter-tribal conflicts or foreign threats. The tribal militias also could be mustered in support of or against the central government during domestic disturbances. This nation-in-arms capability helped the country survive when the central government collapsed, or the state army disintegrated in the face of foreign invasion.

The primacy of tribal and local loyalty among the soldiers impaired the army's commitment to the government cause. The state army was often crudely organized and led, inadequately armed, poorly trained, meagerly paid, and badly fed. Such an army was hardly capable of standing firm in the face of a determined foe. However, the same soldiers would fight with utmost determination with their kinsmen in their own space under the leadership of their local chiefs. A British observer of the Afghan society, Edward Hensman, wrote in 1881:

> The Afghan does not lack native courage, and in hill warfare he is unrivaled, so long as it takes the shape of guerrilla fighting. But once he is asked to sink his identity and to become merely a unit in a battalion, he loses all self-confidence and is apt to think more of getting away than of stubbornly holding his ground as he would have done with his own friends led by his own chief [2].

In such a socio-political environment, the state armies were faced with two major challenges: creating national loyalty among the soldiers that would surpass their tribal allegiance; and providing the military units with the skills to fight effectively in both counterinsurgencies and conventional wars. The response to both challenges was slow and unsteady. The pace of progress was linked to social and political development, expansion of government influence through economic modernization, and the availability of resources to increase the army's professional effectiveness.

Many attempts by the Afghan leaders and outside partners have been made in the past to reduce the reliance of the army on local militia or foreign supporters with mixed results. The British occupiers in the nineteenth century and the Soviets in the 20th took measures to "Afghanize" the security through building formal state institutions to enhance their indigenous capacity. In 1840 the British occupiers raised state army and levies to balance off the influence of tribal chiefs. The effort, however failed to create a viable counterbalance to the chiefs. Nor did it improve the ability of the state to sway the mostly independent tribal lords. Instead, the hasty change alienated the chiefs without winning the devotion of the paid militia to the king or to their British officers. There were two known ways for the puppet Afghan king Shah Shuja and his British supporters to rule the country: through cooperation of the chiefs or destroying their power by an alternative non-tribal force. The latter was only possible by permanent occupation of the country by a significantly large British force which was neither feasible nor acceptable for its prohibitive costs and the risks of facing a drawn-out war with the tribes. A similar "Afghanization" policy was initiated by the Soviet occupiers in 1980–1989 with attempts to broaden the popular base of the Moscow-backed Kabul regime through a policy of reconciliation and coopting tribal militia and some Mujahedin field commanders [3].

While the inconsistent and poorly resourced practices failed to ensure the army's self-reliance and resourcefulness the ad-hoc policy reinvigorated the militia culture with a long-term influence following the departure of foreign supporters of the Afghan government. It also provided perennial challenges to the authority of the government army. Following the first Anglo-Afghan war (1839–1842), it took the Amir of Kabul, Dost Mohammad, the whole length of his 20-year rule to reduce and control the power of local strongmen and militia leaders who fought against the British occupiers during the wars and expected to be rewarded in power sharing-which would weaken the authority of formal state institutions [4].

Following the second Anglo-Afghan war (1878–1880), it took Abdur Rahman twelve years to rein-in many power contenders who were real and pretended heroes of the war. "Every priest, mullah, and chief of every tribe and village considered himself an independent king," he wrote, "So long as they were the rulers, the King could not do justice in the country" [5]. The Amir did not spare even those national leaders who had led popular resistance against the British invasion when these same leaders opposed him.

Modernization of the Afghan army that began in the first quarter of the twentieth century was a slow and incremental process. The national army became more attached to the government and acquired a solid institutional identity as the country underwent a process of integration through a nationwide education system, economic progress, and political development. However, whatever stability that was achieved during the 50 years of peace (1929–1978) under a legitimate government was gradually unraveled after the Soviet invasion and the pursuant civil war (1978–2001) which disintegrated the regular Afghan army and caused its degeneration into factional militias. The key question faced by the builders of a new Afghan army in 2002 was whether to construct a new military from scratch or to transform existing factional militiamen into government soldiers.

The empowerment of non-statutory forces following the Soviet invasion (1979–1989) was more complicated. Politicization of the anti-Soviet resistance split social communities along ideological lines while ideologies failed to resist ethnic divisions. Internally, traditional society was split along different party lines (both Communists and Mujahedin factions) and externally it became polarized through affiliation with the international competition of the Cold War system. This resulted in factionalization of the resistance and the emergence of new local leaders with access to outside resources who challenged the authority of any post-Soviet state security institutions.

2 Political Context of Building State Institutions in Afghanistan

In May 2002, when American Green Berets began training the first group of Afghan soldiers for the new Afghan National Army (ANA), I published an article in *Parameters* (Autumn 2002). I noted that this was the fourth time in 150 years of Afghanistan's history that the country was recreating its military establishment after its total disintegration caused by foreign invasions or civil wars [6].

In all previous cases, it took the consistent effort of decades, not years, under stable and legitimate governments with strong leadership to turn tribal militiamen into government soldiers through a process of integration of multi-ethnic recruits into a cohesive professional organization.

The process was always influenced by prevailing political and social conditions. Building a national army in 2002 had to be intertwined with the creation of a legitimate government, creation of relevant state institutions, economic reconstruction, and demobilization of factional militias. All these efforts were hard to coordinate and synchronize while fighting terrorism and a growing insurgency at the same time.

Few reconstruction tasks have proved more difficult than building indigenous security forces in war-devastated Afghanistan. While state security institutions were destroyed during the war numerous factional militias and non-state armed groups emerged because of an extended period of foreign invasion (1979–1989) and civil war (1992–2001) involving internal armed factions with extensive foreign links. The Taliban regime defeated many warring factions but it too failed to establish a viable state and state security institutions. Other challenges included the lack of basic infrastructure, low economic capability, the prevalence of a culture of corruption, the primacy of factional loyalties and a massive illiteracy problem. Meanwhile the need to build national security forces while fighting terrorism and insurgency generated competing demands that were hard to reconcile.

The political context where the development of ANDSF took place was shaped by several factors including the chaotic post conflict-situation inside the country; the nature of international military intervention and its policies to rebuild the post-Taliban Afghanistan while fighting a counterterrorism/insurgency war; and available resources, institutional capacities, and ways they were used.

Domestically, the breakdown of central authority, during more than two decades of conflict and violence, stimulated a socio-political transformation that vitalized non-state patronage networks under the leadership of regional commanders who often invoked ethnic references to legitimize their leadership. The country became politically fragmented, economically bankrupt, socially atomized, and regionally the vortex of proxy wars waged by rival regional powers who fiercely competed for influence in war-torn Afghanistan. Afghanistan, one of the poorest countries in the world had suffered heavy institutional, economic, social, and political destruction during a lengthy period of war and violence with little resources, poor infrastructure, and a rough and land-locked geography. This situation fueled ethno-regional competition for power and resource distribution.

Meanwhile, the 2001 military invasion of Afghanistan by the United States and its allies, amidst an ongoing civil war in Afghanistan, was not primarily aimed at fixing the failed Afghan state through military action and stability operations. It was an accidental war with unforeseen consequences. Had the 9/11 terrorist attacks in the United States by-Taliban-hosted al-Qaeda not happened, the U.S. military move against the Taliban and al-Qaeda in Afghanistan would be hardly imaginable. The U.S.-led military invasion targeted one side of the civil war (Taliban regime) and its in-country support network (al-Qaeda) in close alliance with the second party in the civil war. In contrast to the Soviet invasion of Afghanistan in 1979, the U.S. operation was launched with minimum American and allied ground troops helping the local anti-Taliban militia forces (represented mainly by the Northern Alliance) to conduct the ground attack supported by the coalition military and technological edge [7].

The co-option of resurrected anti-Taliban Afghan militia in the campaign inevitably empowered them after the fall of the Taliban. This regime was removed from power but not decisively defeated or reconciled and still had the potential to return to the battlefield. Nor was the al-Qaeda fully defeated, although its leadership and network were driven out of Afghanistan across the border into difficult-to-access tribal areas of Pakistan. The situation in the country looked like a post-conflict environment but in fact it was a more complex situation. Consequently, the fall of the Taliban regime, was not a transition from war to peace or from conflict to post-conflict situation, but a new phase in the long-standing conflict. Many al-Qaeda terrorists and Taliban forces who had snuck across the border into Pakistan regrouped there and established new bases for cross-border attacks.

Finally, the false assumption that the conflict was over, and victory was achieved had a profound effect on the establishment and development of Afghanistan security forces. The defining factors in determining the size and capacity of the indigenous security forces were based on the assessment of short-term political and security conditions ignoring the potential of strategic changes in the area and the dynamics of ongoing conflict. Several issues clouded the long-term vision for institution building when, on May 1, 2003, U.S. Defense Secretary Donald Rumsfeld announced that major combat in Afghanistan was ended.[1] The official word on the end of combat was mostly motivated by political considerations rather than the military situation on the ground.

Meanwhile, competing demands of fighting terrorism and building democracy hampered institution building. Enlisting the militia commanders and warlords as coalition allies in fighting terrorism hindered the development of formal democratic institutions since they were not interested in the development of formal state institutions. Aiding the non-state partners in counter-terrorism effort empowered them at the expense of formal state institutions. Consequently, the development of the Afghan National Defense and Security Forces (ANDSF) experienced fluid political and security conditions as it progressed under shifting political context causing changing assumptions and priorities that impacted the army's mission, size, structure, and capacity at various times. As a recent Special Inspector General for Afghanistan

[1] Rumsfeld: Major combat over in Afghanistan, CNN.com. World, Thursday, May 1, 2003.

Reconstruction (SIGAR) report noted, "it was not a consistent 20-year effort but twenty, one year reconstruction efforts" [8].

2.1 A National Institution or an Attachment to Outside Partners?

In the past 20 years, the Afghan National Security Forces (ANSF) had come a long way transforming from an odd assortment of factional militias into modern security institutions with professional capacity and political loyalty to a unified state. It was a long journey over a bumpy road entailing significant achievements and avoidable failures with lasting impacts that shape the final status of the country's security institutions—A force of immense capability to face ongoing security challenges while still bearing the consequences of many years of narrow strategic focus, institutional confusion, underfunding, and uncoordinated development approaches.

The United States, the main sponsor of the effort, visualized the project to serve two main purposes:

- To build an effective alternative to the expansion of international security forces to police the war-devastated country.
- To augment the multilateral struggle against terrorist activity in the region by indigenous forces

Nether vision totally materialized. The growing insurgency forced a gradual surge in the number of US and NATO troops, while the U.S. military's aid and reliance on factional militias in the counterterrorism effort empowered them at the expense of formal state institutions including the ANDSF. These rival factional militias were integrated into the government system but continued to respond only to their faction leaders, often instigating turf battles at the expense of public security.

The framework for international assistance to build Afghan National Army (ANA) and Afghan National Police (ANP), known as the Security Sector Reform (SSR) was formally established at a security donors' conference in Geneva in April 2002. The reform program consisted of five pillars, each supported by a different donor state: military reform (US); police reform (Germany); the disarmament, demobilization, and reintegration of ex-combatants or DDR (Japan); judicial reform (Italy); and counter-narcotics (UK). While all these pillars of reform and development were interconnected, they were pursued by different actors with varying levels of commitment, resources, priorities, and procedures. It was mostly a "burden sharing" arrangement than an integrated development program. "No country or agency had ownership of the ANDSF development mission. Instead, ownership existed within a NATO-led coalition and with temporary organizations" [9]. The development process was assisted by constantly changing staff and rotating military and civilian advisors. Such a constant personnel turnover hampered continuity and institutional memory. "The result was an uncoordinated approach that plagued the entire mission" [9].

Meanwhile, the persistent insurgency led to a strategy where the United States opted to "militarize the police as a localized defense force" rather than encourage a law enforcement organization. On the one hand, assumptions that Afghanistan faced a post conflict status, Germany, the main supporter of police reform planned for a long-term professional development program that would take years to reach fruition. On the other hand, the United States police assistance which was motivated by the need to deal with immediate security, "prioritized rapidly increasing the quantity of police officers in the ANP over the quality and sustainability of police training. This resulted in poorly trained police being sent into communities. For example, DOD pushed to increase the ANP force strength from 62,000 to over 120,000 police, while hastily deploying poorly trained local auxiliary forces to fight on the front lines" [10]. Meanwhile, the legacy of the Soviet-led reconstruction of the Afghan security institutions in 1980s, which resulted in over militarization of the police force which strongly influenced the US sponsored rebuilding of the ANP.

One argument that I often heard from our international partners was the assumption that training of police is less costly that soldiers when the former is used as fighters. In fact the development of police as professional law enforcement elements and protection of the population was much costlier and much longer-drawn than training the soldiers. Consequently, the fighting police deployed in thousands of checkpoints failed to win the trust of the population by protecting the citizens through community policing. Instead, the poorly trained police in many cases turned into corrupt and abusing elements. The deterioration of law and order across government-controlled territory perpetuated the culture of impunity. Such behavior created a gap between the police and the population which in many cases was filled by the insurgents.

Special police units, such as the Afghan National Civil Order Police and the Border Police, proved themselves able to operate effectively, but, as a whole, the policing sector—formal units and informal militia—remained plagued by corruption and abuse of authority [11]. Faced by rising insurgency the military needed to targets the insurgents and terrorists to achieve immediate security, while the police had to target the population to attain long-term security through winning the cooperation of the public. In both cases, it was essential that any use of force should have served a political purpose aimed at delegitimizing the insurgents and legitimating the government. Consequently, at the end of the day both the police and ANA had been configured to fight an internal insurgency. However, the ANA became larger than it needed to be to protect the country, and the police had been militarized in a way that made it largely unfit for law enforcement, especially in rural areas [11, pp. 37, 38].

Other factors that hindered progress and affected capacity building included insufficient donor investment, political influence of non-state armed groups and warlords and insufficient reform in the ministries of defense and the interior which were haunted by the influence of factional interest groups not committed to reform. The absence of shared vision and strategy in the security sector reform blurred the distinction between means and ends. Means too often defined goals, tactics too often drove strategy, supply too often determined demands, and short-term necessities too often took precedence over long-term priorities [12]. The Chairman of the US Joint Chiefs of Staff, Admiral Michael Mullen, precisely summarized the challenges facing the international community in Afghanistan when he told the US Congress in 2007 that "in Iraq, we do what we must... In Afghanistan, we do what we can" [13].

A recent SIGAR assessment finds the length of the U.S. commitment to build the ANDSF was "disconnected from a realistic understanding of the time required to build a self-sustaining security sector—a process that took decades to achieve in South Korea. Constantly changing and politically driven milestones for U.S. engagement undermined its ability to set realistic goals for building a capable and self-sustaining military and police force." The assessment also notes that The U.S. approach to reconstructing the ANDSF lacked the political will to dedicate the time and resources necessary to reconstruct an entire security sector in a war-torn and impoverished country. As a result, the U.S. created an ANDSF that could not operate independently, milestones for ANDSF capability development were unrealistic, and the eventual collapse of the ANDSF's was predictable."[2]

More importantly, the ANA was built in the image of a western military institution mostly as an attachment to the coalition forces rather than a sustainable indigenous force. The key features of traditional Afghan military culture include fierce loyalty to national/tribal cause, faith in strong leadership, and ability to operate with agility and offensive action. A military too dependent on US enablers' support, had incrementally lost its self-confidence to operate independently. It often saw U.S. air support as an indispensable element to fight against the enemy.

While the lightly equipped Taliban followed the Afghans traditional dynamic way of war, the ANDSF, in contrast, was trained as a force not to win but not to lose. Deployed in thousands of small security outposts in a defensive posture, the units were exposed to insurgents' attacks who chose the time and place of their strikes. Consequently, major sweeping counterinsurgency operations had to be conducted by forces brought temporarily to the affected areas only to be reclaimed by the insurgents after the striking forces left. The ANDSF showed an elevated level of professional effectiveness when they took the initiative to act aggressively against insurgents. But only 15% of ANA (Special forces corps) was trained and equipped to achieve this capability. History has taught more than once that, in an insurgency, the counterinsurgent could lose-if it fails to win. Hence, continued government loss of territory to the insurgents over time is a clear indicator of failing to win.

While the Taliban based their narrative on fighting a foreign invasion, masking their own dependence on foreign support, their sanctuaries in Pakistan and ties with notorious global terrorist networks, the Afghan government failed to counter the narrative by discrediting the insurgents for their proven failed policies and extremist ideology connected to global terrorism. The counter narrative insufficiently focused on how the Afghans overthrew the Taliban previous regime with the help of foreign forces who were willingly received as rescuers.

Consequently, the ANA, remained dependent on partner countries' support, both professionally and psychologically. Until the 2012 transition, the size, composition and mission of the ANA was shaped mostly by the requirements for U.S. strategy

[2] SIGAR 22-22-IP [9], p. 30.

of fighting terrorism rather than by the exigencies of creating a self-reliant Afghan national army. Even the fast-paced numerical force generation of the ANDSF during the transition period, left little time to develop capabilities of the ANA to operate independently.

The race to add battalions and police units as part of the exit strategy left the ANDSF with limited maneuverability, fire power, aviation support, intelligence capacity, logistic capacity, and command and control aptitude. These limitations were major liabilities. Plainly, the main goal of the security transition was meeting the deadline set for the exit strategy rather than building a self-reliant indigenous defense and security operation. Consequently, despite its major achievements under extremely trying conditions, the ANDSF continued to depend on international assistance.

In an early 2021 net assessment of the ANDSF versus the Taliban, looking at the size, material resources, external support, force employment, and cohesion, the Taliban forces were seen in a slightly advantageous position. The study concluded that while the ANDSF were of larger size and enjoyed vast technical superiority they were almost entirely reliant on external powers for sustainment [14]. An earlier assessment suggested that "without a peace deal, the further withdrawal of U.S. forces... will likely shift the balance of power in favor of the Taliban" [15]. In the words of one observer "It is a tale of two armies, one poorly equipped but highly motivated ideologically, and the other nominally well-equipped, but dependent on NATO support, poorly led and riddled with corruption" [16].

3 The Dynamics of Civil-Military Leadership

The ANDSF were not built in a vacuum. The domestic political context strongly influenced the security forces' cohesiveness and commitment to mission. A fractured government plagued by corruption, interest groups' predation, nepotism and leadership incompetence undermined morale, motivation, and faith in the leadership of a force that otherwise fought bravely. Afghan government legitimacy was helped more by strong international financial backing, particularly U.S. support of Kabul, than by the fraudulent Afghanistan presidential elections.

One of the main problems was the lack of nationally oriented leaders that were competent in managing and coordinating national security affairs. For more than eight months before the fall of the government, the Minister of Defense was out of the country for medical reasons. The ministry was on autopilot run by a deputy, a decent man, but lacking any professional security education or experience. As noted by the bipartisan Afghanistan Study Group assessment, despite the significant efforts by Afghans and the international community to build state institutions and create democratic accountability, by becoming "the Afghan state's lifeline" the international community has "paradoxically undermined its objective of enabling an effective social compact between the government and its citizens. Instead, as every Afghan presidential election since 2004 has shown, the international community, and the

United States in particular, have become major, albeit often ineffective, local political actors" [17]. The pattern consistently influenced the leadership of the Afghan state where political leaders failed to develop a workable national security strategy that could assume responsibility for nationwide security following the withdrawal of U.S. forces. The Afghan leaders for long had misperceptions about the withdrawal of U.S. forces and counted on open-ended foreign support to ensure the survival of the regime.[3] Even when Washington announced the calendar of American forces' pullout, President Ashraf Ghani continued to assume that the U.S. government did not intend to leave.

Further, internal division within the government led to fragmentation of authority and uncoordinated operation of the army and police. "Institutions are bypassed as personal relations become the means of communication on policy issues."[4] Micro-management of the ANDSF by the top leadership and an endemic cronyism undermined institutional development of the defense and security establishment. Meanwhile the restrained delegation of command responsibilities down the chain of command disincentivized the field commanders to take bold decisions and initiate timely actions in the interest of combat agility. The problem faced by the civilian authority, as Samuel Huntington noted, is to decide on the ideal amount of control. Too much control over the military could result in a force too weak to defend the nation, resulting in failure on the battlefield. Too little control would create the possibility of failure of the government [18].

The ANDSF's low space-to-force ratio resulted in an overextension of forces, impeding operational capability and tactical agility, and often precluding holding those areas cleared of enemy forces. ANDSF operations primarily consisted of covering population areas through a network of security posts. While the insurgents acted with tactical agility, choosing the time and place for hitting individual government posts, the ANDSF fought with little tactical cooperation between the checkpoints network and often with no immediately available quick reaction forces to join the battle.

During the 2015 fighting season, the network of isolated ANA and ANP security posts suffered heavily. Many of the posts of ten to fifteen men were attacked by often dozens of well-armed militants.[5] The Afghan Uniform Police particularly suffered in this regard, because they were often deployed in small teams in remote areas without adequate combat means, supplies, and training. In some cases, control of the area was contested by local strongmen connected to influential officials in Kabul. According to a Pentagon assessment, as of September 2015, the ANP devoted more than half of its total end strength of approximately 147,000 to checkpoints and fixed sites. ANP leaders seemed reluctant to consolidate force due to the influential politicians' demand for police presence in all communities. The absence of checkpoints even in areas that could not be guarded against insurgents were often perceived as giving up space to the Taliban. The deployment of ANA and ANP units to a wide network of

[3] I (author) learned this from many discussions I had with top Afghan leaders since 2014.

[4] Author's interview with General Helaluddin Helal, Kabul, July 2019.

[5] Author's interview with several ANDSF officers and political leaders, Kabul, August 2015.

fixed security posts and checkpoints hampered maneuverability and impeded force concentration against the enemy at the right time and place [19].

Plans to change the ANDSF operational posture from being defensive to offensive and enhancing their resilience and steady improvement to achieve more than a stalemate against the insurgency across the country, were not fully implemented. Such a shift that could significantly avoid overextension and improve the space-to-force ratio was ignored for political reasons. Instead of prioritizing areas of troops concentration, the obsession of a politically inspired need to have "government presence everywhere" weakened the ANDSF control as it had to hold every district and village with small contingents. Seeking to be strong everywhere caused weakening everywhere. The number of committed insurgents in Afghanistan was often five times less than that of the ANDSF, and the latter were better trained and equipped than insurgent forces. The Taliban advantages were their agility and choice of time and space to hit government targets.

The ANDSF fought hard under difficult command and logistics conditions and suffered heavily. Some estimates indicate that up to October 2018 60,000 Afghan soldiers and police have been killed, close to 40,000 civilians have lost their lives in the violence [20]. In January 2019, Afghan President Ashraf Ghani said more than 45,000 members of the country's security forces had been killed since he became leader in late 2014. But corruption remained a "key vulnerability" in ANDSF combat power and combat readiness. The theft of resources like fuel, funded through the Afghan Security Forces Fund, underscored this vulnerability. Corrupt ANDSF officials at all institutional levels degraded security, force readiness, and overall capabilities. The high-level corruption that some ANDSF leaders exhibited promoted lower-level corruption. In Afghanistan, the culture of impunity started at the top and then normalized extortion, embezzlement, fraud, and other abusive behaviors within the entire system.[6]

Additionally, poor personnel management, uneven training system and poor logistic support, took a heavy toll on troops morale, force retention and combat effectiveness. Rates of desertion differed between regions and time of year but were highest during fighting seasons and in the most insecure or remote areas. The main reasons cited for desertion include poor leadership, insufficient and untimely pay, difficulties accessing pay, the absence or misunderstanding of leave policies, constant combat deployments with little or no leave or training rotations, the lack of casualty care, and inadequate living and working conditions. "Poor leadership, fatigue, lack of training, and hunger were driving the high attrition rate within the Afghan security forces."[7]

Consequently, as a DoD report asserts, "The U.S. and Afghan governments share in the blame. Neither side had the political commitment to doing what it would take to address the challenges, including devoting the time and resources necessary to develop a professional ANDSF, a multi-generational process. In essence, U.S. and Afghan efforts to cultivate an effective and sustainable security assistance sector were

---

[6] SIGAR 22-22-IP [9], pp. 31, 32.

[7] General John Campbell speech at the Brookings Institution, Washington, August 5, 2015.

likely to fail from the beginning. The February 2020 decision to conduct a rapid U.S. military withdrawal sealed the ANDSF's fate."[8]

4 Monumental Shifts in Political-Strategic Context—Beginning of the End

The US-Taliban agreement on February 29, 2020, and President Biden's April 14 announcement to end US military presence in Afghanistan by September 11, put a process in motion that led to the August debacle. A U.S. government report finds that the single most crucial factor in the Afghan National Defense and Security Forces' collapse in August 2021 was the decision by two U.S. presidents to withdraw U.S. military and contractors from Afghanistan, while Afghan forces remained unable to sustain themselves [9, p. 6].

The U.S.-Taliban agreement introduced tremendous uncertainty into the U.S.-Afghan relationship. Many of its provisions were not public but are believed to be contained in secret written and verbal agreements between U.S. and Taliban envoys. Some U.S. analysts believe that one classified annex detailed the Taliban's counterterrorism commitments, while a second classified annex detailed U.S. and Taliban restrictions on fighting. The Doha agreement was predicated on the Taliban guarantees to prevent the use of the Afghan soil by any terrorist groups against the United States and its allies; and enter into peace talks with the Afghan government, in exchange for an American commitment to withdraw its forces from Afghanistan within 14 months.[9] Confident that withdrawal of foreign troops would increase their chances of military victory, the Taliban never showed serious interest in honoring their commitments but only tried to run the clock. Meanwhile, the United States leverage to respond to Taliban lapses incrementally weakened as its military forces continued to withdraw. Further, the legitimacy that the Taliban gained in the Doha process enabled them to win new supporters in the region through increasing diplomatic efforts aimed at improving their image as a nationalist force not interested in expanding their war beyond Afghanistan borders.

After the signing of the U.S.-Taliban agreement, the U.S. military changed its level of military support leaving the ANDSF without a critically important force multiplier: U.S. airstrikes. According to a SIGAR report, in 2019, the United States conducted 7423 airstrikes, the most since 2009 while in 2020, the U.S. conducted only 1631 airstrikes, with almost half occurring in the two months prior to the U.S.-Taliban agreement. The report notes that a former commander of Afghanistan's Joint

[8] SIGAR 22-22-IP [9], p. 30.

[9] Agreement for Bringing Peace to Afghanistan between the Islamic Emirate of Afghanistan which is not recognized by the United States as a state and is known as the Taliban and the United States of America, February 29, 2020, which corresponds to Rajab 5, 1441 on the Hijri Lunar calendar.

Special Operations Command told SIGAR that "overnight… 98% of U.S. airstrikes had ceased."[10]

The Taliban gain in Doha became the loss of the Afghan government and its army which were demoralized, amid a sense of abandonment, which undermined their will to fight with the same fervor of the past 18 years. While ANDSF forces were limited to a defensive posture, the Taliban took advantage of its freedom of movement to launch an undeclared offensive targeting vulnerable ANDSF supply lines. To make thing worse, the Afghan government failed to make strategic preparations to face the challenges of full withdrawal of the foreign forces from the country. They just did not believe the US and NATO had the intention of ending their military presence in Afghanistan. This was what I personally heard from them during my conversation with them during the recent years.

A calendar withdrawal of US forces from Afghanistan was not expected to bring peace but a political settlement based on the stated peace process could create favorable conditions for withdrawal of foreign forces. A political settlement, nor a calendar date, therefore, was to define the withdrawal - Withdrawal not a condition for peace but an end for bringing sustainable peace. There has been a plethora of debates in favor of and against the withdrawal of U.S. forces before achieving a workable peace deal among the Afghans. Despite many indications that show Americans did not want to stay in Afghanistan, the way troops were pulled out, and its immediate disastrous consequences have drawn enormous criticism and disapproval. The Biden Administration has argued that the withdrawal of US troops was agreed upon by the Trump administration and became a political commitment to end the 'for-ever wars." Delaying the withdrawal, it suggested, would cause renewed fighting by the Taliban against US forces that would need deployment of more troops in Afghanistan for no different results. Further, Washington argued that the purpose for what the U.S. forces intervened in Afghanistan was achieved and could be followed up without physical presence in the country.

One often asked question is whether a delay in troops withdrawal could facilitate creation of better political and security conditions on the ground for a non-violent transition to an intra-Afghan political settlement? In its final report, the Congressional mandated Afghanistan Study Group (ASG) suggested that "A responsible and coherent set of U.S. actions could greatly increase the chances of a peaceful resolution to forty years of conflict; a rash and rushed approach could increase the chances of a breakdown of order in Afghanistan and a worsening of this long and tragic war with negative consequences for the region and heightened threats to the security and interests of the United States and its allies." A calendar withdrawal, the report noted "will likely lead to a collapse of the Afghan state and a possible renewed civil war" [21].

The ASG report suggested that for a reasonable delay of withdrawal "the Biden administration can make the case that there has been insufficient time for these negotiations to create the hoped-for conditions under which international military forces could leave Afghanistan by May (2021), as envisaged in the Doha agreement.

[10] SIGAR 22-22-IP [9], pp. 10, 11.

A withdrawal would not only leave America more vulnerable to terrorist threats; it would also have catastrophic effects in Afghanistan and the region that would not be in the interest of any of the key actors, including the Taliban." Obviously, managing the delay necessitated initiation of an active regional diplomacy "to midwife an Afghan political agreement, guaranteed by the international community with U.S. backing, that sufficiently holds in abeyance the region's various and overlapping security dilemmas to allow space and confidence to transform negative competition into positive cooperation or, at least, neutral coexistence" [21, p. 43]

With the retrograde of the US forces (almost completed in July), which also meant departure of the NATO contingents and up to 18,000 contractors, many of whom were providing essential services to the Afghan defense forces, including up to 100% maintenance of the Afghan air force, the capacity of Afghan air assets in running bombing missions and supporting the ground forces and facilitating the operation of the elite special forces significantly degraded. A leading American observer of Afghanistan, Jonathan Schroder, noted "contrary to popular perceptions, in many cases and places, the ANDSF fought valiantly to defend the country. It is true that in the immediate aftermath of the United States beginning its withdrawal, some ANDSF units deserted or surrendered without a fight. But across the country in the months that followed, ANDSF members fought and died in battles against the Taliban" [22].

But the fate of the government was sealed months ago. As the U.S. forces were packing up, the Taliban, with stepped up material support of Pakistan, rushed to the final phase of a military victory. Complication of supply and replenishment of provincial military bases due to the control of roads by the Taliban left ANA and police running out of ammunition and other supplies. finally tribal and community leaders made local peace deals between the advancing Taliban and demoralized and isolated army and police posts. The secrecy around U.S.-Taliban negotiations and the Doha agreement meant there was a lack of official information for the ANDSF. Taliban propaganda "weaponized that vacuum" against local commanders and elders by claiming the Taliban had a secret deal with the United States for certain districts or provinces to be surrendered to them. In a space of ten days virtually all provincial capitals fell to the Taliban with little or no fighting. The flight of the leadership of the Ghani government on August 15, opened the gates of Kabul to the victorious Taliban.

5 The Outlook

The violent return of the Taliban to power ushered in an uncertain future in Afghanistan with potential impact on the region. The country will have to recreate the disrupted state institutions—the fifth time in the past less than two centuries. Given their low governance experience and their international isolation the Taliban lack the capacity to deal with the challenges they face. It was easy to overthrow a weak and divided government but hard to form an effective replacement of the fallen administration that would gain domestic and international legitimacy.

Domestically, the legitimacy of the Taliban government is being challenged by the country's political circles and civil society at large. The exclusiveness of the government, its poor administration, and failure to address the country's deepening economic crisis became the source of growing public rejection of the regime. Without achieving domestic legitimacy, the current Taliban government would have little chance of getting international recognition. The situation will deepen the ongoing humanitarian crisis and further isolate the country.

Meanwhile, the transition from the mentality of fighting to the attitude of running a responsible government is hindered by many constraints. One is the politico-ideological divisions within the movement, the other is the issue of deradicalization of their ranks and files who were stimulated during the war by hostility toward the very socio-political trends they now need to embrace in support of running an effective government. They created so many enemies in the minds of their fighters that inhibit their policies as a government. Their reluctance to widen the base of the government, allow women to work and girls to go to school is caused by their years of preaching that such activities were un-Islamic. They are not ready to lose hardline members by making immediate drastic changes in their wartime policies.

All these trends create doubts in the region on whether the Taliban are ready, capable, and willing to prevent foreign jihadist groups from becoming a threat to the region and beyond. The internal political and ideological disputes amongst the Taliban factions can grow wider. Potential competition for the new leadership between different factions may lead to another internal crisis with the possibility of some dissidents joining the Da'ish and other jihadist groups.

With regards to the military, Afghanistan is again facing the challenge of rebuilding a national army. Although the Taliban have populated the hollow shells of former ANDSF units, the existing military network is neither inclusive nor it is united. The force is composed of several factions who have been fighting and attached to various regions. After spending 20 years fighting as an insurgency, the group finds itself grappling with providing security and delivering on its promised commitment of law and order. The challenge is how the Taliban will defend their administration and civilians by a force trained in rural insurgency war. It is uncertain whether the Taliban will be able to design a new military system through voluntary recruitment or a draft system. An army takes its legitimacy and sustainability from the legitimacy and acceptability of a government. The Taliban government is yet to achieve such legitimacy at home and wide international recognition in the world.

References

1. Jalali AA (2002) Rebuilding Afghanistan national army, Parameters, pp 72–73
2. Hensman E (1881) The Second Afghan War, London, p 329
3. The Russian General Staff (2002) The Soviet Afghan War—how a superpower fought and lost (trans and eds:Grau LW, Gress MA). University Press of Kansas, pp 18–28
4. Jalali (2017) A military history of Afghanistan from the Great Game to the Global War on Terror. University Press of Kansas, pp 156, 173
5. Khan MSM (ed) (1900) The life of Abdur Rahman, the Amir of Afghanistan,vol I. John Murray, London, p 217
6. Jalali AA (2002) Rebuilding Afghanistan national army, Parameters, pp 72–86
7. Jalali AA (2016) Afghanistan national defense and security forces: mission, challenges and sustainability, USIP, p 6
8. SIGAR (2021) What we need to learn: lessons from twenty years of Afghanistan reconstruction, p VIII
9. SIGAR 22–22-IP (2022) Evaluation report, Special Inspector General for Afghanistan Reconstruction (SIGAR), Interim Report, Collapse of the Afghan national defense and security forces: an assessment of the factors that led to its demise, p 23
10. SIGAR (2022) Police in conflict: lessons from U.S. experience in Afghanistan, p xix
11. Afghanistan Study Group Final Report (2022) A pathway for peace in Afghanistan, United States Institute of Peace, p 36
12. Jalali AA (2009) Winning in Afghanistan, Parameters, p 6
13. Gilmore GJ (2007) American forces press, 11 Dec 2007. U.S. Department of Defense
14. Schroden J (2021) Afghanistan's security forces versus the Taliban: a net assessment, CST Sentinel, Combating Terrorism Center at West Point, vol 14, issue 1
15. Jones SG (2020) Afghanistan's future emirate? The Taliban and the struggle for Afghanistan, CTC Sentenal, November/December 2020, vol 13, issue 11
16. Wintour P (2021) A tale of two armies: why Afghan forces proved no match for the Taliban, The Guardian, 15 Aug 2021
17. Afghanistan Study Group Final Report (2021) A pathway for peace in Afghanistan, United States Institute of Peace, p 30
18. Nix DE (2012) American civil-military relations: Samuel P. Huntington and the political dimensions of military professionalism, Naval War College Review, vol 65, no 2
19. Jalali AA (2016) Afghanistan national defense and security forces: mission, challenges and sustainability, USIP, p 22
20. The Watson Institute at Brown University, Human Costs of the Post-9/11 Wars: Lethality and the Need for Transparency (2018)
21. Afghanistan Study Group Final Report (2021) A pathway for peace in Afghanistan, United States Institute of Peace, p 45
22. Schroden J (2021) Lessons from the collapse of Afghanistan's security forces, CST Sentinel, Combating Terrorism Center at West Point, vol 14, issue 8

Ali A Jalali is an Afghan politician, diplomat, and academic. Jalali served as the Minister of Interior from January 2003 to September 2005. He has also been a distinguished professor at the Near East South Asia Center for Strategic Studies at the National Defense University in Washington, D.C.

Restoring and Relinquishing Women's Rights in Afghanistan: Fundamental but Fragile Gains

Bahar Jalali

Abstract The U.S. led military intervention in Afghanistan following the 9/11 attacks in the United States was a catalyst for restoring the rights of Afghan women. As such, the international community played an outsized role in the conception, promotion, and implementation of women's rights programming. This had both positive and negative consequences for Afghan women. At the start of the war, Afghan women's suffering under the Taliban was strategically invoked to garner support for the invasion of Afghanistan. This led to the depiction of Afghan women as archetypal victims whose liberation via Western military rescue garnered geopolitical and geoeconomic currency that often benefitted privileged international workers and elite Afghans instead of the targeted beneficiaries—ordinary Afghan women. Foreign donors overwhelmingly dictated the terms of women's programming without the input of the intended recipients. At the same time, novel approaches and mechanisms were introduced to address gender-based violence including the Elimination of Violence Against Women (EVAW) law, special prosecution units, and women's shelters which gave Afghan women access to new outlets for bringing cases of domestic violence and related issues to the judicial system and finding refuge from abuse. The role of the international community in pushing through domestic reforms around women's rights was decisive. When confronted with domestic opposition to reforms in the area of women's rights, the intervention of Western embassies in Kabul proved critical. While the international community successfully lobbied on behalf of pro-women Afghan activists, the decision by the United States to partner with former jihadis, strongmen, and warlords to remove the Taliban from power led to the re-empowerment of domestic actors who often served as formidable opponents against initiatives to promote women's rights. This made the fundamental gains in women's rights fragile and heavily dependent on continued international support.

Keywords Women · Gender · Reform · War · International community

---

B. Jalali (✉)
Loyal University Maryland, Baltimore, MD 21210, USA
e-mail: bjalali@loyola.edu

Department of History, Loyola University Maryland, Baltimore, MD, USA

A. Farhadi and A. Masys (eds.), *The Great Power Competition Volume 4*,
https://doi.org/10.1007/978-3-031-22934-3_10

The Taliban's swift seizure of power in mid-August 2021 has reversed two decades of progress in Afghan women's rights which initially served as a strategic focal point of the U.S. led military intervention in Afghanistan. Among other restrictive measures, the Taliban have banned teenage girls from attending school as Afghan women bear the brunt of the second iteration of the Taliban regime. Today, Afghanistan is the only country in the world where adolescent girls are deprived of an education. Despite some protests against these draconian decrees by brave Afghan women, the international community, which has issued some verbal condemnations, continues to engage with the Taliban in a manner that normalizes the abnormal. The shift in policy from invoking women's rights in support of the U.S. invasion of Afghanistan to abandoning the cause of gender mainstreaming for the sake of ending America's longest war has had a momentous impact on Afghan women.

Wartime gender advocacy featured prominently in the U.S. led military intervention in Afghanistan following the al-Qaeda terrorist attacks on September 11, 2001, in the United States. Couched in the rhetoric of a 'savior complex,' the liberation of Afghan women from the brutally misogynistic edicts of the Taliban became a cause célèbre of the war. Graphic images of suffering, burqa-clad women, consigned to a life of abject poverty and ensnared in Taliban tyranny were a fixture in global media coverage of the war and served as popular wartime motifs. The restoration of Afghan women's rights following the ouster of the Taliban from power by the U.S. and coalition forces was hailed as a centerpiece achievement of America's engagement in Afghanistan. However, the United States relinquished its earlier resolve to promote women's empowerment when the Trump administration brokered a deal with the Taliban on February 29, 2020, as it bypassed the Afghan government and the very system it had spawned, to end America's longest war. The abrupt withdrawal of U.S. troops and return to power by the Taliban in August 2021 has undone significant, if uneven, progress with respect to women's rights. As the Taliban resurrect their erstwhile gender regime, it comes on the heels of two decades of international efforts to restore the rights of Afghan women.

This Chapter looks at how the critical issue of women's rights became inextricably linked to, and dependent on, foreign support during America's two-decade war in Afghanistan. The way the international community supported and funded programs for women had both positive and negative outcomes. On the one hand, American and NATO intervention facilitated the re-emergence of Afghan women in public life. Novel approaches and mechanisms were introduced in addressing gender-based violence with the support of foreign donors whose leverage often proved indispensable in countering domestic obstacles. Substantial improvements in education and health occurred. A new generation of young Afghan women attained professional mobility and accessed gainful opportunities made possible by the surge in donor funding. On the other hand, the hyper-politicized rhetoric of 'saving Afghan women,' created a victim archetype that acquired geopolitical and geo-economic currency. This, in turn, led to exploitative projects benefitting privileged international workers and local elites instead of reaching a critical mass of Afghan women. Donor driven programs and projects were designed with the view to 'emancipate' women from oppressive sociocultural norms via quick-impact projects in unrealistic and impractical ways without considering historical realities, social context, practical needs, and desires of Afghan women [30]. Other programs focused only on quantifiable outputs that did not have the potential to translate into long-term growth [12, pp. 86–91]. Moreover, the decision by the United States to partner with discredited warlords and jihadi figures who had a track record of imposing restrictions on women's rights led to the empowerment of domestic actors who would prove to be formidable opponents against advancing legislation against gender-based violence. Uneven levels of progress created a chasm between urban women who were the primary beneficiaries of developmental aid and their rural counterparts who were often victims of U.S. military operations.

1 Historical Background

The world today is mostly familiar with the image of Afghan women as archetypal victims. This is largely a consequence of how the deprivation of Afghan women's rights under the Taliban regime was appropriated and amplified when the U.S. invaded Afghanistan in October 2001. However, Afghanistan's antebellum history includes indigenous efforts to address women's rights—both radically and incrementally—and deserves mention to counterbalance popular misperceptions of the country as a place where gender repressive regimes are the norm and promoting women's rights is alien to its culture. In her meticulously researched book, *The Pitfalls of Protection: Gender, Violence and Power in Afghanistan*, Torunn Wimpelmann points out that gender relations in Afghanistan are not determined by religion, culture, or other fixed social attributes but are contingent on situational politics [39, p. 49].

The history of formal rights for women in Afghanistan can be traced to the reign of Amir Abdur Rahman (1880–1901), who is considered the founder of the modern Afghan state [21, p. 7]. He passed numerous laws that sought to bring women and family under closer government regulation. All marriages had to be registered with the authorities, a maximum limit was placed on bride-price, underage (prepuberty) marriage (when against the will of the girl) and levirate (marrying a widow to her deceased husband's brother) were prohibited. Amir Abdur Rahman also supported inheritance and divorce rights for women in line with the provisions of Islamic law. Another of his reformist measures included a law permitting women to sue their husbands for alimony or divorce in cases involving cruelty or lack of support [16, pp. 138–139]. His modernist grandson, King Amanullah (1919–1929), would go much further in enacting legal reforms in the Family Code of 1921, which undertook to regulate marriages and engagements. He established the first schools for girls and sent the first group of Afghan women abroad to study. During his reign, King Amanullah promulgated 140 regulations called *nizamnamas*, as well as Afghanistan's first Constitution. Among the *nizamnamas* were several versions of a new marriage code, first published in 1920. A more exhaustive version was published in 1923, requiring the registration of all marriages. Polygamy was made subject to the court's approval, and marriages in which the bride had not yet reached puberty were banned outright [27, pp. 66–91]. These were radical steps in comparison to later legislation in the decades that followed.

Many of King Amanullah's female family members, including his wife, Queen Soraya, were instrumental in his reform program, which went beyond legal transformation of women's public and private roles. Queen Soraya gave public speeches calling on Afghan women to educate themselves so they could serve their newly independent nation, and she managed the establishment of the girls' schools, a government-published women's magazine, and a new association through which women could petition for lawful treatment by their husbands [23, p. 94]. King Amanullah failed in rapidly modernizing Afghanistan as he was overthrown in 1929. However, he left a durable legacy that his successors modified while some of his reforms survived his immediate downfall. Among his major legacies was the establishment of formal education for girls.

After a brief interregnum, King Amanullah was replaced with Nadir Shah, who inaugurated the Musahiban dynasty (1929–1978). The reign of the Musahiban family witnessed a more gradual and selective approach to developing Afghanistan which resulted in a slow but steady path towards modernization ([31], pp. 97–98). In 1959, the government of prime minister Daoud abolished mandatory veiling [9, pp. 531–532]. King Zaher Shah, who was a figurehead ruler upon ascension to throne in 1933, finally exercised actual power when he inaugurated the Decade of Democracy which lasted from 1963 to 1973. A liberal Constitution, which served as the touchstone for the birth of democracy in Afghanistan, was passed in 1964 and granted women the right to vote [35, pp. 561–562]. In 1965, four women were elected to parliament—Raqia Abu Bakr, Masuma Ismati Wardak, Khadija Herari, and Anahita Ratibzad [24, pp. 323–327]. The first woman to hold a cabinet post, Kubra Nurzai, was appointed as Minister of Public Health during the same period [15, p. 7]. Moreover, the 1964 Constitution established an independent Supreme Court in Kabul, with authority to review all lower-level decisions as well as administrative power over the courts [35, p. 594]. The 1964 Constitution introduced another new institution, that of the attorney general to investigate and prosecute crimes. The attorney general's office was to be independent of the executive power of the government, reporting only to the executive. The judicial branch was not to interfere in its activities. Compared to prior legal provisions, the 1964 Constitution favored statutory law over Islamic jurisprudence. According to some observers, this essentially made Afghanistan a secular state, even while paying lip service to Islam [9, p. 583].

By the late 1960s and early 1970s, urban women had been entering the workforce and higher education in growing numbers for more than three decades. Initially working as teachers, nurses, and secretaries, women were employed in public administration, at the universities, in parliament, and in the courts, though they were few and rarely held high positions. While social reforms were state sponsored initiatives rather than resulting from women's mobilization, by the late 1960s, a few events foreshadowed the more substantial mobilization of women during the period of communist rule. For example, in 1968, hundreds of women participated in street protests a proposal by conservative members of parliament to prohibit unmarried women from studying abroad (Zulfacar, cited in Wimpelmann, p. 49). In 1970, several thousand women demonstrated in front of government ministries after a series of assaults on schoolgirls by a man opposed to the nascent changes to women's position in urban areas (Emadi, quoted in Wimpelmann [39, p. 35]).

In 1973, Prince Daoud ended the Decade of Democracy and abolished the monarchy by overthrowing his cousin, King Zaher Shah, in a military coup. A republic was proclaimed, and political liberalization was terminated. Although the country returned to a more authoritarian style of governance, Daoud's second stint in power (he was previously prime minister form 1953 to 1963) witnessed developments with regards to women's legal status and protection. In 1975, separate family courts were established in Kabul, Kunduz, and Kandahar to deal with issues related to family law. They were intended to give women and female judges easier physical access to the courts. In 1976 and 1977, Daoud enacted new civil and criminal codes by decree respectively. The 1977 Civil Code abolished child marriage for girls under the age of fifteen, introduced some restrictions on polygamy, and special provisions for divorce, which had previously been regulated by reference to the Hanafi school of Islamic law [20, p. 152].

The end of the Old Regime in the bloody coup of 1978 brought the Afghan Communists, People's Democratic Party of Afghanistan (PDPA), to power. The new regime, seeking to transform Afghanistan socially and economically along socialist lines, issued a notorious ruling called Decree Number 7, Dowry and Marriage Expenses. It banned underage and forced marriage, as well as excessive wedding celebrations and dowries, and it imposed a punishment of six months to three years in prison for violators. This decree was highly provocative to religious and tribal groups and caused a public uproar that galvanized support for the movement of Islamist resistance to communist rule known as the Mujahedin (Saikal 2006, pp. 188–189).

Despite a systematic campaign to revoke the membership of polygamous husbands and women who had married as second wives, there was overall limited progress under the PDPA with regards to how the justice system treated women. There is little indication that the family courts took on revolutionary or reformist positions during communist rule. Most of the caseload dealt with divorces which the courts did not grant without the husband's consent, even when there was evidence of domestic violence. Despite this enduring conservatism, women who worked as justice officials at the time remember a justice system much less interested in pursuing moral crimes in comparison to subsequent regimes. No women were prosecuted for running away from home, as they would be during the Mujahedin, Taliban, and Karzai governments. Adultery was prosecuted only with caution [39, p. 37].

The collapse of the PDPA brought the Islamist Mujahedin to power which had been supported by the United States against the Soviet-backed Afghan communist regime. After seizing control of Kabul in April 1992, an assorted group of Mujahedin leaders formed a government and proclaimed Afghanistan as an Islamic republic for the first time in history. The new government of Burhanuddin Rabbani issued an edict calling for the veiling of women, prohibiting women from working in offices and radio and television stations, and closure of schools for women (which were labeled as centers of debauchery and immorality). This was followed by a Supreme Court decree in 1994 which stated that women should not leave their homes unless absolutely necessary and should not wear attractive or revealing clothing [25, p. 71]. It was during the Mujahedin era that the practice of incarcerating "runaway" women first began. As Kabul fell to infighting among various Mujahedin factions and a civil war broke out, there was little opportunity to carry out judicial administration of any sort [39, p. 38].

The anarchy and decentralization of the civil war era gave rise to the Taliban, who quickly took control of most of the country and made their extreme version of Islamic justice and order the bedrock of their claim to legitimacy. The Taliban was unlike other political movements seeking to impose authority in Afghanistan. This was due to the exclusively clerical origin of its leaders and the refugee origins of its followers, as well. The bulk of its rank and file were born and bred in Pakistan who had never seen Afghanistan or experienced life there. Educated in the conservative madrasas of Pakistan, Taliban policy represented a counter-modernity vision rather than a return to the past. The Taliban's Islamic Emirate of Afghanistan saw itself as implementing an Islamic order—a novelty in the history of Afghanistan. It was inspired from the Taliban's notions of an Islamic golden age rather than a re-establishment of Afghan traditions [4, pp. 255–256]. The Taliban's restrictions on women's movements and visibility amounted to a curfew on women. Apart from health workers who were allowed to treat only female patients, women were banned from working, and girls were prohibited from attending school. Women were only permitted to leave their houses for religious reasons wearing the all-enveloping burqa and accompanied by a male relative.

The Taliban never adopted a constitution but declared their commitment to sharia and issued a number of laws. They founded a notorious "vice and virtue" religious police modeled on and reportedly funded by Saudi Arabia, who often violently and arbitrarily enforced the Taliban's edicts regarding religious observance and the complete seclusion of women. The Taliban carried out flogging and public execution of women—an unprecedent act in recent Afghan history. To the overwhelming majority of the Afghan population, the segregation imposed by the Taliban was unparalleled. A significant number of home schools set up in response to the ban on female education [17, p. 2].

2 Restoration of Women's Rights: 2001–2021

The reinstatement of Afghan women's formal rights following the U.S. led military intervention and ouster of the Taliban regime was legally enshrined in the 2004 Constitution. The Constitution was a significant achievement of the Western-led reconstruction of post-Taliban Afghanistan. The 2004 Constitution, which once again declared Afghanistan as an Islamic republic, included provisions to safeguard women's rights. Article 22 stated that 'citizens' of Afghanistan—whether man or woman have equal rights and duties before the law. In addition, quotas were set up for female representation in the government—women would be guaranteed roughly 25% of the seats in the lower house of Parliament and 17% of the seats in the upper house. Moreover, women were granted two seats in each provincial council (Thier, p. 571). Although the Constitution made frequent references to human rights, it contained components from across the broad spectrum of Afghanistan's past legal traditions: the Islamist orientations of the Mujahedin and Taliban era, the emancipatory goals of the 1964 Constitution and communist government, and the strong executive that had been a constant feature of the country's institutional design. This made it open to contradictory interpretations like many other constitutions. However, in the absence of security and the rule of law, the expansion of women's formal rights did not translate into substantive benefits.

In 2005, the first parliamentary elections resulted in a legislature with almost 30% female members—a percentage higher than in many Western countries, though this was due to the quotas set forth in the 2004 Constitution. However, female members of parliament were for the most part unable to establish alliances that could pursue gender issues—or, in some cases, they were not interested in doing so. Many female parliamentarians were beholden to powerful male patrons or sought to distinguish themselves as go-to champions of women's rights for Western embassies. In 2017, I was contacted by a journalist from Radio Free Europe when the parliament did not approve of the nomination of Narges Nehan for the post of Minister of Mines and Industries—the only female cabinet nominee to go up for approval at that time. Nehan's rejection by parliament was widely criticized for a number of reasons, including gender discrimination [34]. Nehan was kept on in the post in a caretaker capacity. I pointed out to the Radio Free Europe reporter that the most significant impediment was the lack of institutionalized support for women's entry and effective participation in politics. While the Afghan government had publicly stated its commitment to women's empowerment, much of this was symbolic rather than a reflection of tangible and genuine progress.[1]

In 2015, I founded the first undergraduate Gender Studies program in the history of Afghanistan (Friends of AUAF [13]) and reached out to First Lady Rula Ghani for support in launching the initiative. She not only declined the invitation but distanced herself from the project.[2] Although President Ashraf Ghani appointed young women to high profile governmental and diplomatic posts, it was perceived as largely cosmetic to garner Western support and please the donor community. Many of my students who enrolled in Gender Studies courses felt that official support for women's rights programs and the political participation of women in politics was largely due to international pressure placed on the Afghan government. Given Afghanistan's dependence on foreign aid, the limited visibility of women in public roles was perceived by many as designed to showcase a façade of assurance towards the goal of empowering women. It did not reflect a burgeoning support for women's rights at the state level. In 2010, while speaking to several female parliamentarians in Kabul, they all concurred that one of the greatest challenges they faced in their jobs was perceptions by male colleagues that they were morally compromised women. It

[1] Email exchange with Frud Bezhan from Radio Free Europe, December 5, 2017.

[2] Email exchange with Rula Ghani, March 10, 2015.

was even said that some male parliamentarians accused them of using their positions to promote prostitution.[3]

The lack of domestic support for women's rights was linked to the re-empowerment of reactionary forces in Afghanistan via U.S. and allied military strategy. While the United States and its NATO partners emphasized the improvement of women's lives in their efforts to rebuild Afghanistan, the decision to partner with Afghan war criminals, military commanders, and strongmen from the Mujahedin era in order to remove the Taliban from power and for counterterrorism operations witnessed the rehabilitation and re-empowerment of figures who were not receptive to women's rights initiatives or were products of the jihadi era that saw a reversal of the gains made by Afghan women under the Old Regime. The focus on capturing military opponents led to alliances with armed commanders and strongmen, alliances that ran counter to attempts to monopolize the use of violence and build a unified state. Pledges to eliminate corruption and to support human rights and good governance often had to yield to short-term political stability and intelligence gathering. The shock of 9/11 and the sense of political urgency to respond quickly led the administration of George W. Bush to opt for a bombing campaign and partner with the only anti-Taliban force on the ground in Afghanistan, which was the Northern Alliance [19, p. 459]. Out of expediency, the United States chose to partner with the Northern Alliance—led mainly of Mujahedin figures—and headed by Burhanuddin Rabbani who was the United Nations recognized president of Afghanistan when the attacks on America occurred. As outlined above, it was Rabbani who introduced restrictions on women's rights before the rise of the Taliban. During the Mujahedin era when civil war led to a state of total lawlessness, sexual violence was weaponized as an instrument of war and rapes were carried out with impunity. As already noted, it was this unruly situation that contributed to the emergence of the Taliban.

Moreover, the international support that was thrown behind Hamid Karzai to head the new political order further strengthened the position of conservative forces as he tried to placate conservatives. In 2012, Karzai supported a 'code of conduct' put forth by a group of clerics that would among other restrictions allow husbands to beat their wives unless there was a "sharia-complaint" reason [2]. Moreover, as will be noted below, he signed the controversial Shia Personal Status Law which was, in many ways, contradictory to the Elimination of Violence Against Women (EVAW) law, which he signed at the same time. This meant that on the one hand, the international community was using its leverage to uphold women's rights, but, on the other hand, was elevating domestic forces who were not receptive to gender mainstreaming programs or were hostile to it. This created a scenario where those who sought to advance feminist legislation and address gender-based violence in Afghanistan often turned to the international community for support to overcome powerful domestic opposition instead of pursuing intra-Afghan negotiations.

[3] Interview with four female members of parliament, Kabul, March 2010.

3 The Victim Archetype as a Wartime Motif

The social construction of Afghan women as the victim archetype was part of the broader strategic calculus of the War on Terror. As such, it was a symbolic wartime motif that essentialized Afghan women as perennial victims and decontextualized their lived experiences. As donor funding for women's programming grew during the war, the grand narrative of Afghan women's suffering and liberation via Western military rescue was at times exploited by privileged members of the foreign expat community in Afghanistan as well as elite Afghans. This sometimes conflicted with the purported goal of improving the lives of Afghan women while flat out endangering them at other times. Afghan women suffered immeasurably over decades due to war-related violence and related trauma. However, the one-dimensional image of women as archetypal victims that emerged after 9/11 was devoid of context and authenticity. It was a simplistic representation of victimhood designed to mobilize public opinion and muster support for the war strategically while often producing material gain for a privileged few, who capitalized on the trope of suffering and liberation.

In the wake of the 9/11 attacks on the American homeland, the Taliban's draconian measures against women emerged as a major ideological plank of the case for war. On November 17, 2001, First Lady Laura Bush made an extraordinary radio broadcast from her ranch in Crawford, Texas, during which she stated that "the fight against terrorism is also a fight for the rights and dignity of women" [14, pp. 142–143]. In the same vein, the wife of the British Prime Minister, Cherie Blair, was depicted in a London newspaper as being on a campaign to liberate Afghan women [37]. These public appeals by the spouses of mega powerful world leaders catapulted the plight of Afghan women onto the global stage and triggered a flurry of debate on the correct approach in responding to gender based discrimination and violence in Afghanistan. Thus, the U.S. led military intervention, created an image of Afghan women as suffering victims in need of saving from Taliban brutality via Western military rescue. Almost instantly, the grand narrative of Afghan women as the victim archetype emerged as a form of graphic propaganda in support of the war.

A number of scholars have a critiqued the tactical framing of Afghan women's victimhood for wartime expediency by citing it as neocolonialist, imperialist, and out of sync with women's lived experiences [8, pp. 137–160; 3, pp. 112–133]. In her seminal work, *Do Muslim Women Need Saving*?, Laila Abu Lughod points out the centrality of Afghan women's rights to the politics of the War on Terror. Abu Lughod urges caution about placing feminism and secularism only on the side of the West. She has written about the challenges faced by Middle Eastern feminists when Western feminists initiate campaigns that make them vulnerable to local denunciations by conservatives [1, pp. 44–45]. Similar trends took place in Afghanistan after the U.S. invasion when efforts by prominent Afghan female politicians to secure the passage of the Elimination of Violence Against Women (EVAW) law in parliament met with criticism by powerful jihadi parliamentarians as an attempt to impose alien Western values on Afghan society.

One of the most glaring examples of showcasing Afghan women's victimhood to justify continued American military presence in Afghanistan occurred in 2010. A *Time* magazine cover story featured a young Afghan woman, Bibi Ayesha, whose nose had been cut off by her Taliban husband and in-laws. The disturbing photo of Bibi Ayesha's disfigured face was juxtaposed with the headline "What Happens if we Leave Afghanistan?"—implying that women would be the first victims. The shocking photograph, which was selected as World Press Photo of the Year, unleashed a firestorm of controversy and *Time* was criticized for exploiting Bibi Ayesha's mutilation for political messaging. It was widely pointed out that the presence of NATO countries in Afghanistan had not prevented the mutilation of Bibi Ayesha. She had been photographed in a shelter in Kabul run by an American organization with a large local staff, Women for Afghan Women (WAW). She was waiting there to be sent to the United States for reconstructive surgery funded by generous donors and the Grossman Burn Foundation. The media frenzy surrounding the photograph led WAW to shield Bibi Ayesha from publicity, eventually preventing all interviews and photographs. By then they were sheltering her in New York, hoping she would recover enough from her trauma for surgery to take place.

The managing editor for *Time* defended the decision to publish the photograph and stated that he was not taking sides but would "rather confront readers with the Taliban's treatment of women than ignore it." WAW was founded by expatriate Afghans in New York in 2001. It was very adept at fundraising, lobbying, and organizing high-profile events in the United States. Questions had long been raised regarding its media policy. The fact that WAW allowed journalists to publish photographs of residents of women's shelters was especially criticized given that Afghan women who allow the circulation of their photos often face stigmatization in conservative communities. The publication of Bibi Ayesha's photo coincided with an increase in targeted operations including frequent night raids and the detention and killing of suspects, which led to massive fear and anger among many Afghan groups [1, p. 28].

In this case, the mutilated face of Bibi Ayesha was on full display as cannon fodder to stoke debate about the implications of withdrawing troops from Afghanistan. The award-winning photo brought acclaim to the South African photographer, Jodi Bieber, who captured Biba Ayesha's defaced visage in this now infamous photo. Bibi Ayesha's story, on the other hand, was stripped of context and in some Afghan circles, it was viewed as a national humiliation. The case came to characterize the powerlessness of the Afghan people vis-à-vis the international community. It epitomized the 'salvation discourse' which depicted Afghan women as quintessential victims while promoting the career of a privileged international worker.

In *The Carpetbaggers of Kabul and Other American-Afghan Entanglements*, Jennifer Fluri and Rachel Lehr examine how the grand narrative of saving Afghan women acquired geopolitical currency and generated profit and professional gain for privileged international workers while discounting the practical needs of Afghan women. The currency of the Afghan woman as a poverty-stricken victim in need of Western liberation was exchanged for donor funding. Much of the gender-focused programming in Afghanistan was donor driven. Most large donors did not incorporate the input or needs of Afghan women. While the grand narrative of suffering became commodified to generate funding for grants, fellowships, training programs, and other types of programming, the practical needs of Afghan women—who were the targeted beneficiaries—were ignored. Few Afghan women had direct operational control over programs or funds that proposed to help or assist them. Most organizations employed privileged international workers in leadership positions, which included foreign passport-holding (and international salary earning) Afghan expatriates and returnees. These jobs were dependent on the existence of the international community in Afghanistan, rather than the sustained efforts of local organizations [12, pp. 9–17].

The Kabul Beauty Academy, also known as the Beauty Academy of Kabul, is a famous example of an Afghan women-focused project that received significant media attention and corporate funds in the early years of Western intervention. Mary MacMakin, a U.S. citizen, longtime resident of Kabul, and founder of Physiotherapy and Rehabilitation Support for Afghanistan (PARSA), first came up with the idea of creating a beauty school in Kabul after the fall of the Taliban. Her vision intended to encourage women's enterprise by providing venues for women to offer beauty services to other women for social occasions. MacMakin partnered with Terri Grauel, a New York-based hairstylist, to solicit financial support and product donations from beauty industry corporations. The project also received support from the Afghan government's Ministry of Women's Affairs. This project, which was designed to help women, simultaneously traded on the geopolitical currency of "saving" Afghan women to generate funds and support. As a result, one of the privileged international women involved in this project traded on the currency of Afghan women to generate her own personal wealth [12, pp. 60–61].

The Kabul Beauty Academy received widespread media coverage in popular U.S. women's magazines such as Vogue (also a sponsor) and a documentary was made about it When the funding for the school ended, Debbie Rodriguez transitioned the school into a salon that catered to privileged international workers. This project, which was intended as a project to develop local-Afghan women-to-women sustainable enterprise, turned into an American-run business that catered to temporally bounded incomes of well-paid internationals and wealthy Afghans. The most overt example of capitalizing on the currency of Afghan women can be found in Debbie Rodriguez's pseudo memoir, titled *The Kabul Beauty School: The Art of Friendship and Freedom*, or alternatively titled *Kabul Beauty School: An American Woman Goes behind the Veil*, which received a movie option with Columbia Pictures. Rodriguez describes various acts of "saving" in the book as well as her own experiences living in Afghanistan. She identifies herself as the founder of the Beauty School, which is disputed by most of the women involved with the initial project [10].

In 2005, I met Debbie Rodriguez and years later interviewed one of the women who she trained at the Kabul Beauty Academy and later employed as a beautician when it transitioned into a salon. This Afghan woman, whose name will not be mentioned for privacy reasons, was affiliated with the Kabul Beauty Academy and Oasis Beauty Salon from 2004 to 2007. She mentioned that her life was endangered when her photos were published in the book written by Rodriguez. As of 2017, she had tried unsuccessfully to be evacuated from Afghanistan as the publicity from the book led to her receiving threats from her in-laws.[4] In her book, Rodriguez exploits the victim trope to describe the extreme suffering of Afghan women at the hands of Afghan patriarchy while depicting herself as a foreign liberator. Rodriguez had instructed all the Afghan beauticians working at the salon to recount their stories but had told them that none of their photos would be included for security reasons. When the book was released and the beauticians' photos were published, the women protested that their lives were at risk. Debbie Rodriguez left Afghanistan after the publication of her book and the salon closed.

Another highly publicized women's program funded by USAID was the Promote program, which was launched in 2015. It was hailed as the world's biggest program ever designed exclusively for female empowerment. But, according to the Special Inspector General for Afghanistan Reconstruction (SIGAR), it was a failure and a waste of taxpayers' money. The project was originally budgeted at $280 million and was supposed to help 75,000 Afghan women get jobs, promotions, apprenticeships and internships. In 2018, one of the few concrete results cited in a study of the project was the promotion of 55 women to better jobs. But the report said it was unclear whether the program could even be credited for those promotions [29].

Many people who rushed to Afghanistan after the fall of the Taliban were interested in the U.S. national response and rescue mission in Afghanistan. Some of these expats viewed their participation as either a duty or a way to assist Afghan women who were imagined as similar victims of 9/11-style terrorism. In other cases, individuals saw the economic opportunity afforded by trading on the geopolitical and geo-economic currency associated with Afghan women. Gender identities and conflict-driven dislocations neither can nor should be reduced to arbitrated geographies, narrow depictions of places and people, and dramatic characterizations of suffering or liberation. Everyday lived experiences are much more complex and multifaceted than suggested in mediated representations [12, p. 62].

Concocting a sensationalized image of Afghan women as victims in need of saving by Western armies was a wartime motif that lost currency once the Trump administration entered into direct negotiations with the Taliban. The U.S.-Taliban deal that was signed on February 29, 2020 normalized the same insurgents whose oppression of Afghan women had been intensely invoked as justification for the U.S. led military intervention. The deal that the United States signed with the Taliban left the future of Afghan women completely up to the outcomes of the intra-Taliban negotiations and battlefield developments. I spoke to NPR and mentioned that there was consternation about how the deal re-invented the Taliban as viable political actors given their dismal record on women's rights [38]. A year before the signing of the deal, which one-time Trump National Security Adviser, H.R. McMaster later dubbed as a "surrender agreement," [26], Cheryl Benard penned an incendiary piece in *The National Interest* that condescendingly instructed Afghan women to take charge of

[4] Interview with former student of Kabul Beauty Academy and employee of Oasis Beauty Salon, Kabul, 2017.

their own fate and not expect the world to rescue them [6]. Gone were the days when liberating Afghan women was presented as a strategic rationale for military intervention. The fact that Benard's husband, Afghan-American diplomat Zalmay Khalilzad, was leading the U.S.-Taliban negotiations at the time, was viewed by many Afghan women's rights activists as not only patronizing but a sign that the strategic calculus had shifted. Women's rights were given short shrift during the negotiations to make way for so-called peace with the Taliban. Shortly after the Taliban take-over of power in August 2021, Benard wrote another piece for *The National Interest* criticizing undue concerns about the future of human rights in Afghanistan by referring to it as "hysteria" in what was a dubious effort to depict the Taliban as having evolved [7].

4 Novel Approaches to Gender Based Violence: EVAW, Special Prosecution Units, and Women's Shelters

The most significant developments that took place with regards to addressing gender-based violence were the law on Elimination of Violence Against Women (EVAW), the Special Prosecution Unit for Crimes Against Women, and the establishment of women's shelters. All these initiatives, which were designed to specifically address violence against women were novelties in Afghanistan. None of these innovations transformed gender relations as women continue to be wards of their households and communities with historically little recourse to justice and protection outside of these domains. Nevertheless, the introduction of these novel procedures with Western support and funding amplified the issue of gender-based violence in Afghan society at large and gave women access to procedures and apparatuses that resulted in their limited empowerment. It often took the heavy involvement of the international community and exertion of considerable diplomatic pressure backed up donor funding to ensure the realization and sustainability of these measures. In the case of the women's shelters, it was such pressure that allowed them to operate independently of the Afghan government.

The indifference of the Afghan government to addressing violence against women as well as the intersection of the law with the social context of Afghanistan as it relates to gender and kinship were major contributing factors to low prosecution rates and limited progress. However, this should not deflect from the significance of introducing such protective mechanisms. The history of Afghanistan has shown that sustainable progress and lasting change regarding women's rights is best achieved when it is done gradually. This was demonstrated from 1929 to 1978 when incremental and selective modernization resulted in the passage of the liberal 1964 Constitution. That charter led to universal suffrage and emergence of women in parliament and other areas of public life without external assistance or encouragement. However, in the post-2001 era, women's rights were inextricably linked to international assistance and foreign donors often made aid contingent on the delivery of quick-impact projects which were neither realistic nor responsive to the needs of Afghan women. Thus, the success of these innovative processes should not be viewed solely based on number of prosecutions, but that is how the donor community approached them. In fact, donors would often cut funding or threaten to decrease further aid if certain benchmarks were not met [39, pp. 78].

The EVAW law was unprecedented in Afghan history, listing twenty-two acts of violence against women and mandating punishments for them. It also required the government to take specific actions to prevent violence and protect victims. The EVAW was regarded as important because unlike the existing Penal Code, it designated rape as a crime distinct from consensual adultery, provided considerably stricter punishments for forced and underage marriage, and criminalized a number of violations of women's civil rights. In general, the law was regarded by many as an important instrument of advocacy, indicating that women were independent holders of rights, and that the Afghan state had an obligation to protect them from abuses at the hands of their families. Although the existing Penal Code covered crimes such as beating and murder, those who supported the new law argued that there was a propensity in legal practice and more broadly in Afghan society to view such acts as crimes only when they were committed by people outside the family.

The application of the EVAW law, which was decreed by President Hamid Karzai in 2009 and reconfirmed by his successor, President Ashraf Ghani, in 2018 was partial and tenuous. The legal status of the law was uncertain. It was neither approved nor rejected by Parliament, and it appeared to be in contradiction to other laws, namely the Shia Personal Status Law. The final version of the latter law appeared to allow underage marriage, made a woman's right to marry dependent on their fathers' and grandfathers' consent, and constructed a marital relation in which wives were supposed to submit to sexual relations on demand—or risk forfeiting claims of maintenance from their husbands. The EVAW law as it was decreed did signal a potential reform of both women's subject position and the reach of state realms versus that of kinship. Had the decree been enforced in full by the courts, it would have resulted in a substantial increase in the power of the government in overseeing and regulating sexual and gender relations, in underwriting women's protection and equal rights in marriage, and in punishing transgressions.

The EVAW law encountered significant opposition from the former jihadi parties who were dominant in Parliament [28]. It was widely believed that their leaders were able to coerce and intimidate parliamentarians. As the law was reviewed in parliament, efforts were made to amend it in a direction that would safeguard fathers' and husbands' prerogatives over women—by allowing fathers to marry off minor daughters and by establishing husbands' right to polygamy and beating. However, women's right to maintenance was confirmed. Unlike similar legislation put forth by former Afghan rulers such Amir Abdur Rahman and King Amanullah, the EVAW law represented the force of external actors—namely the United States and Western embassies. It did not reflect an attempt by the Afghan ruler to curtail the power of kinship groups and modernize society by emancipating women from familial control. It was obvious from attempts to implement and promote the EVAW law that a group of foreign diplomats, activists, and pro-women justice officials had pushed it through with guarantees of funding and the infrastructure of aid agencies [39, pp. 75].

The U.S. embassy repeatedly discussed the EVAW law with Sarwar Danesh, who was Afghan Minister of Justice at that time. In one diplomatic cable, it was remarked that the U.S. wanted to see cabinet approval of the legislation prior to the arrival of the U.S. Ambassador-at-Large for Global Women's Issues. While the U.S. did not want to take public credit for the EVAW law as it would tarnish its status as a foreign initiative—it did engage in the type of law reform where isolated, symbolic, and sometimes hollow achievements obtained through informal pressure and negotiations was pursued over more fixed, long-term, gains requiring broader coalitions. Unlike the not too disparate legislation put forth by kings such as Amir Abdur Rahman and Amanullah, the EVAW law did not represent the attempt of a ruler to limit the power of kinship groups. This was not an effort to modernize society by freeing women from the control of their families. It was clear from attempts to promote and implement the law that there was a group of foreign diplomats, activists, and pro-women justice officials, backed by the funds and infrastructure of aid agencies, who played an outsized role in pushing through gender reform. Although such an assemblage of actors worked with Afghan state institutions, this did not mean that there was one, clear, sovereign national power ruling over a single public domain. The fact that President Hamid Karzai signed two contradictory laws—the Shia Personal Status Law and EVAW law as presidential decrees—on July 19, 2009, exemplifies the multiplicity of social realms in post-2001 Afghanistan. Karzai had offered two constituencies to each have its own law—the progressive EVAW law for Afghan feminists and their international allies and Shia Personal Status law to the Shia clergy [39, p. 79].

The Mujahedin, who were rehabilitated to power as partners to the U.S. invasion that overthrew the Taliban regime, presented themselves as guardians of Islam. This was a powerful way to demonstrate their nationalist credentials and distance themselves from the 'infidel' military and diplomatic presence. Ironically, although they returned to power through Western military force, they managed to increase their clout by exploiting popular resentment against Western military operations. The Western-led military reconstruction of Afghanistan entailed strong support for women's rights, but it also empowered forces—namely jihadis and warlords with notorious human rights records—who would use their renewed influence to create obstacles in the way of advancing gender related reform. This created a scenario whereby a disconnect emerged between foreign support for women's rights on one hand, and powerful domestic resistance to it, on the other hand. This in turn increased the dependency of Afghan gender reform advocates on Western backing.

5 Special Prosecution Units

The enactment of the EVAW law led to the establishment of special prosecution units which were intended to serve as a mechanism for the law's application. The implementation of the EVAW law increasingly became the focus of donor and international engagement concerning women's rights in Afghanistan. Most of the special prosecution units were supported by international aid. The first prosecution unit was established in Kabul in 2010. By the end of 2015, there were special units for crimes of violence against women at the prosecutor's offices in twenty of the thirty-four of the country's provincial capitals. Female prosecutors, dedicated resources, and increased visibility of the prosecution units were all designed to make this new branch of the legal system more accessible to women and more accountable to activists and donors. Reliable data on number of prosecutions became difficult to determine due to inadequate record keeping. In 2014, after extensive Western pressure, the Ministry of Women's Affairs, provided a report based on data from twenty-six provinces. It suggested that 18% of all cases registered with the prosecution were referred to courts and that 13% led to a criminal conviction. Independent research on the prosecution units has confirmed these trends [39, pp. 88–89].

There was a considerable gap between how the prosecution unit in Kabul functioned and those in the other provinces. In Kabul, it was relatively easy to register a case at the special prosecution unit for Violence Against Women (VAW). Complainants would submit a formal written petition after which prosecutors would proceed to open a file without any specific requirements regarding evidence or the seriousness of the claim. In Kabul, prosecutors would not berate women for bringing their private matters to outsiders. The prosecutors would also make a systematic effort to bring in suspects. Complainants were given a summoning letter to hand over to the local police, who would deliver the letter to the suspect. If the suspect failed to show up at the prosecution, the police would be sent to identify his (or, less often, her) whereabouts and bring him in. As part of gathering evidence, prosecutors questioned complainants and suspects. When questioning female complainants, efforts were made to establish facts and a sequence of events. In cases of beating, complainants were asked if they had any physical marks and, if so, refer them to the department of forensic medicine. When interrogating male suspects, prosecutors' line of questioning would include a mixture of fact-finding and reproaching. On the other hand, in many of Afghanistan's more rural provinces, simply registering a case of gender violence at the VAW unit was a difficult process. In rural provinces, prosecutors were all male who held an attitude that VAW cases did not belong in the legal system at all, and that the stains left on the family and the women's honor dictated that such cases should be solved within the family, with the women remaining within their marriages or families. Another reason women encountered challenges in registering VAW cases outside of Kabul was when perpetrators resorted to bribery or used their political connections to influence prosecutors [39, p. 90].

While the VAW unit in Kabul did not function according to donor expectations, its creation marked a significant step for women's rights given the social context that Afghan women found themselves situated in. More than 85% of cases registered at VAW unit in Kabul were closed prior to trial. This was often because prosecutors did not want to pursue cases. Prosecutors suggested to women that it was better to reach an agreement with their abusers than to be left without a home and an income. In some cases, it was victims or their families who did not want their petitions to proceed to court. Some women used the prosecution unit in Kabul without any intentions on seeing their husbands or other family members convicted or imprisoned. Such women sought assistance for obtaining a divorce or were attempting to get wayward spouses to return to the marital home, and still others were seeking a government sanctioned guarantee that the violence would cease. In other cases, the prosecution units became an outlet for families in conflict with other families or for the restoration of reputation and honor. For example, complaints of rape and abuse might be filed with the intent of pressuring the defendant's family in matters of bride-price or compensation for elopement.

Prosecutors, women, and their families negotiated within a pattern of gender norms in which rehabilitating the family unit and assimilating sexual indiscretions into the framework of conventional marriage ranked supreme. The overall functioning of the special prosecution units thus worked to reaffirm the notion that acts of violence against women were family problems rather than criminal cases belonging in the mainstream legal system. The special prosecution units operated at the nexus of donor expectations, government indifference, and lack of attention to wider relations that viewed women's independent status outside of the family unit as inherently transgressive. This does not take away from the fact that for many women and their families, the VAW unit provided a novel recourse in the face of abuse, desertion, or neglect. Even if the VAW unit served as a medium for negotiation between complainants and defendants rather than an apparatus for application of the law, it was able to at least raise the specter of criminal prosecution and, in this manner, sometimes helped women secure a better deal than what was available elsewhere. The threat of prosecution was a sort of bargaining tool for some women. In these situations, the withdrawal of cases by complainants should not be seen as a sign of weakness or as an imposition on women by prosecutors or their husbands and families. The prosecution unit provided women with a new form of leverage which, relatively speaking, strengthened their bargaining position [39, p. 87]

The low prosecution and conviction rates of cases of gender violence were disappointing to many supporters of the EVAW law. It was widely viewed that the Afghan government was not committed to the law's enforcement or even pandering to conservative constituencies. This was partially the case. It was certainly true that to a certain degree the special VAW units reflected an attempt to address the problem of missing national will by placing part of the legal implementation under direct global management. But an exclusive emphasis on a lack of political will does not explain women arriving at the prosecution—not as autonomous individuals seeking the dispensation of justice—but as social beings entrenched in and constrained by social relationships. Thus, the actual objectives of women or their families might not be the execution of the law, but, rather, the negotiation of a better deal within the confines of those relationships. Similarly, legal officials operated within cultural boundaries and material realities and, like the women, did not always view the ideal outcome as enforcement of the law [5, pp. 41–75]. Therefore, a more holistic analysis must consider that laws and justice are filtered through an array of social relations especially when there is a substantial gap between legal frameworks and the lived structures of gender and kinship. It makes sense to look at legal institutions as a space where rights and obligations are negotiated, and where the law serves as one reference point instead of one overarching framework. In many instances, the prosecution unit served to empower women in a limited sense—by affording them leverage in pursuing objectives other than punishing the perpetrators according to the law [39, pp. 106–107]. Moreover, the extent to which Afghan women and their families accessed the prosecution units as a recourse to solve their domestic disputes challenged prior scholarly conclusions that Afghan notions of privacy and shame prevented them from turning to the formal justice system unless absolutely necessary [39, p. 101].

6 Women's Shelters

Perhaps the most radical initiative in addressing gender-based violence to emerge in Afghanistan via the U.S. led invasion of Afghanistan was the establishment of women's shelters. These were spaces designed to accommodate women who left their homes and families due to abuse. From 2001 to 2014, more than two dozen women's shelters were established in the country. These shelters provided a living space for women and girls who could not—or for security reasons—did not want to live with their families. It was highly uncommon, and downright unheard of, for women to live on their own outside of the family unit. Therefore, the concept of the women's shelters challenged government practices and long-held gender norms that cast women who lived independently of their families as deeply subversive and even criminals. They also problematized the family home as a site of potential abuse, from which a women had a legitimate right to seek protection. The women's shelters operated through donor funding and donations from abroad. As they relied on international networks and resources, they partially functioned as transnational institutions. This enabled the shelters to operate fairly autonomously of the Afghan government on many counts and to challenge and prevent efforts by Afghan government officials to regulate and supervise them. The international leverage and resources that the shelters were able to assemble became a source of friction to many Afghan officials and conservative commentators who viewed them as foreign-sponsored entities that challenged conventional gender norms which recognized women as wards of their families [39, pp. 108–109].

At the heart of the women's shelters were "runaway women" who were viewed as having committed a "moral crime" for leaving their familial abode. Human rights advocates who visited Afghan prisons after the U.S. led invasion discovered that most female detainees were held for reasons related to "moral crimes." The majority were detained for a variety of offenses related to family law such as refusing to live with their husbands, refusing to marry a husband chosen by their parents, or for having run away from either the parental or marital home. These women did not have access to lawyers, had no information on their rights, and were left to languish in jail until their relatives intervened [22, p. 42]. In 2015, I visited the Kabul women's prison and interviewed women who were incarcerated for running away (farar az manzil) and the majority indicated that they were attempting to escape forced marriages. A couple of women mentioned that their families were trying to coerce them into prostitution. One woman said she felt safer in prison than out in society and feared that her family would harm her if they got the chance.[5]

Being incarcerated in Afghanistan for running away is a gender-specific predicament. Women who wander outside the home, especially unaccompanied, are suspect as the public domain is seen as a man's world. Thus, only women can "escape" from the home. In essence, running away as a gender-stratified crime or subversive act is based on a gender order in which women are considered legal minors. Women are the wards of household heads, husbands, and families. Moral crimes are a broad category; it encompasses both premarital and extramarital sexual intercourse (zina), and the vaguer crime of running away. While the Penal Code made zina a punishable crime for both sexes, there was no reference in the legal canon to any crime of running away. As previously mentioned, the Mujahedin started the practice of incarcerating runaway women. In the 1990s, the Taliban would go even further by mandating that women could only leave home if escorted by a male guardian and only if fully veiled. This radical Taliban imposition essentially usurped kinship power. After 2001 following the ouster of the Taliban regime, government policies on women's mobility became much less restrictive. However, women were detained, charged with, and convicted of running away from home (farar az manzil). During the Karzai presidency, the number of women incarcerated for moral crimes was several hundred at any given time (Wimpelmann, pp. 110–111). The retention of the practice of incarcerating runaway women should come as no surprise given the renewed empowerment of the Mujahedin in the U.S. led military intervention in Afghanistan. But it also reflects the failure of the Karzai government to make a total break with a policy that was never in place in pre-war Afghanistan but was a legacy of the Mujahedin and Taliban periods.

The first shelter was registered in Kabul in 2003 and at first was a response to the deportation of single Afghan women from Iran. They were a novelty in Afghanistan. The advent of shelters addressed the need of accommodating women who had been dislocated by the upheavals of war. By 2014, there were about thirty shelters in nineteen provinces. There were no shelters in the more unstable and conservative parts of the country in the east and south. In these areas, women faced the difficult choice of either enduring abuse or entering an illicit relationship [39, pp. 114].

The shelters were run by NGOs and completely funded by international aid. The U.S. government and UN Women were major funders. In 2011, the Afghan government tried to bring the women's shelters under its direct management through the Ministry of Women's Affairs, which was established in 2001 shortly after the U.S. led invasion. The government established a set of regulations that determined who could be admitted to women's shelters as well as other criteria. Included in the list of stringent guidelines for admission to women's shelters were virginity tests for women and girls who had been compelled to leave their house. These regulations

---

[5] Interview with women incarcerated for running away from home in Kabul Women's Prison, March 2015.

provoked an international campaign to keep the administration of the women's shelters outside Afghan government control. In addition to women's rights organizations including the Afghan Women's Network—the umbrella network for women's organizations—Western politicians and diplomats were vocal in expressing concern about the attempt by the Afghan government to regulate the shelters according to criteria that would expose women to scrutiny regarding their chastity. Three U.S. Senators wrote to the Afghan president reminding him of American assistance to Afghanistan and calling the regulations a "grave mistake." The assistant secretary of state of the United States followed in a more subtle tone expressing concern over the regulations. As the international controversy escalated, President Karzai succumbed to the pressure and backed down. The shelters remained autonomous, and the international pressure mounted on the Afghan government proved to be the decisive factor. The shelters' success in blocking government attempts to nationalize them demonstrated the consolidation of an effective transnational alliance around women's rights in Afghanistan.

For the Afghan government, attempts to take charge of managing the shelters was a matter of reinforcing familial sovereignty over women as well as concerns about the possible consequences of unsupervised women. It was about the ability of society to enforce modesty and propriety, which in the context of Afghanistan, meant that women could not have an independent existence outside of the family unit. The government saw the women's shelters less as safe houses and more as centers of potential debauchery [39, pp. 121–128]. For the international community and Afghan women's rights activists who succeeded in maintaining the autonomous status of the shelters, they served as safe spaces for women fleeing abuse who had nowhere else to go. In the case of the women's shelters, the lack of political will on the part of the Afghan government to put the plight of vulnerable women above social mores meant that international support was critical for their sustainability and survival. Without solid internal backing and dependence on external donor support, the women's shelters, like the EVAW law and special prosecution units, were a fundamental but fragile gain for Afghan women. The return of the Taliban to power has witnessed the closure of the shelters amid a renewed clampdown on women's rights [32].

7 The Outlook

The role of the international community in promoting gender reform proved indispensable. The loss of this critical support for women's rights in Afghanistan has been overshadowed by the efforts of the international community to engage with the Taliban. Despite their continued repression of Afghan women's rights, the Taliban continue to be normalized through invitations at high-profile platforms, VIP treatment, and diplomatic engagement. Verbal condemnations by the United Nations and other international agencies regarding the Taliban's treatment of Afghan women will remain hollow and impotent responses to their exceptional violations of human rights. As the Taliban impose alien customs on Afghan women such as mandating the wearing of face veils for female journalists, it is important to recall that Afghanistan has a prior history of modernizing regimes who were not encouraged or assisted by Western support or money. Thus, any notion that reforms in the area of women's rights is not in keeping with Afghan traditions is not borne out by the historical record[6] [36].

It has been over 300 days since the Taliban banned teenage girls from school. Taliban policies that deprive teenage Afghan girls of an education have once again reportedly led to the establishment of secret schools [18]. The humanitarian crisis has led families to resort to selling their daughters, some of whom are babies and children [11]. The clampdown on women's mobility, employment, clothing, and other restrictions have reportedly led to an increase in suicides among women [33]. Continued engagement with the Taliban by the international community should emphasize women's rights with the potential for tangible impact. In the rush to leave Afghanistan, the United States relinquished significant leverage over the Taliban during the so-called peace talks. This emboldened the Taliban and disempowered pro-women activists. The international community must once again utilize its significant leverage to help restore the rights of the women of Afghanistan, as it did before.

References

1. Abu-Lughod L (2013) Do Muslim women need saving? Harvard University Press, Cambridge, MA
2. Associated Press (2012) Karzai backs restrictive code for women. The Guardian
3. Ayotte KJ, Husain MA (2005) Securing Afghan women: neocolonialism, epistemic violence, and the rhetoric of the veil. NWSA J 17(3):112–133
4. Barfield T (2010) Afghanistan: a cultural and political history. Princeton University Press, Princeton, NJ
5. Basu S (2006) Playing off courts: the negotiation of divorce and violence in plural legal settings in Kolkata. J Legal Pluralism Unofficial Law 38:41–75
6. Benard C (2019) Afghan women are in charge of their own fate. The National Interest
7. Benard C (2021) What do the Taliban really want. The National Interest
8. Berry K (2003) The symbolic use of Afghan war in the war on terror. Humboldt School Soc Relat 27(2):137–160
9. Dupree L (1973) Afghanistan. Princeton University Press, Princeton, NJ
10. Ellin A (2007) Shades of truth: an account of kabul school is challenged. New York Times
11. Engel R, Joselow G, Mengli A (2021) Desperate for cash, Afghan families are selling young daughters into marriage. NBC News
12. Fluri JL, Lehr R (2017) The carpetbaggers of Kabul and other American-Afghan entanglements. The University of Georgia Press, Athens, GA
13. Friends of the American University of Afghanistan (2015) AUAF Launches Afghanistan's First Gender Studies Program. [online]. Available from https://friendsofauaf.org/news/auaf-launches-afghanistans-first-gender-studiesprogram/. Accessed 25 June 2022

[6] In December 2021, then prime minister of Pakistan, Imran Khan, stated that educating girls is not part of Afghan culture.

14. German L (2008) Women and the war on terror. Fem Rev 88:140–149
15. Ghosh H (2003) A history of women in Afghanistan: lessons learnt for the future or yesterdays and tomorrow: women in Afghanistan. J Int Women's Stud 4(3):1–14
16. Gregorian V (1969) The emergence of modern Afghanistan: politics of reform and modernization 1880–1946. Stanford University Press, Stanford, CA
17. HRW (2001) Afghanistan, humanity denied: systematic violations of women's rights in Afghanistan, vol 5, no 3. Human Rights Watch
18. Hadid D, Qazizai F (2022) Secret schools enable Afghanistan's teen girls to skirt Taliban's education ban. NPR
19. Jalali AA (2017) A military history of Afghanistan: from the great game to the global war on terror. University Press of Kansas, Lawrence, KS
20. Kamali MH (1985) Law in Afghanistan: a study of constitutions, matrimonial law, and the judiciary. Brill, Leiden
21. Kandiyoti D (2005) The politics of gender and reconstruction in Afghanistan. United Nations Research Institute for Social Development (Occasional Paper), Geneva, pp 1–31
22. Lau M (2003) Islamic law and the Afghan legal system. In: Cotran E (ed) Yearbook of Islamic and Middle Eastern Law 2001–2002. Kluwer International, The Hague, pp 27–44
23. Majrooh PA (1989) Afghan women between Marxism and Islamic fundamentalism. Soc Central Asian Stud 8:87–98
24. Ministry of Information and Culture of Afghanistan (1965) Tarikh-i Afghanistan Dar Dawrah-yi Hukumat-i Intiqali. pp 323–327
25. Nawid SK (2007) Afghan women under Marxism. In: Moghadam VM (ed) From patriarchy to empowerment: women's participation, movements, and rights in the Middle East, North Africa, and South Asia. Syracuse University Press, Syracuse, pp 58–72
26. O'Connell O (2021) We defeated ourselves: trump's national security adviser says Pompeo signed surrender agreement with Taliban. Independent
27. Poullada LB (1973) Reform and rebellion in Afghanistan in Afghanistan, 1919–1929: King Amanullah's failure to modernize a tribal society. Cornell University Press, Ithaca, NY
28. Ruttig T (2006) Islamists, leftists—and a void in the center. Afghanistan's political parties and where they come from (1902–2006). Konrad Adenauer Stiftung
29. SIGAR (2018) Promoting gender equity in national priority programs (promote): USAID needs to assess this $216 million program's achievements and the Afghan government's ability to sustain them. 18–69 Audit Report, September 2018. Special Inspector General for Afghanistan Reconstruction
30. SIGAR (2021) Support for gender equality: lessons from the U.S. experience in Afghanistan. Lessons learned program, February 2021. Special Inspector General for Afghanistan Reconstruction (SIGAR)
31. Saikal A (2006) Modern Afghanistan: a history of struggle and survival. I.B. Tauris
32. Shaheed M (2021) Taliban closure of domestic abuse shelters, voice of America [online], p 1. Available from https://www.voanews.com/a/taliban-closure-of-domestic-abuse-shelters-leaves-thousands-at-risk-expertssay/6349979.html. Accessed on 2 July 2022
33. Synovitz R (2021) Empty shell: extreme depression, suicidal thoughts haunt Afghan women under Taliban rule. RFE/RL Radio Azadi
34. TOLO News (2017) Nehan rejected due to discrimination, say women
35. Thier JA (2006/2007) The making of constitution in Afghanistan. New York Law School Law Rev 51:558–579
36. Tu J (2021) Imran Khan slammed for inferring that not educating girls is part of Afghan culture. Women's Agenda
37. Ward L (2001) Leaders wives join propaganda war. The Guardian
38. Welna D, Dwyer C (2020) U.S. signs peace deal with Taliban after nearly two decades of war in Afghanistan. NPR (Weekend Edition Saturday)
39. Wimpelmann T (2017) The pitfalls of protection: gender, violence, and power in Afghanistan

Dr. Bahar Jalali teaches the history of the Modern Middle East. Her area of expertise is Afghanistan. She focuses on modernization and women's rights in the twentieth century under the Old Regime. She is currently finishing a manuscript on the Decade of Democracy from 1963 to 1973 in Afghanistan when an urban educated intelligentsia spearheaded democratic reforms and a liberal Constitution. Born in Afghanistan, she fled the country as a child after the Soviet invasion. In 2009, she returned to Afghanistan to work at the newly opened American University in Kabul where she taught History of Afghanistan and founded the first Gender Studies program in the history of the country. She recently started an online protest campaign calling for protection of women's rights and preservation of cultural heritage in Afghanistan. This campaign has received extensive coverage in the international press. It has also triggered wide- ranging discussions on protection of human rights and cultural heritage in Afghanistan under the Taliban regime.

Was There Ever a Plausible Effort for "Peace Negotiations" in Afghanistan and What Went Wrong with the United States' Approach to Afghanistan

Thomas H. Johnson

Abstract The primary purpose of this chapter is to discuss the implications of the "peace talks" with the Taliban for participating in and signing this agreement with the U.S. and, more importantly, why the United States was unsuccessful in Afghanistan. In so doing, this chapter will assess the Taliban's strategy during the last few years of the war; examine the efforts of "formal" peace negotiations, and; explore why the United states lost in Afghanistan. A major conclusion of this chapter is the U.S. went into the war in Afghanistan with an inadequate understanding of the country, its people and its neighborhood. A misguided commitment to U.S. interests prompted mission creep that made America hang around indefinitely in Afghanistan. The U.S. can no longer be a great power if its people do not choose to carry the burdens of leadership that great powers must carry. The U.S., racked with extremism and racial challenges, disagreement over whether its citizens should be vaccinated against an infectious disease, challenges to its democracy, and unwillingness to join to do anything, has also lost almost every significant war since World War II. I deeply regret what has come to Afghanistan, but also the world, which may only be able to see U.S. perfidy as the profound failure of a declining power.

Keywords Afghan Taliban · Taliban strategy · attempted peace negotiations with the Taliban · U.S. defeat in Afghanistan

1 Introduction

The primary purpose of this chapter is to discuss the implications of the "peace talks" with the Taliban for participating in and signing this agreement with the U.S. and, more importantly, why the United States was unsuccessful in Afghanistan. Some of the explicit questions that will be addressed by this chapter are:

T. H. Johnson (✉)
Department of National Security Affairs, Naval Postgraduate School, Monterey, CA 93943, USA
e-mail: thjohnso@nps.edu; Thjohnso52@gmail.org

A. Farhadi and A. Masys (eds.), *The Great Power Competition Volume 4*,
https://doi.org/10.1007/978-3-031-22934-3_11

Assess the Taliban's strategy in the last few years of the war;
Examine the efforts of "formal" peace negotiations in Afghanistan, and;
Explore why the United States lost in Afghanistan.

While sporadic and ineffective efforts to end the conflict in Afghanistan via negotiations took place soon after the U.S. intervention and eventual occupation of the country, in February 2018 "serious negotiations"[1] began between the United States, led by Zalmay Khalilzad and Taliban representatives with no representation from Kabul. These negotiations eventually led to a "peace agreement" between the U.S. and the Taliban that refers to itself as the Islamic Emirate of Afghanistan signed in Doha, Qatar. This was not a "peace agreement" and it will be discussed in-depth below. Many analysts[2] define the "peace agreement" much more than a phased planned for the US to withdraw for Afghanistan—a goal of the Taliban for nearly two decades.

Before explicitly addressing these explicit research questions, it is useful to briefly discuss the background of the Taliban and the Afghan conflict to lay a context for the research that follows.

The Taliban's Origins and Background.[3]

The *Talib* ("seeker of knowledge or student" in Arabic) has been an important fixture in Afghan society ever since Islam was introduced in the seventh century to the present-day. For centuries Talibs travel the countryside as ascetics, often living off the land and tithings from Afghan villagers, in search of religious "truth." The *Taliban* (plural of Talib) would later become extremely important part of the Afghan social fabric running religious schools (*madrassas*), mosques, shrines, and various religious and social services, and serving as *mujahideen* when necessary. And eventually turning onto a political and military institution.

The political foundations of the Afghan Taliban were directly related to the Soviet invasion (1979), occupation and Anti-Soviet *jihad* (1979–89) and the basic inability

[1] George [1].

[2] Mujib Mahal, "Taliban and U.S. Strike Deal to Withdraw American Troops from Afghanistan," The New York Times, August 29, 2020. Professor William O'Malley called the "Afghan Peace Agreement of February 2020 as the worst diplomatic outcome from international negotiations since Munich 1938 and the Vietnam Paris Agreement Of 1973.

[3] For example, see: Barfield [2], Tomsen [3], Crews and Tarzi [4], Maley [5], Giustozzi [6, 7], Johnson and Narratives [8], Johnson and Mason [9, pp. 71–89], Maley [10], Rashid [11], Khan [12], Marsden [13], Abbas [14], Johnson and Mason [15, pp. 2-14]. Some of the Taliban historical background information contained here can be sourced to: Johnson and Adamic [16]; "What Went Wrong in Afghanistan: A Primer," with Larry Goodson, War Room, Army War College, warroom.armywarcollege. 17 September 2021; "How America Lost Afghanistan," with Larry P. Goodson, *Newlinesmag.com*, August 18, 2021; "Political Legitimacy: Why we are failing in Afghanistan," with Larry P. Goodson, *The Strategy Bridge*, July 2221, https://thestrategybridge.org/the-bridge/2021/political-legitimacy-why-we-are-failing-in-afghanistan. Republished.: *The Historical Dictionary of Afghanistan* with the late Professor Ludwig Adamic, (Rowman & Littlefield, May 2021); "The 2019 Afghan Presidential Election: An Assessment of Problematic Processes and Results," *Afghanistan Journal*, spring issue, vol. 4 no. 1, April 2021, pp. 19-46, and; "The Afghan Taliban and the "Peace Negotiations," An Analysis of Narratives, Stories and Messages: Are the Taliban "De-radicalising?,"in Omar [17]. Book also published in Arabic. Arab Centre, Doha, Qatar.

of former Mujahideen commanders to unify and stabilize Afghanistan post-Soviet withdrawal in 1989. After the withdrawal of the Soviet Union from Afghanistan, the United States also withdrew nearly all support for the Mujahideen (U.S. support, aimed at "bleeding" the Soviets to the maximum extent and making it the Soviet's Vietnam, represented the most expensive covert action to a "rebel" group" in U.S. history).

The Mujahideen were successful in overthrowing the Afghan communists in 1992, but shortly after a violent Civil War erupted amongst various Mujahideen groups, and especially commanders, seeking power in Afghanistan.[4]

The violence and criminality of supposed Mujahideen turned warlords who raped, plundered and extracted from the war-weary Afghan population between 1992 and 1994 resulted in the political formulation of the Taliban that was not only a reaction to the criminal warlords, but also represented a reactionary *Deobandi* (a revivalist movement within Sunni—primarily *Hanafi*) Islamist movement. Many of the early Taliban leaders and soldiers had fought the Soviets in the Yunas Khalis' Hezb-e-Islami (Party of Islam) party (HIK) or *Harakat-i-Inqilab-i-Islami* (Islamic Revolution Movement) lead by Mohammad Nabi Mohammadi. So implicitly the U.S. indirectly added the political beginning of the Taliban, as well as the Taliban becoming a military institution.

One of these Jihadists, who had his own madrassa in Maiwand, Kandahar, was Mullah Mohammed Omar Mujahid (or simply Mullah Omar), who either because of piety, adept politicking, or the fact that he was descended from a major regional subtribe, soon became the leader of the Taliban. Although the Taliban were able to use their Pashtun ethnicity to rally many Pashtun tribes, the initial basis of the Taliban leadership core was Hotaki Ghilzai Pashtuns, Mullah Omar's tribe.

The Taliban were a unique political organization in Afghan history because its leaders were almost exclusively clerics and many of its initial members, especially foot soldiers, were Afghan refugees from Pakistan. While the Pashtun areas had seen charismatic cleric led uprisings in the past, none were imbued with the same unique backstory of the Taliban. The radical ideology of the Taliban, in part, was birthed in the refugee camps on the Pakistan side of the Durand Line—as was the desire to transform a lost, Godless Afghanistan into a land sanctified by the effort of Afghan jihadists.[5]

While living as rebels and refugees, the young men that became the Taliban foot soldiers were significantly influenced by their madrassa teachers whose educations were grounded in aspects of the ultra-fundamentalist Wahhabi and, especially, Deobandi ideologies. In their madrassas, the Taliban had been taught jihad and a narrow, rigid, orthodox interpretation of Islam that did not value extortive and criminal behaviour; therefore, though many fought against the Soviets, the Taliban were sheltered from becoming pawns for the Mujahideen warlords. Instead, due to the excesses of the early 1990s, the Taliban grew increasingly angry at the extortive

[4] See: Tomsen.

[5] For an interesting analysis of the present Taliban consult, See: Giustozzi [18]; and especially: Giustozzi [19].

and extractive nature of the warlords working in and around Kandahar and Helmand provinces.

Like other Afghan groups at the time, the Taliban appear to have taken whatever support Islamabad would give, but the Taliban remained independent and according to Antonio Giustozzi, reflected a peculiar Afghan genius.[6] To argue that the Pakistanis completely controlled the Taliban, or a mere "proxy" of Pakistan, just does not stand up to the empirical data[7] and an overly simplistic view of the conflict in Afghanistan from 1994 to 2021.

Taliban Organization. After taking control of Kabul in 1996 from the Mujahideen, Mullah Omar slowly learned the importance of organization and structure in managing his widespread insurgency. In 2003, Omar worked to prevent similar issues and created a council of confidants to help supervise the growing arms of the Taliban. In *On Guerilla Warfare*, Mao[8] highlights the importance of divisional structure writing:

In guerrilla warfare, small units acting independently play the principal role, and there must be no excessive interference with their activities…Only adjacent guerrilla units can coordinate their activities to any degree. Strategically, their activities can be roughly correlated with those of regular forces, and tactically, they must cooperate with adjacent units of the regular army.[9]

Omar's council appointed leaders for matters regarding finance, military operations, governance, religion, and other important divisions, but allowed basic forces to work somewhat independently.[10] The new organization allowed Omar to oversee the movements of the insurgency but gave him the freedom to focus much of his attention on grand strategy. But a close look at the Taliban regime from 1996 to 2001 will suggest that they were totally ill prepared to administer Afghanistan, especially Afghan urban areas.

After the U.S. invaded Afghanistan after 9/11 on October 7, 2001. From 2001 to 2004, the Taliban utilized a trial-and-error method as they worked to revolutionize their insurgency, mostly in Pakistan. Taliban leadership found tactics and methods that worked with the ideological goals of the insurgency and then adapted them to ensure optimal success. This period of experimentation was integral in the success that the Taliban enjoyed from 2005 to 2020.

Taliban forces were initially outnumbered in Afghanistan, but they were experts of the Afghan landscape and found new ways to exploit their knowledge effectively against counterinsurgents. As Mao advises, the Taliban began to move quickly through the mountainous territories of the country and planned attacks and ambushes that they could control. Meanwhile, U.S. conventional forces moved sluggishly,

---

[6] Giustozzi, Afghanistan's Endless.

[7] Ibid.

[8] According to a Senior U.S. Afghan Analyst and Manager, while in exile in Pakistan, 2002–2004, Taliban leaders and commanders read Mao and Che Guevara in their effort to prepare for a guerrilla insurgent war in Afghaniastan.

[9] Tse Tung [20].

[10] Giustozzi [21].

weighed down by heavy equipment, hindered by poor roadways, and slowed by constant sweeps for IEDs. The slow movements of the counterinsurgents allowed Taliban units to trap conveys, retreat, and quickly move insurgents to their next location for preparation of continued attacks.

Again, following the guidelines set forward in Mao's *On Guerilla Warfare*, the Taliban transformed their organization to meet the growing needs of the insurgency. Mao advises:

The soldier must be educated politically. There must be a gradual change from guerrilla formations to orthodox regimental organization. The necessary bureaus and staffs, both political and military, must be provided. At the same time, attention must be paid to the creation of sustainable supply, medical, and hygiene units.[11]

While the government of Afghanistan attempted to modernize and liberalize, the Taliban set out to revolutionize their structure and ideologies as well. Organizational alterations that began in 2008 transitioned the Taliban away from a patrimonial structure and more towards a centralized structure.[12] This substantial change paved the way for a more uniformed and effective Taliban political divisions. Additionally, the Taliban began to adopt new ideologies in matters such as governance, technology, and public services that would allow them to undermine progress made by the central government in Kabul. These changes, when compared to the Taliban of 1994, illustrate the ideological and political ingenuity of the Taliban from 2005 to 2020.

2 Ideology and Domestic Policies Under Its Rule

Regardless of their stated intentions to create an Afghanistan devoid of ethnic division, the original Taliban were primarily Pashtuns who saw all other Afghan ethnolinguistic groups (e.g., Tajiks, Hazaras, Uzbeks) as enemies. The village mullahs and clerics, who became the leaders of the Taliban, rebelled not only against urban modernity but also conflicted with many of the tenets of *Pashtunwali* (literally the "way of the Pashtun—the unwritten rules that drive and significantly influence a Pashtun's life, honour and conflict resolution, especially in rural Afghanistan).

An early tactic of the Taliban was to attack their enemies' arms depots. Shortly after the formation of the Taliban, Pakistan gave a massive arms depot near Spin Boldak originally "belonging" to HIG to Mullah Omar's Taliban. The Taliban used the heavy weapons from this depot to attack their enemies (Afghan Northern Alliance, officially known as the United Islamic Front for the Salvation of Afghanistan, and HIG). Importantly, the Taliban also used demonstrations of power that had significant psychological impacts on the war weary Afghan population. Throughout 1994 and 1995, the Taliban rapidly gained control or made significant advances in much of

[11] Tse Tung, p. 113.

[12] Antonio Giustozzi, "Military Adaptation by the Taliban: 2002–2011," in: Theo Farrell et al. [22] and Johnson [23, pp. 3–27].

south, southwest, central, and eastern Afghanistan. While much of this could be attributed to military abilities, their success also came from early domestic policies. Not only in the south or southwest but also throughout the nation, Afghans were tired of the warlords' greed and violence, so they gravitated towards the hope and security offered by the Taliban. The Taliban became an almost mythical group as it quickly and clearly defeated Afghanistan's most powerful warlords, often without firing a weapon. And the Taliban developed a brilliant narrative campaign to gain the allegiance of rural Afghanistan via a series of resonating stories.[13]

At some point in time, prior to capturing Kabul, the Taliban changed their stated objectives from bringing peace and stability to establishing a "*pure Islamic state (Emirate)*," which included the creation of a religious police (the Committee for the Promotion of Virtue and the Prevention of Vice), appointing only "pious" Muslims to government positions, establishment of Sharia law and an economy based solely on the Quran and Sunnah. They also strictly enforced the burqa—head to foot hijab—for women, prohibition of any "alien cultural influences," and education that encouraged jihad for all Afghans.

Initially after taking Kabul in 1996, the Taliban created an inner council (the Inner Shura) that maintained primacy and an outer council (or Central Shura) for foreign relations and administrative purposes. This structure proved cumbersome and ineffective for managing a modern nation-state, so in 1999 the Taliban re-established many of the administrative practices of the previous Afghan governments. It's important to note that Mullah Omar was bequeathed the title *Amir al-Mu'minin* ("Commander of the Faithful") after he donned the alleged cloak of Muhammad that was held inside the Mosque of the Cloak of the Prophet Mohammed in front of 1500 Afghan religious leaders in Kandahar on April 4, 1996, only travelled to Kabul—the traditional Afghan capital—on a couple of occasions during Taliban-rule of Afghanistan. He ruled the country from Kandahar in a complex built for him by Osama bin Laden. It is also important to note that the five-person Shura that oversaw the administration of Kabul consisted of Taliban who only spoke Pashto in a city that was primarily Dari-speaking. Quite simply and as suggested above, the Taliban proved to be extremely inept administrators.

The Taliban strictly segregated sexes (known as *purdah*). While this is an established practice throughout much of rural (particularly Pashtun) Afghanistan, Afghan urbanites, especially Kabulis, were extremely concerned when the Taliban pushed the practice onto the entirety of the country.

Even after the change of government structure in 1999, which the Taliban intended in part to enhance foreign relations in a bid to gain international recognition, the Taliban's foreign policies remained woefully amateurish. While only three countries formally recognized the Taliban Regime—Pakistan, Saudi Arabia, and the UAE—Mullah Omar's government, shortly after obtaining control of the nation, realized foreign aid was necessary if the Taliban was to retain their power. Hence, the Taliban government did not reject foreign relations, however, their rigid application of Sharia, often violent and draconian domestic policies, and threatening postures to many

[13] Johnson, *Taliban Narratives*.

regional neighbours made contacts with the rest of the world difficult and unpractical. Nevertheless, the Taliban in Kabul, even over the protest of those in the Taliban heartland of Kandahar, worked to obtain United Nations' and other foreign aid.

A central aspect of the Taliban's domestic policy was to provide a haven to many international and regional jihadi groups. Unlike al Qaeda, most of these groups were not global in structure or ideology, rather they were mostly insurgent organizations looking to the Taliban for tactics, techniques, and procedures in overthrowing their own repressive, un-Islamic, Western-influenced governments. Still, though providing haven was part of Taliban practice and policy, the organization never indicated or behaved in any manner that suggested the Taliban were overly interested in the international jihad or terrorism. Rather they were primarily interested in fighting to gain total control of Afghanistan and impose an Islamic Emirate.

From 1997 to 2001, as the Taliban further consolidated control from the Northern Alliance, they simultaneously failed to achieve international recognition while increasing their radical behaviours. By March 2001, and increasingly influenced, but not controlled, by the most radical of Salafis (al Qaeda), the Taliban rid the country of non-Muslims (excluding Afghan Sikhs and Hindus who the Taliban forced to wear special badges) and destroyed the Buddhist statues of Bamiyan. Military and political success only further motivated the Taliban to push Afghanistan into a state of "pure Islam" grounded in the Wahhabi and Deobandi teachings of their youths (and reinforced by their Salafist, al Qaeda allies).

By 2001, while the Taliban were successful in pushing their stated policies onto Afghans, but they were never successful in creating jobs, infrastructure, and institution development, or brought security to the entire nation. Moreover, the United States and their allies would destroy the Taliban regime in November 2001 because they "harboured" the architects (bin Laden and AQ) who planned and executed the September 11, 2001 attacks. But it is important to understand that the Taliban, including Omar, had no specific knowledge of AQ plans and no Afghans participated in the attack.

In his statement to the nation on the night of September 11, 2001, President George W. Bush outlined US strategy emanating from the 9/11 attacks: to capture or kill those responsible for the attacks—including nation-states providing safe-haven. In his post-attack address to the nation, Bush stated, "We will make no distinction between the terrorists who committed these acts and those who harbour them."[14]

On Sunday, October 7, 2001, the United States with the support of the United Kingdom commenced its Operation Enduring Freedom (OEF) air campaign against AQ and the Afghan Taliban. Within four days the US had exhausted its initial target set. During announcing the commencement of operations in Afghanistan, President Bush again reemphasized the multi-national aspects of the coalition.

While there is no doubt that the Bush Administration was successful in creating a coalition to initially support President Bush's metaphorical Global War on Terrorism, this coalition, quite frankly, was never successful in Afghanistan.[15] While all the

[14] Rhem [24].

[15] See the brilliant book by Whitlock [25].

specifics of this failure are beyond the scope of this chapter, one thing is clear: coalitions should not be viewed as a panacea that can solve all of the problems facing the United States and the West. After 20 years, Afghanistan continued to be at war with itself, a corrupt economic basket case and, most importantly, the Taliban were never close to being defeated, and now IS-K is also conducting operations the country. Many have argued that Afghanistan has become another Vietnam for American foreign policy and prominent to this argument is the proposition that the US coalition in Afghanistan has proven to be an interesting alliance, to say the least.[16]

President Obama, who greatly increased US involvement in Afghanistan during his tenure, claimed during his West Point "Afghan surge" speech of December 2009 that Afghanistan was different from Vietnam because it was conducted by a "coalition" of 43 countries (during the "height" of international involvement in Afghanistan). I would suggest that the coalition did *not* have an overly significant impact on military or political results in Afghanistan and the coalition in Afghanistan was quite different quantitatively as well as qualitatively when compared to the U.S. coalition in Vietnam. As I have argued elsewhere, the U.S. Afghan coalition was basically a throwback to President Bush's "Iraq coalition of the willing" mathematics. The truth is that significantly more foreign troops *fought* alongside the United States in Vietnam than fought with the Americans at the height of their involvement in Afghanistan.[17]

President Trump greatly reduced the US presence in Afghanistan and was dedicated to leaving the country completely.[18] During much of the later part of his administration, Trump pursued "negotiations" with the Taliban. During July 2018 U.S. officials met secretly with the Taliban in their new office in Doha Qatar.[19] Shortly after this meeting, Trump appointed Zalmay Khalilzad to led efforts to negotiate with the Taliban. These negotiations eventually led to the February "Peace Agreement" of February 29, 2020, with the Taliban that included clauses that all NATO and US forces would be withdrawn from Afghanistan within 14 months and the Taliban would prevent terrorists, especially al-Qaeda from operating in Afghanistan and that the Taliban would conduct intra-Afghan talks with the Kabul Regime.

The Taliban had risen to power in Afghanistan based on their ability to adapt, innovate, and overcome seemingly insurmountable obstacles.[20] And as important was Afghanistan's population extreme sickness of decades of war. It's also very important to recognize that the Taliban of 1996, 2001, 2006 and 2020 actually had little in common except for major political goals.

---

[16] Johnson & Mason, "Refighting the Last War".

[17] Johnson and Mason [26].

[18] "Trump says he wants full Afghanistan pullout but sets no timeline," *Al Jazeera*, 26/5/2020, accessed on 17/9/2020, at: https://bit.ly/2ZL0FLu.

[19] Shah and Nordland [27].

[20] Johnson, "Taliban Adaptations and Innovations".

3 Analysis

While this chapter has and will focus on the history, organization, ideology and future of the Taliban and introduced the peace negotiations between the United States and the Taliban, we will now explicitly focus briefly on the Taliban's approach to the negotiations[21] and any implications for the transformation of the Taliban into a political party. Here we will explicitly focus on the research question presented as to why and how the inclination towards negotiations/compromise has happened? While an earlier paper focused on these questions by assessing what the Taliban has overtly stated through the media. Specifically, Taliban statements from Twitter social media.[22]

The data examined were gathered from reams of data from the Taliban's official website—Voice of the Jihad and Taliban media statements on the negotiations. Only Taliban Twitter social media statements was assessed by this earlier analysis.[23]

There are a variety of seemingly narratives and associated stories that are either directly presented or directly mentioned by the Taliban over the period of this study (2018-Present) that deserve comment:

When violence or military operations were announced by the Taliban it seemed to largely coincide with rounds of peace talks since early 2018; one interpretation of this is that it made perfect sense for the Taliban to have posted such accounts to suggest that they are coming to the negotiating table from a position of strength (albeit real or perceived) rather than from of position of weakness.

The Taliban often posted messages focusing on both military and civilian population "realizing the truth" and abandoning their government posts to join the Taliban where the Taliban suggested they were always welcomed graciously, given gifts and could return to a "normal" life.

The Taliban found it advantageous to appear as a united front. But certain events suggested that the Taliban was still facing explicit factional problems that could impact on the "peace negotiations."[24]

The Taliban also placed considerable emphasis on the Islamic Emirate of Afghanistan. One of the main questions has always been the actual level of sincerity the Taliban professed in their desire for peace and freedom of all Afghan citizens. I know of no other peace negations that was not accompanied by a cease-fire between conflicting parties.

The Taliban also focused many posts on the COVID-19 pandemic and how it obviously disrupted all facets of both U.S. and Taliban operations in 2020–2021.

---

[21] While there have been attempts at negotiations especially during the Karzai negotiations, our analyses will focus primarily on the negotiations starting in 2018 culminating with the peace agreement signed between the U.S. and Taliban in February 2020. Afghanistan even held a Peace Jirga in June 2010, but the Taliban refused to participate.

[22] See The Afghan Taliban and the "Peace Negotiations," An Analysis of Narratives, Stories and Messages: Are the Taliban "De-radicalising?," in Omar [17].

[23] Ibid.

[24] O'Donnell [28].

They also focused on what they view as poor efforts by Kabul to protect the people from the pandemic.

One could forcefully argue that real and sincere negotiations historically have been accompanied by a cease-fire between warring parties. While there have been a variety of cease-fires between the Taliban and Kabul especially around Eid, the end of Ramadan, in recent years (2019, 2020), they were usually limited to 3 days and after its end the Taliban immediately returned to combat. Why didn't Khalilzad demand a cease-fire during the negotiations? It is not hard to argue that peace negotiations without a cease-fire are an in fact an oxymoron. And, in fact, that is what they turned out to be.

On the surface it appears that the Taliban basically defined the parameters of the negotiations and what we find particularly disturbing is that Afghanistan is not a colony of the United States, and it was highly questionable for the U.S. to negotiate Afghanistan's future without Kabul's participation. One could also argue that the mere fact that the Taliban were able to hold initial negotiations based on their demands suggest that they were not necessarily interested in holding impartial negotiations with the Afghan Government to end this 19-year conflict.

Several things should be discussed concerning the U.S.—Taliban negotiations and the resulting "treaty." First, Khalilizad rarely discussed his talks with the Kabul. Considering that he was literally negotiating Afghanistan's future without any significant input from Kabul, one can reasonably make the argument that Khalilizad and the U.S. were treating Afghanistan as it was a U.S. colony. This became quite evident when Ghani was extremely disgusted that the U.S. was negotiating issues such as prisoners' release without Kabul's input. Second, the "treaty" resulted in no permanent calls for a total ceasefire in Afghanistan. There have been few historical peace negotiations not accompanied with a cease-fire of combat. "Peace negotiations" without an accompanying ceasefire is an oxymoron. Third, because of the points made above and additional points made below, the agreement looked more like a plan for a phased withdrawal and other foreigners from the country. Finally, intra-Afghan talks did not occur in a few months, as suggested by the agreement. They were eventually held in Doha starting 12 September 2020. And these talks basically turned out to be little more than political theater. It took two months for the parties to agree on procedures and objectives. And once they were agreed upon the Taliban took a two-month break from the talks clearly wanting to see what President Biden's view of the treaty and war would be. The success of the negotiations was preordained to be a failure when a key assistant to Mullah Baradar, the co-founder of the Taliban and chief Taliban negotiator, declared the Taliban had two *nonnegotiable demands* for the talks: the reestablishment of the Islamic Emirate of Afghanistan and reinstitution of the Taliban's radical view of Deobandi Islamic Sharia law.

While the peace negotiations, signed on February 29, 2020,[25] did not include one of the major warring parties participating—the Kabul Government. The Taliban had

[25] The actual agreement was the result of 9 different rounds of negotiations involving Khalilzad and Taliban representatives recently lead by Mullah Abdul Ghani Baradar, head of the Taliban's office in Qatar.

long viewed the Afghan Government in Kabul as merely a "puppet" of the U.S. and refused to have Kabul at the negotiating table with the U.S. The Taliban had brief conversations with former President Karzai in Moscow in February 2019 but these talks again did not involve the Afghanistan Government.[26] As with the lack of a demand for a cease-fire to be held during the talks, it is hard to understand why Khalilzad did not raise the issue of having Kabul participate in the negotiations, even just to be at the table and not allowing to speak so at Kabul had exact knowledge of how the U.S. was negotiating Afghanistan's future. While the February 2020 agreement did have a stipulation that intra-Afghan negotiations were scheduled for March 10, 2020 in Oslo, Norway, the disrupted nature of the 2019 Afghan Presidential Election did not allow for the creation of a Kabul negotiating team.[27] The U.S. also agreed to the Taliban demand that 5000 Taliban prisoners be released and before intra-Afghan talks would take commence.[28] Ghani, who had absolutely no involvement in this agreement, initially rejected it. Direct negotiations were expected in March, according to the U.S.-Taliban deal, but the start had been delayed by disagreements over the swap of 5000 Taliban prisoners for 1000 Afghan security forces. The Afghan government initially opposed the plan, saying that they had played no part in the negotiations, but conceded after much pressure from the Trump administration.[29]

The treaty had four main provisions:

A temporary reduction in violence leading to a lasting cease-fire among U.S., Taliban, and Afghan forces will be part of intra-Afghan negotiations.

The United States agreed to reduce its number of troops in Afghanistan from roughly 12,000 to 8600 within 135 days. All U.S. and other foreign troops would leave Afghanistan within fourteen months (1 May 2021).

The Taliban agreed to start talks with the Afghan government in March 2020. Throughout the negotiating process, the Taliban had resisted direct talks with the government, calling it an American puppet. But the Taliban then indicated that talks were possible, with deputy Taliban leader Sirajuddin Haqqani writing in a *New York Times* op-ed, "If we can reach an agreement with a foreign enemy, we must be able to resolve intra-Afghan disagreements through talks."

The Taliban guaranteed that Afghanistan will not be used by any of its members, other individuals, or terrorist groups to threaten the security of the United States and its allies.

On March 1 Ghani stated that he would reject the prisoner exchange:

> The government of Afghanistan has made no commitment to free 5,000 Taliban prisoners. […] The release of prisoners is not the United States authority, but it is the authority of the government of Afghanistan.[30]

---

[26] Higgins and Mashal [29].

[27] Kermani [30].

[28] Ibid.

[29] Mashal et al. [31].

[30] "President Ghani rejects peace deal's prisoner swap with Taliban," *Al Jazeera*, 1/3/2020, acceseed on 17/9/2020, at: https://bit.ly/2ZH0kJG.

It is very clear that the Taliban of 2022 is not the Taliban of 2001, in fact it appears and can be persuasively argued that the Taliban are at a crossroads since they took over Afghanistan in August 2021. They clearly desire international recognition, but will their domestic policies allow for such recognition? Over the last 20 years they have managed to win the war of attrition and "forced" the U.S. out of the country.

It is clear that political and ideological goals associated with their re-emergence of the Islamic Emirate of Afghanistan as an institution and operationalization of the Taliban's view of sharia law. There is no convincing evidence that the Taliban was willing to voluntarily de-radicalise in any significant way.

4 What Went Wrong in Afghanistan and What Are Their Implications

As Afghanistan's major cities and border crossings fell in early August 2021 shortly after the U.S. withdrew from the country. The U.S. began the hard lessons of intervention and occupation of Afghanistan.[31] We have been Afghan students for all that period. In the 1980s the Soviet Union invaded Afghanistan, killed millions of Afghan citizens, and created the greatest refugee crisis of modern times. During the last 20 years the United States invaded Afghanistan, spent over a trillion dollars, and was part of a war that killed more than 170,000 Afghan citizens. The reasons for this incessant conflict are many, but at the top of the lists of reasons have been foreign invasions and occupations along with severe ethnolinguistic cleavages that collectively fuel significant mistrust among large swaths of the county's populations and politics. Terrorist groups and regional interlopers have also contributed to the Afghan instability and conflict. For the US period in Afghanistan, however, we would add American uncertainty about what we hoped to accomplish there, a regularly changing strategy laid on top of an acute misunderstanding about the reality of the situation, and the added and profoundly unnecessary complexity of the American war in Iraq.

5 The Taliban Takeover of Afghanistan

On November17, 2020, Acting US Secretary of Defense stated that the United States would draw down their troops to 2500 by mid-January 2021:

Acting U.S. Defense Secretary Christopher C. Miller announced plans to halve the number of troops in Afghanistan to 2500 by mid-January, days before President-Elect Joe Biden will be inaugurated. Thousands of troops had already been pulled out following an agreement with the Taliban in February, moving closer to fulfilling President Trump's campaign promise to end the so-called forever wars. The announcement

[31] Craig Whitlock, op. sit.

comes as negotiations between the Afghan government and the Taliban are deadlocked and the militant group continues to launch deadly attacks. NATO Secretary-General Jens Stoltenberg warns that withdrawing troops too early could allow Afghanistan to become a haven for terrorists and the Islamic State to rebuild its caliphate.

President Biden eventually announced that the U.S. would not meet the deadline set under the U.S.-Taliban agreement to withdraw all troops by May 1, but the U.S. would be fully withdrawn by September 11, 2021. The Taliban soon started their offensive to take Afghanistan. The Taliban's final drive to takeover Afghanistan started on May 1, 2021 and was completed by August 15, 2021. The quick takeover followed the Taliban's rapid and aggressive advance. During July and August, the Taliban captured all but two of Afghanistan's provincial capitals and seized border crossings. Afghan security forces refused to engage the Taliban and quickly deserted. The size and capabilities were long purposely exaggerated by both the U.S. and Kabul.[32] Chaos almost immediately breakout and the evacuation of U.S. and allied personnel was terribly handled with 13 U.S. Service Members killed and 170 Afghans were initially killed.

The Taliban took over the presidential palace hours after President Ghani and many of his cronies left the country with millions of U.S. dollars. This was almost a replication of the Taliban taking the Capital in Fall, 1996.

6 What Went Wrong?

America's withdrawal has already produced long-term profound changes for Afghanistan and the entire South and Central Asian region. America's withdrawal ultimately reflects profound failure brought about by historical and social realities, poor political choices, and bad strategy. Many factors play a role here, but six stand out.

First and possibly the most important: the U.S. did not understand Afghanistan, its culture, people, history, ethnic-linguistic cleavages, and many other underlying and critical dynamics of country. *The brilliant book, The Afghanistan Papers: The Secret History of the War* by Craig Whitlock suggested that "the United States had jumped into the war with only a hazy-idea of whom it was fighting"—a fundamental blinder from which it would never recover.[33] Similar statements by nearly all U.S. commanders in Afghanistan echoed how destructive our basic lack of knowledge of Afghanistan and our "mission". Moreover, the U.S. military and Kabul consistently misled their populations and the international community as to its ability to secure the country and defeat the Taliban.[34] Anyone interested in the Afghanistan War and the role of the U.S. should read Craig Whitlock's book.

---

[32] Craig Whitlock, op. sit.

[33] Craig Whitlock, pg. 19.

[34] Whitlock, ibid.

Second, of course, when the U.S. invaded Iraq in March 2003, Afghanistan became the "forgotten war." In 2003 over 300,000 American troops invaded Iraq, while less than 10% of that number of Americans were in Afghanistan. Neither war turned out well for the United States and there was perhaps some utility in being the forgotten war, but certainly Iraq overshadowed Afghanistan for much of the last twenty years, often assisting the U.S. march toward doom, allowing our policy makers, generals, and ambassadors to talk of "stalemates," "progress," and "the light at the end of the tunnel."

Third, the U.S. probably most important major failure was that the U.S. had no idea how to crush an opponent in remote, rural Afghanistan and then put something more stable in place. The literal handful of Afghan specialists, including myself, had few good ideas about what to do, as any political leaders in Afghanistan were generally warlords and criminals while the lucky Afghan exiles were either irrelevant and forgotten or trundling up to the water trough as modern-day carpetbaggers. After beginning our bombing of Afghanistan on 7 October 2001, which quickly eviscerated Taliban and Al-Qa'ida targets, the U.S. began a campaign to destroy the Taliban that went much quicker than expected. By early-November the Taliban regime was destroyed, and the U.S. had to cobble together a political future for Afghanistan through meetings held in Bonn, Germany in late November 2001. These meetings only included the Northern Alliance of northern minority groups that had driven out the Taliban with US help, representatives of the former king, and two small groups of former mujahideen exiles; the Bonn meetings did not include Taliban moderates in the negotiations. The subsequent Bonn Accords allowed the Northern Alliance to claim most of the important Afghan ministries in an "Interim Authority" government that was led by Hamid Karzai of the exiles, while the erstwhile king was frozen out. In June 2002 the Afghans held an Emergency Loya Jirga (Grand Assembly) to select an Interim Administration, with which one of us was intimately involved. The plan was that all warlords would be kept out of the government, but notwithstanding the efforts of the UN election monitors to achieve that goal, the front row seats of honor at the Loya Jirga were full of warlords and criminals, placed there by the Interim Authority. Not surprisingly, the government was quickly filled with all these men. This led to Francesc Vendrell's[35] observation that "[t]he return to power of persons widely despised and dreaded by most Afghans for the atrocities and sleaze that had characterized their rule during the mid-1990s ensured that from the very beginning bad governance and corruption became the norm."

These disastrous early steps, taken by the U.S. that demonstrated that it had the nuanced understanding Afghanistan at all, might have been overcome, but the United States also failed to understand how to rebuild Afghanistan (or Iraq). In a country with little economy other than subsistence agriculture and narcotics production, enormous poverty, politicians who mostly controlled militias as their basis of claiming power, and an ongoing civil war, there was no good plan for rebuilding Afghanistan. The United States turned to the UN to run elections, the international community for

[35] Francesc Vendrell, "What went wrong after Bonn," MEI, https://www.mei.edu/publications/what-went-wrong-after-bonn.

money (often promised but not paid) and forces (which were scattered to different provinces and controlled by the home country rather than the commanding general) and divided the US mission into counter-terrorism (CT) and what eventually came to be called counter-insurgency (COIN), which itself transformed later into a mission to build the Afghan Defense and Security Forces (although the false numbers presented undermined this effort).

The Bonn Accords put in place a political map that was unachievable,[36] but the United States made it worse by helping a Constitutional Loya Jirga in 2003–2004 develop a constitution that created a system of government with so many elections (rare in Afghanistan) that it was almost impossible for them to be held. For example, the district elections required as the basis for one-third of the seats for the upper house of Parliament, were never held. All these elections in a country still controlled by warlords contributed to the tremendous corruption that would dominate Afghan politics for the next twenty years. Moreover, it encouraged tremendously fraudulent elections, most significantly Ashraf Ghani's "victories" in 2019 and 2014 and Hamid Karzai's "victory" in 2009.[37]

Fourth, the U.S. combat performance in Afghanistan failed miserably. Too few soldiers were sent to control a large rural country, and the strategy was divided between counterterrorism (CT) and counterinsurgency (COIN). To find more soldiers and police NATO was asked to join and the hated militias of the warlords were able to survive by transforming into security companies. Collectively, these militias caused significant problems not only in the Afghan military's response but also to a wide variety of political intrigues.

Ever since the 9/11 attack on the United States, the conflict between CT and COIN was almost as problematic as the conflict between the Taliban and US-supported Kabul government. Many of the early CT actions significantly violated Afghan—especially Pashtun—tribal values and codes and helped push many rural Afghans to support or at least aid the Taliban. While it is overly simplistic to refer to rural Afghanistan as simply an honor-revenge society, many Afghans can be greatly influenced by this dynamic. In most of rural Afghanistan Afghan man's honor for a millennium has been defined in a very real sense by how family females who exist in a state of purdah are treated, as well as how their possessions and property are treated. For well over two years, the U.S. raided rural Afghan compounds in the middle of the night to grab and bag prospective "terrorists" and in so doing forced sleeping women

---

[36] See ibid.

[37] For analyses concerning Afghanistan's deeply flawed elections, see: "The Afghan Taliban and the "Peace Negotiations," An Analysis of Narratives, Stories and Messages: Are the Taliban "De-radicalising?,"in Omar [17]. Book also published in Arabic by Arab Centre, Doha, Qatar; "Chaos, Confusion and Fraud: An Examination of Afghanistan's 2018 *Wolesi Jirga* Elections," Journal of Asian Security and International Affairs,, 7(1) pp. 57–100, April, 2020; "The myth of Afghan electoral democracy: the irregularities of the 2014 presidential election," *Small Wars & Insurgencies*, 29:5–6, 2018 1006–1039. Published in Dari (https://aiss.af/assets/aiss_publication/The_Myth_of_Afghan_Electoral_Democracy_The_Irregularities_of_the_2014_Presidential_Election(Farsi).pdf.) and;"The Illusion of Afghanistan's Electoral Representative Democracy: The Cases of Afghan Presidential and National Legislative Elections," *Small Wars & Insurgencies*, 29:1, 2018, 1-37.

in their bedclothes out into the open compound. This is possibly one of the most egregious actions that a non-family member can take against another and immediately caused family members to regain the family's honor by seeking revenge. These were exactly the kinds of actions the Taliban hoped the United States and its allies would pursue. *The overall military goal of the Taliban was to encourage the United States into making such terrible miscues.* Likewise, the Taliban wanted American or NATO maneuver elements to chase a group of Taliban or madrassa students who would then seek the relative safety of an Afghan village and then hope that the U.S. Airforce would drop 500 pounders. The Taliban understood that if an errant weapon killed an innocent Afghan woman and/or child, that village was lost to the United States (and its Kabul allies) forever. Another example is that the United States arrested and transported hundreds of Afghans to Guantanamo and other foreign detention centers under the CT campaign, keeping many of them for years with no charges or ability to be released.

Fifth, the United States and NATO fighting the Taliban from Forward Operating Bases (FOBs) "with Banker's hours" was counterproductive in and of itself. To defeat a group like the Taliban demands near continuous presence in the rural areas that ultimately were the military "center of gravity" of this conflict. The support or at least help of the rural population is critical, and the military generally proved too lazy or ignorant of the desires and needs of rural Afghans, which meant that the Taliban could completely dominate in these critical areas, especially with their narratives and stories that addressed the collective memories of Afghans and were delivered via traditional Afghan distribution efforts. Later the Taliban develop an extremely effective presence on the internet and especially in social media.

Finally, the fifth factor was the broader reality of events happening elsewhere. Two cases stand out, both of which had major implications for the military situation in Afghanistan. First, the Taliban, being predominantly Pashtun, fled across the border into Pakistan, both to Pashtunabad, just outside of Quetta, Baluchistan, and to the Federally Administered Tribal Agencies between Khyber-Pakhtunkwa and the Afghan border. These heavily Pashtun areas, were the old digs of the mujahideen from the 1980s, and once again allowed the Taliban to be influenced, but not controlled, by the Pakistani Military and especially Pakistan's intelligence agency, the ISI. Moreover, it is near impossible to find an insurgency that was defeated that had such an easy refuge in a contiguous state. President Barrack Obama's CT assets found and killed Al-Qa'ida leader Osama Bin Laden in Abbottabad, which is part of Khyber-Pakhtunkwa, Pakistan in May 2011. Pakistan controlled the major access route into Afghanistan as well as sheltering the Taliban, making Pakistan into a complicated "frenemy" of the United States.

7 Conclusion

And so, here we are. The U.S. went to war in Afghanistan with an inadequate understanding of the country, its people and its neighborhood. A misguided commitment to U.S. interests prompted mission creep that made America hang around indefinitely in Afghanistan. The U.S. can no longer be a great power if its people do not choose to carry the burdens of leadership that great powers must carry. The U.S., racked with extremism and racial challenges, disagreement over whether its citizens should be vaccinated against an infectious disease, challenges to its democracy, and unwillingness to join to do anything, has also lost almost every significant war since World War II. I deeply regret what has come to Afghanistan, but also the world, which may only be able to see U.S. perfidy as the profound failure of a declining power.

As for Afghanistan, the likelihood that the U.S. could stand against the realities of world politics and help Afghanistan rise, just would not happen. It was a disaster from the beginning to the end. Many good, inspired, brave, and even noble people from all over America and around the world tried very hard to make Afghanistan a success. Others schemed to thwart all that effort. They won, and we failed.

I am convinced the Taliban would have fought for another fifty years to accomplishment their Islamic, political, and social goals. The "peace talks" were merely political theater based upon the reasonable assumption that two very different US Presidential administrations (Trump and Biden) were keen to get out of Afghanistan.

References

1. George S (2020) U.S. signs peace deal with Taliban agreeing to full withdrawal of American troops from Afghanistan. Washington Post
2. Barfield T (2010) Afghanistan: a cultural and political history. Princeton University Press, Princeton
3. Tomsen P (2013) The wars of Afghanistan. PublicAffairs, New York
4. Crews RD, Tarzi A (eds) (2008) The Taliban and the crisis of Afghanistan. Harvard University Press, Cambridge
5. Maley W (ed) (1998) Afghanistan and the Taliban. New York University Press, New York
6. Giustozzi A (2001) Afghanistan's endless war: state failure, regional politics, and the rise of the Taliban. University of Washington Press, Seattle
7. Giustozzi A (2012) Decoding the New Taliban: insights from the Afghan field. Columbia University Press, New York
8. Johnson TH, Narratives T (2018) The use and power of stories in the Afghanistan conflict. Oxford University Press, Oxford
9. Johnson TH, Mason WC (2007) Understanding the Taliban and insurgency in Afghanistan. Orbis 51(1)
10. Maley W (ed) (1998) Fundamentalism reborn? Afghanistan and the Taliban. New York University Press, New York
11. Rashid A (2000) Taliban: Militant Islam, oil, and fundamentalism in Central Asia. Yale University Press, London
12. Khan RM (2011) Afghanistan and Pakistan: conflict, extremism, and resistance to modernity. Johns Hopkins University Press, Baltimore
13. Marsden P (2002) The Taliban: war and religion in Afghanistan. Zed Books, New York
14. Abbas H (2014) The Taliban revival: violence and extremism on the Pakistan-Afghanistan frontier. Yale University Press, London
15. Johnson TH, Mason WC (2009) Refighting the last war: Afghanistan and the Vietnam Template. Military Rev

16. Johnson TH, Adamic L (2021) The historical dictionary of Afghanistan. Rowman & Littlefield, forthcoming, Lanham
17. Omar A (ed) (2021) Bullets to ballots: transformations from armed to unarmed activism. Oxford University Press and Edinburgh University Press
18. Giustozzi A (2015) The army of Afghanistan: the political history of fragile institution. Oxford University Press, New York
19. Giustozzi A (2019) The Taliban at war: 2001–2018. Oxford University Press, New York
20. Tse Tung M (2007) Mao Tse tung on guerrilla warfare. BN Publishing, Hawthorne
21. Giustozzi A (2009) Koran, Kalashnikov, and laptop. Oxford University Press, Oxford, pp 89–90
22. Farrell T, Osinga F, Russell JA (eds) (2013) Military adaptation in Afghanistan. Stanford University, Stanford
23. Johnson TH (2013) Taliban adaptations and innovations. Small Wars Insurgencies 24(1)
24. Rhem KT (2001) Bush: no distinction between attackers and those who harbor them. DoD News. https://bit.ly/3c7MRzz. Accessed on 17 Sept 2020
25. Whitlock C (2021) The Afghanistan papers: the secret history of the war. Simon & Schuster, New York
26. Johnson TH, Mason WC (2009) Obama's indecent interval. Foreign Policy. https://bit.ly/33ANbmB. Accessed on 17 Sept 2020
27. Shah T, Nordland R (2018) U.S. diplomats held face-to-face talks with Taliban, insurgents say. The New York Times. https://nyti.ms/3kpYbKh. Accessed on 17 Sept 2020
28. O'Donnell L (2020) Factional struggles emerge in virus-afflicted Taliban top ranks. Foreign Policy. https://bit.ly/32DlFFN. Accessed on 17 Sept 2020
29. Higgins A, Mashal M (2019) In Moscow, Afghan peace talks without the Afghan Government. The New York Times. https://nyti.ms/3kqGSc9. Accessed on 17 Sept 2020
30. Kermani S (2020) Afghan conflict: what will Taliban do after signing US deal? BBC News. https://bbc.in/3hEOiH5. Accessed on 17 Sept 2020
31. Mashal M, Faizi F, Rahim N (2020) With delay in Afghan peace talks, a creeping sense of 'Siege' around Kabul. The New York Times. https://nyti.ms/3c7He4o. Accessed on 17 Sept 2020

Thomas H. Johnson is a Research Professor of the National Security Affairs Department at the Naval Postgraduate School (Monterey, California) as well the Director of the Program for Culture & Conflict Studies. For over three decades, Professor Johnson has conducted research and published widely on Afghanistan and South Asia. He is the author of the recent book, Taliban Narratives: The Uses and Power of Stories in the Afghanistan Conflict was co-published by Oxford University Press and Hurst Publishers (London) in 2018 His most recent book is, The Historical Dictionary of Afghanistan with the late Professor Ludwig Adamic, (Rowman & Littlefield, forthcoming, May 2021). The views expressed in this chapter are the author's alone and should not be construed as an official position or policy of the U.S. Government, Department of Defense, or Naval Postgraduate School. The author would like to thank Nathan Herbert for his invaluable research assistance on sections of this chapter. This chapter is partly based on: The Afghan Taliban and the "Peace Negotiations," An Analysis of Narratives, Stories and Messages: Are the Taliban "De-radicalising?," in Omar Ashour (ed.), Bullets to Ballots: Transformations from Armed to Unarmed Activism, (Oxford University Press and Edinburgh University Press, May 2021); "What Went Wrong in Afghanistan: A Primer," with Larry Goodson, War Room, Army War College, warroom.armywarcollege, 17 September. 2021;

"How America Lost Afghanistan," with Larry P. Goodson, Newlinesmag.com, August 18, 2021; and,

"Political Legitimacy: Why we are failing in Afghanistan," with Larry P. Goodson, The Strategy Bridge, July 2221, https://thestrategybridge.org/the-bridge/2021/political-legitimacy-why-we-are-failing-in-afghanistan.

The Afghan Evacuation: A Case Study, Observations and Ethical Strategies Going Forward

Mohammed Moh Kilani

This chapter is dedicated to Celes Eckerman, Melody McCoy, Jenna Ben-Yehuda, Malcolm Phelan, Bryant Jones, Aaron Bratcher, Erica Kaster, Katherine Maher, Rose Jackson, Jeff Phaneuf, Jon Margolick, Jennifer Atala, Ya'ara Barnoon, Chris Purdy, Camille Mackler, Elizabeth Spencer, Maru Sefami, Miriam Magdieli, Heather Wild, Jason Coleman, and Stephanie Rudat.
"We hear you have a plane."

Abstract Over an 18 day period during August 2021, the United States evacuated 124,000 Afghans, but left behind the majority of interpreters and other allies. This chapter will examine the Afghan evacuation, its seven observed components, create a basic metric for evaluation, and propose a strategy built on those observations to solve the Afghan SIV crisis. The proposed strategy is to develop multiple, repeatable points of egress (pipelines) that operate persistently from Afghanistan's neighboring countries to systematically evacuate Afghan allies to long-term host countries in ways that are (a) *measurable*, (b) *statistically significant* and (c) *qualitatively accurate*. Much of the information in this chapter is direct research from the author's experiences before, during and after the evacuation. Recommendations to follow.

Keywords Afghanistan · Evacuation · Operations · Strategy · Geopolitics · Foreign policy · SIVs · Counter-terrorism · Global security

1 Introduction: A Human Perspective

Zahra stood outside the gates of Hamid Karzai International Airport (HKIA) in a ditch for four days along with her husband and children. As a Hazara woman who worked in reproductive health, Zahra was at-risk and in danger from reprisal by the Taliban. She had received numerous verbal and written threats. Like tens of thousands

M. M. Kilani (✉)
Truman National Security Project, Washington, DC 20005, USA
e-mail: mjkilani@gmail.com

A. Farhadi and A. Masys (eds.), *The Great Power Competition Volume 4*,
https://doi.org/10.1007/978-3-031-22934-3_12

of Afghans affiliated with Western institutions, Zahra was desperately attempting to leave the country as the new Taliban regime brutally asserted its dominance.

When the opportunity presented itself in the chaos and Zahra made it to Abbey Gate, manifested on a flight, ready to go—the gate commander stopped her saying "*no US visa, no entry.*" Getting Zahra through the gate required the combined effort of foreign policy and national security intelligentsia leveraging personal relationships with Cabinet-level officials, CENTCOM leadership, a foreign consulate official and a sitting US Senator, intervening to compel the gate commander to let Zahra and her family through. Zahra with family in-tow proceeded to run towards Apron 8, toward her flight, towards safety and a new life. But the complications in the chaos and deep inequality had only begun.

Zahra was taken off the privately funded Airbus A340 by unknown Americans, placed on a military C-17 and flown to Qatar, instead of the original destination of a European country. She was told the A340 was to be occupied by "*Afghan soccer players*"—presumably, a euphemism for high-risk CIA assets who needed exfiltration immediately. Apron 8 in HKIA was rumored to have been under the control of the CIA.

From Qatar, she eventually made her way to the US.

Zahra would prove to be one of the relative few who were able to leave, largely as a consequence of her personal ties to influential American institutions and a little bit of luck. Many Afghans who had worked with the US were not as fortunate. This profound inequality was to become emblematic of the crisis.

2 Catastrophic Failure, Metrics for Success

The deep inequality evidenced in Zahra's story reflects the difficulty of evacuating during the Afghan Noncombatant Evacuation Operation (NEO). The Afghan NEO was catastrophically bottlenecked due to a complete absence of planning, contingent or otherwise, whether via private or governmental means. The NEO itself cannot be called anything other than a disastrous failure; any other description strains credulity, evoking political posturing and the rhetoric of government spokesmen. Consequently, this absence of planning led to a severe lack of operational resources. But what are the minimal criteria for a successful evacuation in an emergent and rapidly evolving crisis?

A successful noncombatant evacuation must include not only quantifiable significance but also a simple measure of qualitative accuracy—in other words, (a) *an overwhelming statistical majority* of (b) *a specific demographic*; i.e., 85% of Special Immigrant Visa (SIV) applicants or residents of a specific location. These success criteria represent a dual metric of success. For that dual metric to succeed, the modalities of evacuation—the operational *what, where, how*—must be planned thoroughly to the very end, accounting for every aspect of the operation.

When applied to the Afghan NEO, the above dual metric fails. Indeed, the US government evacuated 124,000 Afghans–the majority of whom were not SIVs [1].

These Afghans were a random collection of anyone who had 3rd country visas, SIVs, American citizens, Green Card holders, and humanitarian parole applicants who managed to make it into the gates of HKIA, by some means or another. By admission of the US Department of State, the majority of SIVs, a demographic of approximately 70,000 (17,000 SIV applicants, and the remaining 53,000 as dependents) were left behind in Afghanistan [1]. Thus, while quantifiably a success, the NEO should be regarded as a failure because its efforts were applied inaccurately across a cross-section of Afghans fleeing the Taliban.

But if the NEO is regarded as a failure, what were the root conditions that led to such a fully avoidable and visibly chaotic scene of human suffering? What were the consequences and what will they be going forward for US national security concerns?

3 The Upstream Fundamentals: Political Unwillingness, Lack of Planning, Disregarding Expertise

The catastrophic failure of the NEO was largely due to the political unwillingness to address bureaucratic and legal challenges of the SIV program from its inception. One could argue that the failure of the NEO is rooted in the failure of the SIV program, itself rooted in the failure to reform the US immigration system writ large for decidedly political reasons. Contextually, this lack of political will can be seen at two critical junctures: (1) *the unwillingness or inability to prioritize the SIV program over its 13 year lifespan, across multiple administrations of both parties* and (2) *the unwillingness of policymakers to heed the guidance of subject matter experts to begin the evacuation of SIVs six months in advance of arbitrary withdrawal dates due to the assessed, impending collapse of the Ghani government*. This chapter will not address the unwillingness or inability to prioritize the SIV program prior to the withdrawal, as it has been discussed elsewhere. But the lack of political will to proactively support the US government's Afghan allies led to a seemingly complete absence of a transnational evacuation contingency plan, and a resultant paucity of necessary operational resources, chief among them: *time*.

Not only was there seemingly an absence of a plan or planning framework, but the administration disregarded warnings from subject matter experts on Afghanistan, both internal and external to the government [2]. Many of these experts had repeatedly advised for an earlier evacuation response far in advance of the withdrawal. The most prominent examples include the dissent cables from the State Department [3], and the warnings from the CIA that the Afghan government could not hold out against the Taliban advance, with only weeks remaining before total collapse. Externally, a myriad of non-profit organizations and think tanks, working in coalition together, advocated for an earlier evacuation [4].

In particular, Afghan war veterans, human rights organizations, Afghanistan analysts, immigration lawyers and others repeatedly advocated for an earlier evacuation response. This group drew from their collective brain trust—only to be rebuffed by the US policy makers and told the problem was being managed [2]. Was this lack of planning and disregarding of subject matter experts unintentional? Or a deliberate tactic in a domestic political calculus? Or was it simply a stereotypical government response (too slow, methodical and bureaucratic) to a foreseeable crisis?

Whether an intentional political trade-off, or the result of operational decisions built on mistaken calculations, the lack of planning left a vacuum that created chaos. What ensued was a human catastrophe: hundreds of innocent Afghans were killed or wounded from terrorist violence across the country and at the crush at HKIA's gate, in addition to the 13 U.S. Marines killed. Meanwhile, a loose coalition of diplomatic personnel, civil servants, foreign aid workers, military service members, as well as volunteers from the general public coalesced and scrambled to evacuate themselves, their associates and Afghan allies of all sorts, exerting Herculean effort in a disastrous but avoidable situation.

By the time the US government made the formal decision to evacuate on August 15th in the absence of mature planning, the government's lack of agility became apparent [4]. The upstream strategic failure created a tactical and operational Gordian knot of miscommunication, awkward coordination and inappropriate role reversal. This chaotic miasma ranged from US government employees asking external parties for guidance on renting buses in Kabul and how to proceed to the gates of HKIA–to House and Senate offices asking private citizens if they could put specific Afghans on privately-funded planes.

If we posit that the root of the failed evacuation is an upstream lack of political will and planning with the resultant operation constrained from the beginning—what were the components of operational failure and how did they manifest in real time?

4 The Observed Components: Myriad of Operational Constraints, Lack of Resources, Ad Hoc Evacuation

While the dual metric of evacuation success is based on quantitative significance and qualitative accuracy, the operational components are largely quantitative. Any subsequent qualitative problems, like evacuating the incorrect evacuees or risky evacuation measures, are outgrowths of quantitative problems, i.e., a dearth of resources. The primary problem with the NEO—the theme that tinged the entire operation—was a lack of all the necessary components and resources of a successful evacuation. As a consequence of the lack of planning, the entirety of the Afghan evacuation can be characterized as working in a deficit.

This deficit, this lack of resources constrained any ability to evacuate targeted noncombatants accurately in a safe, dignified and organized way. Simply said, there wasn't enough of anything to go around—time, visas, planes, funding, partner countries or points of egress. These resources are generally interdependent and an increase or decrease in one will affect the remainder, e.g., more time allows for more administrative processing or more points of egress and airports allows for more flights, and so on. In the end, crisis management is about managing imperfections, not ideal solutions [5].

4.1 Component 1: Time

First and most important among these diminished resources was time. The fact that evacuation planning began on the day Kabul fell to the Taliban [4] and not months prior, already set the foundation for a high-pressure operation, with increased probability of failure for all actors, public or private. In such a high-pressure and time-constrained environment, all other resource gathering, coordination and execution are races against the clock to meet logistical objectives.

This point becomes increasingly important to private evacuation efforts without access to resources to the same degree of nation-states.

It was precisely this combination of delayed evacuation, a constrained timeline and a target population of 100,000 minimum potential evacuees that created the dangerous bottleneck outside HKIA. The delayed evacuation created exactly what the administration wanted to avoid: the damaging optics of a panicked and out-of-control operation. Such a mass of people could have and did present itself as an open target to violent non-state actors (VNSAs), in particular those with a reputation for savage, indiscriminate killing, like ISIS-K. Only one attack during the NEO can be considered fortuitous, as opposed to a series of attacks, with a potential death toll higher by an order of magnitude. To put it into perspective, the NEO had 18 days from beginning to end, with 124,000 evacuees total, via 218 aircraft, 751 flights and one single point of egress with a daily average throughput of 8571 people.

Yet the SIVs who were in administrative process numbered at roughly 70,000 and ultimately left behind—needed an average of 3888 Afghan departures per day. It can be safely concluded that under the abbreviated timeline from August 15th to the 31st, 2021, it was not possible to process, verify or vet SIVs and their documentation, who had anything less than an issued SIV in-hand. In principle, as evidenced by the raw numbers of the US government-led evacuation, the SIVs could have been *physically* evacuated had an interagency plan been in place to process their paperwork and documentation months in advance. Effectively, processing and evacuating the qualitatively correct subset of Afghans—the SIVs—against the backdrop of tens of thousands of innocent civilians attempting to flee Kabul became impossible. The signal to noise ratio was unfavorable. For the US government, the major constraint was only time—the quantitative factor that rendered the desired qualitative factor less than optimal.

4.2 Component 2: Partner and Host Countries

Secondary to the time constraint was the issue of host countries willing to take in Afghans, SIV or otherwise. Due to the legal necessity of SIV processing occurring outside the United States, SIVs have to be processed in an Immigrant Visa section (IV) in a US Embassy or Consulate. Prior to the fall of the Ghani government and the US withdrawal, Afghan SIVs were processed in Kabul.

With no US diplomatic personnel remaining in the country, SIV processing stopped. In the context of the NEO, all countries surrounding Afghanistan refused to take in any significant numbers of Afghans, citing security as prime motive, with the single exception of Pakistan. The two countries that could not assist the United States in processing the SIVs were Iran, having no US diplomatic presence, and Tajikistan, where the US Embassy did not have an Immigrant Visa section (though this will change starting October 4, 2022). Pakistan's policy, in turn, resulted in hundreds of thousands of Afghans crossing the border.

Naturally, the Immigrant Visa section of the Islamabad post was and remains overwhelmed with SIV processing. It is notable, and again within the theme of a constrained time frame, that the State Department only began to reach out to regional partner countries in July, while the Department of Defense only began to reach out to Qatar in the middle of August. A lack of support from regional partners in turn limited points of egress.

4.3 Component 3: Points of Egress

Points of egress, like airports, were another numerically constrained factor. All evacuation flights during the NEO were either from HKIA in Kabul or Mawlānā Jalāl ad-Dīn Muhammad Balkhī International Airport in Mazar-i-Sharif (MSIA). The majority of flights, as a matter of history, were from HKIA—the operation of which was compromised by the approximately 50,000 Afghans immediately outside and desperate for entry, regardless of visa status or forward pathway.

The remaining two international airports, Ahmad Shah Baba International Airport in Kandahar and Khwaja Abdullah Ansari International Airport in Herat, were seized by the Taliban in August of 2021. This occurrence again underscores the necessity to have begun evacuation months prior to the formal withdrawal. Half the facilities for international travel were available right up to the fall of the Ghani government in mid-August. As a matter of course, more facilities permit more aircraft, and ultimately more flights.

For example, the Department of Defense mobilized 18 passenger aircraft from the Civil Reserve Air Fleet (CRAF) on August 22nd. None of these aircraft flew into HKIA, but only "*from temporary safe havens and interim staging bases*" presumably outside of Afghanistan altogether [6]. Had an evacuation begun 6 months prior, the CRAF could have used all 4 of the international airports within Afghanistan. This point highlights the interdependence of the resources of time, points of egress and aircraft. Aircraft, as the other resources, were another limiting factor.

4.4 Component 4: Civilian Aircraft, Limited Aviation Infrastructure

Transport to and from Afghanistan, like many parts of the developing world, is limited by a smaller aviation infrastructure footprint. The difference is most starkly illustrated by the number of aircraft. The two Afghan national airlines, Kam Air and Ariana Afghan Airlines, had a total of 12 aircraft (7 and 5 respectively), as of September 2021. At an absolute maximum, in an ideal setting, Afghan aircraft could move approximately 2400 passengers at a time. During the NEO, however, a number of these aircraft were non-functional, undergoing maintenance or moved out of the country into Iran for safe keeping. There were 5 or less functional civilian Afghan aircraft during the NEO, reportedly.

This lack of civilian aircraft was compounded by the inability to fly into Afghan airspace and land safely. The NATO decision to allow only military aircraft to land in HKIA, decreased the options for evacuation to three possibilities: (a) *evacuation on a NATO or US military plane from HKIA,* (b) *evacuation on a privately funded charter on a Kam Air or Ariana plane from HKIA or MSIA, or* (c) *fleeing overland to a bordering country.*

The combination of the deteriorating security situation outside the gates of HKIA, particularly after the suicide bombing, the NATO decision to ban civilian aircraft from landing, along with the paucity of Afghan civilian aircraft, minimized the channels for egress. Consequently, only the fortunate who had connections or finalized visas (SIV or otherwise) would have a chance at escape. Even then, the lack of safe or open entry points, and masses of people outside of HKIA, made entry nearly impossible. Those attempting to fly on private planes out of MSIA, then under Taliban control, had similar difficulty with the series of security checkpoints.

4.5 Component 5: Civil Society, Privately Funded Evacuation Efforts

For the civil society and privately funded evacuation efforts, the 18-day time constraint was even more critical. Whereas governments have diplomatic access to other host countries to ask for visas or entry and transit, as well as access to funding, aircraft, cars and buses, civil society and private citizens generally do not. If private citizens do have diplomatic access in potential host countries, this access is usually borne out of personal or professional connections, and generally in small quantities. Here again, we see the theme of deep inequality: Afghans with connections to such civil society groups or influential private citizens had a small chance to leave safely, while other Afghans had a far less likely probability of egress. With the impending total withdrawal on Aug 31st, civil society and private efforts had to build entire logistical pipelines from scratch.

Consider also that a privately chartered flight will generally hold 100–300 passengers and requires each passenger possess approved visas. In turn, this requires processing time for the request to be made to a host country and bureaucratic approval. Private flight organizers would then spend time collecting and collating passenger information. During the Afghan NEO, this process proved to be highly error prone. Private manifests were often submitted for approval with wrongly entered information, expired passports, idiosyncratic manifest templates, or children with no documentation. It is important to note that at any point the host country could require US government approval. This presents a discrete problem wherein anything other than an answer in the affirmative will end the operation entirely—visas, landing and overflight clearances will be canceled. This happened multiple times during the NEO. Canceled evacuation efforts to Uganda and Germany are two examples of such an occurrence.

4.6 Component 6: Funding, Data Intake and Management

Further complications unique to the private evacuations were operational funding as well as data intake and management. For the US government, there were a variety of discretionary funds and appropriations methods to fund the NEO. Similarly, there were data management systems in place for biometrics, even if disparate across departments, bureaus and agencies. Comparatively, private operations had to scramble for millions of dollars virtually overnight from institutions like universities, think-tanks, family foundations or high-value donor networks who had no experience in evacuation.

While the US government had databases for the SIVs, the private evacuations had to build data intake and management systems wholesale. These systems were built on open-source software, typically Google Sheets and Google Forms while navigating potential legal pitfalls of personally identifiable information, global security and immigration. In addition, as many of these efforts were volunteer-driven, everything was done at the speed of trust with no substantive way to vet individuals nor to optimize cybersecurity praxis.

4.7 Component 7: Communication, State Department and USCIS Understaffing

A major resource constraint, second only to time, was communication. Across the span of the 18-day NEO, there was never an overall deliberate, systematic method of communication—neither between various elements of USG (e.g., Congress and the State Department), nor with SIVs and private evacuation efforts. All communication was ad hoc. Official guidance was sparse, and generalized, stemming from an understaffed and overwhelmed State Department. The State Department civil servants, though exerting incredible effort, could not overcome the longstanding problem of understaffing. This understaffing was a political problem, not of their making, that has hampered their capabilities for years—made worse during the Trump administration [7]. Similarly, the United States Citizenship and Immigration Services (USCIS) has been understaffed for several years, contributing overall to the backlog of immigration casework, including SIV applications.

This understaffing cut the legs out from under state agencies when surge staffing was required to coordinate the NEO 24/7. Individuals with specialized knowledge of evacuation, transportation, and diplomatic procedures carried the lion's share of labor and were relied upon too heavily by their respective institutions. There were pockets of self-organized coordination and communication, again built on personal connections to CENTCOM, the NSC, private funders, on-the-ground active-duty US Army Special Forces, and other stakeholders. Two such pockets that successfully organized and evacuated several thousand Afghans were No One Left Behind, an Afghan and Iraqi SIV advocacy organization, and the Truman National Security Project, a left-of-center national security network.

One resource was not missing from the NEO—visibility. The entire world had the ability to watch the simultaneous human and geopolitical disaster for the 18-day duration, as the US ended its longest war. This sustained visibility prompts: (a) questions and conversations around ethical engagement with Afghanistan and with Afghans going forward; and (b) the NEO's lasting effects on geopolitics, both regionally and in the context of the Great Power Competition. What are the opportunities, risks and modalities of engagement with Afghanistan, regionally and globally? What is the ethical path forward?

5 The Downstream Results: Visibility, Harm to US Reputation, Vulnerability of Afghan Allies

An unfinished or unplanned evacuation could imply betrayal, a disrespect for one's allies, incompetence, or unreliability—all of which can be weaponized in the raging disinformation and hybrid wars [8]. This is especially true in context of the Great Power Competition, with VNSAs like the Taliban, ISIS, and Al-Qaeda—all of whom have proven adept at propaganda—as well as adversarial nation-states like Russia, whose superior disinformation tactics need little reference [9].

As with many cases of geopolitical disaster in the era of social media and high visibility of transpiring events, the entire world viewed the difficult and heart-wrenching scenes of the NEO. And it is precisely this uber-visibility that politically, morally and emotionally supercharges the consequent ramifications for Afghans, for Muslims globally, for American domestic politics, and for the general American public. Simultaneously, this visibility throws into high relief the ethical questions of Western engagement in Afghanistan, by extension, the larger Islamic world and very directly relevant in the Great Power Competition.

Additionally, it is incumbent upon policy makers to understand that emergency evacuations, as a subset of migration crises, are likely to be features of geopolitics in the coming years. It is critical that data-driven planning and focused political will are features of any evacuation operations going forward.

5.1 Global Credibility, Realpolitik, Hearts and Minds

Credible reports of reprisal and extrajudicial killings, as well as *"cruel, inhuman and degrading punishments"* by the Taliban against a wide swath of Afghans including interpreters, police, civil society activists, journalists and others continue to surface in the hundreds, implying many more unreported [10]. Taliban commanders have reportedly urged fighters to wait for reprisals as "the attention of the international community is on us right now" and "*only observe those who are acting against us, particularly those government officials and civil society activists who preach against the Taliban.*" Without any real diplomatic or military presence, the US has no immediate means of safeguarding the lives of allies inside Afghanistan at this juncture.

Perhaps the most realpolitik and obvious political consequence to the US is possibly irreparable harm to its credibility as a leader and partner. Global confidence in American leadership has been on a steep decline since the presidency of George W. Bush who presided over another politically driven foreign policy disaster, the falsely-caused Iraq War of 2003 [11]. It requires little explanation that to leave one's allies to this openly-televised, disastrous and lethal *fait accompli* will only make it harder to recruit for critical operational partners like interpreters, human intelligence assets, or even political partners in current and future conflicts—whether said conflicts are hot, cold, low-intensity or hybrid.

If the issue of evacuation of Afghan SIVs and other allies remains unresolved, global perceptions of the US will ossify into one of two interpretations: *a lack of integrity or a lack of operational competence.* In light of the globally-unified response to the crisis in Ukraine, a third, and possibly worse, interpretation is simple racism—a notion thoughtlessly promulgated by editorializing media pundits and news anchors [12]. The arguments against this interpretation of racism are increasingly fragile, and ring hollow, even to the American public, and more so to the developing world, the locus of many global security issues.

As the US continues to engage in counterterrorism against an ever-growing landscape of VNSAs in 85 countries—many of them Muslim-majority—damaged credibility becomes a talking point and recruiting tool for adversaries as well as an operational hurdle, obstructing mission success. This landscape is now increasingly complicated by the Great Power Competition and the usage of VNSAs as proxies in Africa, Asia, LATAM and the Near East, where hybrid, low-intensity conflict merges seamlessly with disinformation warfare.

Who on the international stage will refrain from hedging their bets should the United States seek their aid? How will potential partners—whether interpreters, potential intelligence assets or even heads of insecure, precarious nation-states—assess and calculate personal, communal, or national risk of failed engagement on the behalf of the United States? Simply said, why should potential allies put their faith in the US? More importantly, what policies and narratives are to counter this dangerous perception of unreliability and abandonment?

5.2 *Public Service and Moral Injury*

Moral injury is a parallel harm to US domestic political credibility in the eyes of the American public, whether private citizens or public servants. Like global decline in confidence of American leadership, the American public's trust in government has been in persistent decline [13].

Additional politically driven foreign policy mistakes cannot conceivably add to trust in government and will invariably lead to increased political nihilism, feeding alt-political movements across the spectrum [14]. Military service members—and by extension members of the intelligence community and diplomatic corps—experienced grave moral injury and disillusionment at the decidedly unnecessary disaster of the NEO.

This only compounds the continued lack of interest in younger Americans wanting to serve in government or the military, as evidenced by a continuing drop in recruitment and retention numbers, only partially explainable by low earning potential. This bipartisan failure to mobilize commensurately and appropriately—if not rectified—will only continue to harm public confidence in US political institutions, leadership and public service.

5.3 *Past and Future Evacuation Events*

It is important to note there have been multiple operations where the US successfully and necessarily evacuated noncombatant allies, like interpreters, to safety. In 1975, the Ford administration evacuated 130,000 Vietnamese allies to Guam during Operation New Life, for screening before resettling them in the US, through the Interagency Task Force (IATF) on Indochinese Refugees [15]. Similarly in 1996, Operation Pacific Haven (OPH) evacuated 6600 Iraqi Kurdish allies to Guam, escaping from Saddam's reprisals in the north of Iraq [16].

In the case of OPH, the focus of the two segments of the operation (Mission Quick Transit and Operation Pacific Haven and their Joint Task Forces) was transport, housing and processing—safety as a primary concern. OPH was largely a joint effort of the State Department, the Air Force and the Department of Health and Human Services (DHHS) [16]. The effort was focused on bringing the Kurdish allies to safety in the shortest time possible, by transporting them to Incirlik Air Base, then to Guam. Guam, distant from the continental United States, provided (a) security for the Kurds while processing and (b) time for US authorities including the DHHS, the FBI and the INS to administratively prepare legal emigration [16]. Key features here to note: multiple points of egress; third country partners and staging locations; and political will and directives for agencies to host and prepare allies for emigration. All of these components together generated the time needed for processing and yielded safety to the allies.

It stands to reason that evacuation of allies will be necessary again in future conflicts in high-volume, high visibility operations like the fall of Saigon or the fall of Kabul. Less obvious, more strategic thinking is the need for sustained, secure evacuation pathways that are not immediate, emergent crises. Allies who have paid their dues deserve to be commensurately rewarded for their services as a feature and not a procedural afterthought.

To date, there has yet to be enshrined in law an immigration program for US allies, writ large, like the permanent SIV program espoused by No One Left Behind, a non-profit dedicated to the Afghan and Iraqi SIVs.

6 Solving the Crisis: How to Build and Operationalize an Ethical Evacuation Strategy

Up to this point, this chapter has established logical, if somewhat obvious, criteria and metrics for determining a successful evacuation, as well as the key features and necessary operational components. Additionally, this chapter has examined the root causes for operational failure during the Afghan NEO and their interdependence. Similarly, discussed are risks associated with leaving this crisis unresolved, specifically to US global standing and credibility, to the American public's perception of political leadership and to potential impact on future mission success.

What are the opportunities, risks and modalities of engagement with Afghanistan, specifically in relation to the SIVs and other allies, regionally and globally? What is the most mature, ethical, and multilaterally beneficial path forward? Specifically, what is to be done operationally, given the dimensions examined? What is the strategy that yields sustained operational success? How should the obvious metrics of success and underscored necessary operational components be woven together in a coherent strategy that negates or mitigates risk and accounts for geopolitical realities?

The answer is reexamining and implementing the fundamentals and components of a successful evacuation, posited above, in an overarching, data-driven strategy. To be effective, a strategic assessment needs to be translated into an operating policy [5].

7 Generating Solutions, Data-Driven Strategy, Political Will

The current global response to the Afghan evacuation crisis is inadequate, slow, and ultimately dangerous to the Afghans. The HKIA/PIA-Doha pipeline is a chokepoint that requires alternatives. The scope of the problem is far larger than the current scale of evacuations. Evacuating the 70,000 Afghans who are in imminent danger from the Taliban, a failing economy or vulnerable in Pakistan, at the current rate and throughput of 2 flights a month on average would require 16 years, approximately, if uninterrupted. The US government's current response requires a retooling of upstream strategy built on political willingness to create the minimum number of choices to force a successful, sustainable, and scalable evacuation operation.

7.1 Step 1: Expend Political Will, Recruit Partner and Host Countries

Political will is causal to all other components of a successful evacuation, and a fundament to a coherent strategy to reify American credibility, to mitigate risk to Afghan allies and to solve the crisis. As posited above, lack of upstream political will was and remains the root cause of the Afghan SIV crisis. No strategy, no matter how tactically thorough or operationally coherent, can come to mission success if decision makers neglect to empower it by expending the political capital necessary—domestically, and with counterparts around the world. Using Operation Pacific Haven as a partial model, partner and host countries will need to be recruited for two categories, which may not be mutually exclusive: quick transit countries that immediately border Afghanistan and long-term host countries, likely outside Central Asia [16].

Policy makers will need to expend political capital with partner and host countries (Component 2). Specifically, US policy makers—if interested in solving the Afghan SIV crisis in a thorough, coherent, and safe manner as expeditiously as possible—will need to engage in a range of bi-lateral and multilateral conversations with the nations that border Afghanistan. There is a role for Track 1, Track 1.5 and Track 2 diplomacy. These conversations can encompass not only regional security in relation to containment of the threat of VNSAs like ISIS-K and Al-Qaeda—an issue of mutual concern and hence commonality—but also cooperation to maximize points of egress (Component 3) for Afghan allies [17]. It may be that, subservient to realpolitik, human rights and humanitarian assistance—to include the Afghan SIV crisis—may need to be coupled with hard security concerns. The purpose of expending political will at this point is to recruit and secure buy-in from Afghanistan's immediate neighbors, at a minimum for safe passage and transit outwards. In other words, the immediate goal of the quick transit phase is safety of the Afghan allies.

These immediate neighbors of Afghanistan—Tajikistan, Uzbekistan, Turkmenistan, Pakistan, and excluding Iran—can provide safe passage via overland routes along with temporary entry visas for SIVs, P1/2 s and others identified by the US as allies. The US can then either (a) *fly them out from in-country international airports* via *charter flights to another partner or host country to be processed at an US Embassy with a sufficiently staffed IV section* or (b) *process them at in-country IV section in Pakistan and Tajikistan.*

In addition to bilateral and multilateral conversations with Afghanistan's immediate neighbors, who are leery of the threat of ISIS-K and the added socio-economic pressures of potentially tens of thousands of Afghans within their borders, US policy makers will need to engage amenable countries outside of Central Asia or the subcontinent that are willing to assist. Specifically, the US government should cultivate working relationships with 1–6 amenable countries, with the goal of taking in 10,000 or more Afghans each, over a 1–3 year period. Afghan allies will require a range of time for processing, from a few months in the case of READY stage SIVs, to years in the case of P1/2 applicants. While Afghanistan's immediate neighbors may balk at the notion of tens of thousands of Afghans within their borders for extended periods of time, other countries, like Portugal and Argentina, have signaled a willingness for long term hosting, including pathways to permanent resettlement. Latin America may prove to be the best option for the range of needs of Afghan allies, particularly long-term hosting. The overarching strategy is to develop multiple, repeatable points of egress (pipelines) that operate persistently, to systematically evacuate Afghan allies in ways that are (a) *measurable*, (b) *statistically significant* and (c) *qualitatively accurate.*

Additionally, curating and matching the Afghans for specific countries based on status category, i.e., post-COM post-interview approved SIVs to Pakistan and pre-COM SIVs to Portugal, will help with operational simplicity, measurement, transparency, and expectations. During relationship development, small batches of Afghan allies should be tested to show proof of concept for a given pipeline, which can then scale up.

7.2 *Step 2: Match Host Countries with Category-Specific Afghans*

Aside from developing quick transit pathways via Afghanistan's immediate neighbors, diplomatic agreements with long-term host countries must be detailed in relation to the specific sub-demographics within the umbrella term "Afghan allies." e.g., post-COM SIVs, pre-COM SIVs, READY SIVs, P1/2 referrals, humanitarian parole applicants, or Afghan police or army officers who destroyed documents fearing reprisal have different administrative needs, and differing processing timelines. Commensurate with each group's specific forward pathways—or lack of forward pathway—host countries should be aware of potential timelines for processing to the US, as well as contingency options. This yields a contingent question of what happens to those who fail processing, whether for derogatory security reasons, inability to prove employment with the US, or fail to meet humanitarian parole criteria? Does the host country agree to provide asylum?

It may be that some partner countries may only be willing to host late-stage SIVs—like those in READY status—whose cases could process in months, or less than a year. Other countries may be willing to host pre-COM applicants, who will require significantly longer time. Yet others may be willing to provide an open-door policy to all, up to and including permanent resettlement or asylum. It should be clear that there are three general categories of Afghan allies whose situations need to be ethically addressed:

- SIV applicants: Afghans who verifiably worked for the US government or ISAF for a period of one year.
- P1/2 referrals: Priority 1 referrals are for Afghans who were "known to the US Embassy" in Kabul while Priority 2 referrals are Afghans who do not meet the minimum time-in-service for a SIV but who worked as contractors, Locally Engaged Staff (LES), or interpreters/translators for the U.S. Government, USFOR-A, ISAF, or Resolute Support; Afghans who worked for a U.S. government-funded program or project in Afghanistan supported through a U.S. government grant or cooperative agreement; Afghans who were employed in Afghanistan by a U.S.-based media organization or nongovernmental organization.
- Afghan military or police: Often they do not qualify for either SIV or P1/2, or have destroyed their documentation for fear of reprisal; in Pakistan, some of the most vulnerable are Afghan police women who have no documentation. This group will likely require heavy coordination with the Department of Defense and/or intelligence community components to verify identities.

For example, a proposed long-term host country schema could develop along these lines:

- Mexico: 20,000 READY SIVs over a 12-month period

- Argentina: 10,000 READY SIVs over a 6-month period; 15,000 pre-READY (post-COM and pre-COM) over a 24 month period; 10,000 P1/2 referrals over a 24 month period and 5000 Afghan army officers for permanent resettlement
- Portugal: 10,000 READY SIVs over a 12-month period; 15,000 pre-READY (post-COM and pre-COM) over a 24-month period; 10,000 P1/2 referrals over a 24-month period and 5000 Afghan army officers for permanent resettlement.

7.3 Step 3: Develop Specific Pipelines with Maximum Points of Egress from Quick Transit Countries to Host Countries, Activate Civil Reserve Air Fleet

Once the details have been clarified on the specific numbers and type of Afghan allies sub-groups each host country is willing to accept, then the details of the pipelines should be developed. For each quick transit country, border entry points need to be established in partnership with host governments, where Afghan allies can be screened, and temporarily housed before going forward to flights. To maximize egress from quick transit countries to host countries, all usable airports should be engaged for these operations. In other words, we alleviate the constraint of limited aviation infrastructure, Component 4.

If we take Tajikistan as a potential example, there are at least three international, civilian airports that can serve as points of egress for this strategy, with appropriate planning and landing clearances for CRAF aircraft. Specifically:

- Dushanbe International Airport
- Bokhtar International Airport
- Kolab Airport
- Khujand International Airport.

Similarly, we see in Uzbekistan multiple international airports that could serve as points of egress in a coherent, scaled strategy to bring Afghan allies to safety. With proper diplomatic and operational arrangements for landing clearances, the US could leverage the following locations, and this list is non-exhaustive:

- Bukhara International Airport
- Ferghana International Airport
- Navoi International Airport
- Samarkand International Airport
- Urgench International Airport.

If the US were to engage just these airports, in addition to using PIA in Islamabad, it would have multiplied the points of egress by a factor of 10. If they were to engage the airports in the remaining neighbors of Turkmenistan, and the remaining international airports in Pakistan, the points of egress are 20 times that of the NEO. Similarly, we have to alleviate the other part of Component 4—the constraint of limited aircraft—by engaging the Civil Reserve Air Fleet.

Using the Civil Reserve Air Fleet, there would be a pool of 413 aircraft for international operations from 24 US airlines [18]. Using civilian aircraft in this manner avoids the optics of using military aircraft in foreign countries. With 10 or more points of egress with potentially 413 international aircraft, it is well within reason that if 120,000 Afghans could be transported in 18 days during the NEO with 218 aircraft, 751 flights and one single point of egress—HKIA—then a well-designed and well-resourced 30–90 day plan could transport the 70,000 Afghan SIVs and other allies to host countries, to process in safety [19].

7.4 *Step 4: Convene Joint Task Forces to Execute*

Once quick transit and long-term host countries have agreed to cooperate with clear timelines, metrics and target demographics and the CRAF has been activated, US policy makers should convene interagency joint task forces (IJTFs) for the purpose of operational execution as in Operation New Life and Operation Pacific Haven. IJTFs should be divided into two categories: quick transit and host country. For example, building on the previous theoretical schema:

- IJTF Quick Transit Tajikistan
- IJTF Quick Transit Uzbekistan
- IJTF Quick Transit Pakistan
- IJTF Host Country Mexico
- IJTF Host Country Argentina
- IJTF Host Country Portugal.

A Quick Transit JTF's role is to safely screen, transport from the inside the border to the airfields and organize for temporary housing and care if necessary. A Host Country JTF's is to arrange collaboration with partner NGOs and INGOS in conjunction with host country authorities for reception, care, feeding, long-term housing, transportation for visa processing in addition to final transport to the US, or in the case of failed processing, resettlement in host country.

All of this will require communication and coordination (Component 7) with the Department of State. In particular, the Coordinated Afghan Relocation Effort team (CARE), the Bureau of Consular Affairs (CA), the Bureau of Populations, Refugees and Migration (PRM) will need to properly identify Afghan allies, their status, and location for the purposes of communicating with Afghan allies and operational coordination with the IJTFs as well as host governments.

8 Summary

The Afghan evacuation left a majority of Afghan allies like interpreters behind, due to a failure to plan and execute in due time. This harms American standing and credibility in the world as well as posing a lethal danger to Afghan allies stranded in Afghanistan and Pakistan. The consequent operation and its observed components can be studied to determine (a) *quantitative and qualitative metrics of success* and (b) *to build an overarching strategy to solve the Afghan SIV crisis, and thus mitigate harm to the Afghan allies as well as to American standing and reputation.* The overarching strategy is to develop multiple, repeatable points of egress (pipelines) from Afghanistan's neighboring countries that operate persistently to systematically evacuate Afghan allies to long-term host countries in ways that are (a) *measurable*, (b) *statistically significant* and (c) *qualitatively accurate.*

9 Recommendations

- Create an overarching strategy to evacuate and process Afghan allies involving partner countries, with maximum points of egress, and maximum number of host countries to systematically evacuate in ways that are (a) *measurable*, (b) *statistically significant* and (c) *qualitatively accurate*
- Open Track 1, Track 1.5, and Track 2 diplomatic channels to collaborate and cultivate buy-in with Afghanistan's immediate neighbors for Quick Transit for Afghan allies
- Seek buy-in for collaboration with potentially amenable partner countries for Long-Term Hosting for Afghan allies where they can either be (a) *processed to the US* or (b) *resettled, at a throughput that is in the 1–3 year range; LATAM countries in particular should be considered*
- Develop specific pipelines with maximum points of egress (airports) from Quick Transit countries to Long-Term Host countries, activate Civil Reserve Air Fleet
- Convene Interagency Joint Task Forces to execute evacuation, housing, and processing strategy, to include all relevant stakeholder departments, agencies and bureaus
- Legislate a Permanent SIV program for all US allies in any future or current combat zone.

10 Addenda: No One Left Behind's Permanent SIV Program Recommendation

Recommendation: No One Left Behind is advocating for legislation consolidating the current SIV program for Iraq and Afghanistan, expanding geographic coverage to include those who served the US combat mission abroad in other countries, and formalizing the SIV program as permanent and enduring.

Proposed Legislation

Consolidate SI and SQ SIV programs under the FY2006 NDAA Section 1059 (Afghanistan and Iraq), Afghan Allies Protection Act of 2009 Section 602, and FY2008 NDAA Section 1244 (Iraq) into one permanent program, with the current Afghan program (Afghan Allies Protection Act of 2009 and the Immigration and Naturalization Act, as amended) as the baseline with the following amendments:

Eligibility

- Maintain explicit authority for Iraq and Afghanistan
- Delegate authority to the Secretary of State to expand geographic eligibility to any combat area or foreign state in which a principal alien is harmed, persecuted, or threatened with physical harm in connection with the alien's employment by the United States
- Expand eligibility to those principal aliens who are wounded or seriously injured in connection with or because of the alien's employment and cannot continue service through 1 year
- Expand documentation of the "faithful and valuable service" requirement to either a U.S. Government supervisor or private employment (contractor) supervisor
- Expand surviving spouses and children's eligibility to the pre-Chief of Mission application stage; include those for whom the principal alien's application would have been approved by the Chief of Mission had the principal alien had survived and a petition had been submitted

Employing U.S. Government agency and/or contractor

- Establish a requirement for both the employing U.S. Government agency/department and any contractor to the U.S. Government providing eligible services to establish a database containing the biographical, security screening, employment, and at least one direct supervisor (for any period of time) information of any employed third country nationals who could meet SIV eligibility criteria, as well as updates to that information validating completion of service
- Such a database could be unified between the Department of State, Department of Defense, Department of Homeland Security, and U.S. Agency for International Development (per the June 2020 State OIG report)
- Establish a requirement that the employer U.S. Government agency/department or contractor to the U.S. Government to identify contact information for a direct supervisor to the Department of State for applications denied by the Chief of Mission or designee
- Establish a requirement for the employing U.S. Government agency/department or contractor to the U.S. Government to provide access to the database by the Department of State

- Establish a requirement that any U.S. Government agency that hires contractors employing SIV-eligible aliens designate a point of contact for the Department of State to inquire about employment records and furnish those records when requested

Chief of Mission Approval

- Require the Department of State establish a database in which derogatory information that results in a denial of an application on the basis of: (i) risk assessment, (ii) employment confirmation, and/or (iii) faithful and valuable service, is documented within a Department of State database
- Where an application is denied, require a written decision provided to the applicant stipulate at least the category of information on which the denial is based
- Where a written decision providing the specific (and potentially refutable) basis of such a denial to an applicant is infeasible because of sensitive information, provide the applicant with the contact information for the applicant's supervisor (as tracked in the aforementioned database) and a copy of the written decision, including the derogatory sensitive information where feasible, to that supervisor with the applicant's contact information
- Extend the appeal window to 365 days
- Establish a requirement for the Secretary of State to designate a SIV Coordinator at each embassy located in the geographic areas deemed eligible for the SIV program as aforementioned

Interview

- Explicitly authorize virtual interviews in countries for which there is no U.S. consular presence
- Numerical limitations
- Specify that the total number of principal aliens who may be provided special immigrant status under this section may not exceed 1000 per year.

Application process

- Remove reference to the National Defense Authorization Act

Waiver of medical examination

- Extend authority of the Secretary of State and Secretary of Homeland Security to jointly issue a blanket waiver for medical examinations to permanent SIV program
- Require the Department of Homeland Security to furnish information of authorized medical providers who can provide examinations in the United States

Wait times and protection

- Codify calculation of average wait time with a uniform method
- Establish a requirement for the employing U.S. Government agency/department to offer protective services to applicants if the application period extends beyond the nine month period required for completion of steps under the control of the Departments of State and Homeland Security.

References

1. Ward A (2021) 'Majority' of Afghan SIVs left behind, State Dept. official estimates. POLITICO. [Online]. Available at: https://www.politico.com/news/2021/09/01/afghanistan-sivs-left-behind-state-dept-508327. Accessed 1 Aug 2022
2. Packer G (2021) Biden's betrayal of Afghans will live in infamy. The Atlantic. [Online]. Available at: https://www.theatlantic.com/ideas/archive/2021/08/bidens-betrayal-of-afghans-will-live-in-infamy/619764/. Accessed 1 Aug 2022
3. Salama V (2021) WSJ news exclusiveInternal state department cable warned of kabul collapse. WSJ. [Online]. Available at: https://www.wsj.com/articles/confidential-state-department-cable-in-july-warned-of-afghanistans-collapse-11629406993. Accessed 1 Aug 2022
4. Senate Foreign Relations Committee Minority (2022) Left behind: a brief assessment of the Biden administration' s strategic failures during the Afghanistan evacuation 2. [Online]. Available at: https://www.foreign.senate.gov/imo/media/doc/Risch%20Afghanistan%20Report%202022.pdf
5. Kissinger H (2012) Years of renewal. Simon & Schuster, London
6. United States Department of Defense (2021) Department of defense activates civil reserve air fleet to assist with Afghanistan efforts. U.S. Department of Defense. [Online]. Available at: https://www.defense.gov/News/Releases/Release/Article/2741564/department-of-defense-activates-civil-reserve-air-fleet-to-assist-with-afghanis/. Accessed 1 Aug 2022
7. Department of State Office of the Inspector General (2021) Inspector general statement on the department of state's major management and performance challenges. Department of State Office of the Inspector General. [Online]. Available at: https://www.stateoig.gov/system/files/oig-ex-22-01_508.pdf. Accessed 1 Aug 2022
8. Nossiter A, Abed F (2021) The propaganda war intensifies in Afghanistan as the Taliban gain ground. The New York Times. [Online] Available at: https://www.nytimes.com/2021/08/08/world/asia/afghanistan-government-propaganda.html. Accessed 1 Aug 2022
9. Walla K (2021) Before the Taliban took Afghanistan, it took the internet. Atlantic Council. [Online]. Available at: https://www.atlanticcouncil.org/blogs/new-atlanticist/before-the-taliban-took-afghanistan-it-took-the-internet/. Accessed 1 Aug 2022
10. United Nations Assistance Mission in Afghanistan (2022) UN releases report on human rights in Afghanistan since the Taliban takeover. UNAMA. [Online]. Available at: https://unama.unmissions.org/un-releases-report-human-rights-afghanistan-taliban-takeover#:~:text=UNAMA%27s%20report%20details%20extrajudicial%20killings. Accessed 1 Aug 2022
11. Pew Research Center (2008) Global public opinion in the bush years (2001–2008). Pew Research Center's Global Attitudes Project. [Online]. Available at: https://www.pewresearch.org/global/2008/12/18/global-public-opinion-in-the-bush-years-2001-2008/. Accessed 1 Aug 2022
12. Bayoumi M (2022) They are 'civilised' and 'look like us': the racist coverage of Ukraine. The Guardian. [Online]. Available at: https://www.theguardian.com/commentisfree/2022/mar/02/civilised-european-look-like-us-racist-coverage-ukraine. Accessed 1 Aug 2022

13. Pew Research Center (2022) Public trust in government: 1958–2022. Pew Research Center—U.S. Politics & Policy. [Online]. Available at: https://www.pewresearch.org/politics/2022/06/06/public-trust-in-government-1958-2022/. Accessed 1 Aug. 2022
14. Van Green T, Doherty C (2021) Majority of U.S. public favors Afghanistan troop withdrawal; Biden criticized for his handling of situation. Pew Research Center. [Online]. Available at: https://www.pewresearch.org/fact-tank/2021/08/31/majority-of-u-s-public-favors-afghanistan-troop-withdrawal-biden-criticized-for-his-handling-of-situation/. Accessed 1 Aug 2022
15. Taft JV (1975) Indochina refugees—interagency task force documents. [Online] https://www.fordlibrarymuseum.gov. Department of State. Available at: https://www.fordlibrarymuseum.gov/library/document/0164/1505189.pdf
16. Jones Jr RW (2008) A second chance: operation Pacific Haven. Veritas J Army Special Oper Hist. 4(3). [Online]. Available at: https://arsof-history.org/articles/v4n3_op_pacific_haven_page_1.html. Accessed 1 Aug 2022
17. Senate Foreign Relations Committee (2021) Joint statement from SFRC chairman Menendez, chairs of foreign affairs committees of the G7 member states on Afghanistan|United States Senate Committee on Foreign Relations. [Online] Senate.gov. Available at: https://www.foreign.senate.gov/press/chair/release/joint-statement-from-sfrc-chairman-menendez-chairs-of-foreign-affairs-committees-of-the-g7-member-states-on-afghanistan. Accessed 1 Aug 2022
18. United State Air Force (2019) Civil reserve air fleet. Air Force [Online]. Available at: https://www.af.mil/About-Us/Fact-Sheets/Display/Article/104583/civil-reserve-air-fleet/. Accessed 1 Aug 2022
19. US Transportation Command (2021) Afghanistan evacuation support. Ustranscom.mil [Online]. Available at: https://www.ustranscom.mil/cmd/neo.cfm. Accessed 1 Aug 2022

Mohammed Moh Kilani is a policy professional, advocate and strategist, with a bias for action, and a decade of experience in international affairs and tech. During the Afghan Evacuation, he evacuated several hundred people out of Kabul and Mazar-i-Sharif, one of the first to catalyze private flights, with a wide variety of stakeholders. Moh has published research on tech, national security and economics through the Office of the Director of National Intelligence, the University of Washington and the Washington State Department of Commerce. Kilani is a Security Fellow at the Truman National Security Project, a Carnegie Council for Ethics in International Affairs Fellow, a member of the Pacific Council on International Policy, and a New America MENA National Security & Foreign Policy Next Generation Leaders 2020 Honoree. He is currently Director of Evacuation at No One Left Behind, a 501(c)(3) dedicated to Afghan and Iraq SIVs.

Mission Creep on Repeat: Deconstructing U.S. Strategy in Afghanistan

Michael Kugelman

Abstract With U.S. forces making final preparations to leave, the Taliban seized power on August 15, 2021, setting off a chaotic chain of events. Helicopters evacuated U.S. diplomats from the embassy. Fearful Afghans rushed to the airport in Kabul. Several were so desperate to leave that they grabbed onto the wings of American aircraft as they took off. U.S. military personnel struggled to hold back surging crowds outside the airport, and dozens died in a bomb blast. An attempt to target the perpetrators with a drone strike instead resulted in the killing of 10 innocent civilians. Back in Washington, the Biden administration faced a bipartisan backlash for the botched final phase of the U.S. exit. The decision to withdraw was announced back in April, and it had been implemented relatively smoothly until that point. Some commentators have compared those chaotic two weeks in August to America's humiliating final days in Vietnam (Packer in Biden's Saigon. The Atlantic, 2022). They stunned U.S. allies and partners and gave propaganda ammunition to America's rivals and enemies. One U.S. military leader has even speculated they may have emboldened Russian President Vladimir Putin to launch his invasion of Ukraine six months later ("US European Command Leader Testifies on National Security Challenges," Tod Wolters, Testimony to House Armed Services Committee, C-SPAN, March 30, 2022, https://www.c-span.org/video/?519012-1/us-european-commander-assistant-defense-secretary-testify-war-ukraine).

Keywords Afghanistan · US strategy · Military · Counterterrorism · Counterinsurgency · Development · Lessons learned

The final two weeks of the U.S. withdrawal from Afghanistan will long be remembered, and for all the wrong reasons.

M. Kugelman (✉)
Wilson Center, Washington, DC 20004-3027, USA
e-mail: Michael.Kugelman@wilsoncenter.org

A. Farhadi and A. Masys (eds.), *The Great Power Competition Volume 4*,
https://doi.org/10.1007/978-3-031-22934-3_13

1 14 Days in August: A Microcosm of Nearly 20 Years in Afghanistan

America's failures during its final days in Afghanistan can be chalked up to something that also explains its failures more broadly in Afghanistan: The lack of a good strategy. What became apparent, several months after the Biden administration announced its decision to withdraw, is that the White House did not properly plan for the possibility of the least-likely but worst-case scenario becoming a reality: A Taliban takeover before the end of the U.S. withdrawal.

In July, President Joe Biden had said a Taliban takeover "is not inevitable," and that "the likelihood there's going to be the Taliban overrunning everything and owning the whole country is highly unlikely."[1] Barely a month later, the Taliban controlled Kabul and nearly 100% of the entire country—more than the roughly 90% it held when in power in the late 1990s. The Afghan military had collapsed, and government leaders had fled the country. On August 16, Biden had to eat his words. "This did unfold more quickly than we had anticipated," he admitted.[2]

When the worst-case scenario emerged, the administration had no plan, because it seemingly hadn't properly prepared for the possibility that such a scenario may emerge. There was no overarching strategy to ensure that a proper U.S. response could be implemented, no matter what the reality on the ground. Instead, the policy process succumbed to confusion and chaos. This lack of a strategy had costs not just for policy, but also for credibility and above all human lives. And it prevented Washington from achieving its goal of a smooth and responsible final withdrawal.

The last two weeks of the U.S presence in Afghanistan were a microcosm of the previous nearly 20 years: Strategic failures produced bad results that precluded the possibility of achieving objectives. This essay examines U.S. strategy during America's time in Afghanistan. It argues that U.S. strategy failed because it was inconsistent, lacked a long-term outlook, and didn't properly match resources to goals. The essay also considers why there weren't more efforts to develop a better strategy. It argues that the absence of a sustained policy focus on the war, a lack of consensus among top leaders, an insufficient willingness to acknowledge mistakes, and an absence of public accountability all gave war planners little incentive to course correct. Additionally, the essay discusses how poor U.S. strategy contributed to American failures in Afghanistan by highlighting the deleterious impacts on core U.S. goals such as promoting development, fighting the Taliban insurgency, and combating terrorist groups. Finally, the essay offers a series of lessons learned from nearly 20 years of strategic failures in Afghanistan.

---

[1] "Remarks by President Biden on the Drawdown of U.S. Forces in Afghanistan," White House, July 8, 2021, https://www.whitehouse.gov/briefing-room/speeches-remarks/2021/07/08/remarks-by-president-biden-on-the-drawdown-of-u-s-forces-in-afghanistan/.

[2] "Remarks by President Biden on Afghanistan," White House, August 16, 2021, https://www.whitehouse.gov/briefing-room/speeches-remarks/2021/08/16/remarks-by-president-biden-on-afghanistan/.

This essay does not mean to overlook other factors that contributed to failure in Afghanistan. In particular, Afghan governments—through their corruption, poor leadership, inability to address severe challenges within the Afghan military, and so on—certainly deserve blame as well. However, such considerations lie outside the scope of this essay, which focuses specifically on U.S. strategic failures.

2 Defining a Successful War Strategy: Theoretical and Doctrinal Roots

The basic definition of a strategy is a plan or policy meant to achieve an objective. In war, strategy relates to the use of available resources to attain battlefield goals. For Carl von Clausewitz, the nineteenth century Prussian general and military theorist whose views remain deeply influential in the study of war today, battlefield strategy is about identifying a clear goal that will remain consistent throughout the conflict; moving expeditiously to marshal and deploy the resources to achieve that goal; and then implementing that goal without getting distracted by other concerns.

Clausewitz laid out these ideas in his classic study, *On War* [2]. On the importance of identifying a clear objective, he wrote: "No one starts a war—or rather, no one in his senses ought to do so—without first being clear in his mind what he intends to achieve by that war and how he intends to conduct it." Similarly, "Strategy is the use of the engagement for the purpose of the war. The strategist must therefore define an aim for the entire operational side of the war that will be in accordance with its purpose."

On the need to move quickly and in a focused fashion to prepare for implementing a strategy, he wrote that one of the "basic principles" that "underlie all strategic planning" is acting "with the utmost speed." Additionally, "strong determination in carrying through a simple idea is the surest route to success. The winning simplicity we seek, the simplicity of genius, is the result of intense mental engagement."

And on the need to implement a strategy without getting distracted, he wrote: "The talent of the strategist is to identify the decisive point and to concentrate everything on it, removing forces from secondary fronts and ignoring lesser objectives."

Years later, during the Ronald Reagan administration, a new view of war strategy would emerge that featured some elements of Clausewitz's ideas, and that would later be embraced by some of the U.S. officials that launched the intervention in Afghanistan. In 1984, Defense Secretary Casper Weinberger gave a policy speech meant to set limits on the use of force [3]. He argued that a war should be waged only as a last resort; when there are assurances of public support at home; and when vital U.S. national interests are at risk. Such a war should use decisive force to pursue clearly defined objectives, with the full intention of winning, and there should be frequent assessments about whether the use of force remains necessary.

These views, known as the Weinberger Doctrine, would later be articulated in similar form by Colin Powell, a top aide to Weinberger in the Reagan administration.

Powell later served as Chairman of the Joint Chiefs of Staff and Secretary of State in the George H. W. Bush and George W. Bush governments, respectively. One of the most detailed articulations of the Powell Doctrine came in a *Foreign Affairs* essay he published in the early 1990s [4]. It called for the use of overwhelming force to achieve clearly defined objectives. Its components were visible during the 1990–1991 Gulf War, one of the first conflicts to feature "shock and awe" displays of U.S. military power, which was deployed to pursue the specific goal of compelling the withdrawal of invading Iraqi troops from Kuwait.

These theories and doctrines have their detractors. An emphasis on overwhelming force—even if meant to be waged only briefly, until narrowly defined objectives are achieved—invites concerns about scorched-earth tactics and threats to civilians. Also, many of Clausewitz's ideas can be shrugged off as anachronistic, given that wars in recent decades—with their emphasis on non-state actors, subconventional tactics, and advanced forms of weaponry and technology—are so different from European wars of the nineteenth century. Still, the basic principle underlying these ideas remains simple and compelling: Develop a clear and consistent objective, stick with it, and use available resources to achieve it.

3 US Strategy in Afghanistan: A Muddle and a Mismatch

With the exception of the war's first phase, when American forces targeted al-Qaeda leaders and their Taliban hosts, U.S. objectives in Afghanistan were often unclear and inconsistent. Successive administrations changed objectives. They also failed to achieve a workable match between objectives and the resources used to achieve them.

Initially, Washington had a very specific and focused strategy: Target al-Qaeda in Afghanistan, and remove their Taliban hosts from power. This effort, in some respects, reflected Clausewitz's principles and the Powell doctrine (Powell was Secretary of State during the war's initial years). It entailed heavy uses of force to carry out a specific objective meant to safeguard a vital U.S. interest (protecting the country from terrorism). During the early months of the war, the American public offered robust support for U.S. actions in Afghanistan. Congress showed its backing with a joint resolution authorizing the use of force.

3.1 In Search of Consistency

However, after U.S. forces achieved those initial goals, American strategy in Afghanistan went wobbly and never recovered. In the words of Chas Freeman, a former U.S. ambassador and senior foreign service officer:

> Carried away by our success, in 2002, we began moving the goalposts this way and that. There was no debate about why this was necessary or appropriate. Soon, no one could offer a coherent explanation of why we were fighting in Afghanistan. You can't accomplish a mission when you don't know what it is. We violated the basic principles that must guide all wars: set clear objectives, stick with them, and stand down and let the diplomats go to work when they are achieved [5].

The Bush administration did not make plans to withdraw U.S. forces, even though it had achieved its stated objectives. In December 2001, soon after the Taliban's ouster, U.S. representatives participated in a conference in Germany to negotiate the creation of the first post-Taliban government. The Taliban were not invited, and the Bush administration declined to explore opportunities for reconciliation talks with the group. These U.S. moves would help drive the Taliban's decision to launch an insurgency in the coming years. In effect, there was no good U.S. strategy for dealing with the Taliban after their ouster. Consequently, U.S. actions paved the way for the reemergence of the very group that American forces had come to Afghanistan to remove.

The decision to invade Iraq in March 2003 took Washington's eye off the ball in Afghanistan at the worst possible moment, as the Taliban was regrouping and preparing to launch its insurgency. In May, Defense Secretary Donald Rumsfeld declared an end to major U.S. combat operations in Afghanistan, but there were no plans to remove U.S. forces—and the Taliban were starting to roll out their insurgency. In 2005 and 2006, with Taliban forces now actively fighting U.S. forces, Washington signed new agreements that entailed closer cooperation with the Afghan military. And it scaled up its effort to target the Taliban.

To the extent that the Bush administration had a strategy, it is often associated with counterterrorism-focused goals. Such objectives guided its activities during the initial months of the war, and the administration's rhetoric hit on tough-on-terror themes. It spoke of America and Afghanistan as allies against terrorism, it vowed not to give in to "murderers and extremists," and it emphasized going on the offensive and other kinetic activities.

However, in an indication of the Bush White House's lack of strategic clarity about Afghanistan policy, President Bush also spoke of very different, and much more ambitious, goals in Afghanistan. "We're working hard in Afghanistan," he said in an April 2002 speech. "We're clearing minefields. We're rebuilding roads. We're improving medical care. And we will work to help Afghanistan to develop an economy that can feed its people without feeding the world's demand for drugs."[3] That same year, the administration announced the establishment of Provincial Reconstruction Teams (PRTs), groups of military and civilian officials that sought to facilitate more stability in local communities through strengthening governance and service delivery.

In his 2002 speech, Bush said peace would be achieved in Afghanistan by helping develop the country. But just a few years later, the Taliban insurgency was in full force—in part because soon after announcing ambitious and expensive plans for

[3] "President Outlines War Effort," Remarks by the President to the George C. Marshall ROTC Award Seminar on National Security, Virginia Military Institute, Lexington, Virginia, April 17, 2002, https://georgewbush-whitehouse.archives.gov/news/releases/2002/04/20020417-1.html.

developing Afghanistan, the Bush administration shifted attention away from state-building and counterinsurgency in Afghanistan to its invasion of Iraq.

The Obama administration's Afghanistan strategy was similarly muddled. In a 2009 speech outlining his Afghanistan policy, Obama, like Bush early in his term, emphasized counterterrorism goals. "I want the American people to understand that we have a clear and focused goal: to disrupt, dismantle and defeat al Qaeda in Pakistan and Afghanistan, and to prevent their return to either country in the future. That's the goal that must be achieved."[4] But in the same speech, he emphasized the need to help build up the Afghan army so that it could become self-sufficient and capable of leading counterinsurgency efforts against the Taliban. He also announced the need for more U.S. civilian specialists in Afghanistan. "We need agricultural specialists and educators, engineers and lawyers. That's how we can help the Afghan government serve its people and develop an economy that isn't dominated by illicit drugs."[5]

In December 2009, Obama would famously authorize a military surge into Afghanistan even as he announced that U.S. troops would begin drawing down 18 months later. For the next few years, U.S. forces waged a counterinsurgency effort meant in part to weaken the Taliban enough to make them receptive to reconciliation talks. But the Taliban would only get stronger, and especially after Obama ended the U.S. combat mission at the end of 2014, putting Afghan forces in charge of a war effort that they lacked the capacity to lead from the front. The new U.S. mission, focused on training and advising Afghan forces, failed to prevent major gains by the insurgency. After announcing in 2014 that he would pull all troops by the end of 2016, Obama later decided to keep 5500—and eventually 8400—deployed through the end of his term, in order to provide support to Afghan forces.

Donald Trump became the third consecutive president to initially describe the U.S. strategy in Afghanistan as counterterrorism-focused. But unlike his predecessors, he ruled out other objectives:

> It is up to the people of Afghanistan to take ownership of their future, to govern their society, and to achieve an everlasting peace. We are a partner and a friend, but we will not dictate to the Afghan people how to live, or how to govern their own complex society. We are not nation-building again. We are killing terrorists…Victory will have a clear definition: attacking our enemies, obliterating ISIS, crushing al-Qaeda, preventing the Taliban from taking over the country, and stopping mass terror attacks against Americans before they emerge.[6]

In this 2017 speech outlining his Afghanistan policy, he came closer than any U.S. leader since Bush to articulate a clear and focused strategy. But then, like all his predecessors, he went in a different direction. He decided he wanted to leave Afghanistan, and so he pursued talks with the Taliban, resulting in a 2020 agreement

---

[4] "Remarks by the President on a New Strategy for Afghanistan and Pakistan," White House, Office of the Press Secretary, March 27, 2009. https://obamawhitehouse.archives.gov/the-press-office/remarks-president-a-new-strategy-afghanistan-and-pakistan.

[5] Ibid.

[6] "Remarks by President Trump on the Strategy for Afghanistan and South Asia," Fort Myer, Alexandria, Virginia, August 21, 2017, https://trumpwhitehouse.archives.gov/briefings-statements/remarks-president-trump-strategy-afghanistan-south-asia/.

for all U.S. forces to depart by May 2021. Trump later vowed to have all U.S. troops home by Christmas 2020, but his administration eventually changed course and left office with 2500 soldiers still deployed.

President Biden would honor the agreement with the Taliban to complete a full withdrawal. To his credit, in speeches announcing and defending that decision, he acknowledged that Washington had lacked an underlying strategy for years and that there was little reason to stay given that the initial terrorism-focused objectives had been achieved. "We delivered justice to bin Laden a decade ago," he said in April 2021. "And we've stayed in Afghanistan for a decade since. Since then, our reasons for remaining in Afghanistan are becoming increasingly unclear, even as the terrorist threat that we went to fight evolved." In July, he said: "We did not go to Afghanistan to nation-build. And it's the right and the responsibility of the Afghan people alone to decide their future and how they want to run their country."[7]

There was a sad irony in Biden announcing that the achievement of initial counterterrorism-focused objectives warranted a withdrawal. Islamic State-Khorasan (IS-K), a terror group that didn't exist before U.S. forces arrived in Afghanistan, staged dozens of attacks in Afghanistan during the latter years of the war. It carried out one of its deadliest-ever attacks during the final days of the American troop presence. IS-K capitalized on the administration's inability to plan for the possibility of a Taliban takeover prior to the final withdrawal, targeting large and panicked crowds of people outside the Kabul airport who were desperately trying to flee the country to escape the Taliban.

3.2 *Too Many Discrepancies*

U.S. strategy in Afghanistan wasn't only a muddle. It was also a mismatch, and on multiple levels. Its goals were too ambitious to be achieved with the resources at hand. For example, U.S. strategy identified development goals that would have required far more people, money, time and expertise than were available. Counterterrorism objectives exceeded deployed personnel and intelligence capacities [6].

The opposite problem also prevailed: U.S. strategy sometimes allocated far too many resources to achieve goals, resulting in undesirable outcomes. Development assistance poured into a country that lacked the absorptive capacities to manage it, generating new opportunities for the very corruption that U.S. officials sought to reduce. Overwhelming force was used to achieve counterterrorism goals, which killed terrorists but also large numbers of civilians, alienating local communities and benefiting Taliban insurgents.

Some international relations scholars believe that foreign policy priorities accounted for another U.S. strategy mismatch in Afghanistan. One reason for U.S.

[7] "Remarks by President Biden on the Way Forward in Afghanistan," White House Briefing Room, April 14, 2021, https://www.whitehouse.gov/briefing-room/speeches-remarks/2021/04/14/remarks-by-president-biden-on-the-way-forward-in-afghanistan.

failure in Afghanistan, they argue, was that too many resources were invested in efforts to pursue goals—such as the protection of human rights—that didn't rank high on the list of U.S. foreign policy priorities [7].

Other experts have argued that the problem was less an inconsistent or ever-changing strategy, and more an inconsistent and ever-changing set of ways and means used to pursue a strategy that was more consistent than is often given credit for (for example, the goal of ensuring Afghanistan never becomes a sanctuary for terrorists that threaten America remained constant throughout the war). In these experts' view, it's not that the goalposts were shifting, but that balls kept getting kicked in different directions and didn't come close to clearing goalposts that never shifted an inch. Dan Fisher, an analyst with the office of the Special Inspector General for Afghanistan Reconstruction (SIGAR), a Congress-mandated independent entity charged with studying U.S. reconstruction efforts in Afghanistan, speaks of a "constancy of the ends compared to shifting ways and means…we had sort of a reasonably clear…over-arching goal that held constant over time but different degrees of…commitment as to what we were willing to do and to achieve that goal and how we wanted to do it" [8].

Ultimately, a strategy can't be described as effective if, even though it's consistent, the means used to pursue it are not. Additionally, when experts speak of what they believe to be indications of a consistent U.S. strategy, they actually identify several different strategies—such as curbing terrorism in Afghanistan, or creating democracy in Afghanistan—thereby undermining the notion of consistency.

National security scholar Nazanin Azizian takes an even more skeptical view. Azizian, who has served as a U.S. official, contends that the ways, means, and ends were all problematic during the course of the war. "Failure to achieve strategic goals is often attributed to the imbalance between one or more variables of a strategy," she argues, lamenting a "loss of equilibrium between the 'ends' sought and the 'ways' and 'means' allocated to achieve those ends" in Afghanistan. The original envisioned "end" was to eliminate al-Qaeda, but it would morph into an "over ambitious state-building effort that was not sufficiently planned, coordinated, and resourced." The "ways" used to pursue these shifting ends also changed and featured a "whole-of-government approach" that "was not always coherent, aligned, and coordinated." And the necessary "means," she contends, were underestimated, leading to insufficient resources made available for pursuing Washington's goals [9].

3.3 *From the Horse's Mouth*

The most searing indictment of failed U.S. strategies in Afghanistan comes from the very officials charged with implementing and managing those strategies. *The Afghanistan Papers*, a book by *Washington Post* reporter Craig Whitlock based on previously unpublished interviews that SIGAR conducted with U.S. and NATO officials from 2014 to 2018, reaches a simple conclusion: The people overseeing the war were as much in the dark about the war as those trying to understand it from the outside [10]. And that is because, in Whitlock's words, the interviewed officials believed Bush and Obama "failed in their most important task as commanders in chief—to devise a clear strategy with concise, attainable objectives" [11].

According to those interviewed, there were no answers to the most basic questions: Who were U.S. soldiers meant to be fighting and why, what was the timeframe for the U.S. war effort, and what would or should victory look like. In 2007, with 50,000 U.S. troops on the ground, Lt. Gen. Dan McNeill, then the top NATO commander "tried to get someone to define for me what winning meant, even before I went over, and nobody could. Nobody would give me a good definition of what it meant" [11]. Perhaps part of the problem is that these questions weren't even addressed at the top. Nicholas Burns, the Bush administration's ambassador to NATO, said that "after 2003 and 2004…I can't remember us ever saying, should we still be there? Are we being useful? Are we succeeding?…I've often wondered did we make a mistake…in not deciding strategically if there was going to be an endpoint…We probably should have asked those questions more consistently by 2005 and 2006" [11].

What also stands out about *The Afghanistan Papers* is how those at the very top at times didn't appear to care that there were fundamental problems with America's strategy. In his interview, Gen. David Richards recalled an exchange with Rumsfeld in 2006, when Richards was NATO commander in Afghanistan. Richards told Rumsfeld a lack of resources was leading to NATO struggles in southern Afghanistan: "Rummy said 'General, what do you mean?' I said, 'We don't have enough troops and resources and we've raised expectations.' He said 'General, I don't agree. Move on'" [11].

4 What Accounted for Strategic Failure?

Why was U.S. strategy in Afghanistan inconsistent, muddled, and poorly matched with resources? And, if U.S. officials like Rumsfeld knew that something was wrong and that U.S. strategy wasn't working, why were more efforts not made to course correct? Why were more efforts not made to define the mission more clearly and consistently? In effect, why the long-term mission creep? Such questions may take years to answer conclusively, but a few initial possible explanations come to mind.

4.1 Absence of Sustained Policy Focus

Despite a nearly 20-year U.S. presence in Afghanistan, there was little sustained policy focus on fighting the war. And without that sustained policy focus, there was little incentive to think about long-term strategies. The Bush administration allowed itself to be distracted by Iraq. In October 2002, only a year after U.S. forces entered Afghanistan, President Bush admitted he didn't know the name of the commanding general in Afghanistan and said he had little desire to speak with him [11]. In May 2003, soon after U.S. forces entered Iraq, Rumsfeld effectively declared mission accomplished in Afghanistan: "We have concluded we're at a point where we clearly have moved from major combat activity to a period of stability and stabilization and reconstruction activities. The bulk of this country today is permissive, it's secure."[8]

It wasn't only about distractions, but also about a consistent desire to depart. John Sopko, the Special Inspector General for Afghanistan Reconstruction, believes that a hankering to head home constitutes one of the few clear and consistent U.S. strategies in Afghanistan: "The one constant strategy was that none of the administrations over those 20 years wanted to be there and one of their strategies was 'we're going to get out.' So, whether it was Bush, Obama, Trump, or ultimately Biden… it influenced a lot of the day-to-day on-the-ground strategies." [12].

Obama announced his timeline for a troop drawdown at the same time that he laid out his surge plans, and there was frequent messaging during the Obama and Trump administrations about reducing troop levels. Stepped-up counterinsurgency efforts were meant to weaken the Taliban and create the conditions for talks and an eventual withdrawal. And when the Taliban wasn't weakened, the Trump administration pursued and concluded withdrawal talks with the Taliban anyway. One of the main findings of a SIGAR report from May 2022 was that U.S. personnel in Afghanistan were under constant pressure to pursue and achieve short-term goals in order to demonstrate progress, presumably in order to better justify an eventual withdrawal.[9]

For policymakers and lawmakers in Washington, constantly keeping an eye on the exits in Afghanistan can likely be attributed in part to a war that became increasingly unpopular with the American public the longer the conflict raged on. A 2017 *Politico* poll found that only 17 percent of Americans strongly supported troop increases in Afghanistan [13]. Nearly 40% wanted U.S. troops numbers to decline. A Gallup survey released in July 2021, weeks before the final withdrawal, found that nearly 50% of Americans believed military involvement in Afghanistan was a mistake [14]. In contrast, a Gallup survey from the early days of the war found nearly 90% of Americans supporting military action [15].

Another reflection of the lack of a sustained policy focus in Afghanistan was the short tours of many of those serving there, both civilian and military officers, and especially the military advisors meant to provide expertise and strategic insights to senior leaders on the ground. Although the entire war played out under only four U.S. presidents, there was unrelenting turnover on the ground in Afghanistan. With people being cycled in and out so quickly, it was difficult to foster an environment conducive to long-term strategic thinking. Admiral James Stavirides, NATO's supreme allied commander from 2009 to 2013, wrote soon after the withdrawal that "We had 20 1-year wars. The lack of continuity and sense of 'I've just got to make it to my departure date' hindered strategic coherency." [16].

---

[8] "Rumsfeld: Major Combat Over in Afghanistan," CNN.com, May 1, 2003, https://www.cnn.com/2003/WORLD/asiapcf/central/05/01/afghan.combat/.

[9] "Collapse of the Afghan National Defense and Security Forces: An Assessment of the Factors That Led to its Demise," Special Inspector General For Afghanistan Reconstruction, Interim Report, May 2022, https://www.sigar.mil/pdf/evaluations/SIGAR-22-22-IP.pdf.

4.2 Lack of Consensus at the Top

One can also argue that the absence of a clear strategy in Afghanistan was a consequence of indecision and disagreements between and within different U.S. administrations. According to government officials interviewed by SIGAR, there was no consensus on what the core objective should be in Afghanistan—kill terrorists, strengthen women's rights, transform Afghanistan into a U.S.-style democracy, or even seek to recast power balances in the region.[10] In 2015, a USAID official likened U.S. policy to "a present under the Christmas Tree for everyone. By the time you were finished you had so many priorities and aspirations it was like no strategy at all. If you have 50 priorities then you don't have any priorities at all." [17].

Indeed, disagreements played out over the course of the war. Though the Bush administration's initial wartime goal was narrowly focused on targeting al-Qaeda, there were fundamental differences between key officials about the proper response to the 9/11 attacks. Deputy Secretary of Defense Paul Wolfowitz called for a wider campaign in other countries, including Iraq, while Colin Powell argued for a more limited approach focused on Afghanistan.[11] Powell prevailed initially, but Wolfowitz would triumph later on—with deleterious consequences for the war effort in Afghanistan.

Meanwhile, there were differences within the Obama administration about the president's troop surge, with Vice President Joe Biden preferring a smaller troop footprint. Trump initially wanted out of Afghanistan before changing course, only to revert to his original instinct. Different administrations vacillated between preferences for counterinsurgency and counterterrorism, and between security-focused goals and more development-oriented objectives. And there were frequent u-turns on decisions about troop numbers. The result was a lack of strategic clarity.

4.3 An Unwillingness to Hear Bad News

U.S. officials in Washington knew the war wasn't going well. So why weren't there more efforts to course correct and devise a sharper, clearer strategy? After all, as Azizian notes, "Changes in strategy must be made courageously when assessments show policies are no longer working."[12] The problem, some U.S. officials would later admit, was that senior U.S. personnel in Afghanistan often didn't want to hear what wasn't working. They preferred instead to focus on the positives, and on progress—because bad news would have raised concerns that troop levels and other allocated resources were not up for the job [19]. Another reason for the emphasis on positivity, according to SIGAR's conclusions, was the short tours in Afghanistan. Since they

---

[10] Azizian, "Easier to Get Into War Than to Get Out."

[11] For a detailed account of these Bush White House differences over war policy, see James Mann [18].

[12] Azizian, "Easier to Get Into War Than to Get Out.".

were so brief, military personnel felt pressure to show success before their time was up.[13]

This helps explain why senior military leaders so often fixated on highlighting progress, repeatedly proclaiming that "we're turning the corner" and drawing derision from critics that accused them of sugarcoating failure. This ultimately meant that key officials on the ground preferred to discount bad news that, if actually addressed and rectified, could have created the conditions for more strategic successes. Bob Crowley, a U.S. counterinsurgency adviser in Afghanistan in 2013 and 2014, said that "when we tried to air large strategic concerns about the willingness, capacity, or corruption of the Afghan government, it was clear it wasn't welcome" at U.S. military headquarters.[14]

The long pattern of top U.S. officials putting a happy face on bad news may have contributed to the Biden administration's surprise at the Taliban's rapid takeover. After years of not fully acknowledging just how serious the corruption and morale problems were within the Afghan military, U.S. authorities may not have been prepared to accept the possibility that Afghan security forces could collapse so quickly.

4.4 Lack of Public Accountability

Another possible explanation for the lack of an effective strategy was that there wasn't enough external pressure on U.S. officials to do better. Once the Bush administration turned its attention to Iraq, the public's focus was diverted as well. U.S. media coverage of the war would decline, too. A telling moment about media coverage occurred one day in June 2018, when images surfaced on social media of a Pentagon official giving a press briefing on Afghanistan to a room that was practically empty.[15] Several journalists defended the low turnout, noting that many reporters preferred to watch the livestream. But the small in-person presence was still striking, given that this was during a pivotal moment in the war, when efforts were being made to launch talks with the Taliban. One would have expected more journalists would have wanted to be in the room asking questions.

This diminishing focus on the war extended to Congress as well. The number of Congressional hearings on Afghanistan plunged dramatically from more than 50 during the 111th Congress, which coincided with the height of the troop surge in 2009 and 2010, to 6 during the 115th Congress in 2017 and 2018. The number of such hearings fell exponentially after 2009 and 2010, reflecting Congress's decreasing

---

[13] Sopko, video interview with the author.

[14] Whitlock, "At War With the Truth.".

[15] Idrees Ali (@idreesali114), "There are exactly four journalists at the Pentagon briefing on Afghanistan," Twitter post, June 27, 2018, https://twitter.com/idreesali114/status/1011988426504187905?lang=en.

focus on the war after the end of the combat mission and as troop numbers dwindled—even though thousands of troops were still deployed and billions of dollars were still being spent.[16]

Additionally, while many Americans were unhappy about the war during its second decade, many others were indifferent or even unaware of it altogether. A Rasmussen Reports survey in July 2018 found that one in five likely American voters, a demographic one would expect to be fairly well informed about public policy, didn't think the war in Afghanistan was still happening—and that another 20% were not sure.[17] Many Americans—perhaps unable to understand why U.S. troops were still in Afghanistan after so many years—simply didn't want to be bothered about it. As scholar Dominic Tierney put it in an *Atlantic* essay in 2015, "Raising the topic of Afghanistan these days is like mentioning mortality. There's a profound desire to change the subject." [20].

During its final years, the war in Afghanistan, once viewed in Washington as the good and necessary war, became the invisible and forgotten war. With relatively little media coverage, public debate, and Congressional hearings, war planners were not subjected to much accountability. Additionally, the political risks of not focusing more on getting strategy right were not terribly high. Unlike the Iraq War, the war in Afghanistan was not a galvanizing issue within U.S. society and politics. Most Americans had little skin in the game; the lack of a military draft means that only a small percentage of Americans had direct links to the conflict. In effect, American leaders had relatively few political incentives to get strategy right in Afghanistan.

5 How Strategy Failures Undermined Core U.S. Goals

Over the course of the war, U.S. objectives were often inconsistent and unclear, but three broad aims did stand out at different times: Development, counterinsurgency, and counterterrorism. U.S. officials sought to create a more stable and prosperous country including self-sufficient institutions; they emphasized degrading if not defeating the Taliban; and they focused on targeting terrorists. The lack of a clear, consistent, and sustained strategy precluded the achievement of all three core goals.

---

[16] This data is based on a review of all Congressional hearings over the war period. These hearings are posted by the U.S. Government Publishing Office's Govinfo service at https://www.govinfo.gov/app/collection/chrg/106/senate/Committee%20on%20Governmental%20Affairs. Not all of the hearings focused exclusively on Afghanistan. In some cases, Afghanistan was one of several topics featured in a hearing.

[17] "Do Voters Know We're Still At War With Afghanistan?" Rasmussen Reports, July 30, 2018, https://www.rasmussenreports.com/public_content/politics/current_events/afghanistan/do_voters_know_we_re_still_at_war_with_afghanistan.

5.1 Botched Development Efforts

In 2013, Dov S. Zakheim, the Pentagon's coordinator of civilian programs in Afghanistan from 2002 to 2004, published a revealing account of U.S. policy in Afghanistan during the George W. Bush era [21]. He argued that the Iraq War distraction prevented Washington from building on its early military triumphs in Afghanistan to achieve successes in development and reconstruction. In effect, in great part because the Bush administration turned its attention to Iraq, the absence of a new U.S. strategy in Afghanistan, following Washington's successful mission to target al-Qaeda and oust the Taliban, meant that America squandered an opportunity to leverage its military successes into post-conflict achievements.

Indeed, instead of trying to reconcile with a defeated Taliban from a position of strength and plan an exit strategy following the achievement of its original objectives, Washington blundered into a deeply flawed nation-building project. America sought to transform Afghanistan, a nation that couldn't be more dissimilar from the United States, into a mirror image of itself, through the development of new state institutions—including a national military—and infrastructure projects modeled on those in America.[18] This wasn't the first time U.S. officials tried to transform a conquered land into its own image: It was also attempted in the Philippines a century earlier, and with equally problematic results.[19] But that didn't stop America from trying again, this time in Afghanistan—even after similar attempts at U.S.-style nation-building in Afghanistan decades earlier, during an era of relative peace, didn't pan out either.[20]

Additionally, U.S. civilian and military personnel were guided by American experiences in other countries that had little bearing on Afghanistan. U.S. military efforts to strengthen the Afghan military were often viewed through the lens of America's time in Iraq. During the earlier years of the war, considerations about approaching the challenge of narcotics in Afghanistan were influenced by U.S. drug-fighting policies in Colombia. A major criticism of the U.S. experience in Afghanistan is that many war planners didn't have a strong understanding of Afghanistan—and one reflection of that was the inaccurate belief that what happened in far away countries could neatly and successfully be replicated in Afghanistan.

Another example of poor strategy impacting development goals was the use of funding. With U.S. foreign policy officials focusing much of their fiscal attention on Iraq, Congress appropriated less than $40 billion between 2001 and 2009, and more than half of that was used on the Afghan military. This didn't leave nearly enough to achieve Washington's ambitious development goals, including an Afghanistan Marshall Plan proposed by Bush in 2002. And yet, the money that was available was often used in profligate fashion, resulting in immense amounts of funding being

[18] See SIGAR [22].

[19] See Peter May [23].

[20] Chandrasekaran's *Little America* describes U.S.-funded modernization programs in Afghanistan—from irrigation projects to American-style living communities—that began in the 1940s but failed to achieve their intended goals, such as more agricultural production.

lavished on a single development project, and reflecting the tendency for U.S. officials to throw large amounts of money into projects in the hopes that this would maximize their chances of success. Afghan officials lacked the capacity to manage all this funding, meaning that corruption flourished and projects were often unsustainable.[21] Corruption expert Sarah Chayes has produced extensive research on how U.S. policy enabled and deepened graft in Afghanistan more broadly, including through empowering corrupt warlords [24]. Afghan officials, including former spy chief Mohammad Stanekzai, would later say that "inconsistencies" in U.S. strategy encouraged corruption.[22]

SIGAR analyst James Cunningham uses the metaphor of sedimentary rock to describe the lack of strategic planning in managing financial resources in Afghanistan: Each year brought a new layer of appropriated resources, with each year's allocation piled on top of the previous year's, with little thought of how these different layers were all meant to be managed. "We never had a long-term posture on how we were actually going to get there," he says [25].

Finally, in line with US strategy on the whole, development policy goals were inconsistent. The most prominent example is counternarcotics. Multiple administrations recognized the challenges posed by a powerful drug industry that not only helped fund the Talban insurgency, but also dominated the economy and crowded out licit industries and potential growth sectors. The Bush administration wanted to eliminate the industry by spraying poppy fields with chemicals—a plan influenced by U.S. actions in Colombia several years earlier [26]. William Wood, Bush's ambassador in Kabul in 2007 and 2008, was previously the U.S. envoy in Colombia. And then, in a dramatic policy about-face, the Obama administration would choose to focus instead on creating alternate livelihood programs.[23]

Afghan officials rejected the first plan, and Washington never implemented it. The alternate livelihoods policy did enjoy some fleeting success, in the sense that some new legal growth industries would emerge in the years that followed. Afghanistan's telecommunications sector, for example, would take off in a big way, until heavy Afghan government taxation policies cut into its profits and departing foreign troops and contractors badly reduced demand [27]. Poppy production, meanwhile, remained a dominant economic input. New annual harvest records were broken during the final period of the war.[24] For many poor farmers, poppy remained the best and most profitable choice.

---

[21] Azizian, "Easier To Get Into War Than to Get Out.".

[22] See "Expert Panel Reflects on Afghanistan's Future at William & Mary Law School Event," William & Mary Law School, March 30, 2022, https://law.wm.edu/news/stories/2022/future-of-afghanistan.php.

[23] "Afghanistan's Narco War: Breaking the Link Between Drug Traffickers and Insurgents," Report to the Senate Committee on Foreign Relations, U.S. Government Printing Office, August 10, 2009, https://www.govinfo.gov/content/pkg/CPRT-111SPRT51521/html/CPRT-111SPRT51521.htm.

[24] "Last year's Record Opium Production in Afghanistan Threatens Sustainable Development, Latest Survey Reveals," United Nations Office on Drugs and Crime, May 21, 2018, https://www.unodc.org/unodc/en/frontpage/2018/May/last-years-record-opium-production-in-afghanistan-threatens-sustainable-development--latest-survey-reveals.html.

5.2 Counterinsurgency Failures

Some of these development failures also delivered blows to another core goal, that of counterinsurgency. The corruption that emerged from so many unregulated aid dollars strengthened a core Taliban narrative: The U.S.-aligned Afghan government is corrupt and not worth supporting. The Taliban, in recruitment drives, often depicted themselves as a better alternative to the government of the day because they were not corrupt. To be sure, Washington's development efforts did result in genuine success stories, especially more infrastructure and improved women's access to education and health care. But Washington didn't do Kabul, and U.S. counterinsurgency activities, any favors by implementing wasteful aid programs that exacerbated graft and delivered propaganda victories to the Taliban.

Additionally, the Taliban benefited in at least one case from Washington's strategy of rebuilding Afghanistan in America's image. U.S. officials believed that empowering Afghan police would help bring much-needed law and order to local communities, as they understood this as a goal of policing in the United States. But in Afghanistan, where many police are corrupt and abusive, such a move undermined development, angered local communities, and gave more propaganda ammunition to the insurgents. The discontent of local communities also provided recruitment opportunities for the Taliban. "In our mind, in our mental imaging and strategy we think through," explains SIGAR's Cunningham, "the police are good guys when they walk around ... [but] in Afghanistan, they're actually the abusers and so our strategy actually created more discontent at a sub-national level [that] the Taliban was able to exploit."[25]

Another indication of poor U.S. strategy impacting counterinsurgency goals was that not enough thought was put into how to build a truly comparative advantage into the Afghan military, relative to the Taliban. Stavrides admits that the U.S military "missed out on opportunities to evolve our weaponry and phase in less traditional and more agile equipment earlier, and to train the Afghan forces into something more like the Taliban itself." Capitalizing on such opportunities, he suggests, would have given Afghan soldiers a greater capacity to fight the Taliban.[26]

Furthermore, the uncertainty surrounding future U.S. intentions in Afghanistan, given the lack of clarity about U.S. strategy, prompted a hedging policy by both Afghans and their Pakistani neighbors, and with deleterious implications for U.S. interests in Afghanistan. According to the scholar Stephen Biddle, many Afghans, and especially the political elite, hedged against the uncertain future role of America in Afghanistan by using corrupt practices to amass riches, ensuring they'd have enough wealth to leave the country following a U.S. departure and potential Taliban takeover [28].

Biddle also notes that Afghanistan's neighbor Pakistan allied itself with the Taliban as a hedging strategy, so that Islamabad could be assured of having leverage over and access to the Taliban, the most powerful non-state actor in Afghanistan, in

[25] Cunningham, video interview with author.

[26] Staviridis, "How the 20-Year War in Afghanistan Changed the U.S. Military.".

the event of a U.S. withdrawal [28]. Pakistan provided safe havens to the Taliban leadership and medical assistance to Taliban fighters during much of the war. Hedging considerations may also account for the more limited forms of military support provided to the Taliban by Iran, though Tehran was undoubtedly also motivated by the fact that Iran and the Taliban shared a common enemy in America [29]. Both external backers made the Taliban stronger, complicating the U.S. goal of degrading the insurgency.

Pakistan allied itself with the Taliban for other reasons, too. It had been close to the group since it formally emerged in the 1990s (it grew out of the anti-Soviet Afghan mujahideen forces that emerged in the 1980s). Additionally, Pakistan viewed the Taliban as a useful asset against Pakistan's rival India, which was no friend of the Taliban and enjoyed close partnerships with non-Taliban governments. However, Pakistan's need to hedge offered a sharp disincentive to ever consider easing its close ties to the Taliban. In this sense, the lack of a clear U.S. strategy in Afghanistan ensured an enduring relationship between the Taliban and their top external backer.

The Pakistan problem would loom large for America throughout the war. Pakistan provided sanctuary to the Taliban, but because U.S. forces depended on Pakistani airspace and roads to convey military supplies in and out of Afghanistan, Washington never resolved to launch strikes on Taliban sanctuaries in Pakistan (though on some rare occasions it did target top Taliban leaders there, including supreme leader Mullah Mansour in a drone strike in 2016). Also, in the early years of the war, Washington benefited from Pakistani counterterrorism assistance, including arrests of al-Qaeda leaders in Pakistan, and from the approval of Pakistan's military for America to use Pakistani territory to conduct drone strikes.

And yet, America never developed a regional strategy that would have enabled it to rely less on Pakistan, and that may have reduced the costs of taking a tougher line on Pakistan that could have helped America better achieve its counterinsurgency goals. For example, South Asia security scholar C. Christine Fair has argued that had Washington been able to improve relations with Iran, it could have considered that country, which borders Afghanistan to the west, as a potential candidate to host U.S. military supply routes [30]. Despite their enmity, the U.S. and Iran did briefly cooperate in efforts to curb al-Qaeda during the early years of the war—until a hardening Bush administration position toward Tehran made additional collaborations untenable. Additionally, Tehran played a key role in negotiations at the December 2001 Bonn conference that led to the formation of the first post-Taliban government [31].

5.3 Counterterrorism Struggles

The United States achieved its initial goal of degrading the al-Qaeda presence in Afghanistan. However, in 2015, more than a decade after U.S. forces entered the country, a previously nonexistent group, Islamic State-Khorasan (IS-K), formally announced its presence. While it is difficult to gauge whether the group would have emerged if there was not a war going on, conflict conditions—lawlessness, violence, the absence of the state's writ—certainly provided an enabling environment for a group that thrives on those very conditions. In this sense, a poor U.S. strategy, having contributed to an ongoing and intensifying war, helped facilitate the emergence of a major new terrorist threat.

Between January 2015, when the parent Islamic State formally announced the arrival of IS-K in Afghanistan, and the U.S. departure in August 2021, IS-K carried out hundreds of attacks in the country. According to one study, there were a whopping 365 IS-K attacks in Afghanistan in 2021 alone, compared to 85 in 2020.[27] That the number of IS-K attacks more than quadrupled during the last year of the U.S. military presence in Afghanistan attests to the failures of U.S. counterterrorism actions. U.S. and Afghan forces largely used kinetic tactics to manage the IS-K threat, with U.S. air strikes and other heavy firepower targeting IS-K fighters. Such operations—including, famously, the use in 2017 of America's biggest non-nuclear weapon on an IS-K cave hideout—did little to curb the threat.

Some scholars have rightly noted that counterterrorism activities can only be successful if accompanied by some counterinsurgency tactics—especially the "winning hearts and minds" activities that generate the necessary public and political support for counterterrorism activities. This didn't happen in Afghanistan. The IS-K threat would intensify in the years after 2014, when the U.S. military presence had shrunk and the mission had transformed from combat to training and advising Afghan troops. Additionally, the Trump administration relaxed the rules of engagement for U.S. forces that had been tightened in the Obama era, which resulted in more robust uses of forces in counterterrorism activities and surges in civilian casualties.

Consequently, U.S. forces did not have the numbers or mandate to focus on the hearts and minds goals that they had pursued earlier in the war. In the words of Trump, they were there "to kill terrorists." In the Trump era, heavy-handed counterterrorism operations may have killed many bad guys, but they also may have had a radicalizing effect that produced new bad guys. On the whole, America's counterterrorism strategy failed to include a broader counterextremism component meant to identify and curb the drivers of radicalization and terrorism. It's easy to kill terrorists, but it's much harder to kill the ideology that fuels them—and especially in a conflict zone, with a limited mission and personnel on the ground.

6 Lessons Learned

The U.S. war in Afghanistan marked the last of a century-plus-long succession of extended American military interventions. In Afghanistan as well as in Iraq, Vietnam, Haiti, and the Philippines (among other places), U.S. forces struggled to

[27] "Summary of ISIS Activity Around the Globe in 2021," Meir Amit Intelligence and Terrorism Information Center, Israeli Intelligence Heritage and Commemoration Center, January 2022, https://www.terrorism-info.org.il/app/uploads/2022/01/E_008_22.pdf.

address a range of challenges, from unwelcoming publics and corrupt leaders to ferocious insurgencies and unfamiliar terrain. The American military intervention in Afghanistan also marked the last of a series of foreign invasions of that country stretching from the Soviets and British to the ancient Greeks—all of which struggled as well.

In many of these cases, mistakes and failures of earlier interventions were not properly taken into account. And yet, even though the current administration is keen to avoid future military adventures abroad, considering the lessons learned from the U.S. experience in Afghanistan is critical. One can't rule out another American military conflict sometime down the road, whether one tied to a confrontation with China over Taiwan or disputed territories in the South China Sea, or with Russia over NATO countries in Eastern Europe. In any such conflict, an effective U.S. strategy would be essential. And that will require understanding the lessons learned from the troubled experience with U.S. strategy in Afghanistan.

Know what you want. It is instrumental that a war strategy not only be clear and consistent, but also agreed upon by the entirety of the government overseeing it. A lack of consensus can lead to policy u-turns that breed confusion and ultimately embolden the enemy. It can also lead to internal distractions that take critical political bandwidth away from focusing on implementing the strategy.

Keep your eye on the ball. A strategy can only be effective if it remains top of mind for those that developed and implemented it. The war in Iraq marked a turning point for the war in Afghanistan because U.S. officials became distracted by Iraq at a pivotal moment in Afghanistan, with American forces still on the ground but with no clear mandate following the achievement of their original objectives. Additionally, it is difficult to maintain a clear focus on executing a war strategy if the commander in chief is also thinking about an exit strategy—as was the case with Presidents Obama and Trump.

Align goals with resources. Too often in Afghanistan, there were mismatches between means and ends. Officials laid out development objectives without enough funding to achieve them. Too much money was lavished on basic infrastructure projects. It is more prudent policy to formulate objectives based on available resources, rather than to announce objectives and try to generate the requisite amount of resources to achieve them. At the same time, judiciousness in the disbursal of financial resources can be advantageous, and help move the needle forward on attaining key goals, such as reducing corruption.

Plan for the worst and take bad news seriously. It's essential to plan for all possible battleground contingencies, including those that may appear to be the most unlikely. War is unpredictable. A good strategy will feature a range of policies that can be implemented in response to any possible scenarios, including worst-case ones—such as a Taliban takeover prior to the completion of the troop withdrawal. In addition, when top leaders are informed that the war effort isn't going well, reassessments and policy reviews—and not mere shrugs and a fixation on projecting positivity and progress—are imperatives.

Agree on the basics and communicate them to those overseeing them. When diplomats, military officers, and aid workers didn't have definitive answers to simple questions about war—why are we here, who is our enemy, and so on—strategy implementation on the ground can become a very dysfunctional process, and it also risks morale problems for those on the ground. Senior leaders in Washington must ensure that their representatives on the ground are properly informed—and if even senior leaders don't have answers to those basic questions, that should sound an alarm about the need for a strategy rethink.

Develop a historically and socially informed approach. One reason why U.S. strategy failed in Afghanistan was that war planners repeated mistakes from the past—including the development of infrastructure projects more suited for the United States than Afghanistan—and made policy decisions based on realities in other countries that have few similarities to Afghanistan. The views of scholars and other area specialists outside government should be considered as essential inputs to help inform the policy planning and strategy development process. And at the least, care should be taken to ensure that top leaders are aware of the costs of not taking more historically and socially informed approaches.

Demand accountability. When an American war is not making regular headlines and doesn't feature prominently in debates in Washington, political leaders have less incentive to course correct when the war's strategy is not going well. When they do feel pressure, the political costs of not changing a bad strategy will likely be much higher. The American public—including civil society, the media, and research and policy analysis organizations—should ensure proper focus and coverage on a war that involves thousands of deployed soldiers and billions of dollars spent. And Congress should ensure a steady set of hearings that enable senior officials to be asked tough questions about war strategy.

7 Conclusion

In two of President Biden's final policy speeches on Afghanistan, he strongly implied that the United States lacked a clear strategy to justify keeping boots on the ground. In his April 2021 address announcing his decision to withdraw, he said that "our reasons for remaining in Afghanistan are becoming increasingly unclear."[28] Then, in a speech on August 31—the day after the last U.S. soldier left Afghanistan—he said "we must set missions with clear, achievable goals—not ones we'll never reach."[29]

[28] "Remarks by President Biden on the Way Forward in Afghanistan.".

[29] "Remarks by President Biden on the End of the War in Afghanistan," White House Briefing Room, August 31, 2022, https://www.whitehouse.gov/briefing-room/speeches-remarks/2021/08/31/remarks-by-president-biden-on-the-end-of-the-war-in-afghanistan/.

It took nearly 20 years for a U.S. commander in chief to admit what had long been the elephant in the room. But by then, it was too late. Tens of thousand of Afghans had perished in the war, and 2455 U.S. soldiers had lost their lives. At least $2.3 trillion had been expended on the war [32]. And American soldiers had completed their withdrawal from a country now controlled by the same group they had ousted nearly two decades earlier, and threatened by a terror group that didn't exist when U.S. forces first entered Afghanistan.

This, then, may be the most sobering illustration of U.S. strategy failures in Afghanistan: Once the withdrawal was complete, even America's initial achievements—degrading terrorist threats and ousting the Taliban—had gone up in smoke.

References

1. Packer G (2022) Biden's Saigon. The Atlantic. https://www.theatlantic.com/magazine/archive/2022/03/biden-afghanistan-exit-american-allies-abandoned/621307/
2. von Clausewitz C (1976) On war (Trans. Howard M, Paret P). Princeton University Press, Princeton, NJ
3. Weinberger C (1984) The uses of military power. Remarks to the National Press Club, Washington, DC, https://www.pbs.org/wgbh/pages/frontline/shows/military/force/weinberger.html
4. Powell C (1992/1993) U.S. forces: challenges ahead. Foreign Affairs, Winter
5. Freeman C (2021) How should we deal with our failure in Afghanistan?. Remarks to a Panel of the Rhode Island National Guard. https://chasfreeman.net/how-should-we-deal-with-our-failure-in-afghanistan/
6. Azizian N (2021) Easier to get into war than to get out: the case of Afghanistan. Harvard Kennedy School, Belfer Center, Paper. https://www.belfercenter.org/publication/easier-get-war-get-out-case-afghanistan
7. Winnefeld JA, Morell MJ, Allison G (2020) Why American strategy fails: ending the chronic imbalance between ends and means. Foreign Affairs. https://www.foreignaffairs.com/articles/united-states/2020-10-28/why-american-strategy-fails
8. Fisher D (2022) Video interview with author
9. Azizian Easier to get into war than to get out
10. Whitlock C (2021) The Afghanistan papers. Simon & Schuster, Boston
11. Whitlock C (2019) Stranded without a strategy. Washington Post. https://www.washingtonpost.com/graphics/2019/investigations/afghanistan-papers/afghanistan-war-strategy/
12. Sopko J (2022) Video interview with author
13. Shepherd S (2017) Poll: voters split on Afghanistan increase, oppose shutdown for border wall. Politico. https://www.politico.com/story/2017/08/30/afghanistan-troops-trump-poll-242146
14. Brenan M (2021) Americans split on whether Afghanistan war was a mistake. Gallup. https://news.gallup.com/poll/352793/americans-split-whether-afghanistan-war-mistake.aspx
15. Newport F (2001) Public opinion of the war in Afghanistan. Gallup. https://news.gallup.com/poll/9994/public-opinion-war-afghanistan.aspx
16. Stavirides J (2021) How the 20-year war in Afghanistan changed the U.S. military. Time. https://time.com/6090623/afghanistan-us-military-lessons/
17. Clayton T (2020) *The Washington Post*'s 'Afghanistan papers' and U.S. policy: main points and possible questions for congress. Congressional Research Service R46197. https://sgp.fas.org/crs/misc/R46197.pdf

18. Mann J (2004) The rise of the Vulcans: the history of bush's war cabinet. Penguin Books, New York
19. Whitlock C (2019) At war with the truth. Washington Post. https://www.washingtonpost.com/graphics/2019/investigations/afghanistan-papers/afghanistan-war-confidential-documents/
20. Tierney D (2015) Americans have forgotten the war in Afghanistan. The Atlantic. https://www.theatlantic.com/international/archive/2015/06/afghanistan-war-memory/396701/
21. Zakheim DS (2013) A Vulcan's tale: how the bush administration mismanaged the reconstruction of Afghanistan. Brookings Institution Press, Washington, DC
22. SIGAR Collapse of the Afghan national defense and security forces: an assessment. Chandrasekaran R (2013) Little America: the war within the war for Afghanistan. Vintage Books, New York
23. May P (1980) Social engineering in the Philippines: the aims, execution, and impact of American colonial policy, 1900–1913. Greenwood Press, Westport, CT; Karnow S (1990) In our image: America's empire in the Philippines. Penguin Random House, New York
24. Chayes S (2007) The punishment of virtue: inside Afghanistan after the Taliban. Penguin Books, New York
25. Cunningham J (2022) Video interview with author
26. (2009) Afghanistan's Narco war: breaking the link between drug traffickers and insurgents. Report to the Senate Committee on Foreign Relations, U.S. Government Printing Office. https://www.govinfo.gov/content/pkg/CPRT-111SPRT51521/html/CPRT-111SPRT51521.htm
27. Shevory K Once a bright spot, Afghan telecoms face unsustainable losses. New York Times. https://www.nytimes.com/2016/04/09/business/international/once-a-bright-spot-afghan-telecoms-face-unsustainable-losses.html
28. Biddle S (2011) Long term goals for Afghanistan and their near term implications. Testimony before the U.S. Senate Foreign Relations Committee. https://www.foreign.senate.gov/imo/media/doc/Biddle%20Testimony.pdf
29. Kugelman M (2016) What was Mullah Mansour doing in Iran? Foreign Policy. https://foreignpolicy.com/2016/05/27/mullah-mansour-iran-afghanistan-taliban-drone/
30. Ayres A (2018) Five questions with C. Christine fair on India, Afghanistan, and Iran. Asia Unbound Blog, Council on Foreign Relations. https://www.cfr.org/blog/five-questions-c-christine-fair-india-afghanistan-and-iran
31. Slavin B (2009) Bitter friends, bosom enemies: Iran, the U.S., and the twisted path to confrontation. St. Martin's Griffin, New York; Dobbins J Negotiating with Iran: reflections from personal experience. Wash Quart 33(1):149–162
32. Crawford N (2021) Calculating the cost of the Afghanistan war in lives, dollars, and years. Conversation. https://theconversation.com/calculating-the-costs-of-the-afghanistan-war-in-lives-dollars-and-years-164588

Michael Kugelman is director of the South Asia Institute at the Wilson Center in Washington, DC. He is grateful to Mahdi Surosh and Hamna Tariq for their research assistance.

Afghanistan: A Case Study in Self-defeat

Ambassador Hugo Llorens

> How pleasant it is, 'Says the Roman poet', to listen to a storm from the safe shelter of the shore.
> —Marc Bloch, Strange Defeat

Abstract The Taliban's seizure of Kabul on a beautiful summer afternoon of August 15, 2021, is reminiscent of the German Army's triumphant march into Paris in June 1940. The Taliban's campaign, like the German Panzer onslaught that humbled the French Army of 80 years ago, was as much about the spiritual and psychological collapse of the Afghan state as the military prowess of the Taliban war machine. While most experts and analysts (myself included) predicted an inevitable Taliban victory in the wake of the precipitous U.S. and NATO withdrawal, almost no one foresaw the rapidity of the result. Although this chapter will touch on some of the tactical decisions made by various actors hastening the collapse, the focus will be on the broader strategic decisions contributing to the final outcome.

Keywords Afghanistan · Self-defeat · NATO withdrawal · Taliban

Obviously, the true significance of the Taliban victory is also much less about the defeat of the Afghan Armed Forces, or the demise of the Islamic Republic of Afghanistan, and more about the military and political defeat of the United States and its closest allies at the hands of a fourth-rate military and political organization. The U.S. defeat, like the French in 1940, was not foreordained, but ultimately was the product of a lack of statesmanship and vision, at critical points, by America's most senior leaders (principally Presidents Trump and Biden). The Taliban's seizure of Kabul on August 15, 2021 did not have to happen and in many ways can most aptly be described as a self-defeat by the United States.

U.S. leaders' scuttling of our position in Afghanistan has, is, and will yield a series of primary and secondary order consequences detrimental to both U.S. interests and to the stability of the South Asia region. First and most immediate is the dramatic

A. H. Llorens (✉)
Ambassador Hugo Llorens and Global Partners LLC, Marco Island, Fl 34145, USA
e-mail: hugo.llorens@ambassadorhugollorens.com

A. Farhadi and A. Masys (eds.), *The Great Power Competition Volume 4*,
https://doi.org/10.1007/978-3-031-22934-3_14

humanitarian plight confronting the people of Afghanistan caused by the complete collapse of the Afghan economy and financial system. The economic and financial conflagration has resulted in a 50% decline in GDP, a sharp spike in unemployment, and a dramatic rise in poverty. Above all, the grim economic situation has resulted in a major shortfall of food with the World Food Program estimating most Afghans are not getting enough to eat, and millions, including children, are confronting starvation. In addition, there are the immediate political effects of the removal of an admittedly flawed and somewhat dysfunctional government, but one firmly pro-Western and committed to liberal and democratic rule. The result is a very large chunk of the best human capital has fled, and Afghanistan is now ruled tyrannically by an aberrant clique on behalf of a narrow rural Pashtun constituency. Meanwhile, the vast majority of the people, urban Pashtuns, Tajiks, Uzbeks and Hazaras are oppressed and relegated to second-class status. What is worse, the Taliban's ascension to power has resulted in a total reversal of the major political, economic, health and education gains achieved by Afghan women in the past two decades. The harsh reality is Afghanistan today has become, once again, one of the most brutal, backward, least free and autarchic societies on earth. President Biden's bold assertion that the major issue of our time is the confrontation between democracy and autocracy rings hollow juxtaposed with his reckless and completely unnecessary decision to abandon the Afghan people.

In terms of core national security issues, America is far less safe today than prior to August 15, 2021. The Taliban has seized billions of dollars of sophisticated U.S. and NATO military hardware of every type and variety. Unable to absorb such a large cache of weaponry, much of it will be acquired by our enemies—state and non-state. The Taliban victory and our defeat has been hailed by every militant and jihadists group around the world. It has raised their morale and reaffirmed their long-held faith that America does not have the stomach for a protracted fight against terror. In practical terms, by withdrawing from Afghanistan, we have lost a precious counterterrorism platform in Central Asia and removed from the equation a political, military and intelligence network in Kabul completely allied to us and committed to our efforts to prevent groups like Al-Qaeda, ISIS-K and others from attacking America and its allies. Afghanistan, once again, is ungoverned space as it was pre-September 11. With thousands of the most hardcore jihadists freed from Afghanistan's high security prisons, these groups will reconstitute and expect that they will knock loudly on our door at some future date.

America's failure in Afghanistan is also a major setback in terms of our long-term objective of being a decisive trade and investment player in South-Central Asia. Afghanistan is remarkably well-endowed in natural resources. It's mineral and energy potential are immense, and U.S. Geological Survey studies estimate easily recoverable mineral wealth (what we know is there) at $1 trillion. The U.S. withdrawal opens up Afghanistan for Chinese capital to take the lead in the long-term control of Afghanistan's vast resources, to include copper, gold, lead, silver and some of the richest rare earth mineral resources in the world, including potentially the largest lithium reserves in the world.

Last but not least, there are and will potentially be geopolitical ramifications to the U.S. and NATO debacle in Afghanistan. The U.S. retreat in South Asia sends a message of weakness to our potential adversaries in China, Russia, Iran, and North Korea. While Putin's views on Russia's historic territorial claims in Europe may be unrelated to Afghanistan, the Taliban's victory likely emboldened the Russian dictator and the timing of the invasion less than six months after the spectacular Kabul collapse may be far more than coincidence.

No doubt, as anyone who has bothered to read the literature on the subject or served there knows, the United States made numerous large-scale mistakes in Afghanistan. While President George W. Bush deserves great credit for his decisive leadership in the wake of the September 11 attacks, and he has my admiration, many analysts and historians have rightly pointed out possibly the first major U.S. strategic mistake was the decision to take our eye off the Afghan ball and vector to Iraq. It is not so much about whether toppling Saddam Hussein was or was not the right move in the big scheme of things, but simply at a critical juncture the President and his advisors' relative neglect of Afghanistan provided a crucial interval for the defeated Taliban forces to lick their wounds in their Pakistan sanctuary, prepare and eventually launch a renewed insurgency in southern and eastern Afghanistan. Probably most consequential was the Bush Administration's easing of pressure on Pakistani President Musharraf. Seeking to achieve strategic depth in Pakistan's confrontation with India, Musharraf hypothesized America's focus on Iraq and the Middle East suggested Afghanistan was of relatively less importance. The final result was his decision to step back from the Western-backed government in Kabul, and cynically cast his lot once again with a renewed Taliban insurgency. Pakistan's decisive shift towards the Taliban caught a distracted U.S. by surprise as we blindly transitioned from "The Liberation" (2001–05) to the "Bogging Down" phase (2005–09) of our experience in Afghanistan. Over the next two decades, Pakistan's military, particularly its intelligence and special operations branch (ISI), would play a pivotal role in protecting the Taliban leadership, and arming, training, basing Taliban fighters, and treating their wounded.

While working the Afghanistan portfolio hard, President Barrack Obama, a leader who also merits respect, is not immune from fault in Afghanistan. Many of his critics cite with some merit his approach was erratic and inconsistent. In the 2008 Presidential campaign, candidate Obama asserted Afghanistan was "the good war" and Iraq the "bad one." He vowed to focus the vast array of U.S. power on achieving success in Afghanistan. Once in office, however, he dithered before ultimately approving his military advisors' recommendation for a major "surge" in our military presence. Yet, despite supporting the expansion of the war, he was throughout plagued with doubt and sent conflicting messages and made contradictory decisions. By the end of his first term, Obama was already pulling the plug on his surge and committed to Afghanize the conflict and implement a blueprint and timetable for a U.S. withdrawal. Like his predecessor, President Obama failed to consistently and forcefully pressure Pakistan to end its support for the Taliban.

Again, in our two decades of involvement the U.S. government and bureaucracy writ large, myself included, committed a series of major errors. Certainly, our tolerance for corruption was a primary and costly error. In our zeal to get the job done, we breathlessly handed out extraordinary sums of money on a broad spectrum of military and development initiatives without creating sufficient controls and audit systems to ensure our tax payers got their money's worth. In addition, we were inconsistent in holding some of our Afghan allies to account for their role in stealing our money. An argument could be made that early in our involvement it was essential to work with the Afghan war lords, believing they could get things done. The reality was the opposite: we empowered a group of rapacious leaders notorious for their ineptitude and corrupt ways. Again, at no point in our two-decade involvement did we do enough to promote into positions of power and influence the mass of technocrats and well-meaning civilian politicians who had returned from exile to help their country rebuild. Among our other litany of missteps included creating an Afghan military too much in the U.S. and NATO image—although I wonder if institutionally U.S. and allies' military training could have realistically been designed to create something far different from our own. Others cite Western support for an Afghan constitution that overcentralized power and control in Kabul. On the economic side, we failed until too late to emphasize a more market-based policy approach aimed at achieving true self-reliance.

All that said, and despite innumerable mistakes by both our senior civilian and military leadership, as well as our bureaucracies, including the wasteful and inefficient management of our resources, much good was achieved in the two decades we were in Afghanistan. Certainly, President Bush made the right call early in understanding that following years of brutal rule by the Taliban, when all opposition was ruthlessly suppressed, the only viable way forward was to support a consensus approach bringing together leaders of all the principal ethnic and tribal groups, and political parties and factions. U.S. support for the UN's convening of all key players to the Bonn Conference (December 5, 2001), and the signing of the Bonn Agreement, ratified by UN Security Council 1383, resulted in the creation of an interim administration, led by Hamid Karzai. In addition, UN Security Council Resolution 1386 of December 20, 2001, established the international Security Assistance Force, or ISAF. These decisions were followed by an emergency Loya Jirga of June 2002, attended by 1550 delegates (including 200 women), representing Afghanistan's 364 districts. The Loya Jirga reaffirmed the role of President Karzai as Chairman of the transitional government. In the interval, ISAF and the international community established Provincial Reconstruction Teams (PRTs), and the international donor community and the western allies re-established their diplomatic missions in Kabul and began to implement large scale development and economic assistance programs to rebuild Afghanistan. These were heady and optimistic days—a period I call "the Liberation" phase of our involvement in Afghanistan (2001–2005). The most significant political event of the Liberation phase was the convening of the constituent assembly in January 2004 and the enactment of a liberal-democratic constitution. In the context of the times, and with memories fresh of the horrors the Taliban had inflicted on the Afghan people, as well as the Taliban's nefarious role in backing

Osama Bin Laden and Al-Qaeda, it is hard to imagine back then anyone seriously advocating negotiations with Mullah Omar, or in any way seeking to reincorporate the Taliban into the broader political system.

The critics have it wrong as well in asserting the Bush Administration's nation building effort and support for a democratic Afghanistan were naïve and misplaced. In my view, President Bush's basic approach was appropriate at the time and rather than "nation build" the Administration was seeking to provide urgently-needed humanitarian help, as well as financial assistance to restore economic growth and achieve some level of social stability. Similarly, on the political side no one in the Bush or Obama teams I worked with ever deluded themselves into believing Afghanistan could evolve into some kind of Vermont town-council democracy. What officials did believe is a more liberal and more tolerant political order was the best way for Afghans to manage a very complex ethnic, tribal, religious and gender social tapestry.

What no one can deny is in the 20-year period following the toppling of the Taliban regime in December 2001, the United States and its allies helped the Afghan government and people recover from an utterly catastrophic situation and achieve a major socio-economic and political transformation. In these two decades, the nation's GDP grew five-fold and per capital incomes four-fold. The international development assistance led to the rebuilding of large swathes of the countryside devastated by the Soviet invasion, and cities destroyed in the subsequent mujahadeen civil war. Bilateral and multilateral donors invested in road construction and rural electrification. The public works investments in irrigation networks contributed to a major expansion in the cultivation of basic grains and a wide variety of cash crops, such as fresh and dried fruits, vegetables and nuts. Education was an area of unprecedented progress. In 2001 under Taliban rule, there were 600,000 children in the system (K-university), and women were barred from attending school. By contrast, in the past two decades Afghanistan experienced dramatic progress. As late as 2021, there were 8,500,000 Afghans in the school system, and 40% were women. A new generation of educated Afghan men and women were on the cusp of replacing the old generation and leading the country to a more modern and prosperous future. Health was another area of great gain. In 2001 there was no health care system to speak of. Following the overthrow of the Taliban, and thanks to generous international support, by 2021 60% of Afghans were a 45-min walk to the nearest health clinic. Underscoring the robust expansion of medical care were the dramatic drops in infant and maternal mortality rates in Afghanistan. Communications is another area of revolutionary change. Under the Taliban, Afghanistan was one of the most isolated/least connected countries in the world. After 2001, the adoption of a new telecommunications law and regulatory regime led to massive private investment in the sector. Prior to the return of the Taliban there were 22 million cell phones in Afghanistan, a country with a population of 32 million. I recall in my trips into rural Afghanistan often seeing women in full bourka walking down a dirt road and speaking on a cell phone—a true information revolution! Best of all, the adoption of a liberal constitution in 2004 led to significant expansion of individual rights and freedoms for the citizenry, particularly in urban areas. Many benefited, but none more than women who were empowered to study, work, own businesses, lead NGOs, and even serve as parliamentarians, cabinet ministers and ambassadors. Indeed, Afghanistan became one of the freest nations in Central Asia and the Afghan media was one of the most vibrant and its investigative journalists some of the best in the region. In the past two decades Afghanistan became a real country with a President, a parliament, and a judiciary. Although flawed there were elections, and Afghanistan had a thriving and vibrant civil society. In sum, despite being subjected to a relentless Taliban terror campaign, the Afghan people were committed to creating a better world for themselves, but they needed time and sustained international solidarity. All of that is no more.

President Bush was essentially correct in supporting a more open and liberal Afghan society and helping the Afghans rebuild their nation, although his focus on Iraq had serious consequences. Likewise, President Obama, despite the mixed signals, was equally right in reaching the conclusion by the end of his first term that we needed to gradually wind down our combat role and help the Afghan National Security Forces (ANSF) take the lead in combatting the Taliban insurgency. The fact was despite a major expansion in U.S. and NATO ground and air forces in the period 2009–12 (peaking at 140,000 U.S. and NATO), and an intensification of offensive and counter-insurgency (COIN) operations, U.S. military commanders could not convince the President they had a clear path to victory. With the U.S./NATO surge costing the U.S. tax payer as much as $140 billion per year in combined military-diplomatic-intelligence-development efforts, annual casualty levels in the hundreds, and with no end in sight, the President prudently opted to find a way out and begin phasing out the U.S. presence.

I arrived in Afghanistan for my first Ambassadorial-level assignment in the summer of 2012 to coincide with the ending of "the Surge" (2009–12) and the very start of the Afghanization phase (2012–16). Prior to arriving in Kabul, I had extensive consultations in Washington with senior National Security Council, State and intelligence community officials, as well as with Members of Congress (House and Senate) and their senior staffers. In meetings with Administration officials, I found little appetite for continuing our military engagement. Many Administration officials, particularly on the civilian side, had lost faith in our partnership with President Karzai and did not believe the cost in blood and treasure of our effort in Afghanistan was worthwhile. Our instructions, were clear: the U.S./NATO surge was coming to an end, and we were embarking on a policy to quickly transition away from a U.S. and NATO lead role in combatting the Taliban insurgency to the ANSF.

In Kabul, I was one of 5 U.S. Ambassadors. With a staff of 8500, including U.S., local Afghan, Third Country Nationals, and contract personnel, the U.S. diplomatic presence in Kabul and across Afghanistan was the largest Embassy operation in history. Just as there is only one CEO in a private firm, diplomatic missions are led by a single Ambassador. The one-Ambassador maxim was broken for several years in Afghanistan, as in Iraq, partly because of the extraordinary size and complexity of our diplomatic presence. However, likely the most relevant reason related to our need to effectively coordinate our activities with our U.S. and NATO military counterparts, which at the time included several hundred Flag-rank officers (Generals and Admirals). Militaries are enormously rank conscious institutions, and our ability to engage symmetrically with these senior military officers could most effectively be done with equivalent Ambassador-ranked officials. Obviously, there was still one senior U.S. Ambassador, or Chief of Mission, who retained the "Extraordinary and Plenipotentiary" powers, and was the President's personal representative to the Afghan government. Our Chief of Mission was one of America's best career diplomats, Ambassador Ryan Crocker. Of the other Ambassadors, I held the position of Assistant Chief of Mission, the third ranking official in the Embassy. I had a full plate of responsibilities, which included running the day-to-day of the Embassy's management, human resources, budget, and logistics, as well as a vast security apparatus. As the Embassy's Chief Operating Officer, the administrative piece, was a full-time, 7-day per week job. Yet, I also had the lead policy portfolios with oversight and responsibility for directing our Political and Political-Military sections, as well as Public Affairs (media, information, and cultural issues) and our Consular operations (American Citizen Services and Visas). The Deputy Ambassador, Ambassador Jim Cunningham, shared the "Front Office" role and was Ambassador Crocker's chief strategic advisor and alter ego. The other members of the Ambassadorial leadership team included Ambassador Steve McFarland, who managed a vast "Rule of Law" portfolio involving our counterdrug and law enforcement policy and programs, and Ambassador Hilda "Bambi" Orellana, who directed the economic/commercial/agriculture policy team, and the largest USAID economic development assistance program in the world. Ambassador Crocker, and Ambassador Cunningham who replaced him, were first-among-equals, but as an Ambassadorial leadership team the five of us ran things in a most productive and collegial manner.

The key diplomatic document paving the way for the withdrawal of most U.S. forces and the handing of lead responsibility to the Afghans was the Strategic Partnership Agreement (SPA). Ambassador Crocker devoted the lion's share of his 2011–12 tour in Kabul to personally negotiate the SPA with President Karzai and National Security Advisor Dadfar Spanta. Cocker delivered a tour de force performance concluding the agreement in a period of tense relations between Washington and Kabul. President Obama and Karzai signed the SPA on May 1, 2012 in Washington. As part of the agreement, the U.S. designated Afghanistan a Major Non-NATO Ally and made a series of broad commitments to provide long-term political, diplomatic, military and economic support for the Islamic Republic of Afghanistan. The SPA also paved the way for the launch of negotiations for the conclusion of a Bilateral Security Agreement (BSA) with our Afghan counterparts replacing the Status of Forces Agreement. The BSA would define the long-term military and security relationship between the United States and Afghanistan. As the Ambassador responsible for the Political-Military portfolio, I was the senior U.S. official in Kabul leading our team in the BSA negotiations in close partnership with our Washington-based lead negotiator Ambassador James Warlick. During my term in Kabul (2012–13) we were able to negotiate 95% of the BSA text, which included everything from basing rights, to customs, to legal jurisdiction over our servicemembers stationed in Afghanistan. A final agreement was never concluded in my tour due to President Karzai's unwillingness, for domestic political reasons, to formally agree to a long-term presence of a modest number of U.S. military advisors and a small special forces contingent. The BSA was not formally signed until President Ashraf Ghani assumed office at the end of 2014, but once enacted it did serve its purpose of defining the rules for the U.S. military presence and operations during the Obama, Trump and Biden Administrations (2015–22).

Soon after my arrival in May 2012, I recall a one-on-one working lunch with U.S. Forces and NATO Commander General John R. Allen (U.S. Marine Corps). An exceptionally bright and competent military officer, Allen aptly described his role of leading his forces in the fight against the Taliban, Al-Qaeda and other terrorist forces at the same time we were beginning our phased withdrawal, and intensifying our training advisory role with the Afghan military, as something akin to "flying and building a plane at the same time." Despite the difficulty of the fight Allen expressed confidence our military successes had diminished the Taliban. He also had faith our ramped-up NATO Training Mission (NTM-A/CSTC-A) was on track to increase the capabilities and professionalism of the ANSF, both the army and police elements. He was particularly pleased with the progress made creating highly capable Afghan Special Forces that could take the fight to the insurgents. He lamented the high levels of corruption within both military and civilian sectors, and noted professionalizing the officer corps was necessary to gradually eradicate much of the "predatory war lord culture" of the Afghan military—a key to a successful outcome. Based on my Washington consultations with the U.S. State Department's Political Military Affairs Bureau's I conveyed our concerns about the slow-pace of the development of the Afghan Air Force, and mentioned the importance of the Pentagon completing the adjudication of the bidding contract to procure a ground attack aircraft for the Afghan Air Force. We both agreed enhancing the Afghan military's logistical capability was also vital.

It is important to underscore during the surge hundreds of civilian U.S. diplomatic, development, counternarcotics (DEA), and intelligence officers were deployed and embedded with U.S. and NATO forces in every one of Afghanistan's 34 provinces. These dedicated men and women, the very best of the U.S. Foreign and Civil Service, were integral to U.S. and NATO counter-insurgency operations. Our people were involved in managing ties with Afghan provincial, district and tribal leaders, reporting on fast-breaking political and economic developments, collecting and analyzing intelligence, and implementing a wide range of rural development projects. I had the privilege of traveling widely and visiting our superbly talented and motivated people on the frontlines of the war serving in large Provincial Reconstruction Teams (PRTS) and small isolated Forward Operating Bases (FOBs) in all regions of Afghanistan, including Kandahar, Helmand, Farah, Nangarhar, Balkh, Herat, Bagdhis provinces. We were very proud of them and several of our best people were killed in the line-of-duty during my time there.

In nearly all of the venues outside Kabul, security for our civilian staff outside of the Embassy in Kabul was primarily the responsibility of our military. Obviously as U.S./NATO forces withdrew, our people would also have to be pulled out from the field. One of our most importance functions at the Embassy was to smoothly coordinate with our military colleagues the sequenced, safe and secure withdrawal of official civilians. We also devoted much time and attention working with the U.S. State Department's Bureau of Management on a blueprint for the U.S. official civilian presence in Afghanistan post-Surge. In extensive discussions back-in-forth with both State and the White House, the Embassy made the case for a "1 + 4" diplomatic presence. We proposed sustaining a large Embassy in Kabul (the center), with smaller contingents in the north, south, east and west. Specifically, in the west this would include maintaining the U.S. Consulate in Herat (formally opened soon after my arrival in June 2012). In the north, we advocated standing up a small diplomatic presence in the city of Mazar-i-Sharif (Balkh province) housed within the perimeter of the German army-air base there. We also suggested deploying small U.S. diplomatic contingents to be co-located in military and intelligence facilities in Kandahar (in the south) and Jalalabad (in the east). In the wake of the attack on the U.S. diplomatic compound in Benghazi, Libya (September 2012), extreme risk-aversion seemed to take hold of the Obama Administration, and the result was a reluctance to commit to anything beyond the existing Embassy and Consulate in Herat. The Administration quickly closed the Herat consulate following a Taliban bombing of the consulate's front gate in September 2013.

Ultimately, the decision made by the Obama State Department resulted in bringing all of our diplomatic, development and official civilians back to our Embassy in Kabul. I always believed it was a mistake to completely withdraw all U.S. official civilians from the field. While we maintained a gigantic Embassy in Kabul our ability to do our diplomatic work and manage our vast development program portfolio was seriously hamstrung by not being able to have more direct access to provincial and district officials and civil society outside the capital. We made creative efforts to maintain open lines of communication with the provinces using technology, local contractors, and receiving visiting Afghan delegations from the field. Nevertheless, I always believed the White House and State Department's zero tolerance for risk with our diplomatic staff was a signal of our weakness. Our meek approach to risk was also a psychological blow to our many Afghan allies outside Kabul and was possibly perceived as the first signal that we might eventually abandon them.

In terms of our job in Kabul at the time I was there and beyond, we successfully completed the transition to an Afghan-led military effort by the end of 2014. On the military side, we completed the withdrawal of 85% of our troops and retrograde (bring back home) vast stores of military equipment. Our post-Surge military presence (U.S. and NATO) consisted chiefly of NATO headquarters in Kabul, the large Bagram Air Base 30 miles north of Kabul, and major U.S., Italian, and German bases in Kandahar, Herat, and Balkh provinces, respectively. There were a number of other smaller military and training facilities, and the intelligence community maintained a reduced but vital footprint outside of Kabul.

On the political side the most significant event of the post-Surge period were the 2014 Presidential elections. These elections, both the first and second rounds, were marred by allegations of large-scale fraud. Ultimately, the National Elections Commission declared Ashraf Ghani the winner. The runner-up, Dr. Abdullah Abdullah, refused to concede. While the UN conducted an extensive audit of the results, it took the personal intervention of U.S. Secretary of State John Kerry to broker a deal resulting in the creation of a Government of National Unity (GNU). Ashraf Ghani assumed the Presidency on September 29, 2014, but he and Abdullah agreed to a power-sharing arrangement giving Abdullah powers somewhat akin to a Prime Minister. While many in Afghanistan favored a winner-take-all result, and criticized the U.S. imposition of a governance formula, Secretary Kerry's role was decidedly a positive one. In a situation where the results of the elections were in dispute, creating a government led by the two leaders the Afghan people believed had garnered the most votes, gave these flawed elections some measure of legitimacy, and allowed for the peaceful transfer of power from President Karzai to his elected successors Clearly, in a nation facing an existential threat from a radical Taliban insurgency, an effort by the Afghan political class to attempt to work and govern together was the only path open to them. Creating an inclusive government with broad ethnic, tribal and gender representation was a key political element then, as it is now.

President Ghani and Chief Executive Abdullah were at the helm of the GNU on January 1, 2015 when the transition was formally completed and the Afghan military assumed principal responsibility for defending their country. As usual, the Afghans continued to bicker among themselves. Largely driven by patronage networks seeking to benefit from government service, the powerbrokers pushed Ghani and Abdullah into a state of constant intrigue, maneuver and conflict. Despite the myriad threats facing the country, the political class devoted more time to seeking their own personal advantage then to doing what was right for the country. The adverse situation was compounded by a significant weakening of the economy. The massive international military and development presence across Afghanistan during the Surge years had brought an extraordinary financial bonanza, not only in Kabul, but to virtually every part of the country. The COIN strategy in those years led to the doling out of massive sums of cash, and the implementation of billions of dollars of infrastructure projects of every kind. While much good was achieved the rush to disburse led to huge, unnecessary waste of resources, and limitless opportunity for a predatory political class to steal money. While in the heady years of the Surge the Karzai government managed an economy that was growing 10–12% per year, the withdrawal of foreign military and civilian personnel sent the economy into reverse gear. The misfortune for the GNU was they assumed office at the exact moment when the boom years had ended. Despite the adverse economic conditions, Ghani and Abdullah deserve immense credit during their period in office in lessening the proportion of the Afghan budget financed by international grants. The credible management of the nation's public finances led to a significant increase in public revenue to cover as much as a third of Afghanistan's current account budget. The result was international reserve levels were stable, as was the currency (the Afghani), and the nation stayed in the good graces of the IMF and the Multilateral Development Banks, including the World Bank and Asian Development Banks. So, while times were difficult, as they always are in Afghanistan, macro-economic stability was maintained and there was a path charted on the road to self-reliance. Most importantly, and despite the growing pains, the Afghan military, now significantly on its own, was performing reasonably well.

Following my tour in Afghanistan in 2013, I was assigned as the Principal Officer in Sydney, Australia. Sydney was a dream assignment for me and my wife (Lisett) and a wonderful "golden handshake" to end my 35-year career in the U.S. diplomatic service. In June 2016, I was coming to the end of my time in Australia, and had filed my retirement papers effective September 30, 2016. I received a message from Secretary Kerry informing me of his interest in having me return to Afghanistan and lead the U.S. Embassy in Kabul in the final months of the Obama Administration and the first year of the incoming Administration. By June of 2016, it was evident the Democratic nominee would be former Secretary of State Hillary Clinton and the GOP standard bearer, real estate magnate Donald Trump. Not being a political person, my one query to the Secretary was whether I was being asked to go to Afghanistan to pull-the-plug on our involvement in Afghanistan. I underscored despite the difficulties and complexities of our mission in Afghanistan, I remained convinced it was a worthwhile venture. I noted if the task was abandonment, I was the wrong person to send to Kabul. At the most senior levels of the U.S. State Department the view was if Clinton won the election, she would remain fully committed to Afghanistan. Likewise, while recognizing Trump's campaign to end "the endless wars," the high-level view was GOP Congressional leadership and party foreign policy establishment's staunch support for our presence in Afghanistan would influence Trump to be responsible. Ultimately, I accepted the job based on reasoned Administration assurances.

Several months later the State Department's "D Committee" approved my assignment to Afghanistan as a "Special Chargé d'Affaires," or Acting Ambassador. I arrived in Kabul in the first days of December 2016. I had a very smooth and productive overlap with Ambassador Mike McKinley, a superb professional and an old friend dating back to our Junior Officer days in La Paz, Bolivia in the mid-1980s. Mike was extraordinarily generous with his time and helpful getting me up to speed on the situation in the country. He conveyed his continual frustration with Ghani and Abdullah's failure to work in common purpose. In his final weeks he also dealt with the political tremors caused by Afghan Vice President Dostum's violent assault of an opponent. We both agreed the Dostum case provided an opportunity for the GNU to reaffirm its support for human rights, the rule of law, and to crackdown on the impunity culture of the warlord class, of which Dostum was a prominent member.

I assumed leadership of the U.S. Embassy in Kabul in the second half of December 2016. The Christmas holidays and the Obama-Trump transition provided an opportunity for me to focus in-house and get a good grip and feel for the core human resource, budget, security, and infrastructure issues. While we were no longer deployed in the field, my Embassy was still the largest in the world and only marginally smaller than at the height of the Surge. We still had a huge staff of 7200 U.S., Afghan, Third Country and contract staff representing nearly two dozen U.S. government agencies. I devoted most of the first couple of weeks meeting with our Embassy sections and agencies and made a point of holding the sessions in their own offices, rather than summoning them to the Ambassador's office. I also scheduled a series of town hall meetings with all our people. In all of these sessions, my purpose was to discuss my leadership style, convey what I expected from each and every member of our team, provide the policy context, shed some light on the Washington transition, and listen to employees' views and concerns. Again, we had an excellent professional team on the ground in Afghanistan committed to loyally serving our elected leaders irrespective of their party or ideological affiliation.

Another key objective in my first two months on the job was to close out the Afghanistan portfolio for President Obama and his Administration. President Obama had devoted the better part of the previous eight years on this vital national security issue. He knew the country, the issues and the players extremely well. I recall in the final weeks of the President Obama's tenure, we scheduled a lengthy farewell secure video conference between the President and National Security Advisor Susan Rice (from the White House Situation Room) and President Ghani, Chief Executive Abdullah and I in the Presidential Palace. I will not go into the details of the discussion, since it is not appropriate and much of what was discussed was classified. I can say it was a very substantive policy exchange with the President providing some final advice to the Afghan leaders and it was also a very human and emotional conversation attesting to the fact that statesmanship and diplomacy is above all a personal and at times intimate craft. I appreciated the President graciously encouraging Ghani and Abdullah to work closely with me and to seek my advice and close counsel in the weeks and months ahead.

Based on my Washington consultations prior to arriving in Kabul and subsequent contact with an extensive network of Washington insiders in government, the think tank community and the Hill, I learned President-elect Trump remained deeply skeptical about our involvement in Afghanistan. I used this fact as leverage with Ghani and Abdullah. In my initial round of individual and collective meetings in December–January, I conveyed to both of them that President-elect Trump continued to believe our adventure in Afghanistan had been disastrous for the United States and his gut-instinct was to end our involvement. I stressed the negative image created by the petty in-fighting within the GNU, the greed of the patronage networks, and excessive corruption could only strengthen President Trump's conviction on the importance of cutting our losses and leaving. I urged them to demonstrate to the incoming Administration that the GNU was functional and could engage the new Trump team in a productive relationship. Although unstated, I had the clear impression Trump's electoral triumph had surprised and rattled them to some degree. No doubt, Ghani and Abdullah had my attention and understood the need to do things better. The three of us agreed to have joint meetings every Wednesday afternoon to review political, military, economic and intelligence developments, work through differences and seek to find common cause. To their credit, and despite the continued differences driven by their own supporters, during the entire year (2016–17) I served in Kabul, we never failed to have our two-hour Wednesday session. Those sessions proved invaluable to clear the air, avoid the worse of the miscommunication, and mitigate the damage caused by the poisonous schemes and mischief of what I called, "the Palace eunuchs" of the Arg and Sepidar Palaces.

Another top priority for me was to quickly establish strong working ties with my military counterpart, USFOR-A and NATO Resolute Support Commander General John "Mick" Nicholson. For me, as U.S. Ambassador, there was no more consequential relationship than Nicholson's. Nicholson was a brilliant strategic thinker and military commander, and had done multiple tours in Afghanistan and so understood the nuance and complexities of the place as well as anyone. Fortunately, we were able to connect personally and professionally. In those transition months, the two of us devoted many hours in one-to-one conversation and working lunches to review our options. Nicholson and I shared very similar views on the importance of Afghanistan in the U.S. strategic chess board, but also understood the challenges and opportunities. We both agreed on the need to give our "field" views to the incoming President and his key National Security officials as soon as feasible. We concurred on the importance of offering a fresh approach on Afghanistan that was "sustainable" from a "blood and treasure" perspective. Nicholson and I brought our Embassy and USFOR-A Leadership teams together in a series strategy, policy and program sessions. Based on these intense discussions the Embassy drafted a Country Team Afghanistan strategy options paper. We shared our draft with General Nicholson and his command team and encourage their comments and clearance. While our document was more focused on the diplomatic, political, economic, development, intelligence and counterdrug issues, the military component very much reflected General Nicholson's perspective, which I fully concurred with. Likewise, Nicholson prepared a military strategy document that had our Embassy input. These strategy documents were drafted in the transition phase, and both were submitted to Washington within weeks of President Trump's inauguration. In my case, the "Country Team" document was sent electronically at a very high level of classification and given the most limited access to the Secretary of State and the National Security Advisor. Copies were given to an extremely small group of senior policymakers at State and the NSC. Nicholson's document was similarly classified and had strictly limited distribution to include the Secretary of Defense, the Chairman of the Joint Chiefs of Staff and if I remember correctly, possibly the CENTCOM Commander. I gave a copy of my paper to Nicholson to share with Secretary of Defense Mattis and Chairman Dunford, and they shared their paper with our people.

The main point here is we were able to influence the senior decision makers very early on. More importantly the leadership at State, Defense, the National Security Council and the intelligence community could see the diplomatic and military teams on the frontlines of the fight were fully joined-at-the-hip on what needed to be done as to strategy, policy and programs. Our approach from the field contained some basic points: First the importance of implementing our approach on Afghanistan in a broader South-Central Asia regional context, as well as bringing in the key stakeholders from outside the region (EU, UK, China, Russia, Saudi Arabia). Of our many recommendations front-and-center was taking a persistently tough and unambiguous line with Pakistan. On the military piece, we advocated giving the military commander on the ground the full authorities of using our small ground and air contingents as needed. The U.S./NATO Commander needed to have the tactical flexibility and freedom to fully utilize our forces in a very fluid battlefield. Unfortunately, for most of the Obama Administration micro-management, or the "ten-thousand-mile screwdriver," stifled initiative and at times paralyzed and demoralized our effort. We also advocated shifting to a more market-based economic policy and development approach. For too long, the main economic policy question for Afghanistan was how the state would manage and direct the huge flows of international donor assistance. The notion of having the right policy formulae to more efficiently allocate resources and help those Afghans who were actually creating wealth needed to be our priority. We also proposed a limited but more effective counterdrug approach focused on restricting the money flows to the Taliban generated from the opium/heroin trade.

Our field contributions provided our senior leadership in Washington with working documents that could help kickstart the discussion within the NSC on how to devised a new strategic approach on Afghanistan. Obviously, Secretary Tillerson, Secretary Mattis, Chairman Dunford and NSA McMaster were climbing a steep hill in attempting to convince a very skeptical President on the need to stay in engaged in Afghanistan. Ultimately, Nicholson and I were lucky three of the key players (Mattis, McMaster and Dunford) were all highly expert on Afghanistan. In the case of my boss, Secretary Tillerson, although he had little substantive knowledge of South Asia, he quickly grasped what was at stake and proved to be a constructive player in making the case to the President, along with the others, on the vital importance of staying on in Afghanistan, but doing it in an affordable and sustainable way.

Ultimately, thanks principally to the Mattis-Tillerson-McMaster triumvirate, a reluctant President Trump agreed to unveil the new South Asia Strategy in a speech on August 21, 2017. It was one of the President's best foreign policy addresses of his Administration. While conveying his frustration with America's longest war, he made the case for staying the course in Afghanistan. He advocated a 'conditions-based' approach bringing together the diplomatic-military-development and intelligence pieces. The President made clear the U.S. would no longer stay silent on Pakistan's complicity with the Taliban. As a corollary to the White House South Asia Strategy, NSA McMaster had advocated the adoption of the Afghanistan-U.S. Compact. President Ghani and I inaugurated the Compact in a meeting on August 23. The Executive Committee included CEO Abdullah, Afghan cabinet ministers, General Nicholson and senior members of our teams. The Compact provided tangible programmatic teeth to the President's South Asia Strategy. In the first Executive Committee session, the Afghan and U.S. working leads highlighted the completion of the Compact's four matrices focusing on Governance, Economics, Peace and Reconciliation, and Security. The Governance matrix had a main anti-corruption component. The economics piece aimed to shift the policy approach more in a market-based direction and focused on generating more private sector investment. The Peace and Reconciliation matrix underscored our success was ultimately premised on creating the conditions on the ground for the launch of Afghan-owned and Afghan-led negotiations ultimately leading to a political settlement and an end to the conflict. Finally, the security piece had many components, but included giving the commander more tactical control of our forces, and further strengthening the capability of the ANDSF.

In the wake of the President's launch of the South Asia Strategy, I was upbeat about our ability to turn the tables on the Taliban, and believed we were shaping the ground for ultimate success. Throughout 2017, we had received extremely productive and reassuring visits from Tillerson, Mattis, McMaster and Dunford. All four senior U.S. officials, as well as numerous Congressional Delegations, including a critical last visit by Senator John McCain in July 2017, sent a consistent message of support for the GNU, but one premised that there was no blank check and that we were looking for major improvements on governance and anti-corruption. On the anti-corruption side, we were successfully closing one of the longstanding corruption nodes by using digital technology to sharply reduce the problem of "ghost soldiers" and the theft of ANDSF fuel. On our side we were working to close a similar problem with the Afghan police forces, as well as working with SIGAR to radically reform and monitor the Afghan government's procurement and public contracting system. On Pakistan the intelligence suggested the Pakistani government and the military high command had been taken by surprise and unnerved by President Trump's unorthodox, but highly effective public exposure of Pakistan's damnable role in supporting the Taliban insurgency. For the first time in many years, the Pakistanis were on the defensive, and the pressure was only going to increase.

Above all, I was convinced we had created a sustainable model for the President consistent with the conditions-based nature of our strategy. While at the height of the Obama Surge the U.S. was spending as much as $140 billion per year in the combined military, diplomatic, development and intelligence components, under our approach in 2017, our total resource outlays were under $25 billion. To put this figure in perspective, total U.S. defense spending is $750 billion per year, so our aggregate military-diplomatic-development-intelligence investment was marginal, accounting for less than 3% of total U.S. military outlays. Yet, while our military resource contributions in Afghanistan had been massively reduced, we had learned our lessons and were achieving success on the battlefield. On the military side, our operating premise was while there was no clear-cut military victory to be had, at a minimal cost we could support our Afghan partners to keep terror groups like Al-Qaeda and ISIS-K in permanent check, while ensuring the Taliban could never achieve a military victory. First, the President's decision to restore Nicholson's authorities allowed him to skillfully use our residual ground and air forces in a manner that truly enabled our Afghan military allies who were doing 95% of the fighting and 99% of the dying. For example, our small Special Forces contingent of no more than 2500, achieved strategic success in making more effective their Afghan counterparts. These were particularly useful in several instances when the Taliban were seeking to isolate and potentially seize several provincial capitals. In every attempt the insurgents failed badly and took huge casualties. Likewise, Nicholson's decision to redeploy our 6000 advisors down to the battalion level made Afghan army units significantly more effective and ensured our military assistance was not siphoned off by corrupt officers. Equally, Nicholson was given the freedom to utilize our small air force in Afghanistan of no more than 120 combat aircraft to good effect. Our air contingent included two squadrons of F-16s, a couple of squadrons of armed Predator drones, Apache and Blackhawk armed helicopters and a few 'Puff the Magic Dragon AC-47' gunships. In 2017, this very modest air corps unleashed a huge volume of free and precision-guided ordinance on the enemy. The guiding military principle was simple: for the Taliban to achieve strategic success on the battlefield required them to mass their forces. Unfortunately for them, under the model we developed any attempt by the Taliban to maneuver in large formations to seize a strategic objective, such as a provincial capital or large military facility, would invariably result in the launch of a few aircraft from our string of U.S. and NATO bases who within a short span of time were dropping free and precision-guided munitions to devastating effect. During a field visit to Helmand, I saw for myself the destructive power of Predator gunships on a large Taliban concentration attempting to attack the outskirts of Lashkar Gah. There was not much left of them.

As my agreed one-year tour came to a close in early December 2017, I was extremely confident about our situation in Afghanistan. We had a South Asia Strategy approved by our President and we were implementing numerous program aspects through the Compact. As was the nature of Afghan politics, serious tensions continued between Ghani and Abdullah, mostly over patronage appointments. A good part of my job was to intermediate between the two, try to mitigate the differences, and keep everyone focused. On the positive side, Ghani and Abdullah essentially agreed on policy substance and both remained committed to a liberal-democratic system of government. For example, during my tenure, and at our insistence, Ghani and Abdullah had marginalized Vice President Dostum, a serial troublemaker with a criminal past. Ghani and Abdullah barred the notorious warlord from accessing the Vice President's office complex, and he spent most of his days in purgatory-exile in Turkey. Again, corruption continued to be a major problem, but some of the largest nodes of corruption were being reduced or closed. Meanwhile, the GNU had appointed reformists in key economic policy positions, and the World Bank, Asian Development Bank, USAID, the EU other donors were advocating significant policy reforms aimed at strengthening private sector development, including bolstering the export capacity of a number of core industries such as marble, fresh and dried fruits, nuts, saffron, carpets and much more.

Politically, a priority was assisting the Afghans prepare for more open and transparent Presidential and Congressional elections in 2019. On the military side, we had taken a battlefield situation of an eroding stalemate at the end of 2016, to one where the ANDSF was increasingly gaining capability and confidence. The Afghan Special Forces contingent was on track to be increased by a factor of three, the military's leadership and logistical capability was improving, and the Afghan air force, particularly the introduction of the Super Tucanos ground attack aircraft was making a tactical difference on the battlefield. Above all, with the Afghans in the lead, and the U.S. and NATO a completely residual force of no more than 15,000 U.S. (and an equivalent number of NATO), plus a small number of aircraft, we had inflicted huge losses on the Taliban enemy. In very late 2017, just prior to my departure, there were estimates of Taliban combat losses (killed, wounded, captured, deserted) totaling 17,000, and the intelligence indicated Taliban infantry morale was at a low ebb. In stark contrast, U.S. losses in 2017 were only 25. Yes, 25 too many, but still a fraction of annual U.S. Army training accidents. We had finally learned the lessons from our past mistakes and developed a sustainable model from a blood and treasure perspective. This approach would give us the time and space to create conditions with Pakistan and the broader South Asia region that could result in the start of a dialogue between the Afghan government and the Quetta Shura. The successful peace negotiations between the Kabul government and warlord Gulbuddin Hekmatyar and his

Hezb-i-Islami in April 2017 signaled the way the wind was blowing and offered a possible template for the Taliban and their patrons in Islamabad to consider.

Sadly, all of the gains achieved in 2017 were thrown to the winds. President Trump, a man operating on gut instinct, and woefully ignorant and unread as to the reality in Afghanistan, resisted his experts and scuttled his own policy. While there are numerous reasons for the departure of McMaster, Tillerson and Mattis, I have no doubt Afghanistan played some role. On innumerable occasions these three dedicated public servants pushed back on the President's views on Afghanistan, while facing sniping and smear campaigns by anti-establishment/isolationist elements at the heart of Trump's base.

Once these individuals were removed, President Trump assumed direct ownership of our policy in Afghanistan. The President's appointment of Ambassador Zalmay Khalilzad to lead the Administration's reconciliation efforts was a good example. His decision to direct Ambassador Khalilzad to negotiate directly with the Taliban was a significant mistake in my view, and a major departure from the three previous Administrations' position. By excluding our allies in Kabul, we were proving the Taliban's assertion as to the illegitimacy of the Kabul government. It justified the Taliban's long held view the only people they needed to talk to was the puppet master in Washington. What was worse, the Trump Administration showed little interest in supporting successful Presidential elections in September 2019. A positive elections result might still have led to the creation of another coalition government and given our friends in Kabul added strength through the power of the ballot box. Instead, the Trump Administration revealed its disdain for the Islamic Republic of Afghanistan's constitutional system by advocating (through Khalilzad) the creation of an interim government.

U.S. negotiators held a reported nine negotiating sessions leading to the signing of the infamous U.S.-Taliban Agreement of February 29, 2020. Again, in my nearly 40 years of work in the U.S. diplomatic service I cannot conceive of a worse diplomatic deal. In essence Khalilzad negotiated with Mullah Baradar America's capitulation to the Taliban. A quick read of the accord will tell you that there is one hard and fast condition in the February 29, 2020 agreement, the complete and total withdrawal by May 1, 2021 of all U.S. forces from Afghanistan. The other elements of the accord, including the Taliban's ties to Al-Qaeda, the issue of mitigating violence, and the Taliban's agreement to participate in talks with political elements in Kabul, were all nebulously written and amounted to nothing. The obvious result is the Taliban continued to maintain its close ties to Al-Qaeda, was emboldened and increased the pace of its military operations, and at no time engaged the Kabul government and other political representatives in good faith political negotiations. Adding insult to injury was the U.S. insistence the Kabul government release 5000 Taliban prisoners. The U.S.-Taliban Agreement is the 'Original Sin' of U.S. failure. At the time, in my contacts with Afghan friends and colleagues in Kabul, civilian and military, it was obvious to all the U.S. had moved from the "South Asia Strategy" phase (2017–18) and entered the 'Abandonment or Betrayal' phase (2018–21) of our experience in Afghanistan. The psychological blow on the government and political class in Kabul, the business community and the security forces were devastating.

If it can be imagined, President Trump actually contributed further to destabilizing the U.S. situation in Afghanistan by seeking to exploit such a sensitive and complex issue for his own political gain. Throughout the 2020 elections campaign, Trump repeatedly made clear his intention (through tweets and statements) to completely withdraw U.S. forces irrespective of what the Taliban or anyone else did. The President's call for bringing all U.S. troops home by Christmas 2020 was a prime example, and led to the resignation of Secretary of Defense Esper. Ultimately, the President's military advisors convinced him U.S. military lives would be endangered by such a rapid rush for the exits. In the end as the Trump Administration came to an end, the U.S. force component was only 2500. In 2016–17, we had done copious number-crunching on the minimal number of troops we could operate in Afghanistan and carry out basic security, military advisory work and maintain a minimal level of combat (Special Forces) capability. The bare minimum number of U.S. forces was 8500. Clearly, 2500 placed the U.S. in a precarious position in Afghanistan.

President Biden's inauguration on January 20, 2021 gave some people hope the situation in Afghanistan could be salvaged. I was not one of them. I recalled how during the Obama Administration then Vice President Biden had had been the most unabashed advocate for a full and unconditional withdrawal from Afghanistan. Once he assumed office, President Biden went through the motions by permitting a full inter-agency policy review, but in retrospect it is clear the President's mind had long been made up. I can say from having spoken to many in the policy community the advice given to the President by his professional military, diplomatic and intelligence advisors was to pause and avoid a precipitous decision to leave Afghanistan. The view was that even if the preference was to end our presence, we should wait at least until the end of the year (and fighting season) to determine if the Taliban was serious about negotiating with Kabul. While our NATO/coalition partners were not consulted on whether to stay or leave, in discussions with senior Biden Administration officials had with our closest allies in London, Paris, Berlin and Rome, the unanimous message was to stay the course in Afghanistan.

So, rejecting the advice from his own professional staff and our allies, on April 14, 2021 President Biden formally announced his fateful decision to totally withdraw all U.S. forces. He agreed to postpone the departure by a few months beyond the May 1 date stipulated in the U.S.-Taliban Agreement. Upon listening to the President's decision, it was obvious to me the Taliban were poised for victory and were going to plant their flag once again in Kabul and Kandahar. Like most Afghan policy hands, I never expected the collapse would happen so soon. My own judgment was by the end of the fighting season in late 2021, the Taliban would be poised or seized key provincial capitals and begin to envelope and choke Kabul. My own view was the collapse would likely happen by the spring–summer of 2022. Clearly, the morale effect of our own capitulation to the Taliban and betrayal of our allies was far more significant than many of us could appreciate from afar. In the end, though, while President Trump contributed immensely to the debacle, President Biden has to assume the greatest responsibility for our defeat in Afghanistan. I have no doubt President Biden's legacy will be tarnished by his strategic blunder in Afghanistan. As we began this chapter, I am also convinced historians will place the humanitarian, security and geopolitical consequences of the calamity very much on Biden's shoulders. And speaking of historians, I am reminded of a conversation I had with my good friend General McMaster in Kabul in April 2020. He pointed out, and I agreed, if we ever got everything wrong in Afghanistan and the situation collapsed, it would be completely unnecessary and a self-defeat. A most comparable example, he asserted, was contained in historian Marc Bloch's history of France's dramatic collapse in 1940, Strange Defeat. I could not agree more!

Ambassador Hugo Llorens is a leader, executive and advisor with vast international experience. He is a two-time Ambassador (Afghanistan and Honduras), Deputy Ambassador (Argentina and Spain), and Consul General (Vancouver and Sydney). Ambassador Llorens has 36 years of diplomatic experience serving in the United States, Latin America, Canada, Europe, Asia/Pacific, Europe and Central Asia. He has earned numerous awards for superior performance and achievement from six Presidents, including leading the largest Embassy in the world in Kabul with a staff of 7200 U.S., Afghan and Third Country National employees representing 22 U.S. government agencies. His skills include directing/leading inter-agency policy teams, overseas and in Washington, strategic planning and communications, security and crisis management, conflict resolution, and negotiations. Ambassador Llorens has decades of public speaking and media experience on a diverse range of national security, foreign policy and trade, investment, energy and financial issues. He is an author of numerous articles and think pieces covering foreign policy and grand strategy. He is married to his wife of 37 years, Lisett Aparicio Llorens. They have two sons, Andrew and Dirk, and a grandson, Cody.

A Comparative Study on Iraq and Afghanistan: Two Nations, Similar Journeys, Different Destinations

Arman Mahmoudian

Abstract For the last twenty years, US and its allies have struggled with their mega-project of nation-building in Afghanistan and Iraq. Both countries, with the support of the US entered to a new era that neither Afghani nor Iraqi jihadi revisionists could tolerate. The resistance of extremist elements escalated when they realized that the nation-building process involves all actors, including those ethnic-religious groups who were considered "others" by Saddam and the Taliban's regimes. Thus, as political inclusion increased, the attacks against the ex-others grew, especially against Shia communities in both nations. Although, when it came to ethnolinguistic minorities, Afghanistan and Iraq experienced different scenarios. In Iraq, non-Arab speaking Kurds maintained a relatively safe environment and successfully defended themselves from ISIS aggression, and later turned to be a reliable ally for US and international actors in the fight against ISIS, but in Afghanistan, non-Pashtun speaking minorities fell to the hand of Taliban to suffer from its tyranny, once again. This outcome is not the only difference between Afghanistan and Iraq; the other difference is the prelude. In Iraq, ethnic-minority Kurds have been given the right of autonomy, and hence, for more than a decade, they had the privilege of building up a local system for themselves. Thus, when ISIS raided Iraqi Kurdistan, the Kurdish Regional Government (KRG) had a pre-established system and infrastructure to counter ISIS. However, the Afghani nan-Pashtun minorities were denied such a privilege; thus, in the absence of foreign power and central government, they failed to defend themselves. As Iraqi Kurds had no intention of leaving their destiny in the hands of Saddam's regime and chose to fight, today's Afghani non-Pashtuns are slowly but gradually choosing the same path, fighting for the right of "self-determination," which warns us that in the near future, Afghanistan might experience what Iraq went through in the mid-1990s, the Iraqi-Kurdish Civil War. Eventually, decades of bloodshed between the Iraqi central government and Kurds declined in 2005, when Baghdad recognized the autonomy of KRG. The historical experiences across the world show us that Afghanistan has two choices either choosing the "Iraqi remedy" and implementing a federal system in which "All are Afghans but separate,"

A. Mahmoudian (✉)
University of South Florida, Tampa, FL 33620, USA
e-mail: Armanm@usf.edu

A. Farhadi and A. Masys (eds.), *The Great Power Competition Volume 4*,
https://doi.org/10.1007/978-3-031-22934-3_15

or reliving the "Yugoslavian experience" and witnessing the birth of a new "-Stans" from falling Afghanistan.

Keywords Afghanistan · Iraq · Kurds · Non-Pashtun · Arabs · Farsi-speakers · Taliban · Islamic State of Iraq and Syria (ISIS) · Al-Qaeda · Insurgency · State-system · Identity · Constitutional law

1 Introduction

One might ask why there should be a comparative study between two nations that are 1417 miles (2221 km) apart. Well, Afghanistan and Iraq are indeed not neighbors of one another, but their lack of geographical proximity should not overshadow their excessive similarities. For instance, both countries have mostly Muslim populations; 99.7% of Afghanistan's population[1]; and 97% of Iraq's[2] are Muslims.

Also, both Iraq and Afghanistan are sharing a feature that has been impacting their destiny from the very beginning of their time, "Plural Society." Both Iraq and Afghanistan, in terms of linguistic, religious, and ethnic identity, are strongly diverse. In terms of ethnolinguistic demography, Iraq's population is mostly divided between major ethnic groups, Arab-speaking Arabs, who constitute about 75% of the population, and Sorani-Speaking Kurds, who make up between 15 and 20% of the country's population.[3] Also, the country is home to numerous small ethnic minorities, including Turkman, Assyrians, Armenians, Roma, Black Iraqis, Yezidis, Shabak, and Chaldeans.[4]

In addition, religious-wise, the Iraqi community is heavily polarized between Shias, who make up 55–60% of the population, and Sunnis, who represent 35–40% of the country.[5]

In this context, Afghanistan's community is even more diverse. In terms of ethnic demography, Afghanistan is populated by five major ethnic groups: (1) Pashtuns (42% of the population), (2) Tajiks (27% of the population), (3) Hazara (9% of the population), (4) Uzbek and Turkmen (6%[6] of the population), and Aimaqs (3% of the population).[7]

Linguistic-wise, Afghan people are divided into three major groups: (1) ~ 49% of the population are Farsi/Dari-speaking Tajik, Aimaqs, and Hazaras, (2) ~ 11%

[1] The Cultural Atlas team [1].

[2] U.S. Department of State [2].

[3] Minority Rights [3].

[4] Ibid.

[5] Religious Politics in Iraq [4].

[6] Some sources claim Uzbeks constitute 9% and Turkmen represents 3% of population which is equivalent to total of 12% of population. See below: Country Reports [5].

[7] World Population Review [6].

of the population are Turkic-speaking Uzbeks and Turkmen, and (3) ~ 40% of the population are Pashto-speaking Pashtuns.[8]

Despite Linguistic and ethnic diversity, Afghanistan is also polarized when it comes to the religious fate of the country; about 84–89% of the country are Sunnis, and 10–15% are Shias,[9] about 90% of them are Farsi/Dari-speaking ethnic Hazaras.[10]

So, it is fair to argue that the absence of a homogeneous community is another similarity between Iraq and Afghanistan. However, having a heterogenous society by nature is not less valuable than a homogenous one. In fact, a plural society can have many advantages, but in order to have a prosperous and progressive diverse community, a nation needs to have to experience various stages of cultural and social development, known as the "nation-building process."

2 The Big Struggle: Nation-Building

Nation-building is a process in which independent societies endeavor to construct an identity (national identity) that would unify the inhabitants of a specific territory under the rule of their own government. In order to process successful nation-building processes, nations need to establish a functional national institution and also create national symbols/paraphernalia. The functional institution will help the country to maintain necessary administration duties, such as security, judiciary, and other bureaucratic responsibilities, which all together sustain the survival of the country. In addition to all mentioned mundane activities, a functional institution will develop the most important factor in the nation-building process, the "education system.[11]"

The education system is the major actor in providing the community with skills and knowledge. It is via the schools that members of the community will learn about their history, the achievement of their ancestors, and the roots of their culture. All of these together build what is known as the "public mindset,[12]" or people's understanding of "themselves" and "others," which forms the bases of their identity. Education systems, by expanding the mindset of "us versus others," creates a sense of nostalgia among the member of the society toward their place of habitant (county), which will inspire people's loyalty toward their territory/country.[13]

Also, education provides the population with a coherent understanding of their language and its literature. It is needless to say that most of the population will learn at least two out of four basic language skills at school, writing and reading. Playing a major role in the advancement of the public's language skills makes education

[8] World Atlas [7].

[9] United States Commission on International Religious Freedom [8].

[10] U.S. Department of State [9].

[11] Idris et al. [10].

[12] Ibid.

[13] Paglayan [11].

a vital actor in the nation-building process. Simply put, language is the primary means of communication, and it is via the divulgence that inhabitants of a territory can advance a sense of association which is one of the cornerstones of the nation-building process.[14] In light of the importance of language education, it is necessary for nations to have a functional national institution that can advance public education in which values of all groups of citizens would be represented, and all members of the community would receive an equal education.

Along with education, national institutions play another significant role in nation-building, and that is promoting "national symbols" such as the national flag, coat of arms, national mottos, national anthems, national colors, seal/stamps, and national emblem. However, one might ask what the significance of these symbols is; well, it is true that at first glance, national symbols might not seem crucial, but in fact, they have an essential place in nation-building since these visual aids are representing the "idea of the home/homeland." Basically, national icons remind people of their "motherland," its "values," and its "history". It is in light of such an important function that national symbols have been used to inspire the sense of patriotism among citizens, and their "pride for homeland" (national pride). Therefore, we can conclude that national paraphernalia is one of the factors that give spiritual meaning to the society's territory and turns "a land" into "a motherland"; as Benedict Anderson, in his *Imagined Communities*, argued; "nations were built around the common imagination of citizens based on symbols and practices like maps, media and the census.[15]"

Both national education and national symbolism can only arise through a functional and stable institution. In this context, we can safely argue that there is a quid pro quo relationship between successful nation-building and the existence of a stable and functional state. In other words, without an efficient and steady national institution, neither nation-building nor the national identity would be constructed.

In this vein, the state should be functional enough to maintain a solid and progressive education system and run its propaganda machine to create a universal national symbolism. Additionally, the national institution should have a "consistent existence" to be able to establish a coherent environment for the nation-building process. Simply put, creating national icons and developing a national education would take generations; ergo, nations, at least for a considerable period of their history, should have a national institution that had been free from foreign occupation and domestic chaos to naturally proceed the nation-building process.

This precondition for a successful nation-building process brings us to the historical obstacles of Afghanistan and Iraq. Neither of these countries had the chance to experience a steady and constructive nation-building process. In fact, neither Afghanistan nor Iraq had the time to undergo the nation-building process. Iraq gained its independence in 1932, and until then, the country was under the rule of a series of foreign powers, starting from the Persian domination in 539 BC and ending with the fall of the British Mandate in 1932. Comparably, Afghanistan, except for a brief period in the eighteenth and nineteenth centuries (1706–1879), during which the

---

[14] Karna [12].

[15] Anderson [13].

territory of today Afghanistan was under the control of rival local tribal actors, for most of its history was either under foreign rule or was a protectorate of a foreign power. Eventually, Afghanistan gained its independence in 1919. Ergo, neither of these two countries ever found the chance to practice stable nation-building.

Foreign interventions and domestic rivalry not only deprived Afghanistan and Iraq of an unwavering nation-building process but also disturbed the "natural" path of creating national boundaries and identity. Simply put, what we today know as the territory of Iraq and Afghanistan is the outcome of the grand designs of foreign colonist powers. Iraq's today border was drawn by the Sykes-Picot agreement between France and the United Kingdom in 1916.[16] An agreement that basically put different ethnic and religious groups together to form a newborn nation while they had never experienced the idea of ruling a land together and only had the history of coexisting with one another under foreign dominations (British, Turks, etc.)

Similarly, Afghanistan's current border owns its existence to more than half a century of foreign planning. A long-lasting endeavor which began in 1830 with the political and diplomatic competition between Russia and Britain. The Russo-Anglo rivalry, also known as the Great Game, was fueled by Britain's fear of the Russians' expansion in India via today Afghanistan, and Russia's fear of a British invasion of Central Asia from Afghanistan.[17] Ergo, both great powers launched campaigns to turn Afghanistan into a shield/buffer zone between them and the other party.

In light of this great power competition, Great Britain forced the Persian Qajar dynasty to withdraw from its claim on Herat and facilitated the formation of the "Emirate of Herat," Britain's new buffer zone to protect India from Russian ambition.[18] The Emirate of Herat, later in 1863, was conquered by Dust Muhammad Khan and turned into a new piece of today Afghanistan. Evidently, the British design was completed in 1893 by the implementation of Durand Line,[19] which settled the southern boundaries of Afghanistan and completed the territorial development of Afghanistan. The Durand line divided the Pashtun and Baloch between Southern Afghanistan and Northern Pakistan. This division escalated the unbalanced ethnic demography in Afghanistan and tied up the political destiny of Afghanistan and Pakistan together, which up to this day, continues to have an impact on Afghanistan.

Now, one might ask why "foreign interruption" in the formation of borders in Iraq and Afghanistan matters. To answer this question, we must keep in mind that plural societies, ethnically diverse nations, can only thrive with the help of ethnoreligious harmony,[20] which is also known as "national coexistence."

Tolerance is a skill that only can be mastered by exercise and patients, and only can be achieved over a long-time period of practice. Nations are not exception from this rule either, and its members/citizens require time to acknowledge their differences, respect their differences, and learn how to resolve their disputes peacefully and

[16] Dodge [14].

[17] Rezun [15].

[18] Ingram [16].

[19] Lambah [17].

[20] Sidanius [18].

violence-free,[21] and it only can be achieved after centuries of experience. In this context, heterogeneous (plural) nations will need more time than homogenous nations since in heterogeneous countries, there are more actors from different identities and backgrounds that need to be accustomed to the idea of cultural coexistence. Ergo, heterogeneous societies require a long period of practice to establish a solid national coexistence among all groups of the community, but if this process is interrupted by destabilizing forces which could be foreign invasion or domestic chaos, the national coexistence will not take place, and as a result, different groups of the country instead of endeavoring to "rule the country with each other" and "for each other," will race against each other to "rule the country over each other."

Unfortunately, both modern Iraq and modern Afghanistan did not find a solid chance to experience this process since both have a relatively short history of independence, and even that short period has been interrupted by waves of domestic chaos. For instance, Afghanistan's nation-building process was interrupted by domestic conflicts such as the 1928 Civil War between the Pashtun Shinwaris and Tajik Saqqawisit, the Collapse of the Tajik Saqqawisit establishment in 1929, the Kuhistan rebellion (1930), the 1973 Afghan Coup, the Saur coup (1978) and three other generations of Afghan civil wars (1989, 1992, 1996). Similarly, Iraq has been dealing with various domestic bloodsheds, including the series of Shia uprisings (1935, 1991, 1999), six generations of Coups (1936, 1941, 1958, 1963, 1963, 1968), and two generations of Iraqi-Kurdish War (1961–1970, 1947–1975) and Kurdish uprising of 1991 (Table 1).

What makes Afghanistan and Iraq more similar to one another is the fact that if ethnicity and religion were not the major players in most of their internal conflicts, they were among the involved factors. Simply put, these conflicts either were caused by ethnic rivalries like the Kuhistan rebellion in Afghanistan and the Kurdish uprising in Iraq or led to division of the different ethnic groups into various blocs against one another, such as Afghanistan's third civil war (1992) or Iraq's 1991 uprising. The involvement of ethnicity and religion in these clashes excessively undermined Iraqi and Afghani attempts to achieve national coexistence; simply put, the units of their societies, instead of experiencing national harmony, built up long memories of hostility and resentment toward one another.

However, internal pressures were not the only factor that undermined the nation-building in Afghanistan and Iraq; external stresses from foreign powers actively undermined both nations' chances of successful nation-building. For example, Iraq, only a few years after its independence, in 1941 was invaded by Great Britain, leading to the occupation of the country, removal of Rashid Ali al-Gaylani from power, and reinstalment of the pro-British Hashemite dynasty. Also, the Iranian 1979 Revolution strongly affected the Shia-Sunni relations in Iraq and led to the second Shia uprising in 1979 (-1980) against Sunni Saddam Hussein of Iraq. Later the war with Iran (1980–1988) escalated the Shia-Sunni gap in Iraq since many Iraqi Shias actively or covertly supported Iran in the fight against Iraq. Also, the war deepened the tension between Baghdad and Iraqi Kurdistan since major Kurdish parties sided with Iran.

[21] Kriesberg [19].

Table 1 Timeline of major domestic crises in Iraq and Afghanistan: from their independence to the US invasion

Afghanistan's domestic crisis: from its independence to US invasion of Afghanistan	Iraq's domestic crisis: from its independence to US invasion of Iraq
The Khost Rebellion (1924–1925)	The First Shia Uprising (1935–1936)
The First Afghan Civil War (1928)	The Yazidi Revolt (1935)
The Collapse of Saqqawisit Establishment (1929)	The Bakr Sidqi Coup (1936)
The Kuhistan Rebellion (1930)	The Golden Square Coup (1941)
The Shinwari Revolt (1930)	The Kurdish Barzani Revolt (1943)
The Pashtun Ghilzai Rebellion (1938)	The 14 July Coup (1958)
The Afghan Tribal Revolts (1944–1947)	The Mosul Uprising (1959)
The First Hazara Revolt (1945–1946)	The First Iraqi-Kurdish War (1961–1970)
The Second Hazara Revolt (1949)	The Ramadan/February Coup (1963)
The Afghanistan Coup (1973)	The November Coup (1963)
The Saur Coup (1978)	The 17 July Coup (1968)
The Herat Uprising (1979)	The Second Iraqi-Kurdish Civil War (1974–5)
The Second Afghan Civil War (1989–1992)	The Second Shia Uprising (1979–1980)
The Third Afghan Civil War (1992–1996)	The Kurdish Rebellion (1980–1988)
The Taliban Uprising (1994)	The Third Shia Uprising (1991)
The Fourth Afghan Civil War (1996–2001)	The Fourth Shia Uprising (1993)

Additionally, Iraq's invasion of Kuwait led to the Persian Gulf War (1990–1991), which created chaos in the country and resulted in the 1991 uprising. Also, the US active supports of Kurds in the Iraqi-Kurdish Civil War (1995) led to the collapse of Baghdad's control over Iraqi Kurdistan. At last, the bombing of Iraq in 1998 by US-UK joint forces dismantled most of Saddam's offensive military capability and encouraged the 1999 uprising (Table 2).

Similarly, Afghanistan has suffered from destructive foreign impacts, especially from its northern neighbor, Russia. The Communist Revolution of 1917 changed the destiny of Afghanistan, since slowly but surely, the Soviet Union turned into a major actor in Afghanistan's political spectrum. Initially, Afghanistan and the Soviet Union were on good terms, and actually, their growing relations led to the spread of the anti-Communist revolt of Basmachi into Afghanistan, which led to the Red Army's intervention in Afghanistan in 1929. However, after WW II and the emergence of the Cold War between the US and Soviet Union, Russia changed its view on Afghanistan from a friendly neighbor to a potential sphere of influence. In this context, Moscow found Afghanistan as a gateway to the Indian subcontinent.[22] Therefore, Kremlin launched its propaganda campaign to export Communism to Afghanistan, which evidently led to the 1978 Coup and the takeover of the People's Democratic Party

[22] Hauner [20].

Table 2 Timeline of Iraq's major international crises: from the independence to the US invasion

External pressures on Iraq	Their Impact on Iraq's domestic stability
The Anglo-Iraqi War (1941)	The Barzani Revolt (1943)
The Islamic Revolution of Iran (1979)	The Second Shia Uprising (1979–1980)
The Iran-Iraq War (1980–1988)	The Kurdish Rebellion (1980–1988)
The First Persian Gulf War (1991)	The Shia-Kurdish Uprising (1991)
US Support of Kurds in Iraqi-Kurdish Civil War (1995)	Withdrawal of Baghdad from Iraq Kurdistan (1995)
The US-UK Joint Bombing of Iraq (1998)	The 1999 Uprising

of Afghanistan (PDPA), a pro-USSR Communist Party. In light of Moscow's guidance, the PDPA Implemented series of forceful socialist reforms which faced backlash from the public and led to generations of revolts, resulting Soviet Union of Afghanistan in 1979. The Russian military intervention to support PDPA lasted for ten years and completely disrupted Afghanistan's nation-state platform. To resist the Soviet's invasion, all ethnic groups of Afghanistan started to arm themselves to force Russians to withdraw from Afghanistan. The resistance forces succeeded in their fight, and eventually, in 1989, the Soviet Union withdrew its troops from Afghanistan, and hence the chaos began. In the absence of a mutual enemy, the Soviet Union, Afghans faced the inevitable question of who was entitled to run the country. Unfortunately, since previous experience has left an immortal wound on national trust and coexistence, different ethnicities who already were armed started to form an alliance against other groups, resulting in three generations of civil wars in Afghanistan (Table 3).

Ergo, we can argue that neither Iraq nor Afghanistan had peace of mind from either harmony among domestic factions or powerful foreign actors. Therefore, they did not find the chance to develop the idea of national harmony, which played a great role in the construction of today's ethnoreligious conflicts in both countries. Conceivably,

Table 3 Timeline of Afghanistan's major international crises: from the independence to the US invasion

External pressures on Afghanistan	Their impact on Afghanistan's stability
The Russian Revolution (1917)	The Basmachi Revolt (1929)
Cold War (1947–1991)	Afghan Communists Takeover
Soviet Invasion of Afghanistan	The Mujahideen Insurrection (1980s)
Soviet Withdrawal from Afghanistan	Three Generation of Civil Wars (1989–2001)

once Americans set their foot on Afghanistan and Iraq's soil, they started the long-lasting Nation-building project in Iraq and Afghanistan,[23] which today's events in both countries prove that the project did not turn out as well as it was expected.

Speaking of Americans' journey in Iraq and Afghanistan brings us to the last similarity between Iraq and Afghanistan which is the fact that both Iraq and Afghanistan, for about two decades, were at the top of global news. On 11 September 2001, the terrorist attacks mapped Afghanistan, al-Qaeda's host country, at the center of international attention. Two years later, on 19 March 2003, Saddam's weapons of mass destruction and its support of terrorism brought Iraq once again to the top of the breaking news. Both countries were invaded by the United States, and immediately after the invasion, both turned into bloody battlefields between US and extremist jihadi groups.

In Afghanistan, the US had to encounter the Taliban and its allies' insurgency, and in Iraq had to face waves of attacks from terrorist cells of Ba'ath loyalists, Naqshbandi Army, al-Qaeda, the Islamic State of Iraq and Syria (ISIS), and their allies. The forces of insurgency might have different names, still, they all had the same agenda, challenging the change and, more importantly, resorting to the previous status quo, in which the Sunni-Pashtuns of Taliban had the upper hand in Afghanistan, and the Arab-Sunnises of the Ba'ath regime were the ruler of Iraq. Hence, waves of attacks on communities and minorities who used to be recognized as second citizens began. In Afghanistan, the Taliban forces launched various attacks against Farsi-speaking groups, and in Iraq, the Sunni Jihadi targeted Shia communities wherever they could. In this context, attacks against these ethnic groups grew as ISIS advanced its territorial campaign in Iraq in 2014, and the Taliban began its offensive campaign in 2015.

3 Two Nation; Various Forces of Insurgency but Similar Modus Operandi

The Taliban militants, after losing the full-scale war to the US collation, retreated to their "safe zones" in Pakistan to regroup and mobilize the remaining of their troops.[24] Hidden in their safe zone, Taliban slowly but gradually started to establish training camps across the Afghanistan-Pakistan border and develop multiple routes to bring supplies and jihadists from Pakistan.[25] Thus, after building their Jihad setups, the Taliban started their insurgency by launching terrorist attacks such as suicide attacks, rocket attacks, and guerilla deployments. These attacks, specifically, targeted NGOs, UN personnel, Afghani civilians, military personnel, and coalition forces, which all were icons of the "new and changed Afghanistan."

---

23 Dobbins et al. [21].

24 Baldaufl and Tohid [22].

25 Tohid [23].

Similarly, in Iraq, right after the fall of Saddam, the Ba'ath regime's loyalists, Fedayeen Saddam and Naqshbandi Army, and all other militias, including the Supreme Command of Jihad and Liberation (SCJL), who rather the previous status-quo in Iraq than the new one. Ergo, misoneism's forces started hiding in their sanctuaries across Iraq's borders with Syria.[26] Similar to the Taliban, the radical Iraqi militants, especially al-Qaeda, began to smuggle weapons and guerilla fighters to Iraq from the neighboring country, Syria, which, according to US officials, was the safe haven of al-Qaeda.[27] In Iraq, the insurgency forces also targeted US forces, Iraqi government personnel, and ethnoreligious groups, mainly Shia communities who, to Jihadists, were considered as others. The terrorists in Iraq used the same modus operandi that the Taliban used in Afghanistan; suicide bombing, rocket attacks, guerilla attacks, roadside and car bombing.[28]

In both Afghanistan and Iraq, neither US Coalition nor the terrorists could defeat the other party, however the US-backed central-governments maintained control over most of their territory; but in the meantime, the Jihadi forces managed to survive and stay active. Although in 2011, when the official withdrawal of US forces from Iraq and Afghanistan began, the destabilizing attacks started to grow. Thus, as of 2011, Iraqi and Afghani people have witnessed waves of mass terrorist assaults designed to challenge the status quo. In this context, similarly, both Taliban and ISIS raided their target country from unstable neighboring countries, Pakistan and Syria.

Evidently, in 2014, as the gradual withdrawal of US troops from Afghanistan and Iraq grew, both Taliban and ISIS increased their attacks which led to the ISIS occupation of Mosul and the Taliban's significant gains after the Helmand Offensive.

4 Different Predators, Different Targets, but Similar Target Selection

ISIS and the Taliban primarily targeted areas with similar demography to their bases. In this context, ISIS captured the city of Mosul, the second-largest city of Iraq with about 1.4 million residents, most of whom were "Arab-Sunnis," and the Taliban chose Kunduz and Helmand provinces which both were mostly populated by Pashtuns.[29]

To understand ISIS and Taliban's logic in identifying targets, we should be mindful of similar characterizations between these targets:

1. *Geographical Proximity*
   Helmand and Kunduz were close to the mountainous Waziristan and desert terrain of Baluchistan in Pakistan, which were the Taliban's traditional place of refuge.[30]

---

[26] Howard [24].

[27] U.S. House of Representative, Committee on Homeland Security [25].

[28] Anti-Defamation League [26].

[29] Helmand Provincial Overview [27].

[30] Dorronsoro [28].

Similarly, Iraq's Mosul was neighboring Syria's Deir-al-Zour Province, which at the time was ISIS's major stronghold. Thus, from their point of view, attacking these regions was convenient and safe; since it was easy and quick to mobilize and march their forces toward these targets, and also since these areas were close to their stronghold, in case of defeat, they could easily retread to their safe places.

2. *Strong Familiarity with the Regions*
   Abu Ibrahim al-Qurashi, the former caliph of ISIS and ISIS's second-in-command at the time who led the ISIS raid on Mosul, was born and grew up in the city of "Tel Afar," which was only 39 miles far from Mosul.[31] Thus, it is safe to argue that while he was commanding the invasion of Mosul, he had the 'home-field" advantage. In a similar fashion, the Taliban's 2015 offensive was led by Ibrahim Sadr, the military chief of the Taliban at the time, who was born and grew up in the south of Helmand province.[32] Therefore, one can argue that warlords' personal familiarity played a great role in their decisions, or at least gave them the confidence to take such a risk because they were confident that they knew all ways of "getting away" from the consequence.
3. *Counting on Local Sympathy*
   We should keep in mind that based on the Sharia law, jihad against infidel tyrants is mandatory, ergo, jihad against Americans whose by the forces of the insurgency were seen as "infidels" (Kafir), and US-backed governments which radical militants consider them as "tyrants" (Zaalim), were matter of religious "mandate" (Wajeb Sahrie).

   Simply, Iraq was led by a Shia government which ISIS saw as infidels,[33] and comparably in the Taliban's view, Afghanistan was under "military occupation," which is tyranny against Islamic community (Ummah) and must be defeated.[34] In this context, the Taliban and ISIS targeted regions where they were hoping their locals would share their point of view and support their Jihad, ergo, Taliban targeted Kunduz and Helmand with Pashtun Sunni population, and ISIS raided the Arab Sunni populated Nineveh province.
4. *Potentials for Recruitments*
   In addition to counting on the partial local support, it sounds reasonable to argue that the forces of insurgency might have hoped that locals in these regions would join their offensive campaign to advance the Jihad. However, up to this moment, we did not see any valid evidence suggesting that locals helped radical militants' operations. Although, what we know is that both Taliban and ISIS strongly attempted to recruit locals, and in this vein, ISIS established various centers all across Mosul to recruit youths, and a number of locals, either out of fear or because of brainwashing, have joined ISIS.[35] Similarly, the Taliban has been

---

[31] Amir and Sa'id Abdal-Rahman al-Mawla [29].

[32] Muzhdah [30].

[33] Kadivar [31].

[34] Semple [32].

[35] O'Driscoll [33].

actively recruiting fighters from Helmand with the help of its active radical Islamic schools, semi-Madrasas.[36]

5 Two Attempts to Return, One Failed, the Other Succeeded, but Why?

Despite all similarities between ISIS (2014) and the Taliban's offensive campaigns (2015), they met different endings. From the Northern Iraq Offensive that ISIS eight years ago launched, almost nothing has left except a few minor terrorist cells hidden in the deserts. However, in contrast to the ISIS campaign, the Taliban's insurgency, after two decades of perusing violence, succeeded in its mission. Now, the question is why ISIS, which at its peak had more than 40% of Iraqi territories under its occupation, failed to establish the Islamic Caliphate,[37] but the Taliban, after years of struggle, succeeded in re-setting up the Islamic Emirate of Afghanistan once again.

To answer this question, we should start looking in a new direction, the differences between Afghanistan and Iraq in their struggles with radical insurgency, which can be categorized into two major groups.

5.1 Foreign Factors

Once ISIS raided Iraq, the general understanding among international actors was that ISIS must be stopped. In this vein, the United States launched "Operation Inherent Resolve" (OIR) to defeat the ISIS Northern Iraq offense.

US military action against the ISIS began on August 8, 2014, when President Barack Obama authorized strikes against the Islamists. This decision was dictated by the need to protect American personnel in the city of Erbil (Iraqi Kurdistan) and Yezidi Kurds who fled from the Islamists in the Sinjar mountains. The overall direction of the operation of the US forces is carried out by the US Central Committee of the Armed Forces, whose area of operational responsibility includes the Middle East. The command headquarters is located at MacDill Air Force Base (Florida, USA) under the commander of General L. Austin. According to official statements by the White House, the goals of the operation were to undermine the military potential of terrorists, prevent the expansion of their influence in the region and subsequently defeat them.[38]

In this context, to counter ISIS, Washington decided to create a new international coalition that could act against ISIS. Ergo, on September 5, at the NATO summit in

[36] Azami [34].

[37] Wilson Center [35].

[38] The White House President Barak Obama [36].

Wales (UK), the head of the US State Department, John Kerry, invited[39] the heads of the Foreign and Defense Ministries of Australia, Great Britain, Germany, Denmark, Italy, Canada, Turkey, and France to join the fight against ISIS.

In addition, on September 8, 2014, the State Department announced that more than 40 countries, including members of the Arab League, are ready to join international efforts to counter Islamists.[40] Some of them expressed their readiness to contribute to the war against ISS through military operation or by providing financial aid, intelligence, logistical support, or political sympathy.

In light of the international cooperation, evidently, a combined Air Force of the United States, Australia, Belgium, Great Britain, Denmark, Netherlands, and France began strikes against ISIS facilities and armed units in Iraq. Also, US Navy warships, carriers of sea-based Tomahawk cruise missiles, were deployed to the Red Sea and the Persian Gulf to launch missiles and air strikes against ISIS.[41]

In addition to air strikes, the US and UK, the leading forces of the coalition, launched limited tactical operations "on the ground," such as the British special forces (SAS) mission to eliminate 200 British jihadis in Iraq,[42] or US special forces operations for hostage rescue and training missions.[43]

To successfully advance their operation, the anti-ISIS coalition received logistic, financial, and intelligence support from 20 countries, including Austria, Albania, Hungary, Israel, Ireland, Spain, Italy, Canada, Kuwait, Luxembourg, New Zealand, Norway, Republic of Korea, Slovakia, Turkey, Czech Republic, Switzerland, Sweden, Estonia, and Japan.[44]

Thanks to international support and political determination, US and its allies, during their fight against ISIS, carried out a total of 13,582 air strikes against ISIS in Iraq,[45] which wholly disoriented ISIS's mission and dismantled most of its offensive capabilities.

However, the US and its allies were not the only foreign elements that significantly appeared in the fight against ISIS. Surprisingly when it came US traditional enemy in the region, the Islamic Republic of Iran and its Shia Axis decided to fight on the very same side that the US and its allies were standing on.

Iran was one of the first countries immediately engaged in the war against ISIS. Only a week after the beginning of the ISIS Northern Iraq offensive, Iran deployed 2 thousand well-trained IRGC troops to protect Baghdad against ISIS's potential invasion.[46] In this context, Iran's elite Quds Force and their commander at the time, Major General Qasem Soleimani, was assigned by Ayatollah Khamenei to develop the military intervention against ISIS.

---

39 Reuters [37].

40 U.S. Government Publishing Office [38].

41 The Washington Post [39].

42 Kentish [40].

43 CNN [41].

44 U.S. Government Publishing Office [42].

45 BBC World News [43].

46 Chulov [44].

In addition, Iran mobilized its sources to arm a mega-militia of Iraqi Shias with about 70 thousand manpower,[47] the Popular Mobilization Forces, which played a crucial role in pushing ISIS back. Iran was not the only enemy of the US that accompanied the international coalition in the war against ISIS, Russia, although to a lesser extent, was another actor that tried to contribute to fighting against ISIS. In this context, in 2015, Russia, Syria, Iran, and Iraq formed a security coalition (RSII), with two operation headquarters in Damascus and Baghdad. The coalition was assigned to share intelligence on ISIS and exchange ideas between parties on the next steps in countering ISIS.[48]

In the context of the newly established framework, Russian air forces and Iran-backed Shi'a militias started confronting ISIS across the Iraqi Syrian border.[49] This strategic move severely undermined ISIS' logistic capacity to transfer its fighters from Syria to Iraq and vice-versa.[50]

Given to international support of the Iraqi government in its fight against ISIS and the fact that a cross-coalition of traditional enemies (West vs. Iran/Russia) were fighting for the same end, it is safe to argue that Iraq had the highest level of support that a country can ever have in a war. Hence, after three years of bloodshed, on 9 December 2017, Iraq rightfully declared victory in the fight against ISIS.

Now that we covered foreign factors behind Iraq's victory over the ISIS insurgency, the question is, did Afghanistan have the fortune of having international support as Iraq had? To answer this question, we should keep in mind that there are few differences between Iraq and Afghanistan. First, unlike Iraq, in Afghanistan, except for a brief period after the 9/11 attacks, we have never witnessed a cross-coalition between the US and its rivals against the forces of insurgency, the Taliban. As a matter of fact, in the last decade, we have been observing a growing competition between the US/-backed coalition and Iran and Russia in Afghanistan. Each of these actors was trying to expand relations with specific factions in Afghanistan in order to establish an influence in the country.

For instance, after the outbreak of the Syrian civil war in 2011, Iran, which traditionally had solid relations with Shia Afghan Hazara, actively started recruiting Shia Hazaras to deploy them to Syria. Iran's recruitment of Shia Afghans grew to the extent that in 2014 Iran had formed a specific battalion for Afghans, named Liwa Fatemiyoun, with about 20 thousand manpower.[51] Additionally, for the last 15 years, there have been numerous speculations regarding the Taliban and Iran's relations to the extent that from early 2004 US military officials have been accusing Iran of providing the Taliban with a place of refuge, training programs, and arm supplements.[52] The guesswork on Iran-Taliban relations continued to the point that by

[47] Global public policy institute [45].

[48] Mullen [46].

[49] AL Jazeera [47].

[50] Mahmoudian [48].

[51] Jamal [49].

[52] Meyer [50].

2019, Western sources were arguing that Iran had expanded its relations with the Taliban on all fronts, including economic, security, and political.[53]

Similar to Iran, Russia also tried to expand its ties with various political parties. Although, in the beginning, Moscow tried to establish a friendly relationship with the Kabul Government, even from 2002 to 2007, Russia provided Afghanistan with military-technical assistance worth more than $200 million, including, an air defense system to protect the Kabul airport, communications equipment, automotive equipment, and repairing 8 helicopters and 4 military transport aircraft of the Afghan Air Force. Also, Russian armed force trained more than 300 Afghan servicemen.[54] Additionally, in 2007, Russia wrote off Afghanistan's previous debt worth of $1 Billion to the USSR.[55]

However, despite all developments in interstate relations, Moscow started to take a new direction which signaled Russia's interest in developing relations with non-state factions inside Afghanistan. In this context, in 2009, Russia hosted the "Russia-Afghan Forum," in which many Kabul Pashtun opponents with ties to the Taliban attended. The participation of the pro-Taliban figures escalated Afghanistan officials' suspicious of Russia's interest in finding its own Taliban, the Kremlin's Taliban.[56] In light of establishing ties with the Taliban, in 2019, Moscow held the "Afghan Peace Talks," in which Taliban delegations and Afghan politicians attended to discuss the future of Afghanistan and a new political platform that can bring peace among all factions in Afghanistan, however, the conference had one prominent absent participant, the Afghan Government.[57]

In addition to the lack of universal consensus and the great power competition in Afghanistan, the other difference between Afghanistan and Iraq is that in Afghanistan, not only have we not witnessed a cross-coalition between international powers, but also, we have seen waves of tension and disagreement among the US-backed coalition, specifically between Pakistan and the United States. During the first decade of the US presence in Afghanistan was accompanied by strategic cooperation with Pakistan against the jihadist forces, and Pakistan granted the US the airfield use of its military airport to carry out airstrikes and logistic operations; it also handed over about 400 Qaeda suspects to the United States.[58] The cooperation grows to the point that the US announced Pakistan as its major non-NATO ally.

However, the golden days of cooperation moved toward meeting its end in 2008, when Admiral Mike McConnell, the director of NSA at the time, accused ISI, Pakistan's Inter-Services Intelligence (ISI), of informing the jihadist forces of US aerial attacks which enabled them to skip many of US's attacks.[59] In addition, the on and off military incidents between the US military and Pakistan military across

[53] Tabatabai [51].

[54] Андреевич, Мендкович Никита [52].

[55] Radio Free Europe [53].

[56] Андрей, СЕРЕНКО [54].

[57] Higgins and Mashal [55].

[58] CBC News [56].

[59] Goldberg and Ambinder [57].

the Afghanistan-Pakistan border led to frequent closures of US/NATO supply routes from Pakistan to Afghanistan, which raised the cost of intervention for the US, imposed pressure on Afghanistan economy, and escalated the tension between the two countries. Eventually, in 2011, at the beginning of the second decade of the US war in Afghanistan, the decline of American-Pakistani cooperation began. From early 2011, official statements and media speculation were indicating that the US was developing suspicious of Pakistan's growing secretive relationship with the Taliban, which reached its new peak in May 2011, when the US spotted Bin Laden's hideout less than a mile from Pakistan Military Academy in a neighborhood populated mainly by retired military officers.[60] Finding out that Bin Laden was living under the nose of the Pakistani Army for about six years definitely left an immortal wound of distrust to the US-backed coalition in Afghanistan, and hence US efforts to stabilize Afghanistan.

Next to division in US-Coalition, the other significant difference between Afghanistan and Iraq was the international community's semi-flattery with the Taliban. From the early 2010s, many courtiers started to have a direct discussion with the Taliban, including:

1. Qatar's decision to host the Taliban's "political office" in Doha in 2012.[61]
2. Iran-Taliban negotiations in 2013.[62]
3. The official high-level bilateral discussions with Pakistan in 2016.[63]
4. Negotiations with China took place in China between the Taliban's delegation from the Doha office and Chinese diplomats in 2016.[64]
5. The Moscow Peace talks in 2018, which was the first international event that the Taliban was invited to attend, and representatives of Iran, Russia, China, India, and Central Asian nations were other participants of it.[65]
6. The Uzbekistan-Taliban security talks in 2018.[66]
7. Iran-Taliban Bilateral Talks in 2018.[67]

Despite the result of these negotiations, we can argue that, in general, they gave the message to the Taliban and its base that the international community is leaning toward recognizing the Taliban as an inescapable part of Afghanistan's political spectrum. A semi-recognition that boosted the Taliban's morale and encourage them to wait out the hard days of confrontation. However, the most major game-changing event occurred in 2018 when the US began its negotiation with Taliban in Doha,[68] on 29 February 2020, after 18 months of negotiation under the lead of Zalmay Khalilzad, negotiations

[60] Froelich [58].

[61] Sheikh and Greenwood [59].

[62] Graham-Harrison [60].

[63] Voice of America News [61].

[64] The Diplomat [62].

[65] BBC News [63].

[66] Reuters [64].

[67] Gul [65].

[68] Basit [66].

Table 4 International and foreign actors' response to ISIS invasion of Iraq (2014) and Taliban's latest insurgency (2021)

International actions	Iraq versus ISIS (2014)	Afghanistan versus Taliban (2021)
Diplomatic	Universal consensus against the ISIS	Isolated and bilateral negotiation with the Taliban
Cooperation	Cross-coalition of traditional rivals	Great power competition
Military	NATO Airstrikes + Iran involvement on the ground + Russia/Syria intelligence support	No major action been taken
US coalition	United	Division between NATO and Pakistan
US/NATO actions	13,582 air strikes	Evacuation

resulted in the US-Taliban deal, also known as the Agreement for Bringing Peace to Afghanistan.

One might argue that the US-Taliban deal was a "knock-out" hit on the Afghans' morale since the Afghan Government was entirely excluded from the deal, and the spirit of the agreement was signaling that the during the process of NATO troops withdrawal from Afghanistan, US would not engage with Taliban, unless they threaten NATO troops while they are evacuating Afghanistan.[69] Therefore, on the one hand, the Afghan National forces found themselves left alone in their struggle against the Taliban, and on the other hand, the Taliban forces found a sense of relief from the US threat and start their mobilization to march toward Kabul.

In addition to psychological impact, the US-Taliban deal had another effect that changed the destiny of Afghanistan, the absence of US military support. Unlike Iraq, where the US and its allies launched thousands of aerial attacks against ISIS, in Afghanistan, while the Taliban was advancing its 2021 offense, the NATO forces not only did not take any actions but also expedited their evacuation. This neutralized a significant portion of the Afghanistan Army which was used to operate on the ground while NATO jets were supporting them over the skies. In addition, NATO's rush to evacuate created "psychological chaos" among Afghanistan National Forces and disoriented them from their mission, while the Taliban's forces' spirit was boosted by seeing NATO's withdrawal. Hence, unlike Iraq, Afghanistan failed in its fight against the forces of insurgency (Table 4).

In light of what has so far been explained, we can safely argue that, unlike Iraq, where we witnessed a "universal consensus" on supporting Baghdad against the ISIS insurgency, in Afghanistan, Kabul did not have such support in confronting the Taliban's 2021 offenses which led the fall of Afghan Government. However, one might argue there are a few reasons behind the lack of international support for Afghanistan's latest struggle with the Taliban:

[69] Welna and Dwyer [67].

First—ISIS attacked Iraq about ten years after the US invasion of Iraq, and for most of that decade, Iraq was dealing with civil wars and ISIS's first insurgency. Therefore, in the eyes of international actors, the Iraqi government was not expected to be able to defeat ISIS single-handedly. Unfortunately, this was not quite the case for Afghanistan. The Taliban 2021 offense took place two decades after the US invasion of Afghanistan, which is not outrageous to say that is quite enough time for a state to achieve security independence. On top of this, Taliban launched the 2021 offense with about 80,000 barefooted militants who at most were equipped with Ak-74 and RPG-7, while the Afghan Army had 300,000 manpower equipped with the worth of $83 billion West-made weapons and trained by the mightiest armies.[70] Ergo, when the Taliban 2021 offense began, the idea of launching a new military campaign or even postponing the withdrawal was not relatively justified to the public in the West. Although these reasons do not mean that the Afghan army was not willing to sacrifice all it had to protect the country, the collapse of Kabul was the result of wrong political decisions in Afghanistan and abroad.

Second—Unlike ISIS, which was pursuing a global jihad that triggered many terrorist attacks against civilians across the world, the Taliban neither was seeking to establish an Islamic Caliphate beyond its traditional border nor handled any terrorist operation beyond the Afghanistan border. In a way, the Taliban successfully managed to introduce itself as an "isolationist" force of insurgency with no international ambition. Obviously, these political maneuvers of the Taliban cooled off the temptation among foreign actors to intervene in the 2021 offensive.

Last but not least, in 2014, when ISIS raided Iraq, in the absence of the disoriented and shocked Iraqi Army, there was another dependable factor engaging on the ground that the West could rely on it, the Iraqi Kurdistan Regional Government (KRG). KRG forces were among the first units that confronted ISIS and appeared as a strong obstacle against the ISIS ambush. These troops were well trained and equipped, and thanks to their previous cooperation with the US/UK against Saddam, they were well-known and trustworthy. The presence of KRG forces and their resistance which later was accompanied by the Shia militants, bought time and also motivation for the international actors to step in Iraq's War against ISIS. Unfortunately, in 2021, when the Taliban ambushed toward Kabul, Afghanistan was lacking such an element, and in the absence of the Afghan Army, other factions neither had the readiness nor the material capability to fill the vacuum which eased the Taliban's takeover in Afghanistan. But why Afghanistan was deprived of such a factor is a question that its answer will take us to the next section and the most significant difference between Afghanistan and Iraq, the system of governance.

[70] Deutsche Well [68].

5.2 Domestic Actors

It is not only the ending that varies Afghanistan from Iraq, but also their beginning has been different. The post-Taliban Afghanistan and post-Saddam Iraq began their new days under different systems of governance. The former was set to be ruled by a heavily centralized government, and the latter was ought to begin its journey under a federal structure in which the Kurdish regions in the north were granted autonomy. Ergo, in Afghanistan, all courses of action could only be decided by the capital, but in Iraq, there has been a division of authority between the two cities of Baghdad and Erbil. Hence, in Afghanistan, once Kabul fell, the rest of the country collapsed with it too, but in Iraq, even when the Iraqi Army lost its disciple and was disoriented after the fall of Mosel, Erbil remained organized and began to confront ISIS. This distinction brings us to the root of the contrast between the Iraqi and Afghan systems, their constitution.

The designed system of governance for Afghanistan marked its beginning on 26 January 2004, when the constitution of the Islamic Republic of Afghanistan was formally ratified. Despite Afghanistan's heavily diverse society, the new constitution began by stating that Afghanistan's system of governance is unitary (centralized) and its territory is indivisible.[71] However, to restore the trust of national unity, especially among Farsi-speaking minorities, which were the particular target of the Taliban, the new constitution made the gesture of recognizing Dari as the second official language of the country, next to Pashtun and other local languages as the third in their own region.[72]

Although, despite the progressive implications of Article 16, the rest of the constitution was set in a way that majorities (Pashtun) would make all crucial calls in the country. Since according to Article 64, the new constitution left the monopoly of controlling armed forces and national security policies, terminating the national assembly, appointing attorney generals, Supreme Court Justices, and supervising the constitution itself to the office of President.[73]

On top of handing a great deal of power and authority to the Presidential Office, the new constitution barred ethnic minorities in Afghanistan from local self-governance since, according to the constitution, the provisional governors who were the extension of the executive branch in provinces were only appointed by the majority-elected President.[74] The expansion of central government authority over local politics might have been seen as a set-back to minorities, especially Farsi-speaking Tajiks who, prior to the US invasion of Afghanistan, under the lead of the Northern Alliance, had a self-claimed system of leadership over a notable part of the north-east of the country (about 10% of Afghanistan territory). Mostly-Tajik populated Northern Alliance achieved and sustained this territorial control through 15 years of struggle

---

[71] The constitution of the Islamic Republic of Afghanistan [69]. Article 1.

[72] The constitution of the Islamic Republic of Afghanistan. (2004). Article 16.

[73] The constitution of the Islamic Republic of Afghanistan. (2004). Article 64.

[74] Mukhopadhyay [70].

with the Taliban, but after the establishment of the Islamic Republic of Afghanistan, they lost their privilege of "making the last call" in the region to Kabul.

In this context, after 15 years of resistance, the North Alliance and other ethnic groups handed their weapons and armors to the central government; hence the safety of their and regions became solely dependable on the might and efficiency of Kabul.

While Afghans were endeavoring to adjust to the new system of governance, on October 15, 2005, the foundation of the state system in Iraq was established by the adoption of a new constitution which was approved by the majority of voters in a general referendum. The new constitution, which was containing 144 articles, was set to a much more extensive range of various socio-political relations, specifically regarding the distribution of power among different ethnic groups. In this context, in the very first article, Iraq was introduced as a "federal state.[75]"

In light of Iraq's previous challenges in establishing a "united and inclusive national identity", the new constitution characterizes Iraq as a multi-ethnic and multi-confessional state. Ergo in Article 4, both Arabic and Kurdish were identified as official languages.[76] Also, according to Article 4, local languages such as the Turkomen, Chaldean, and Assyrian were designated as the official languages in areas where most residents speak these languages.[77]

In addition to recognizing the identity of other ethnic groups, the new constitution was set to heal the major wound to Iraq's nation-building experience, the systematic discrimination against non-Arab groups. In the past, most of the country's sources were spent on Arab-Sunni populated regions which deprived another part of countries of the right of prosperity. Therefore, according to Article 112, the Federal Government assumed the responsibility of distributing the oil and gas revenues fairly among all demographic basis throughout the country.[78]

In this context, to assure minorities that they would have a place in marking the destiny of Iraq, Article 116 identifies Iraq's system of governance as a federal form of government, characterizing it as "democratic" and pays much attention to local government bodies,[79] giving regions greater autonomy. Ergo, Article 117 officially recognizes Iraq's Kurdistan Regional Government as autonomy.[80] In light of recognizing regional autonomy, Article 120 states the regional government is allowed to adopt a constitution of its own and an autonomous mechanism to exercise its authority.[81] To guarantee the survival and functionality of the autonomous mechanism, Article 121 authorities' regional governments to establish their own armed forces, including police, security forces, and guards.[82]

[75] Constitution of the Republic of Iraq [71]. Article 1.

[76] Constitution of the Republic of Iraq. (2005). Article 4.

[77] Ibid.

[78] Constitution of the Republic of Iraq. (2005). Article 112.

[79] Constitution of the Republic of Iraq. (2005). Article 116.

[80] Constitution of the Republic of Iraq. (2005). Article 117.

[81] Constitution of the Republic of Iraq. (2005). Article 120.

[82] Constitution of the Republic of Iraq. (2005). Article 121.

Table 5 Iraq and Afghanistan's Constitution

Constitution	Iraq	Afghanistan
System of Governance	Federal	Centralized
Armed Forces	National + Regional	National
Recognition of Minorities Language as an Official Language	Yes Arabic + Kurdish	Yes Pashtun + Dari
Regional Self-Governance	Yes	No
Regional Vanguard	Yes	No
Current State	Effective	Terminated by Islamic Emirate of Afghanistan

Thanks to Iraq's constitutional legitimization of fair distribution of national resources/revenue, and Kurdish regional autonomy, the Kurdish Regional Government found the capability to establish a well-trained armed force, Peshmerga, which by the time of the ISIS invasion had expanded its size to 190,000 of manpower.[83] Peshmerga were one of the first forces that on June 8, only two days after the fall of Mosul, confronted ISIS in the West of Kirkuk and prevented ISIS forces from occupying Kirkuk.[84] Without the Peshmerga, the fall of Kirkuk, one of largest cities of Iraq with more than half a million population, would be inevitable since the shock of the fall of Mosul completely disoriented and paralyzed Iraqi National Army to the extent that once they noticed ISIS advancement toward the city, they fled Kirkuk[85].

The Peshmerga not only filled the vacuum of power that the fleeing Iraqi Army left behind, but also their resistance bought time for international actors and also Shia militants to mobilize their forces to counter the ISIS ambush, which marked the survival of Iraq (Table 5).

Unfortunately, Afghanistan's centralized constitution never left a chance for either existence of regional vanguards like KRG Peshmerga or the chance of formation of non-states armed forces similar to Iraqi Shia Mobilization Forces. Ergo, once the system of leadership in Kabul collapsed and the Afghan National Army fled apart, there were no forces to fill the vacuum of power against the Taliban; hence the entire country, in less than 100 days (92 days), fell into the forces of insurgency. Thus, the Taliban Second Realm again began to impose its discriminatory system of governance against ethnic and religious minorities, especially Shia Afghans and Farsi-speaking Tajik and Hazaras.

[83] BBC World News [72].

[84] BBC World News [73].

[85] Paasche and Gunter [74].

6 Conclusion

The journey for neither Iraq nor Afghanistan has not been finished yet, and as long as these nations exist, their destiny could change. Still, for now, it seems that they are moving toward different destinations, although it is too early to anticipate their final destination. In the meantime, it is crucial to acknowledge what we have learned from the US journeys in Afghanistan and Iraq.

In this context, it is safe to argue that the events in Iraq and Afghanistan reminded us of one of the classical textbooks of history; heterogeneous nations have a long path to process a successful nation-building. However, their path can be further prolonged by the absence of an efficient state that advances a national-wide education and upholds national harmony. In addition to the absence of an efficient state, domestic chaos and foreign power interruptions can deprive nations of experiencing consistent and effective nation-building process, as it did to Iraq and Afghanistan. Although, nations who suffered from the the absence of steady and persistent nation-state building, can still achieve national harmony by a establishing a democratic system t in which all units of society are free and allowed to be involved.

Probably this was one of the major reasons behind US decisions to implement a democratic constitution for Afghanistan and Iraq. Although, the problem with foreign-backed state-building is that its survival and sustainability depend on the presence and support of the foreign sponser. Simply put, as long as the foreign power is present, the newly established democratic regimes can survive domestic chaos, foreign interruptions, and insurgencies, but once the foreign-built security umbrella is removed, the newly established state will fall. The fact that the Islamic Republic of Afghanistan in the absence of American troops collapsed, while the Republic of Iraq won over the ISIS offensive with the help of foreign supporters, including United States, are evidence of this argument.

Thus, the presence of external support is vital. As the disorientation of the Iraqi National Army after the ISIS ambush and the collapse of the Afghanistan National Army in 2021 proved to us that foreign-established national armies would not be the savior of the country in the time of crisis, unless the foreign sponsor actively engages the crisis-management processes.

In the light of what happened in Iraq and Afghanistan, we can categorize external supports into two major categories; a foreign Army and non-State militia like KRG Peshmerga, whose benefits and regional autonomy depend on protecting the prior-the crisis status-quoand hence will "put up a fight" against forces of insurgency to protect their own interest and wellbeing. The role of homemade external support groups, non-state actors, brings us to the most crucial difference between Iraq and Afghanistan, their system of governance.

The impact of the system of governance can be considered as the most important lesson that we have learned from the US journey in Iraq and Afghanistan. It appears that in heterogeneous nation-states with incomplete nation-building, the best system of governance is a federal structure that assures minorities a fair share of power and also enables them to protect themselves against regressive insurgencies. In lack of such a factor, Afghanistan failed and now is heading toward a different destination which does not seem very prosperous.

At this point, it is hard to predict the future of Afghanistan. Still, history has taught us that prolonged tension among ethnic groups in diverse societies would either result in the domination of one faction over other factions or the territorial collapse of the entire nation. The former usually takes place where the central government has a great degree of military capabilities, such as in Russia and Turkey; the latter happens in nations with chaotic and traumatized governments like the "former" Yugoslavia. For now, we have to wait because only time can tell us whether Afghanistan will survive the current crisis or fail, and we will witness the rise of new "-stan nations" in the region.

References

1. The Cultural Atlas Team (2020) Afghan culture. https://culturalatlas.sbs.com.au/afghan-culture/afghan-culture-religion
2. U.S. Department of State (2019) Report on International Religious Freedom, Iraq. https://www.state.gov/reports/2019-report-on-international-religious-freedom/iraq/
3. Minority Rights. (2018). Iraq. https://minorityrights.org/country/iraq/
4. Religious Politics in Iraq (2003) United States Institute of Peace. https://www.usip.org/publications/2003/05/religious-politics-iraq
5. Country Reports. Afghanistan demography. https://www.countryreports.org/country/Afghanistan/population.htm
6. World Population Review (2022) Afghanistan population 2022 (live). https://worldpopulationreview.com/countries/afghanistan-population
7. World Atlas. What languages are spoken in Afghanistan? https://www.worldatlas.com/articles/what-languages-are-spoken-in-afghanistan.html
8. United States Commission on International Religious Freedom (2020) Afghanistan. https://www.ecoi.net/en/file/local/2028946/Afghanistan.pdf
9. U.S. Department of State (2022) Report on International Religious Freedom, Afghanistan. https://www.state.gov/reports/2019-report-on-international-religious-freedom/afghanistan/
10. Idris F, Hassan Z, Ya'acob A, Gill SK, Awal NAM (2012) The role of education in shaping youth's national identity. Procedia—Soc Behav Sci 59:443–450
11. Paglayan A (2021) The non-democratic roots of mass education: evidence from 200 years. Am Political Sci Rev 115(1):179–198
12. Karna MN (1999) Language, region and national identity. Sociol Bull 48(1/2):75–96
13. Anderson B (1983) Imagined communities: reflections on origin and spread of nationalism. Verso, New York, p 168
14. Dodge T (2016) The danger of analogical myths: explaining the power and consequences of the sykes-picot delusion. Ajil Unbound 110:132–136
15. Rezun M (1986) The great game revisited. Int J 41(2):324–341
16. Ingram E (1980) Great Britain's great game: an introduction. Int Hist Rev 2(2):160–171
17. Lambah SK (2012) The durand line. Indian Foreign Aff J 7(1):42–60
18. Sidanius J, Feshbach S, Levin S, Pratto F (1997) The interface between ethnic and national attachment: ethnic pluralism or ethnic dominance? Public Opin Quart 61(1):102–133
19. Kriesberg L (1998) Coexistence and the reconciliation of communal conflicts. In: Weiner E (ed) The handbook of interethnic coexistence. Continuum International Publishing Group, New York, pp 182–198

20. Hauner M (1981) The soviet threat to Afghanistan and India 1938–1940. Mod Asian Stud 15(2):287–309
21. Dobbins J, Poole MA, Long A, Runkle B (2008) Post-9/11 nation-building: Afghanistan and Iraq. In: After the war: nation-building from FDR to George W. Bush. RAND Corporation, pp 85–134
22. Baldaufl S, Tohid O (2003) Taliban appears to be regrouped and well-funded. Christ Sci Monit. https://www.csmonitor.com/2003/0508/p01s02-wosc.html
23. Tohid O (2003) Taliban regroups—on the road. Christ Sci Monit. https://www.csmonitor.com/2003/0627/p06s01-wosc.html?related
24. Howard M (2007) Battle for new leader likely. The Guardian. https://www.theguardian.com/world/2007/jan/01/iraq.michaelhoward
25. U.S. House of Representative, Committee on Homeland Security (2013) Assessing the threat to the homeland from AL-Qaeda operations in Iran and Syria. Serial No. 113-19
26. Anti-Defamation League (2022) Al Qaeda. https://www.adl.org/resources/profiles/al-qaeda
27. Helmand Provincial Overview (2017) Naval Postgraduate School. https://nps.edu/web/ccs/helmand
28. Dorronsoro G (2009) The Taliban's winning strategy in Afghanistan. Carnegie Endowment, pp 7. https://carnegieendowment.org/files/taliban_winning_strategy.pdf
29. Amir M, Sa'id Abdal-Rahman al-Mawla (Deceased) (2022) US reward for justice mission. https://rewardsforjustice.net/terrorist-rewards/amir-muhammad-said-abdal-rahman-al-mawla-2/
30. Muzhdah W (2016) Who is the new Taliban military chief? Deutsche Well. https://www.dw.com/en/who-is-the-new-taliban-military-chief/a-19516155
31. Kadivar J (2020) Exploring Takfir, its origins and contemporary use: the case of Takfiri approach in Daesh's media. Contemp Rev Middle East 7(3):259–285
32. Semple M (2014) Rhetoric, ideology, and organizational structure of the Taliban movement. United States Institute of Peace, pp 8
33. O'Driscoll D (2019) The Islamic State in Iraq. UK Department for International Development—Helpdesk Report, pp 12–14. https://assets.publishing.service.gov.uk/media/5d430bafed915d09dac9bb05/580_The_Islamic_State_in_Iraq.pdf
34. Azami D (2014) How the Taliban groom child suicide bombers. BBC News. https://www.bbc.com/news/world-asia-27250144
35. Wilson Center (2019) Timeline: the rise, spread, and fall of Islamic state. https://www.wilsoncenter.org/article/timeline-the-rise-spread-and-fall-the-islamic-state
36. The White House President Barak Obama (2014) Statement by the president on ISIL. https://obamawhitehouse.archives.gov/the-press-office/2014/09/10/statement-president-isil-1
37. Reuters (2014) U.S. says forms 'core coalition' to counter Iraq militants. https://www.reuters.com/article/cnews-us-iraq-crisis-coalition-idCAKBN0H00TA20140905
38. U.S. Government Publishing Office (2014) United States strategy to defeat the Islamic State in Iraq and Syria. https://www.govinfo.gov/content/pkg/CHRG-113shrg95102/html/CHRG-113shrg95102.htm
39. The Washington Post (2014) U.S. airstrikes target Islamic State militants in northern Iraq. https://www.washingtonpost.com/world/middle_east/islamic-state-militants-seize-christian-town-in-northern-iraq-thousands-flee/2014/08/07/942a553a-1e2b-11e4-ab7b-696c295ddfd1_story.html?hpid=z1
40. Kentish B (2016) SAS in Iraq given 'kill list' of 200 British jihadis to take out. Independent. https://www.independent.co.uk/news/world/middle-east/sas-special-forces-hit-list-iraq-syria-isis-terrorist-attacks-drones-a7400756.html
41. CNN (2015) Pentagon IDs American killed in Iraq hostage rescue. https://www.independent.co.uk/news/world/middle-east/sas-special-forces-hit-list-iraq-syria-isis-terrorist-attacks-drones-a7400756.html
42. U.S. Government Publishing Office (2016) Counter ISIL (Islamic State of Iraq and the Levant) operation and Middle East strategy. https://www.govinfo.gov/content/pkg/CHRG-114shrg25619/html/CHRG-114shrg25619.htm

43. BBC World News (2019) IS fight: US-led coalition says it killed 1300 civilians in Syria and Iraq. https://www.bbc.com/news/world-middle-east-48473979
44. Chulov M (2014) Iran sends troops into Iraq to aid fight against ISIS militants. Guardian. https://www.theguardian.com/world/2014/jun/14/iran-iraq-isis-fight-militants-nouri-maliki?CMP=fb_gu
45. Global Public Policy Institute (2017) Who's Who: quick facts about local and sub-state forces|Iraq After ISIL. https://www.gppi.net/2017/08/16/quick-facts-about-local-and-sub-state-forces
46. Mullen J (2015) Iraq agrees to share intelligence with Russia, Iran, and Syria. CNN. https://edition.cnn.com/2015/09/27/middleeast/iraq-russia-iran-syria-intelligence-deal/
47. AL Jazeera (2015) Iraq liaises with Syria, Russia and Iran to bomb ISIL. https://www.aljazeera.com/news/2015/10/14/iraq-liaises-with-syria-russia-and-iran-to-bomb-isil
48. Mahmoudian A (2021) Russia and Iran's relations in Iraq. London School of Economy. https://blogs.lse.ac.uk/mec/2021/07/02/russia-and-irans-relations-in-iraq/
49. Jamal AS (2018) Mission accomplished? What's next for Iran's Afghan fighters in Syria. War Rock. https://Warontherocks.Com/2018/02/Mission-Accomplished-Whats-Next-Irans-Afghan-Fighters-Syria/
50. Meyer H (2009) Iran is helping Taliban in Afghanistan, Petraeus says (update1). Bloomberg. https://web.archive.org/web/20110814093743/http:/www.bloomberg.com/apps/news?pid=newsarchive&sid=aru5H2YB1Tv8&refer=india
51. Tabatabai AM (2019) Iran's cooperation with the Taliban could affect talks on U.S. withdrawal from Afghanistan. The Washington Post. https://www.washingtonpost.com/politics/2019/08/09/irans-cooperation-with-taliban-could-affect-talks-us-withdrawal-afghanistan/
52. Андреевич, Мендкович Никита (2012) Политика России в Афганистане в 2001–2011 гг. Перспективы. https://www.perspektivy.info/rus/desk/politika_rossii_v_afganistane_v_20012011_gg_2012-02-17.html
53. Radio Free Europe (2006) Russia to forgive Afghanistan's Soviet-Era debt. https://www.rferl.org/a/1065243.html
54. Андрей, СЕРЕНКО (2009) На «линии Дюранда» появились «русские талибы». фганистан.Ру. https://afghanistan.ru/doc/14815.html
55. Higgins A, Mashal M (2019) In Moscow, Afghan peace talks without the Afghan Government. The New York Times. https://www.nytimes.com/2019/02/04/world/asia/afghanistan-taliban-russia-talks-russia.html
56. CBC News (2006) Pakistan hands hundreds of suspects to U.S. for money: report. https://www.cbc.ca/news/world/pakistan-hands-hundreds-of-suspects-to-u-s-for-money-report-1.620799
57. Goldberg J, Ambinder M (2011) The ally from hell. The Atlantic. https://www.theatlantic.com/magazine/archive/2011/12/the-ally-from-hell/308730/
58. Froelich P (2021) What Osama bin Laden's compound looks like today. The New York Post. https://nypost.com/2021/08/14/what-osama-bin-ladens-pakistani-compound-looks-like-today/
59. Sheikh MK, Greenwood MTJ (eds) (2013) Taliban talks: past, present and prospects for the US, Afghanistan and Pakistan (PDF). In: DIIS Report 2013:06. Danish Institute for International Studies, p 14. https://www.files.ethz.ch/isn/162384/RP2013-06-Taliban-Talks_web.jpg.pdf
60. Graham-Harrison E (2013) Afghan Taliban send delegation to Iran. Guardian. https://www.theguardian.com/world/2013/jun/03/afghan-taliban-send-delegation-iran
61. Voice of America News (2016) Afghan Taliban confirms delegation visiting Pakistan. https://www.voanews.com/a/afghan-taliban-confirms-delegation-visiting-pakistan/3304525.html
62. The Diplomat (2016) What to make of China's latest meeting with the Taliban. https://thediplomat.com/2016/08/what-to-make-of-chinas-latestmeeting-with-the-taliban/
63. BBC News (2018) Afghanistan war: Taliban attend landmark peace talks in Russia. https://www.bbc.com/news/world-asia-46155189
64. Reuters (2018) Afghan Taliban delegation visits Uzbekistan to talk security, power lines. https://www.reuters.com/article/us-afghanistan-taliban/afghan-taliban-delegation-visits-uzbekistan-to-talk-security-power-lines-idUSKBN1KX056

65. Gul A (2018) Iran hosts bilateral talks with Taliban on Afghan peace. https://www.voanews.com/a/iran-hosts-bilateral-talks-with-taliban-on-afghan-peace/4722655.html
66. Basit A (2020) The US-Taliban deal and expected US exit from Afghanistan: impact on South Asian militant landscape. Counter Terrorist Trends Anal 12(4):8–14
67. Welna D, Dwyer C (2020) U.S. signs peace deal with Taliban after nearly 2 decades of war in Afghanistan. NPR. https://www.npr.org/2020/02/29/810537586/u-s-signs-peace-deal-with-taliban-after-nearly-2-decades-of-war-in-afghanistan
68. Deutsche Well (2021) Why the Afghan army folded to the Taliban. https://www.dw.com/en/why-the-afghan-army-folded-to-the-taliban/a-58889507
69. The constitution of the Islamic Republic of Afghanistan (2004)
70. Mukhopadhyay D (2016) Provincial governors in Afghan politics. US Institute of Peace
71. Constitution of the Republic of Iraq (2005) Article 1
72. BBC World News (2014) Profile: who are the Peshmerga? https://www.bbc.com/news/world-middle-east-28738975
73. BBC World News (2014) Iraqi Kurds 'fully control Kirkuk' as army flees. https://www.bbc.com/news/world-middle-east-27809051
74. Paasche TF, Gunter MM (2016) Revisiting Western strategies against the Islamic State in Iraq and Syria. Middle East J 70(1):9–29

Arman Mahmoudian is a Ph.D. Candidate in Politics and International affairs at the University of South Florida (USF) and an adjunct faculty at the USF Judy Genshaft Honors College, where he is teaching various courses related to the Middle East, Russia, and International Security. In addition, he is a research associate at the USF Center for Strategic and Diplomatic Studies, where he researches on Iran's regional policy. Mr. Mahmoudian grew up and completed his undergraduate education in Law in Iran and graduate education in International Relations in Russia (Moscow). His Analysis of Regional Security in the Middle East and Russian Foreign Policy has been published by Quincy Institute for Responsible Statecraft, Harvard Journal of Middle Easter Politics and Policy, London School of Economics Middle East Center, Atlantic Council, Middle East Eye, Politics Today, New Arab, Trends Research and Advisory, and the Global Post. Additionally, He has spent a great deal of time providing perspectives regarding the Middle East crisis and the ongoing Russia-Ukraine conflict through media appearances on Al-Jazeera and the BBC.

Fratricide: Lessons Learned from Afghanistan

Anthony J. Masys

Abstract Fratricide (Friendly Fire) continues to be a disruptive and tragic event in military operations. Despite the advent of precision-guided munitions, 'smart bombs', and unprecedented navigational accuracy, friendly fire continues to be prevalent. In the last 20 years we have seen several fratricide incidents emerge during military operations in Afghanistan. For example, on June 19, 2014, an airstrike conducted by a B1 Bomber killed U.S. Special Forces and other American soldiers. It was considered one of the worst friendly fire incidents in more than 12 years of U.S. military operations in Afghanistan. This case resonates with a similar incident from 2002 where a US airstrike resulted in the death of 4 Canadian soldiers (Masys in Fratricide in air operations: opening the black box—revealing the social, 2010). With such friendly fire incidents, investigations often cite poor Situation Awareness (SA) as a major contributing factor (Ministry of Defence (MOD) in Combat identification, report by the comptroller and auditor general HC 661 session 2001–2002, London Stationary Office, 2002). Across the many domains comprised of complex socio-technical systems (aviation, healthcare, military operations…), human error is often cited as a major contributing factor or cause of incidents and accidents. Accident surveys in aviation have attributed 70% of incidents to crew error citing pilot error as the root cause [Woods et al. in Behind human error: cognitive systems, computers, and hindsight, 1994.: 2; Helmreich in BMJ 320:781–785, 2000: 781; Shappell and Wiegmann in Human Factors and Aerospace Safety 1:59–86, 2001: 60]. In a study of accidents among major air carriers, 88% of those involving human error could be attributed to problems with situation awareness (SA); similarly, problems with SA were found to be the leading causal factor in a review of military aviation mishaps (Endsley in Situation awareness in aviation systems. Lawrence Eribaum Associates, NJ, pp 257–276, 1999). In this chapter we 'open the black box' of human error through a systems analysis lens to reveal a complex accident aetiology. This research draws upon the research of Masys (Masys in Fratricide in air operations: opening the black box—revealing the social, 2010) and updates it with insights from recent fratricide occurrences. As noted by Dekker [Dekker in Reconstructing human contributions to accidents: the new view on error and performance, 2001: 3], 'human error is not an

A. J. Masys (✉)
University of South Florida, Tampa, FL 33620, USA
e-mail: anthony.masys@gmail.com

A. Farhadi and A. Masys (eds.), *The Great Power Competition Volume 4*,
https://doi.org/10.1007/978-3-031-22934-3_16

explanation, but is something to be explained'. What is revealed is that our notion of 'teamwork' transcends the traditional human element to include the intersection of the technological and informational domains that shape situation awareness and decision making (Masys in Teamwork and trust: a sociotechnical perspective, 2011). This Actor Network Theory based research initiative points towards the 'hybrid collectif' (Masys in Fratricide in air operations: opening the black box—revealing the social, 2010) as the framework that helps us understand the factors that precipitates the occurrence of fratricide and points to strategies to prevent further incidents through this socio-technical lens.

Keywords Fratricide · Friendly fire · Lessons learned · Afghanistan · Actor network theory (ANT) · Socio-technical system

1 Introduction

Fratricide (Friendly Fire) continues to be a disruptive and tragic event in military operations. Despite the advent of precision-guided munitions, 'smart bombs', and unprecedented navigational accuracy, friendly fire continues to be prevalent. In the last 20 years we have seen several fratricide incidents emerge during military operations in Afghanistan. For example, on June 19, 2014, an airstrike conducted by a B1 Bomber killed U.S. Special Forces and other American soldiers. It was considered one of the worst friendly fire incidents in more than 12 years of U.S. military operations in Afghanistan.[1] This case resonates with a similar incident from 2002 where a US airstrike resulted in the death of 4 Canadian soldiers [1]. With such friendly fire incidents, investigations often cite poor Situation Awareness (SA) as a major contributing factor [2].

Across the many domains comprised of complex socio-technical systems (aviation, healthcare, military operations…), human error is often cited as a major contributing factor or cause of incidents and accidents. Accident surveys in aviation have attributed 70% of incidents to crew error citing pilot error as the root cause [3: 2, 4: 781, 5: 60]. In a study of accidents among major air carriers, 88% of those involving human error could be attributed to problems with situation awareness (SA); similarly, problems with SA were found to be the leading causal factor in a review of military aviation mishaps [6].

In this chapter we 'open the black box' of human error through a systems analysis lens to reveal a complex accident aetiology. This research reports from and draws upon the work of Masys [1] and updates it with insights from recent fratricide occurrences. As noted by Dekker [7: 3], 'human error is not an explanation, but is something to be explained'. Military operations such as those conducted in Afghanistan are characterized by an inherent complexity of interdependent systems and teams working under conditions of VUCA (Volatility, Uncertainty, Complexity, Ambiguity). This is

[1] https://www.businessinsider.com/investigation-finds-june-friendly-fire-airstrike-was-completely-avoidable-2014-9.

particularly evident during military operations involving air-ground cooperation and coordination. Under such conditions, teamwork and trust are a requirement given the conduct of interdependent tasks. This research examines fratricide through the application of Actor Network Theory (ANT) [1]. Yeung [8] notes that much of the work that draws on ANT places its analytical focus on unearthing the complex web of relations between humans and non-humans (technology) within a complex socio-technical system. Through this ANT lens on fratricide, what is revealed is that our notion of 'teamwork' transcends the traditional human element to include the intersection of the technological and informational domains that shape situation awareness and decision making [9]. This Actor Network Theory based research initiative points towards the 'hybrid collectif' [1] as the framework that helps us understand the factors that precipitates the occurrence of fratricide and points to strategies to prevent further incidents through this socio-technical lens.

2 Fratricide Overview

As described in detail in Masys [1], the definition of the term 'fratricide', 'amicicide', 'amicide', 'friendly fire', 'blue on blue' are all terms denoting the action of an accidental death of one's own forces. There are numerous definitions of fratricide in the literature representing national and organizational views.

As reported in the US Joint Publication 1-02 [10: 222] friendly fire is defined as:

> …a casualty circumstance applicable to persons killed in action or wounded in action mistakenly or accidentally by friendly forces actively engaged with the enemy, who are directing fire at a hostile force or what is thought to be a hostile force.

As reported in the UK Joint Doctrine Publication 0-01.1 [11: F10] fratricide is defined as:

> the accidental death or injury which occurs when friendly forces engage their own forces believing either them, or their location, to be an enemy target.

Analysis conducted by The Technical Cooperation Programme (TTCP) under the auspices of Action Group (AG) 13 has made significant contributions to the literature on fratricide. The AG 13 defines friendly fire as:

> …a friendly fire event is the deliberate engagement of non-enemy entities by friendly forces in the belief that the entities are enemy. Entities include both personnel and material [12: 544].

What emerges from the definitions are themes associated with: mistakes, accident and belief regarding target identification.

Syms and Salt [13] compiled an analysis of fratricide cases and noted that air to ground fratricide accounted for 40% of the total for the twentieth century as a whole. Syms and Salt [13] point towards fratricide findings pertaining to a misperception of a decision-maker regarding reality or 'ground truth'. As described in Masys [1:

102], 'decision-making, within the context of a cognitive process situates the blame on the pilot for any outcomes resulting from his/her decision: a decision based on imperfect knowledge and uncertainty. Hence, we are presented with the findings of pilot error as the root cause in the accident aetiology'.

Fratricide incidents are characterized by a complex aetiology. As noted in Gadsden and Outteridge [14: 7] '...incidents rarely (if ever) have single cause. There are often complex interrelationships between contributing factors, which can occur at different levels (strategic, operational, and tactical) and with different levels of impact'. Hence, we apply actor network theory as our systems lens to open up the proverbial 'black box of fratricide' [1] revealing the complex socio-technical nature of fratricide.

3 Actor Network Theory Overview

As described in detail in Masys [1]:

Actor Network Theory (ANT) is a theoretical perspective from the domain of Sociology and has been applied in such fields as information technology, organizational theory, geography, medical anthropology and psychology. This perspective facilitates an analysis of complex socio-technical systems by challenging the way we think of agency, the human and non-human. The strength of ANT lies in the analysis of situations where separation of the social and technical elements is difficult [15]. As a systems thinking approach it embraces a fundamental 'complexity' shift that challenges the traditional paradigm of linearity and reductionism. The methodology associated with ANT shares fundamental principles with other qualitative approaches, such as ethnography. 'Following the actors' [16–18] lies at the foundation of the ANT methodology. As detailed in Masys [1] 'Following the actors' allows the researcher to investigate those actors that have been 'silenced or deleted' and '...to bring them back to light by using archives, documents, memoirs, museum collections' [19: 81]. With this in mind Latour [19: 82] argues that '...if objects are not studied it is not due to a lack of data, but rather to a lack of will'.

The ANT lens reveals an inherent heterogeneity in the system or 'network of actors and the relations' between the actors. As described by Coakes [20: 2], 'Socio-technical thinking is holistic in its essence; it is not the dichotomy implied by the name; it is an intertwining of human, organizational, technical and other facets'. The holistic perspective of ANT thereby makes it well suited to facilitate an examination of the complex socio-technical systems associated with accident aetiology.

4 Case Studies

The analytical framework deployed for this chapter stems from the work of Masys [1] where various fratricide case studies were examined resulting in a theoretical framework that explains the complex aetiology associated with fratricide.

In this chapter we examine 2 fratricide cases that took place in Afghanistan involving air-ground incidents. These cases are:

the 2002 F-16/Canadian soldiers' incident at Tarnac farms

In the evening of 17/18 April 2002, a section from "A" Company, 3rd Battalion, Princess Patricia's Canadian Light Infantry BG (3 PPCLI BG) were conducting a livefire exercise in the vicinity of Kandahar, Afghanistan, when they were mistakenly engaged by two American F-16 fighter aircraft. (Details of this incident can be found at Masys [1]).

the 2014 B1B Bomber and US forces incident at Zabul province in southern Afghanistan

The June 9, 2014, airstrike marked one of the ugliest friendly fire incidents in more than 12 years of U.S. military operations in Afghanistan. A B-1B Lancer bomber dropped its ordnance on five U.S. soldiers, including members of an elite Special Forces team [21].

5 Discussion

The literature is replete with examples of Human Error in complex socio-technical systems. For example, in the aviation industry, human error is reported as the leading cause of both commercial airline crashes and general aircraft accidents. More than 88% of all general aviation accidents are attributed to human error, especially due to loss of control by the pilot during flight. In addition to the attribution of pilot error in the aviation industry, other avenues of human error are also involved including: flight crew members, air traffic controllers, and mechanics or maintenance staff, engineers and managers.

A 1999 Institute of Medicine report highlighted that upwards of 100,000 Americans die each year because of medical errors [22]. Similarly, upwards of 80% of accidents in the offshore oil and gas industry are blamed on human error. According to a new study, human and organisational factors are now widely recognised as the main cause of major oil and gas accidents and initiatives to address risks have only recently been developed.[2]

In understanding human error in such systems, Woods et al. [3: 4], present an argument that human performance is a multidimensional issue that is a function of the context in which an incident takes place; that technology shapes human performance thereby creating new pathways and forms of error and failure; that human performance involves a set of interacting people; that competing goals within the organizational context creates dilemmas shaping accident aetiology; and that the attribution of error is a social judgment rather than an objective conclusion.

[2] https://www.energylivenews.com/2016/07/27/human-error-blamed-for-80-of-offshore-accidents/#:~:text=Image%3A%20Thinkstock-,Around%2080%25%20of%20accidents%20in%20the%20offshore%20oil%20and%20gas,have%20only%20recently%20been%20developed.

To facilitate understanding of human performance/human error in complex socio-technical systems, the application of systems mapping presents a powerful tool/lens [1]. Such an approach enables one to:

Make sense of complexity
Better see and understand system
Clarify relationships between system components
Challenge assumptions and mental models
Identify root causes
Understand feedback loops and patterns
Identify knowledge gaps
To engage various health disciplines and leverage different perspectives
Communicate complexity
Co-create shared understanding of the system.

System mapping (which emerges from the ANT analysis) provides a methodology that explores the various dimensions of the iceberg model (Fig. 1). Explore the complex socio-technical systems associated with military operations through the systems lens, we move beyond an event description of the fratricide event to uncovering the trends, underlying structures and mental models that shape our understanding of fratricide and can shape our design of intervention strategies.

Hence the application of ANT as our systems thinking approach moves beyond just an event description to 'opening the blackbox' of human error in fratricide.

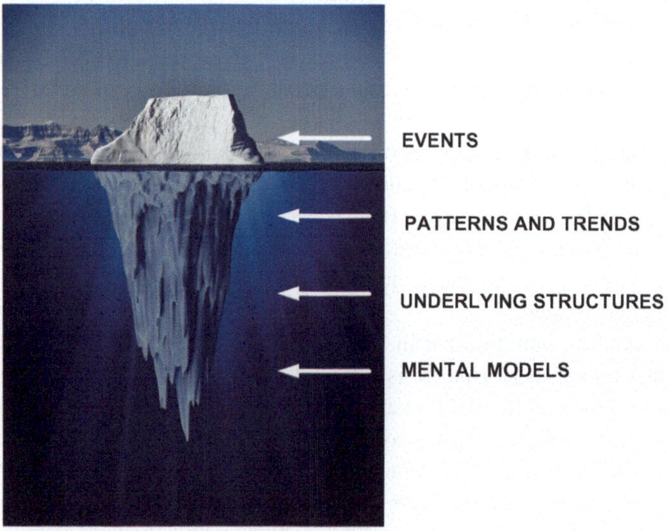

Fig. 1 The iceberg model of systems thinking (adapted from Senge et al. [23: 174])

6 Fratricide Case Study Description

1. **TARNAC Farms 2002**
   As detailed in Masys [1], the Coalition Investigation Board [24] regarding the 17 April 2002 Tarnak Farms Range incident reported that the fratricide was a direct and proximate result of actions taken by the two F16 pilots involved. Based on the evidence presented, given the pilots expectations when he encountered what he believed to be SAFIRE, he misperceived the caliber, trajectory, and distance traveled of the munitions. The investigation revealed that the behavior of the F-16 pilot in flight suggests a perceptual set or mind set regarding the threat associated with surface-to-air fire. It was stated during the course of the investigation that 'when perceptual sets are established, individuals tend to scan the environment for confirmatory cues. Information that would negate what is already believed generally receives minimal to no allocation of attention. Only information that is overwhelmingly contradictory may be sufficient to lead an individual to question current beliefs or hypotheses or to change their overall cognitive assessment of a situation' [CIB 24:52]. It was concluded that the lack of situational awareness exhibited by the F-16 pilot follows from poor planning and preparation combined with problems associated with attention, misperception, and fatigue. The pilot channelized attention and missed important information that could have redirected his course of action.
2. **B1B 2014 Case**
   As detailed in the investigation report and cited in Washington Post [25], on June 9, 2014, a friendly fire incident involving an airstrike marked one of the ugliest friendly fire incidents in more than 12 years of U.S. military operations in Afghanistan. A B-1B Lancer bomber dropped its ordnance on five U.S. soldiers, including members of an elite Special Forces team.

According to the investigation, the fratricide incident was the result of poor communication, inadequate planning and several other mistakes.

A brief summary of the incident as described in Lamonthe [21];

From an altitude of about 12,000 feet, the B-1 bomber was providing what the military calls close air support while U.S. and Afghan ground troops were moving out of the area at the conclusion of their operation.

Muzzle flashes were mistaken for signs of rifle fire from insurgents, in part because the movement of the US Forces was not properly communicated to those coordinating with the B-1 crew.

The B-1 crew said their targeting pods had detected no U.S. marking devices at that location it was decided that targets must by insurgents.

All key leaders in the mission, including the ground commander, JTAC and air crew, thought that sensors on the bomber would be able to see infrared strobe markers. The targeting pod on the B-1B bomber cannot do that, leading the air crew to incorrectly believe there were not troops on the ridgeline they bombed, investigators determined.

The investigation reported that '…had the team understood their system's capabilities, executed standard tactics, techniques and procedures and communicated effectively, this tragic incident was avoidable…'.

7 Analysis: Fratricide Causes/Insights

One of the underlying themes that emerged from these cases and others described in Masys [1] is that fratricide stems from a misperception of a decision-maker regarding reality or 'ground truth' [13]. This lack or loss of Situation Awareness (SA) has been cited by US sources, [FM-1-114, 26: I0, 1], as a primary cause of fratricide characterized by: Target identification errors; Navigation errors; Communications errors; and Weapon errors. The most common cause of fratricide, as reported by Ministry of Defence [2: 7] is a '…lack of Situational Awareness through poor identification and co-ordination of forces, and failures in communication together with inadequate procedures'.

As cited in Masys [1], Gadsden and Outteridge [14: 8–9] details 12 high-level causal categories of failure that characterizes fratricide. These include:

- Command and control
- Procedures
- Equipment/technology
- Situational awareness
- Misidentification
- Physical/physiological factors
- Pre-deployment preparation
- Team work
- Environmental factors
- Communications/information
- Platform configuration
- Cognitive factors.

As noted in Gadsden and Outteridge [14: 7] '…incidents rarely (if ever) have single cause. There are often complex interrelationships between contributing factors, which can occur at different levels (strategic, operational, and tactical) and with different levels of impact'. Situation awareness as noted in the 2 case studies analyzed is rooted in mental models of expectations [1]. These can emerge as 'illusions of certainty'.

ANT Analysis

Systems mapping is facilitated through the ANT lens by neither privileging the human or non-human and views the elements of the network as actors from which facilitates a tracing of the relations that uncover the 'social' that permeates complex socio-technical systems. The research conducted in Masys [1] through the ANT lens reveals 3 domains in this socio-technical system: Physical/Human/Informational that define

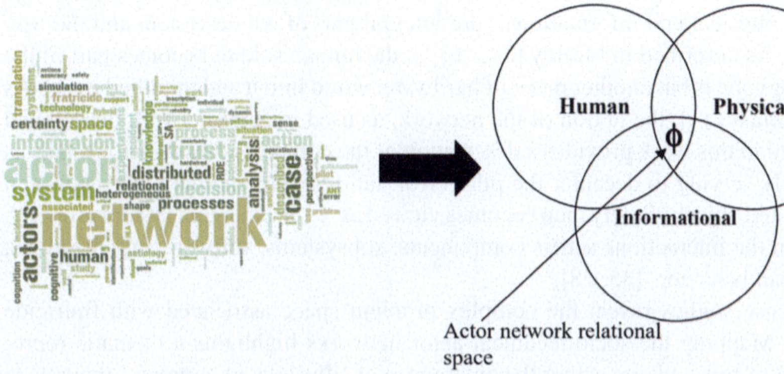

Fig. 2 Thematic analysis of case studies reveals the 'hybrid collectif' [1]

the problem space associated with fratricide. As shown in Fig. 2, these 3 domains are interrelated. The intersection of these domains represented by the symbol φ (Fig. 2) encapsulates the actors that reside within the actor network. It is at this intersection of the domains that the hybrid collectif emerges, whereby the dichotomy between human and non-human (technical) is erased, representing a socio-technical entangled state space [1: 250].

The physical dimensions of the case studies reflected in the technological artifacts such as (strobes, PODS, GPS…) create expectations of behavior of the 'system' that shapes situation awareness and decision making. Cited in Masys [1: 262], Rosen and Rappert [27: 20] report that '…design of artifacts can prohibit certain users or compel particular kinds of uses' and become reflected in the design and implementation of Rules of Engagement (ROE) and supporting architecture that shaped decision making. Emerging from the analysis of the case studies and supported by Viseu [28: 113] '…Objects are not passive containers of human designs and desires. They are actors in that they do things, i.e. by existing they actively shape and transform the character of that which they are part of'. The hybrid collectif reveals the emergence of 'illusions of certainty' [1] created and reinforced by perceptions of redundancy, reliability and capability claims of the system. This complex accident aetiology is consistent with Normal Accident Theory [29], Swiss Cheese model [30] and Normalization of Deviance [31]. With these 2 case studies, new 'technical' solutions (with their inscriptions) become part of the system space. Verbeek [32: 131] in his analysis remarks that 'Artifacts influence the way in which people do things, and this influence could be deliberately inscribed into them'. This is reified in both case studies where '…artifacts co-shape the relational world (network space) by influencing or …giving shape not only to people's actions but also to people's experiences' [32: 139]. As cited in Masys [1], this research reveals that technology, instead of reducing human error, rather changes it and often aggravates the consequences [33]. Fratricide is thereby viewed as a socio-technical phenomenon (as a network of heterogeneous elements relationally interconnected) recognizing that safe and effective flight operations are achieved by the entire socio-technical system as a collective. The actors (human, physical and informational) are integral part of whole system and the way it works. As described in Molloy [34: 16] '…the human soldier becomes part of the technology; he is but another piece of hardware, wired into it and modified by it'. As cited in Masys [1], the notion of the network, as used in actor-network theory, and illustrated in this work provides a description of the complex webs of actor relations effectively serving to decenter the pilot error and to overcome the binary between subjects and objects. Safety then becomes viewed as '…an emergent system property, arising in the interactions across components, subsystems, software, organizations, and human behavior' [35: 28].

The case studies reveal the complex problem space associated with fratricide (Fig. 3). Mapping the socio-technical actor networks highlights a thematic representation of fratricide revealing the emergence of 'illusions of certainty' that reside within the system (Fig. 2).

Exploring the case studies reveals fratricide as a complex aetiology that is 'rhizomal' and de-centered [1]. As described by Boulton et al. [36: 41], 'path-dependency suggests that the detail of pattern and events shapes the emerging future…often many factors over differing scales come together synergistically to shape what happens next'. Action takes place in this 'hybrid collectif' of entangled human and nonhuman actors. As noted in Masys [1], through the hybrid collectif, the event-based 'domino' perspective that characterizes linear event-based models of pilot error/human error disappears revealing a complex temporal and spatial heterogeneity. What emerges from the analysis of the actor network is the notion that time and space are folded (decisions, mindsets, mental models that developed in the past)

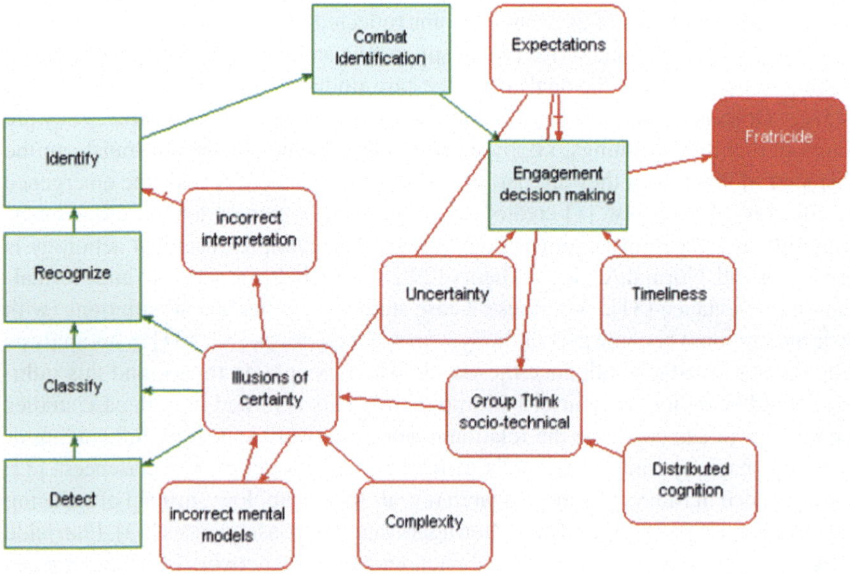

Fig. 3 Thematic representation of Fratricide [1]

are recasted in terms of latent effect/errors as purported by Turner (1978) and Reason [37], in terms of a network schema. Barad [38: ix] writes 'the past is never finished. It cannot be wrapped up like a package, or a scrapbook, or an acknowledgement; we never leave it and it never leaves us behind'.

This emergent rhizomal understanding of fratricide derived from the ANT approach presents key lessons learned that points to the requirement for a paradigm shift. Of note regarding lessons learned is '...finding new ways to help people reflect on mindsets and overarching worldviews, and finding ways to critique some long-held beliefs about the 'way things are', is absolutely critical to their really coming to terms with the world and its complexity' [36: 123). Blamism is not the same as causality. Pilot error is not an explanation, but is something to be explained.

8 Conclusion

Despite the advent of precision-guided munitions, 'smart bombs', and unprecedented navigational accuracy, friendly fire continues to be prevalent. In the last 20 years we have seen several fratricide incidents emerge during military operations in Afghanistan. This case study analysis (leveraging the findings contained in Masys [1]) applies actor network theory to reveal the complex accident aetiology. As detailed in Masys [1] and revealed in this additional case study, fratricide resides within a relational network of actors that creates the possibility for action. In defining the actor network, the root cause of fratricide is 'rhizomal' and emerges as an entangled network of heterogeneous elements, a hybrid collectif. What this reveals is that action, as seen in the case studies, takes place in a 'hybrid collectif' that is comprised of entangled human actors as well as non-human actors in multiple ways. Fundamental to this analysis is the application of systems thinking and system mapping. As reported in Boulton et al. [36: 4], '...we require a radical shift in our perceptions, our thinking and our values. Thinking in systems and more specifically thinking in terms of interconnectedness, is a key element of a broader and more holistic worldview'. This chapter presents an approach that challenges our current mental models regarding of system design and human factors.

Note: For a complete discussion of the analysis of fratricide through the Actor Network Theory lens, the reader is directed to Masys [1].

References

1. Masys AJ (2010) Fratricide in air operations: opening the black box—revealing the social. PhD Dissertation, June 2010, University of Leicester, UK
2. Ministry of Defence (MOD) (2002) Combat Identification. In: Report by the comptroller and auditor general HC 661 session 2001–2002, London Stationary Office.

3. Woods D, Johannesen L, Cook R, Sarter N (1994) Behind human error: cognitive systems, computers, and hindsight. Crew Systems Ergonomics Information Analysis Center (CSERIAC), Wright-Patterson Air Force Base, Ohio
4. Helmreich RL (2000) On error management: lessons from aviation. BMJ 320:781–785
5. Shappell SA, Wiegmann DA (2001) Applying reason: the human factors analysis and classification system (HFACS). Hum Factors Aerosp Saf 1(1):59–86
6. Endsley MR (1999) Situation awareness in aviation systems. In: Garland DJ, Wise JA, Hopkin VD (eds) Handbook of aviation human factors. Lawrence Eribaum Associates, Mahwah, NJ, pp 257–276
7. Dekker S (2001) Reconstructing human contributions to accidents: the new view on error and performance. Technical Report 2001-01, Lund University School of Aviation
8. Yeung HWC (2002) Towards a relational economic geography: old wine in new bottles?. Paper presented at the 98th annual meeting of the Association of American Geographers, Los Angeles, CA. http://courses.nus.edu.sg/course/geoywc/publication/Yeung_AAG.pdf
9. Masys AJ (2011) Teamwork and trust: a sociotechnical perspective. In: Stanton NA (ed) Trust in military teams. Surrey, England, Ashgate Publishing
10. US Joint Publication 1-02 (2001) Department of defence dictionary of military and associated terms. 12 April 2001 (as amended through 19 August 2009). http://www.dtic.mil/doctrine/jel/new_pubs/jp1_02.pdf. Accessed 11 Nov 2009
11. UK Joint Doctrine Publication 0-01.1 (2006) United kingdom glossary of joint and multinational terms and definitions. Edition 7 dated June 2006. http://www.mod.uk/NR/rdonlyres/E8750509-B7D1-4BC6-8AEE8A4868E2DA21/0/JDP0011Ed7.pdf. Accessed 11 Nov 2009
12. Caseley P, Dean D, Gadsden J, Houghton P (2007) Concepts of network enabled capability—safety issues and potential solutions. In: The proceedings of the 25th international system safety conference, Baltimore, Maryland, USA, pp 538–550
13. Syms PR, Salt JD (2004) A catalogue and analysis of amicidal incidents. DSTL/WP08280 Report, 31 August 2004
14. Gadsden J, Outteridge C (2006) 'What value analysis? The historical record of fratricide. Presented at the 23rd international symposium on military operational research, 29 Aug–1 Sept 2006
15. Callon M (1999) Actor-network theory: the market test. In: Law J, Hassard J (eds) Actor network and after. Blackwell and the Sociological Review, Oxford and Keele, pp 181–195
16. Callon M (1986) The sociology of an actor-network: the case of the electric vehicle. In: Callon M, Law J, Rip A (eds) Mapping the dynamics of science and technology. MacMillan Press, London, pp 19–34
17. Callon M (1991) Techno-economic networks and irreversibility. In: Law J (ed) A sociology of monsters? Essays on power, technology and domination, sociological review monograph. Routledge, London, pp 132–161
18. Latour B (1996) On actor-network theory. A few clarifications. Soziale Welt 47:369–381
19. Latour B (2005) Reassembling the social: an introduction to actor network theory. Oxford University Press, Oxford
20. Coakes E (2003) Socio-technical thinking—an holistic viewpoint. In: Clarke S, Coakes E, Hunter MG, Wenn A (eds) Socio-technical and human cognition elements of information systems. Information Science Publishing, Hershey, pp 1–4
21. Lamonthe D (2014) Investigation: friendly fire airstrike that killed U.S. Special Forces was avoidable
22. Kohn LT, Corrigan JM, Donaldson MS (eds) (1999) To err is human: building a safer health system. National Academy Press, Washington, D.C
23. Senge P, Smith B, Kruschwitz N, Laur J, Schley S (2010) The necessary revolution: working together to create a sustainable future. Boradway Books, NY
24. Coalition Investigation Board (CIB) of Inquiry Tamak Farm Final Report (2002)
25. Washington Post (2014) https://www.washingtonpost.com/news/checkpoint/wp/2014/09/04/investigation-friendly-fire-airstrike-that-killed-u-s-special-forces-was-avoidable/

26. FM-1-114 Air Calvary squadron and troop operations: appendix I. Headquarters Department of the Army, 1 February 2000. http://www.adtdl.army.mil/cgibin/atdl.dll/fm/1-114/appi.htm. Accessed June 2005
27. Rosen P, Rappert B (1999) The culture of politics and technology. Soziale Technik 4(99):19–22
28. Viseu AAB (2005) Augmented bodies: the visions and realities of wearable computers. PhD Dissertation. University of Toronto
29. Perrow C (1999) Normal accidents: living with high-risk technologies, 2nd edn. Princeton University Press, Princeton, NJ
30. Reason J (1997) Managing the risk of organizational accidents. Ashgate Publishing, Aldershot, England
31. Vaughan D (1996) Challenger launch decision: risky technology, culture and deviance at NASA. University of Chicago Press, Chicago
32. Verbeek P (2005) Artifacts and attachment: a post-script philosophy of mediation. In: Harbers H (ed) Inside the politics of technology: agency and normativity in the co-production of technology and society. Amsterdam University Press, Amsterdam, pp 125–146
33. Dekker S (2002) The field guide to human error investigations. Ashgate Publishing Ltd., Aldershot
34. Molloy P (2005) Killing Canadians (II): the international politics of the accident. Can J Cult Stud 14(Fall):5–21
35. Woods DD (2006) Essential characteristics of resilience. In: Hollnagel E, Woods DD, Leveson N (eds) Resilience engineering: concepts and precepts. Ashgate Publishing, Aldershot, pp 22–34
36. Boulton JG, Allen PM, Bowman C (2015) Embracing complexity: strategic perspectives for an age of turbulence. Oxford University Press
37. Reason J (1990) Human error. Cambridge University Press, New York
38. Barad K (2007) Meeting the Universe Halfway: quantum physics and the entanglement of matter and meaning. Duke University Press, Durham & London

Dr. Anthony J. Masys is an affiliate Associate Professor and former Director of Global Disaster Management, Humanitarian Assistance and Homeland Security. A former senior Air Force Officer, Dr. Masys has a BSc in Physics and MSc in Underwater Acoustics and Oceanography from the Royal Military College of Canada and a Ph.D. from the University of Leicester. He is Editor in Chief for Springer Publishing book series: Advanced Sciences and Technologies for Security Applications and holds various advisory board positions with academic journals and books series. Dr. Masys is an internationally recognized author, speaker and facilitator and has held workshops on security, visual thinking, design thinking and systems thinking in Europe, Canada, South America, West Africa and Asia. He has published extensively in the domains of physics and the social sciences. Dr. Masys holds 5 national and international academic appointments.

The Spiritual Dimension of Moral Injury and PTSD

Garrett W. Potts and Lily M. Abadal

Abstract This chapter aims to carefully distinguish Post-Traumatic Stress Disorder (PTSD) from Moral Injury (MI) and review the role of the spiritual dimension in the causation, healing, and prevention of both syndromes. We argue that by giving due consideration to this spiritual dimension, the U.S. military can better prepare soldiers to encounter potentially morally injurious events (PMIEs) in deployment and combat.

Keywords Post-traumatic stress disorder · Moral injury · Combat · Troops morale

1 Introduction

Moral Injury (MI) research has gained considerable traction for its potential to aid in the diagnosis and treatment of symptoms coinciding with other mental health concerns. In recent years, MI has been found to frequently coexist with most Post-Traumatic Stress Disorder (PTSD) cases among active duty and retired United States military veterans. The result has been that MI is often conflated with PTSD rather than treated as a separate syndrome. In order to better understand each syndrome and adequately care for the whole soldier, we argue there is a need to (1) clearly distinguish MI from PTSD and (2) consider the spiritual dimension that contributes to their comorbidity. As a result, this chapter will define and clearly distinguish MI from PTSD and explore the spiritual dimension of both syndromes in terms of causality, healing, and prevention. We argue that by giving due consideration to this spiritual dimension, the military can better prepare soldiers to encounter potentially morally injurious events (PMIEs) in deployment and combat.

G. W. Potts (✉) · L. M. Abadal
Department of Religious Studies, The University of South Florida, Tampa, FL 33620, USA
e-mail: garrettpotts@usf.edu

L. M. Abadal
e-mail: lmking2@usf.edu

A. Farhadi and A. Masys (eds.), *The Great Power Competition Volume 4*,
https://doi.org/10.1007/978-3-031-22934-3_17

2 PTSD: Definition and History

In order to best distinguish MI from PTSD, we will begin with a very brief overview and history of PTSD and then summarize the diagnostic criteria for PTSD and its associated symptom clusters. As Garcia [12] outlines, PTSD was first included in the *DSM-III* in 1980 but had a much longer conceptual history in psychiatry, beginning in the late nineteenth century. It was British railway passengers that inspired the first inquiry into what we now call PTSD. Post Office employees who had suffered crashes on the poorly-constructed lines were suffering from sleep disturbances, chronic pain, and fear of railway travel—a phenomenon then called 'railway spine.' Lasiuk and Hegadoren [25] describe the effects of this phenomenon and the questions it raised as follows:

It was during these early debates concerning the etiology of railway spine that many fundamental questions about the nature of psychological trauma began to arise. Is the disorder organic or psychological in origin? Is it the event itself or the subjective interpretation of the event that is the source of the trauma? (p. 16).

These questions are important ones. Unfortunately, sustained focus on finding answers to them did not follow—at least not right away.

Van der Kolk et al. [44] and Herman [14] clarify that serious investigation into the lasting and life-altering effects of trauma has received intermittent attention and focus in psychiatry. However, the research of Myers [30, 31], Da Costa [9], and Kardiner [17] focused specifically on trauma in military contexts and laid the groundwork for PTSD's ultimate inclusion in the DSM. In fact, it is research focusing on trauma in military contexts that ushered forth a renewed interest and heightened emphasis on understanding PTSD.

The most current diagnostic criteria for PTSD entails an exposure or a history of exposure to a traumatic event or events causing patients to identify with some or all of the following symptom clusters: "intrusion," "avoidance," "negative alterations in cognition or mood," and "alterations in arousal and reactivity" [1, p. 271]. Patients who are clinically diagnosed with PTSD have experienced a disturbance of symptoms associated with these clusters, unrelated to the ingestion of any substances, and the symptoms have persisted for at least one month, "causing clinically significant distress or impairment in social, occupational, or other important areas of functioning" [1, p. 272]. Table 1 summarizes common experiences associated with each of the four symptom clusters.

These symptom clusters are essential for advancing our argument—they help clarify a distinction between PTSD and MI, highlighting the fear-based nature of PTSD. Van der Kolk [43] describes the roots of PTSD symptoms quite well:

Ideally, our stress hormone system should provide a lightning-fast response to threat, but then quickly return us to equilibrium. In PTSD patients, however, the stress hormone system fails at this balancing act. Fight/flight/freeze signals continue after the danger is over and... do not return to normal. Instead, the continued secretion of stress hormones is expressed as agitation and panic, and in the long term, wreaks havoc on their health (p. 30).

Table 1 PTSD symptom clusters and associated experiences[1]

PTSD symptom clusters	Common experiences
Intrusion	Intrusive memories, dreams, or flashbacks to events or circumstances that cue the remembrance of a traumatic event
Avoidance	Deliberate efforts to suppress or avoid feelings, thoughts, memories, and/or reminders associated with a traumatic event
Negative alterations in cognition or mood	Inability to remember features of a traumatic event, prolonged negative emotional state and/or negative beliefs about oneself, others, or the world, distorted beliefs about the cause or consequences of a traumatic event, and diminished interest in purposive activities and/or relationships
Alterations in arousal and reactivity	Difficulty concentrating and/or sleeping, hypervigilance and/or exaggerated startle response, irritability, outbursts, and impulsive, self-destructive behavior

As Van der Kolk [43] describes, patients or soldiers suffering from PTSD have an altered stress hormone system. When aroused in the face of threatening stimuli, they fail to return to equilibrium. Moreover, they will take more extreme measures to avoid triggers that elicit memories surrounding the trauma, and they tend to remain vigilant, often in fearful anticipation. They live life as if the trauma is ongoing and enduring.

Research on soldiers returning from Afghanistan has recently brought deeper insights about PTSD to the fore. Chief among these insights is the sheer frequency of the disorder. In studying soldiers deployed in Iraq and Afghanistan, Ramchand et al. [38] found that "PTSD appears to affect between 5 and 20% of previously deployed service members" (p. 28). Some estimate that nearly 23% of those who served in Afghanistan and Iraq following September 11, 2001 experience PTSD symptoms [27]. Deployment is specifically related to an increased risk of PTSD due to the frequency with which soldiers are exposed to both traditional combat events and episodes of extraordinarily abusive violence [38]. Ultimately, increased exposure to trauma yields an increased likelihood of developing PTSD.

Though exposure to trauma is, no doubt, the single most significant risk factor in the development of PTSD, it does not tell the whole story. In other words, exposure to trauma is a necessary condition for the development of PTSD; it is not sufficient. *Understanding PTSD requires making a clear distinction between the traumatic event itself and the subjective experience of a traumatic event.* Included in that subjective interpretation would be the narrative an individual uses to describe an event, attributions of blame or guilt associated with an event, moral evaluations of the event,

[1] The PTSD symptom clusters in Table 1 are reflected exactly as they appear in the APA's DSM-5 [1, pp. 271–272]. The common experiences identified within this table also appear in the DSM-5, but our descriptions of these experiences have been summarized for inclusion in this table.

and how the experience of an event interacts with an individual's unique experiences. For example, witnessing the violent abuse of a non-combatant may affect a victim of child abuse more significantly, given their history of trauma. As Lasiuk and Hegadoren [25] note, an acknowledgment that patients who have PTSD are embodied human beings that uniquely interact with their environments is essential to properly understanding the disorder and treating it.

Furthermore, research shows several existential factors highly correlate with the development of PTSD or are demonstrated to increase the risk of development. For instance, Oren and Possick [33] suggest that "when ideology is congruent with political reality, it may serve as a powerful resilience factor in extreme stress. In contrast, when ideology clashes with reality, it is positively associated with pathological symptoms" (p. 488). In other words, ideology could be a significant risk factor in developing PTSD, especially when those ideological commitments are challenged and repudiated by the political and existential realities surrounding extreme stress or trauma. As Oren and Possick [33] suggest, the specific content of ideological commitments is essential in evaluating their ability to hinder or help to cope following the experience of an extreme stressor or trauma.

More specifically, a significant relationship between PTSD and spirituality has also emerged in the secondary literature [21, 34, 35]. Currier et al. [8] have demonstrated that adaptive and maladaptive spiritual exercises directly correlate with levels of PTSD symptomology.

Veterans who endorsed more daily spiritual experiences, practiced prayer or meditation in a more regular manner, endorsed greater levels of forgiveness (for self, others, and from God or Higher Power), incorporated positive religious coping strategies (e.g., collaborating with God or Higher Power to solve problems, look to divine realm for strength), or were engaged in a church or other community all showed lower levels of PTSD symptomatology at discharge (p. 62).

These findings show the strong correlation between spirituality, a soldier's ability to cope with traumatic stress, and the nature or degree of PTSD symptoms. Some, such as Park et al. [34, 35], argue a reciprocal relationship between spirituality and trauma exists. An individual's meaning-making process informs the way one processes trauma. Conversely, experiences of trauma profoundly impact one's existential position. Traumatic events are moments when big questions about the value of human life, mortality, the existence of God, the reality of human suffering, and so on are encountered—sometimes for the first time.

The research on psychological inflexibility and PTSD also highlights this fact. Psychological inflexibility has been shown to predict the severity of PTSD symptoms in war veterans [27]. Though psychological inflexibility is a nebulous term, drawing upon the work of Bond et al. [3] and Meyer et al. [27] define it as "the rigid dominance of psychological reactions to unwanted internal experiences over contextual factors and personal values in guiding behavior" (p. 384). Essentially, psychological inflexibility occurs when an individual lets negative emotions guide their actions and behavior over their values or goals—a proclivity ripe for the development of cognitive dissonance and mental anguish.

All of this data points to the interesting entanglement of belief—spiritual, ideological, political, or ethical—and the development of pathological symptoms or the prevention of them. It can be deduced from this data that forming balanced meaning-making values is crucial in the formation of soldiers as it informs their subjective conceptualization of traumatic events and subsequent coping with traumatic events. A deeper exploration into the concept of MI will further substantiate this speculation. In the following section, we will now turn toward this exploration of MI.

3 Moral Injury: Definition and History

MI research began in 2003 with the publication of Jonathan Shay's book, *Achilles in Vietnam: Combat Trauma and the Undoing of Character*, to describe the so-called 'character wounds' that veterans and active-duty military service members commonly face. For the purposes of this chapter, we understand MI in light of Hall et al.'s [13] definition, which is derived from a systematic review of the MI literature. They define MI as follows: "Distress (due to guilt, shame, disgust, withdrawal, self-condemnation, etc.) following situations involving moral transgressions" [13], p. 93.

The notion of 'moral transgressions' should be carefully qualified. The sort of moral transgressions that result in potentially morally injurious events (PMIEs) are significant and fundamentally upend a soldier's conception of justice, violate deeply-held convictions, and challenge the notion that there is meaning in the world. For example, Shay [40] argues that soldiers face MI in the wake of betrayals of trust, unjust killing, the falling of a comrade, and so on. Such experiences posit a "violation of assumptions and beliefs about right and wrong," resulting in existential "dissonance and conflict" [26]. Typical symptoms of MI that are associated with moral transgressions include the loss of trust, shame, grief, remorse, the loss of hope, meaning, and/or purpose in life, depression, anxiety, anger, burnout, the loss of belief in God, and the heightened potential for substance abuse as well as other patterns of self-handicapping [4, 13, 16, 21].

While MI research is relatively recent, these phenomenological experiences associated with Hall et al.'s [13] definition are as old as wars themselves. Shay [40], for example, speaks of the 'character wounds' that haunted military personnel in Homer's *Iliad* and *Odyssey* and connected these wounds to similar struggles faced by U.S. veterans for whom MI remains a grave concern. Because Shay's work suggests that Homer's epics can be read as narratives that may have been written, at least in part, "for veterans as a way of communalizing their grief and trauma" and as spiritual expressions of "the universal experience of human beings at war," he maintains that these epics have something to teach us about the experience of MI today and about how to care for those who exhibit its aforementioned symptoms [2, p. 160].

Similarly, Koenig and Al Zaben [21] indicate that a deeper understanding of the MI concept is aided by the existential tension described within Homer's epics and that MI corresponds to the Ancient Greek account of *miasma*, which described the psychological impact of experiencing "moral defilement" on the battlefield (p. 2990).

Table 2 Bi-directional types of MI and their associated causes

MI symptom types	Causes
Self-directed	Something a soldier *does* or *does not do*
Other-directed	Something *done to* or *witnessed by* a soldier

Miasma, Koenig and Al Zaben [21] suggest, is "applicable to any transgression of moral values, whether applied to the perpetrator, the victim, or even the observer" (p. 2990). *Miasma,* therefore, is best understood as an ancient way of describing and extending the research on MI.

Recognizing the enduring patterns of MI from Homer's epics onward and the connection between MI and the Ancient Greek concept of *miasma* allows us to see that MI symptoms can be experienced in multiple directions. The existing literature on MI and the associated scales for its assessment corroborates this. For example, Koenig and Al Zaben [21] indicate that MI scales "typically assess self-directed symptoms (guilt, shame, self-condemnation) and other-directed symptoms (anger toward others, feelings of betrayal)" (p. 2991). That being said, research indicates that MI can be experienced as a result of something that a soldier *does* or *does not do,* something *done to* a soldier, or something *witnessed by* a soldier that causes the soldier to question matters such as meaning, purpose, hope, and transgression. We classify these bi-directional types of MI and their associated causes in Table 2, where we demonstrate in more detail *how* and *why* these four causes specifically correspond to PMIEs in military contexts.

3.1 Something Done by a Soldier

First, MI can result when a soldier *acts* in such a way that directly conflicts with personally held values and commitments. While cognitive dissonance is common when moral agents act in ways contrary to their conscience or deeply held beliefs, there could be a heightened sense of dissonance and guilt that results when individuals are feeling *forced* to act in such a way that is morally compromising and when the matter is serious and grave. The context of war makes both of these circumstances more likely. For example, the hierarchical structure of the military and its emphasis on authority, rank, and duty makes soldiers especially prone to experiencing this. Soldiers may know an order is unjust, realize they don't want to perform the order, and still feel pressured to carry it out despite this self-awareness. The tension between duty and a soldier's personally held convictions is borne out here. What does a soldier do when there is a duty to act in such a way that compromises one's personal values or conscience?

Because modern war is characterized by tactics that "often result in ambiguous, split-second decisions, in which the sanctioned course of action is not readily apparent or results in collateral harm" [18, p. 741], it is particularly rife for PMIEs. Many such tactics appear to be relevant for understanding the context of the war in Afghanistan, where "urban combat and insurgency environments, unmarked combatants, constant civilian threats, and widespread use of improvised explosive devices (IEDs)" were typical [18, p. 741].[2]

3.2 *Something Not Done by a Soldier*

Second, a soldier could experience MI when they fail to act in a way consistent with their deeply held moral convictions, especially when the stakes are high. Another way of understanding this is that "acts of omission" can also be fertile ground for PMIEs [23]. Imagine a soldier who deeply values courage and self-sacrifice but who failed to tend to an injured comrade in order to escape deadly gunfire. There is no fault in saving one's own life; however, the weight of perceived cowardice is heavy. So, failure to act can weigh just as heavily on a soldier, producing the same capacity for shame and regret as an actively performed action.

3.3 *Something Done to a Soldier*

Third, PMIEs can occur when a moral transgression is enacted upon a soldier. There are a variety of instances that could result in this particular sort of MI. For example, soldiers captured as prisoners of war could be subject to extreme torture and struggle to maintain hope amidst intense suffering; soldiers could receive unjust orders or be the recipient of unfair treatment, abuse, and so on. One can imagine the dissonance that would arise when a soldier is, for example, the victim of sexual assault while fighting in a war for 'freedom.' It is challenging to maintain fervor and hope for a cause when the actors in pursuit of it are unjust, especially when the soldier is the victim of injustice.

3.4 *Something Witnessed by a Soldier*

Fourth, a PMIE could result from witnessing an act or continuous behavior that causes a soldier to seriously doubt that the world is just, good, or meaningful. Much of the secondary literature on MI points to a few apt examples—seeing the death of a child, an unarmed civilian, or witnessing the torture of a prisoner of war are just a few examples. Moreover, PMIEs could be significantly more impactful and traumatic when the perpetrator is someone of higher rank, an authority figure, or someone previously thought to be an upstanding comrade. Shay's [41] understanding of MI is

[2] Also see, Forkus et al. [10], Lancaster and Erbes [23], and Lancaster and Irene Harris [24].

particularly sensitive to this reality: "(a) a betrayal of 'what's right'; (b) by someone who holds legitimate authority; (c) in a high-stakes situation" (p. 183). While we are not employing Shay's definition for the purposes of this chapter, it does highlight how much more deeply MI of this sort can cut when an authority figure perpetrates the moral transgression.

In the next section, we summarize the connection between as well as the distinguishing features of PTSD and MI. Following this summary, which is based on our literature survey, we will turn our attention to the impact that MI and PTSD, experienced together, tend to have on the spiritual dimension of soldiers' wellbeing.

4 The Need to Distinguish PTSD and MI

Evidence from emerging research on MI suggests that there is a strong correlation between PTSD and MI, "with nearly 90% of participants [exhibiting PTSD] indicating a rating of 9 or 10 in severity (on a 1–10 scale) for at least one [MI] symptom and 50% indicating this for more than five symptoms" [19, p. 257]. While the statistical significance of PTSD and MI coinciding seems abundantly clear in most cases where PTSD is evident, there remains a need to understand the unique features of each syndrome to address symptoms pertaining to both wholly. Otherwise, there is a risk of conflating these syndromes and consequently neglecting to consider more holistic measures for healing and prevention. This is why we have begun our chapter with an analysis of the unique histories associated with PTSD and MI, *understood as distinct syndromes*.

Indeed, it has been acknowledged that some degree of "definitional overlap" exists between these two terms [19, p. 2990]. Still, most of this overlap can be understood as the result of accounts about PTSD that go beyond the bounds of its cluster symptoms and attempt to grapple with transgressions involving moral and spiritual distress without reference to the MI concept. For example, Wortmann et al. [45] indicate that the link between trauma and PTSD symptoms may be mediated by a moral and spiritual struggle entailing "negative cognitions" related to "understanding or responding to stressful events," and they do this without reference to MI. The result is that their account of PTSD includes additional cluster symptoms associated with heightened PTSD severity that are not captured by the DSM-5, such as "negative beliefs about the self (e.g., I am damaged, inadequate, or unacceptable), the world (e.g., others are dangerous and untrustworthy), and self-blame" [45, p. 2]. Notice how closely associated these themes are with the ancient account of *miasma* and the character wounds typically related to MI. What if, instead of expanding the list of PTSD symptoms to include this additional cluster, Wortmann et al. [45] had argued that emerging evidence supports the fact that MI symptoms mediate PTSD?

Our argument in this chapter is much more closely associated with that alternative way of understanding the mediating role of moral and spiritual struggle in cases of PTSD. The latest research on PTSD and MI supports this conclusion and maintains that they are best understood as "separate syndromes" with a high potentiality for comorbidity [21, p. 2994]. Koenig et al. [19], for example, argue that the two syndromes, although overlapping, may be distinguished as follows:

PTSD is a trauma-based disorder characterized by a cluster of symptoms that include re-experiencing, hyperarousal, avoidance/numbing, and hypervigilance, while MI is a syndrome characterized by psychological and religious/spiritual symptoms of inner conflict (p. 250).

Similarly, Hall et al. [13] posit that PTSD can be distinguished from MI as a "danger- or fear-based disorder" inciting trauma, whereas MI, particularly self-directed MI, is most commonly attributed to the character wounds of "shame or guilt" as well as betrayal and the loss of hope, causing spiritual tension that corresponds to a violation of one's moral convictions (p. 93). Therefore, with the aforementioned research from our survey of the literature on PTSD and MI in mind, we have chosen to distinguish MI from PTSD by focusing in the previous section on the existential tension incurred by military personnel when assumptions about right and wrong are challenged during their service, creating moral dissonance and spiritual conflict about matters of conviction. To distinguish PTSD from MI, we focused in the PTSD section on the fear-based nature of PTSD as opposed to the conviction-based nature of MI. Now in what follows, we attempt to demonstrate how scholars such as Wortmann et al. [45] could better understand the mediating link between spiritual struggle and heightened PTSD symptoms by incorporating the emerging research on MI that supports this link while still maintaining the distinction between these two syndromes.

5 Toward a Better Understanding of the Spiritual Dimension of MI and Its Association with Heightened PTSD Symptoms

Despite the emerging link between the spiritual dynamics of MI and heightened PTSD symptoms, there remains a lack of consensus in the secondary literature regarding how one ought to understand the impact that spirituality and MI have on PTSD symptom severity. In addition to the problematic conflation of PTSD and MI, we suggest that this lack of consensus exists for two other important reasons. First, accounts of the whole soldier are only beginning to emphasize a spiritual dimension that is set apart from the body and mind but which has implications for both. Second, the literature on MI is only beginning to acknowledge the spiritual nature of the existential tension that has been historically associated with MI. We address each of these problems briefly in turn. Following our account of these two problems, we seek to clarify the mediating link between the spiritual struggle of MI and heightened PTSD symptom severity.

5.1 The Need to Emphasize the Spiritual Dimension of Soldiers

In recent years, the Military Health System [28, 29] has made considerable attempts to think holistically about former and active-duty service members' human nature as a way of promoting their wellbeing. They have done this via their Total Force Fitness (TFF) program. Yet, much of the focus of TFF still seems to emphasize how various phenomena impact soldiers' bodies and minds—with the vast majority of the attention being focused on psychological resilience. For this reason, Potts [37] argued that the military ought to revise its two-dimensional understanding of soldiers as inhabiting bodies and minds with a tri-dimensional understanding of soldiers as occupying a physical, mental, and spiritual dimension. Doing so, argues Potts [37], helps us to see and attend to the unique needs that make up soldiers' spiritual dimension, such as the need for "transcendent connection," "meaning/purpose," "moral values/beliefs," and "spiritual practices" like prayer and contemplation which help to mitigate symptoms associated with trauma and PMIEs. Potts [37] demonstrates that once the spiritual dimension is set apart from the mental and physical, it is also easier to see how this core dimension of every soldier impacts all of the other elements of TFF as outlined by the Military Health System [28]. We maintain that this need to emphasize the distinctiveness of soldiers' spiritual dimension and its significance for their spiritual health—not to mention their physical and psychological resilience—remains.

Others have also recently accounted for the frequent neglect of the spiritual dimension of our humanity within military research that attempts to understand the whole soldier. Pearce et al. [36] and Koenig and Al Zaben [21], for example, maintain that there is a spiritual dimension of every person which appears to be minimized or overshadowed by the psychological dimension in explanations of trauma, what mediates it, and how to mitigate it. We agree with these scholars that this minimal understanding of the spiritual dimension of soldiers helps to make sense of the diminished significance of spirituality in general as it pertains to MI research and common symptoms experienced after PMIEs, which we review in the next sub-section.

5.2 The Need to Acknowledge the Spiritual Implications of MI and PMIEs

Much like the need to acknowledge a spiritual dimension to every soldier, there remains a need to recognize what has been described as the "spiritual core" of MI syndrome [21, p. 2990]. While emergent MI research is beginning to reveal that there is a spiritual struggle that must be seen [5, 7, 11, 16, 22], however, earlier MI research often minimized or even excluded this so-called spiritual struggle [8, 26, 32]. Following Koenig and Al Zaben [21], we acknowledge the spiritual core of morally injurious events and assert its relevance in treating comorbid cases of MI and PTSD.

Another way of stating this problem is as follows. While it has been widely recognized that MI symptoms can be experienced in multiple directions (self- and other-directed), it is also crucial to acknowledge the ways that MI syndrome can impact multiple dimensions of the human person—not only the body and mind but also the spirit. Because the psychological dimension tends to overshadow the spiritual dimension in military and therapeutic research, most MI scale measurements have focused on the psychological impact of this syndrome and its concomitant impact on the human body (sleep problems, headaches, physical disability brought about by substance abuse, and so on). Until recently, minimal efforts have been made to understand the impact that MI has on the spiritual dimension of military personnel and what can be done to care for them in this regard.

The gold standard for assessing and measuring self- and other-directed symptoms of MI and the impact of both on all three dimensions of soldiers is the Moral Injury Symptom Scale—Military Version in both its long (45-item) and short (10-item) forms [19, 20]. We focus here on the novel findings that this scale reveals about the connection between MI and spiritual disintegration. In particular, recent research utilizing The Moral Injury Symptom Scale—Military Version demonstrates how MI often causes veterans and active-duty military personnel to struggle with and potentially even lose their religious or spiritual faith and values [19]. For example, following a PMIE, "some may conclude that they live in an immoral world or that they are immoral, irredeemable, and un-repairable" [19, p. 250]. Others who experience PMIEs may struggle to forgive or live with hope for the future [19, pp. 258–259]. PMIEs have especially been associated with a diminished sense of objective meaning in life and a loss of belief in God, as the MI stemming from these events often entails a "life-defining trauma that may challenge one's assumptions about oneself and the world" [18, p. 742]. Findings from this holistic scale measurement for assessing MI suggest that the spiritual underpinnings of this syndrome have implications for former and active duty service members who are coping with trauma.

Connecting The Spiritual Implications of MI to Heightened PTSD Symptom Severity

Where MI and PTSD coexist, Shay [40] indicates that "character wounds" tend to loom, and we argue that the significance of these wounds for one's spiritual well-being is striking. Research suggests that this is particularly acute in cases of MI that are caused by predominantly self-directed patterns of shame and guilt stemming from soldiers' perceived personal transgressions, although evidence also supports the conclusion that other-directed patterns of MI similarly prompt a spiritual struggle to hold onto hope, meaning, and even belief in God or others who are good, which can heighten fear-based cluster symptoms associated with PTSD [21].

While much focus has already been placed on the psychological struggles of PTSD and MI, there remains a need to understand some of the implications of the common spiritual struggles that tend to be associated with these comorbid syndromes. Important to mention here are the findings which suggest that the spiritual trauma associated with PMIEs and MI often exacerbates symptoms of PTSD, particularly

Table 3 The connection between MI, PTSD, and spiritual disintegration

Action →	Moral injury symptoms →	PTSD symptoms →	Spiritual disintegration
Self-directed PMIEs	Spiritual conviction about personal transgressions results in guilt, shame, and self-condemnation	Re-experiencing and ruminating about personal transgressions is associated with depression and the greatest risk for suicidal ideation severity	Soldiers often struggle with the loss of faith, loss of meaning or purpose in life, and exhibit reservations about reconnecting with God and/or spiritual community based on feelings of unworthiness
Other-directed PMIEs	Betrayals and other forms of unethical conduct on behalf of another violate individual spiritual codes, resulting in distrust, anger, and the struggle to forgive them	Ruminating on actions taken by another leads to hyperarousal and hypervigilance in settings where the individual harmed must place trust in others	Soldiers often struggle to hold onto hope, meaning, and faith in God or a belief in others who are good and trustworthy

when the spiritual component of veterans and active-duty service members' questions and concerns go unaddressed [18, 13]. In cases where spiritual trauma is combined with psychological trauma, Pearce et al. [36] indicate that "individuals with spiritual struggles report lower recovery rates from PTSD and a greater need for VA (Veterans Affairs)—approved mental health services" (p. 2). To demonstrate this, in Table 3 we build on earlier research conducted by Bryan et al. [6], Pearce et al. [36], and Kelley et al. [18] to delineate the bi-directional pattern of MI scenarios that tend to trigger PTSD, and we highlight at a high level what the existing research has found as it pertains to the spiritual disintegration associated with each pattern of PTSD that is mediated by either self-directed or other-directed morally injurious convictions.

Given the nature of MI syndrome, which research points to as a frequent mediator of PTSD symptoms and a trigger for greater PTSD severity, we have suggested that it would be prudent for the U.S. military to consider more carefully the spiritual needs of soldiers as outlined by Potts [37]. Moreover, we have argued that the emerging research on the spiritual underpinnings of MI will be helpful to attend to so that the military has a clearer conception of the common ways that PMIEs and MI syndrome trigger spiritual disintegration and prevent healing from traumatic events. Now, in the next section, we turn our attention to the only known spiritually integrated approach for treating comorbid cases of PTSD and MI, which was recently developed by a clinical psychologist, an active-duty military psychologist, a psychiatrist, and a VA chaplain [36]. Rather than reinventing the wheel, it is our hope that highlighting this important and nuanced form of cognitive processing therapy in what follows will prompt the U.S. military to use what has been developed by their own experts.

5.3 How to Alleviate Comorbid Cases of MI and PTSD Holistically

Emerging research on treating cases of PTSD that are mediated by MI suggests that the spiritual dynamics associated with PMIEs pose a significant barrier to recovery from PTSD. Perhaps this is why only about 20–30% of individuals suffering from these comorbid syndromes make a full recovery [36, p. 1]. Given the historically minimal understanding of the spiritual dimension and its implications for overall wellbeing within military culture, the most common forms of therapy that are utilized by the U.S. military to treat these comorbid syndromes have focused on psychologically-based treatments that are primarily functionalist as opposed to spiritual in their approach. Pearce et al. [36] explain this as follows:

Although moral injury is intimately connected with spiritual beliefs and values, these are typically not addressed in secular approaches, not to mention spiritual struggles and loss of religious faith due to trauma, which are typically not addressed at all. Instead, secular approaches focus on thinking errors, dysfunctional cognitions, erroneous underlying assumptions, and rational justifications, not on spiritual resources, spiritual struggles, and the spiritual ramifications of trauma and their interconnection with symptoms of PTSD. *In addition, among approaches that do address spirituality in the treatment of PTSD, moral injury is typically not addressed, particularly for individual treatment* (p. 2).

The clinical methodology that Pearce discusses here favors a primarily functionalist conception of the human person. Thus, the solution is a form of cognitive behavioral therapy that focuses on correcting beliefs and dysfunctional cognitions. Failure to account for the spiritual or existential dimension of a person results in a skewed clinical approach. It fails to acknowledge an individual's need to restore meaning and purpose following the trauma they have endured.

Secular approaches that fit the bill of Pearce et al.'s [36] critique above include all four trauma-focused talk therapy treatments for PTSD that the U.S. Department of Veterans Affairs [42] includes on its mental health awareness website. We list the VA's therapeutic treatments for only PTSD in Table 4—since these appear to be the only available options for individuals facing comorbid syndromes of PTSD and MI combined (in addition to medication):

Indeed, upon reviewing the list of psychotherapeutic treatment options available to veterans and active-duty service members, as they are reflected in Table 4, it seems evident that [36] are correct to assert that neither spirituality nor MI is typically addressed in the treatment of PTSD. Thus, these researchers appropriately conclude that "there is a need for empirically based individual treatments for PTSD that target MI that make explicit use of a patient's spiritual resources, particularly given the evidence that such resources predict faster resolution of PTSD" [36, p. 2]. In what follows, therefore, we reaffirm Pearce et al. [36] conclusion and highlight their revisionist approach to the treatment of comorbid cases of PTSD and MI, which they refer to as "Spiritually Integrated Cognitive Processing Therapy" (SICPT).

Table 4 PTSD treatment options available to veterans and active duty soldiers[3]

3–4 month therapies for PTSD recommended by the VA	Description of treatment
Cognitive processing therapy (CPT)	CPT teaches you how change the upsetting thoughts and feelings you have had since your trauma
Prolonged exposure for PTSD (PE)	PE teaches you to gradually approach trauma-related memories, feelings, and situations that you have been avoiding since your trauma
Eye movement desensitization and reprocessing (EMDR)	EMDR helps you process and make sense of your trauma while paying attention to a back-and-forth movement or sound (like a finger waving side to side, a light, or a tone)
Written exposure therapy (WET)	WET helps you find new ways to think about your trauma and its meaning through writing assignments you complete during sessions. This is a brief, 5-session therapy

As Pearce et al. [36] describe, SICPT differs from CPT in five critical ways. First, it directly targets MI in the treatment of PTSD. Second, it specifically targets MI, "challenging erroneous interpretations of trauma by focusing on cognitive restructuring using clients' spiritual/religious resources (i.e., spiritual beliefs, practices, sacred writings, values, and motivations) to challenge maladaptive thinking patterns" [36, p. 2]. Third, it acknowledges the need for moral repair and seeks to establish that repair through, "the spiritual concepts and rituals of compassion, grace, spiritual guided imagery, repentance, confession, forgiveness, atonement, blessing, restitution, and making amends," [36, p. 3]. Fourth, it encourages patients to seek support through immersion in a spiritual community. Finally, it normalizes spiritual struggles—loss of faith in God, anger with God, shame for sin, etc. [36, p. 3]. Ultimately, SICPT presents a radically different way of understanding the human being as a body-mind-soul composite. In taking MI seriously, it directly addresses the underlying source of dysfunctional narratives that couch a patient's experience of trauma. It does all of this without overly-spiritualizing trauma or dismissing the influence of psychopathology.

Moreover, these scholars have created tradition-specific manuals for utilizing SICPT, which include one for the spiritual but not religious, as well as manuals for Christians, Muslims, Buddhists, Jews, and Hindus. We want to affirm the treatment of comorbid cases of MI and PTSD using tradition-specific manuals of treatment whenever possible. Again, these tradition-specific manuals acknowledge the incredibly unique position of each person who encounters trauma. Truly, these manuals are a refreshing contrast to what we call the 'treatment-in-a-vacuum' clinical orientation. Instead of treating patients as mechanistic responders to objective traumatic events,

[3] The PTSD therapies and descriptions in Table 3 are reflected exactly as they appear on the U.S. Department of Veterans Affairs [42] website.

it positions them as socially and dialogically informed—members of communities with unique spiritual orientations and beliefs.

It is important to note that caring for the spiritual dimension of patients who are suffering from the combined syndromes of MI and PTSD requires healthcare practitioners to realize that "spirituality has the potential to be a positive and protective resource or an exacerbating factor for PTSD or both" [36, p. 2]. In either case, however, it should be noted that our suggestion to care for the religious and spiritual dimension of veterans and active-duty military personnel who suffer from the combined syndromes of MI and PTSD does not run counter to their own wishes. As Koenig et al. [19] indicate—when asked whether they would be open to receiving care for the spiritual dimension of their struggles—"over 70% [of 427 veterans and active-duty military] participants surveyed responded in the affirmative" (p. 263).

That being said, for patients who do not identify as spiritual or religious, Pearce et al. [36] believe attending to the spiritual matters of MI as a way of alleviating PTSD symptoms will not work. However, we diverge from Pearce et al. [36] in this regard. Though arguing for a very particular use and application of the word 'spiritual' is outside the scope of this chapter, we contend that a broad, pluralistic understanding of the term is appropriate here for the identification of what may be considered a dimension that houses universal human concerns. Spiritual matters can include "transcendent connection," "meaning and purpose," "moral values/beliefs" and "spiritual practices" [37]. Thus, while a traditional conception of 'spiritual' is often associated with organized religion or individual beliefs about the transcendent realm, we should use the term more liberally in this context to describe something common to all soldiers on the basis of their human nature. Applying any narrower of a scope to the understanding of 'spiritual' casts aside the reality that there are deep, existential needs present in those who do not identify as spiritual or religious—needs that are readily on display as they suffer from instances of MI. Thus, it would also be beneficial to develop an agnostic SICPT manual for those who neither identify as spiritual nor affiliate with a particular religious tradition.

6 Suggestions for Future MI and PTSD Prevention

Though there is much cause for hope given the increase in spiritually informed treatments of PTSD like SICPT, there is still much work necessary to create spiritually informed prevention modes. This is curious given the research suggesting that ideological commitments and spiritual praxis are major factors contributing to an individual's prospect of recovery and response to PTSD treatment. Additionally, research demonstrating that PTSD symptoms where MI looms are likely to be more severe would also seem to inspire deeper discovery into the prevention of MI—particularly self-directed MI. However, there is little to no research on potential preventive measures. Though a careful consideration of all potential spiritually-informed preventive measures is impossible given the restraints of time and space, we would like to suggest that one way to systematically decrease the severity of PTSD symptoms among military personnel would be to take intentional measures to reduce instances of MI. In what follows, we will detail what preventative MI measures may entail.

6.1 Preparing for Emotional Battles

When a soldier trains in boot camp, they learn how to push themselves to extreme limits, tackling impressive feats of physical and mental fortitude. These practices are in anticipation of what they will face in eventual deployment. They may have to function and make prudent decisions with little to no sleep, while malnourished, and in the face of intense fear. They need to be physically and mentally fit as they encounter these and other significant challenges and difficulties. They train accordingly. Research on MI and PTSD over the last decade has made it apparent that most if not virtually all deployed soldiers will experience PMIEs. The military should address this and attempt to prepare soldiers for that reality in the same way it addresses the physical and mental stressors of deployment—intentionally. What would it look like to prepare and fortify soldiers for their eventual encounter with PMIEs? We have two suggestions: (1) establishing existential security and (2) training in emotional regulation and cognitive framing.

We suggest that establishing existential security contains two parts. First, it requires addressing big questions on the meaning of life, purpose (especially in the context of military conquests), and the systematic exploration of ethical questions beyond the scope of compliance. We contend that failure to confront these fundamental existential questions as central to the human experience compounds the existential tension encountered in trauma and could increase the risk of self-directed MI. Equipping soldiers with a vocabulary to describe their experiences related to life's big questions ensures they are not blindsided by PMIEs during deployment or combat and forced to reconcile their values with the reality of war for the first time. This gives soldiers a voice in naming and explaining the horrors of war *qua* human beings.

The second aspect of establishing existential security involves encouraging and promoting the practice of healthy spiritual exercises. Just as SICPT involves integrating the spiritual praxis as a mode of treating PTSD compounded by MI, we suggest taking up spiritual praxis as a form of strength training for combat and deployment. Examples here include participation in faith-based community groups, prayer, meditation, devotionals, rituals, engagement with sacred writings, yoga, and so on. We obviously do not propose the endorsement of any particular religious tradition here but do suggest the systematic encouragement of religious praxis more broadly, given the positive impact of regular spiritual praxis in healing PTSD. Further consideration of what shape this should take beyond the placement of military chaplains and offering of religious services would be an essential next step for future research to consider.

Now, let us consider our second suggestion—training in emotional regulation and cognitive framing. Our perspective here necessarily follows from our exploration of successful PTSD and MI treatment. If effective healing focuses on repairing dysfunctional cognition of traumatic events, it would make sense to prime soldiers to accept and frame their experiences in healthy ways *before* they happen, thereby better equipping them to narrate their experiences when they do happen. For example, if soldiers were to confront predispositions to shame or avoidance prior to exposure to a PMIE, we suspect it would significantly diminish the likelihood of self-inflicted MI occurrences. Efforts to incorporate training of this sort should not be considered ancillary but necessary given the high probability of exposure to PMIEs that soldiers face. The specific content, structure, and practical implementation mark additional paths for future research.

7 Conclusion

This chapter has covered substantial ground regarding the history, causes, healing, and prevention of PTSD and MI. Our research strongly suggests a clear delineation between PTSD and MI to heal patients suffering from these comorbid syndromes more fully and holistically—by considering their physical, psychological, and spiritual needs. We argue that recognizing the spiritual dimension of MI is an important step in promoting the healing of patients who suffer from severe PTSD symptoms and advocate using SICPT as a promising treatment method that observes the unique spiritual struggles of soldiers and how they can help or hinder response to treatment. Finally, we propose the prevention of MI through establishing existential security prior to deployment and the systemic implementation of training in emotional regulation and cognitive framing. We suggest these preventive measures to better equip soldiers for PMIEs and to help them move through these experiences with an attenuated resilience. In the end, it is our hope that this research, which largely builds on concerns outlined by Potts [37], will prompt further reflection on care for the whole soldier and lead to revisionary insights that can enhance the U.S. military's TFF program.

References

1. American Psychiatric Association (2013) Diagnostic and statistical manual of mental disorders, 5th edn. American Psychiatric Association Publishing, Washington. https://doi.org/10.1176/appi.books.9780890425596
2. Blucher R (2003) Jonathan Shay's Achilles in Vietnam and Odysseus in America. J Mil Ethics 2(2):160–167. https://doi.org/10.1080/1502750310000153
3. Bond FW, Hayes SC, Baer RA, Carpenter KM, Guenole N, Orcutt HK, Zettle RD (2011) Preliminary psychometric properties of the acceptance and action questionnaire-II: a revised measure of psychological inflexibility and experiential avoidance. Behav Ther 42:676–688. https://doi.org/10.1016/j.beth.2011.03.007
4. Brock RN, Lettini G (2012) Soul repair: recovering from moral injury after war. Beacon Press, Boston
5. Bravo A, Kelley ML, Mason R, Ehlke SJ, Vinci C, Redman JC (2020) Rumination as a mediator of the associations between moral injury and mental health problems in combat-wounded veterans. Traumatol (Tallahassee, Fla.) 26(1):52–60. https://doi.org/10.1037/trm0000198
6. Bryan AO, Bryan CJ, Morrow CE, Etienne N, Ray-Sannerud B (2014) Moral injury, suicidal ideation, and suicide attempts in a military sample. Traumatol (Tallahassee, Fla.) 20(3):154–160. https://doi.org/10.1037/h0099852
7. Carey L, Hodgson TJ (2018) Chaplaincy, spiritual care and moral injury: considerations regarding screening and treatment. Front Psych 9:619–619. https://doi.org/10.3389/fpsyt.2018.00619
8. Currier JM, Holland JM, Drescher KD (2015) Spirituality factors in the prediction of outcomes of PTSD treatment for U.S. Military Veterans. J Trauma Stress 28(1):57–64. https://doi.org/10.1002/jts.21978
9. Da Costa JM (1871) On irritable heart: a clinical study of a form of functional cardiac disorder and its consequences. Am J Med Sci 61:17–52
10. Forkus SR, Breines JG, Weiss NH (2019) Morally injurious experiences and mental health: the moderating role of self-compassion. Psychol Trauma 11(6):630–638. https://doi.org/10.1037/tra0000446
11. Frankfurt S, Frazier P (2016) A review of research on moral injury in combat veterans. Mil Psychol 28:318–330. https://doi.org/10.1037/mil0000132
12. Garcia V (2011) PTSD in the military: history, attitudes, and treatment—argument for a paradigm shift. ProQuest Dissertations Publishing
13. Hall NA, Everson AT, Billingsley MR, Miller MB (2022) Moral injury, mental health and behavioural health outcomes: a systematic review of the literature. Clin Psychol Psychother 29(1):92–110. https://doi.org/10.1002/cpp.2607
14. Herman JL (1997) Trauma and recovery: the aftermath of violence—from domestic abuse to political terror. Basic Books, New York
15. Hiraoka R, Meyer EC, Kimbrel NA, DeBeer BB, Gulliver SB, Morissette SB (2015) Self-compassion as a prospective predictor of PTSD symptom severity among Trauma-exposed U.S. Iraq and Afghanistan war veterans. J Trauma Stress 28(2):127–133. https://doi.org/10.1002/jts.21995
16. Jinkerson JD (2016) Defining and assessing moral injury: a syndrome perspective. Traumatology 22(2):122–130. https://doi.org/10.1037/trm0000069
17. Kardiner A (1941) The traumatic neuroses of war. Hoeber, New York
18. Kelley ML, Chae JW, Bravo AJ, Milam AL, Agha E, Gaylord SA, Vinci C, Currier JM (2021) Own soul's warning: moral injury, suicidal ideation, and meaning in life. Psychol Trauma 13(7):740–748. https://doi.org/10.1037/tra0001047
19. Koenig H, Ames D, Youssef NA, Oliver JP, Volk F, Teng EJ, Haynes K, Erickson ZD, Arnold I, O'Garo K, Pearce M (2018) The Moral injury symptom scale-military version. J Relig Health 57(1):249–265. https://doi.org/10.1007/s10943-017-0531-9

20. Koenig H, Ames D, Youssef NA, Oliver JP, Volk F, Teng EJ, Haynes K, Erickson ZD, Arnold I, O'Garo K, Pearce M (2018) Screening for moral injury: the moral injury symptom scale—military version short form. Mil Med 183(11–12):E659–E665. https://doi.org/10.1093/milmed/usy017
21. Koenig H, Al Zaben F (2021) Moral injury: an increasingly recognized and widespread syndrome. J Relig Health 60(5):2989–3011. https://doi.org/10.1007/s10943-021-01328-0
22. Kopacz MS, Connery AL, Bishop TM, Bryan CJ, Drescher KD, Currier JM, Pigeon WR (2015) Moral injury: a new challenge for complementary and alternative medicine. Complement Ther Med 24:29–33. https://doi.org/10.1016/j.ctim.2015.11.003
23. Lancaster SL, Erbes CR (2017) Importance of moral appraisals in military veterans. Traumatol (Tallahassee, Fla.) 23(4):317–322. https://doi.org/10.1037/trm0000123
24. Lancaster SL, Irene Harris J (2018) Measures of morally injurious experiences: a quantitative comparison. Psychiatry Res 264:15–19. https://doi.org/10.1016/j.psychres.2018.03.057
25. Lasiuk GC, Hegadoren KM (2006) Posttraumatic stress disorder part I: historical development of the concept. Perspect Psychiatr Care 42(1):13–20. https://doi.org/10.1111/j.1744-6163.2006.00045.x
26. Litz BT, Stein N, Delaney E, Lebowitz L, Nash WP, Silva C, Maguen S (2009) Moral injury and moral repair in war veterans: a preliminary model and intervention strategy. Clin Psychol Rev 29(8):695–706. https://doi.org/10.1016/j.cpr.2009.07.003
27. Meyer EC, La Bash H, DeBeer BB, Kimbrel NA, Gulliver SB, Morissette SB (2019) Psychological inflexibility predicts PTSD symptom severity in war veterans after accounting for established PTSD risk factors and personality. Psychol Trauma 11(4):383–390. https://doi.org/10.1037/tra0000358
28. Military Health System (2020a) Total force fitness. https://www.health.mil/News/In-the-Spotlight/Total-Force-Fitness. Accessed 27 Nov 2020
29. Military Health System (2020b) Joint chiefs say mind, body, spirit all part of total force fitness. https://okinawa.stripes.com/community-news/joint-chiefs-say-mind-body-spirit-all-part-total-force-fitness. Accessed 27 Nov 2020
30. Myers ABR (1870) On the etiology and prevalence of diseases of the heart among soldiers. J. Churchill, London
31. Myers CS (1915) A contribution to the study of shell shock. Lancet 188:316–320
32. Nash WP, Marino Carper TL, Mills MA, Au T, Goldsmith A, Litz BT (2013) Psychometric evaluation of the moral injury events scale. Mil Med 178(6):646–652. https://doi.org/10.7205/MILMED-D-13-00017
33. Oren L, Possick C (2010) Is ideology a risk factor for PTSD symptom severity among Israeli political evacuees? Ideology as a risk factor for PTSD. J Trauma Stress 23(4):483–490. https://doi.org/10.1002/jts.20540
34. Park CL, Currier JM, Harris JI, Slattery JM (2017) The centrality of meaning in human lives. In: Park CL, Currier JM, Harris JI, Slattery JM (eds) Trauma, meaning, and spirituality: Translating research into clinical practice. American Psychological Association, Washington, pp 15–38. https://doi-org.ezproxy.lib.usf.edu/10.1037/15961-002
35. Park CL, Currier JM, Harris JI, Slattery JM (2017) Trauma, meaning, and spirituality: translating research into clinical practice, 1st edn. American Psychological Association, Washington
36. Pearce M, Haynes K, Rivera NR, Koenig HG (2018) Spiritually integrated cognitive processing therapy: a new treatment for post-traumatic stress disorder that targets moral injury. Glob Adv Health Med 7.https://doi.org/10.1177/2164956118759939
37. Potts GW (2022) Caring for the whole soldier in the wake of COVID-19. In: Farhadi A, Masys A (eds) The great power competition, vol 2. Springer, Cham. https://doi.org/10.1007/978-3-030-90390-9_15
38. Ramchand R, Schell TL, Jaycox LH, Tanielian T (2011) Epidemiology of trauma events and mental health outcomes among service members deployed to Iraq and Afghanistan. In: Ruzek JI, Schnurr PP, Vasterling JJ, Friedman MJ (eds) Caring for veterans with deployment-related stress disorders. American Psychological Association, Washington, pp 13–34. https://doi-org.ezproxy.lib.usf.edu/10.1037/12323-001

39. Ruzek J (2011) Caring for veterans with deployment-related stress disorders: Iraq, Afghanistan, and beyond, 1st edn. American Psychological Association, Washington
40. Shay J (2003) Achilles in Vietnam: combat trauma and the undoing of character. Scribner Press, New York
41. Shay J (2014) Moral injury. Psychoanal Psychol 31(2):182–191. https://doi.org/10.1037/a0036090
42. U.S. Department of Veterans Affairs (2022) PTSD: National Center for PTSD. https://www.ptsd.va.gov/understand_tx/talk_therapy.asp. Accessed 31 May 2022
43. Van der Kolk B (2014) The body keeps the score: brain, mind, and body in the healing of trauma. Penguin soldier, New York
44. Van der Kolk BA, Weiseth L, Van der Hart O (1996) History of trauma in psychiatry. In: Van Der Kolk BA, McFarlane AC, Weisaeth L (eds) Traumatic stress: the effects of overwhelming experience on mind, body, and society. Guilford Press, New York, pp 47–74
45. Wortmann JH, Park CL, Edmondson DE (2011) Trauma and PTSD symptoms: Does spiritual struggle mediate the link? 3(4): 442–452. https://doi.org/10.1037/a0021413

Dr. Garrett W. Potts is an Assistant Professor of Religious Studies at USF. His work focuses on religio-cultural competence, particularly in the areas of Health, Business, and Religion. Dr. Potts conducts research and teaches multiple courses around these topics. Additionally, Dr. Potts has published scholarship on moral injury, servant leadership, social capital, and the pursuit of work as a calling.

Dr. Lily M. Abadal has a Ph.D. in Philosophy and Religious Studies and is a Visiting Professor of Instruction at the University of South Florida. She has a particular interest in moral psychology and applied virtue ethics, especially as it relates to the health care professions.

Afghan Exit Punctuates a Pattern of Broken Promises and Instability in American Foreign Policy

Leif Rosenberger

Abstract The U.S. exit from Afghanistan punctuates a pattern of broken promises and subsequent instability in American foreign policy. Failures in U.S. decision-making have damaged the credibility and resolve of U.S. foreign policy to stay the course. This research uses case studies to translate theories into practice. The study argues that U.S. foreign policy keeps failing because its decision-making process is seriously flawed. Academe provides valuable theoretical insights. At times U.S. bureaucratic infighting results in polarization. At other times groupthink produces conformity and flawed outcomes. Other scholars talk about cognitive theory and the conceptual limitations and errors of leaders. In Afghanistan, U.S. economic exit from the New Silk Road gave China an open playing field to use its Belt and Road to promote its economic power and influence. The U.S. military exit from Afghanistan triggered a humanitarian crisis and economic collapse. This pattern of broken promises and instability is also evident in the mini-case studies on Syria, Iraq, Iran, and Ukraine.

Keywords U.S. · China · Russia · Afghanistan · Syria · Iraq · Iran · Yemen · Ukraine · New Silk Road · Belt and Road · Stabilization · Reconstruction · Territorial integrity · Budapest Memorandum · Decision-making theories · Leadership · Psychological models · Cognitive dissonance · Sociological models · Rational actor · Organizational models · Bureaucratic models · Domestic politics · Groupthink · Credibility · Resolve · Misperceptions · Miscalculations

1 Introduction

In April 2021 President Joe Biden announced that the US military mission in Afghanistan would conclude on 31 August 2021. President Biden made it seem like a simple decision. He said the US military had been in the country for 20 years and

L. Rosenberger (✉)
Army War College, U.S. Peacekeeping and Stability Operations Institute (PKSOI), Carlisle, PA 17013-5234, USA
e-mail: leifrosy@yahoo.com

A. Farhadi and A. Masys (eds.), *The Great Power Competition Volume 4*,
https://doi.org/10.1007/978-3-031-22934-3_18

had achieved all its objectives for the original intervention. Therefore, the president said it was time to leave.

But this decision was more about domestic politics and wishful thinking than thoughtful grand strategy. While this approach to cut and run from Afghanistan may strengthen President Biden's political position at home with the left wing of the Democratic Party, this ill-advised decision threatens U.S. national interests. The U.S. military exit from Afghanistan has predictably triggered a humanitarian crisis, an economic collapse, and a strategic safe-haven for Al-Qaeda and ISIS to plan terrorist attacks on America. President Biden says the U.S. can solve these Afghan problems with a military presence that is "over the horizon" in Central and South Asia. But as former Secretary of Defense Robert Gates would say, over the horizon is really "over the rainbow."

While President Biden claims these negative events were unintentional consequences, his military advisors warned the president about this likely scenario and therefore advised him to keep a small U.S. military presence in Afghanistan. But cut and run is not new to President Biden.

Ten years ago, the Obama-Biden team waived the economic white flag and opted for a U.S. economic exit from the New Silk Road. That ill-advised decision gave China an open playing field in the "great power competition with China [1]. President Biden now concedes this economic exit was a mistake. In fact, he recently asked the G7 and the EU to revive the U.S. Silk Road Initiative. The EU is open to the idea. But they say, "If the US now thinks the New Silk Road was a good idea and did all the feasibility studies and due diligence, why is the U.S. now a de facto bystander?"

While the negative outcomes of these American foreign policy mistakes are well known, what is not so clear to many observers is why this keeps happening. This study explains how flaws in the national security decision-making process leads to a pattern of broken promises in American foreign policy. Adversaries like China and Russia misperceive this pattern and miscalculate their responses by expecting the U.S. to be a paper tiger. This creates instability, over-reaction and increases the risks of crisis and war. To foster more stability and strategic deterrence vis a vis China and Russia, the Biden administration needs to enhance its credibility by showing more resolve. In other words, the U.S. needs to send clearer signals to Russia and China via a pattern of staying the course, militarily and economically.

2 Methodology

While only presidents themselves know for sure what the real reason was why they gave the go ahead for so many foreign policy debacles, scholars have created decision-making models to reconstruct the logical reasoning processes of U.S. presidents and their inner circle. Graham Allison created the rational actor model in his classic study of the Cuban missile crisis [2].

When we try to explain why President Biden acted the way he did, we simplify the process by asking ourselves why "the U.S." decides to send all the troops home from Afghanistan. We imagine the US speaks with one voice and pursues logical goals or objectives. Using this model, the U.S. has perfect intelligence, makes a list of all possible options, weighs up the costs and benefits of all the options and selects that alternative which delivers the greatest benefits relative to cost [3].

But of course, the rational actor model is a convenient decision-making abstraction. In fact, organizational theory says that different organizations are sometimes involved in bureaucratic infighting to get what they want. Organizations can be rigid and resist change. For instance, people in the organizational cultures of the US State Department (STATE) and the U.S. Department of Defense (DOD) sometimes disagree with each other so much that the president struggles to reconcile their differences. In short, when this occurs, there is too much polarization and not enough common ground.

At other times decisions are made in small groups. In this sense, sociological decision-making models also matter. While some small groups try to be even handed and follow the rational actor model, too often social pressures induce conformity and "groupthink." Those dissenters with different perspectives from the groupthink are often excluded from the inner circle. Group members say these dissenters are "not one of us." [4]. For instance, U.S. Secretary of State Cyrus Vance complained that he was excluded from the meeting at which President Jimmy's Carter's gave the go-ahead for an ill-conceived military operation on 24 April 1980 to free American hostages in the Iran Hostage Crisis. Vance claims he was excluded because he was a dissenter and now part of the hawkish groupthink whose ill-advised plan was bound to be a debacle [3: 20–21].

Finally, cognitive decision-making models argue that individuals matter. Even the most intelligent human beings have cognitive limitations. To compensate for their cognitive limitations, they often rely on cognitive shortcuts (such as analogies). For instance, Paul Bremer compared the Iraqi Baath Party and the Iraqi military to Nazi Germany. As a result, he made an ill-advised decision to fire the lion share of these Iraqis. This massive unemployment turned into a full-blown anti-US insurgency. The takeaway is belief systems are remarkably resistant to change. That in turn can cause misperception and miscalculation. Personality and emotions also affect decision-making.

While all the decision-making models cited above provide some useful insights in "decoding" why President Biden decided to withdraw all the US military troops from Afghanistan by the end of August 2021, the cognitive models provide numerous cognitive errors which are especially relevant to President Biden's cognitive style. These include decision-making on impulse [5]. And Leon Festinger talks about the desire of individuals to maintain cognitive consistency and wishful thinking in the face of psychologically uncomfortable new information (or cognitive dissonance).

Robert Jervis' research examines the ways political leaders commonly misinterpret the signals that other leaders intend to send [6]. Jervis would say U.S. leaders like Bremer misperceive the situation in Iraq and miscalculate how to respond. We will also see how Russian President Putin is vulnerable to misperception induced failure to empathize [6]. Ralph White's classic work on empathy is also important here. White would say Bremer failed to put himself in the shoes of the Iraqis in order to better understand their motives [7, 8].

In contrast, General Dave Petraeus had more empathy. He could see that the lion share of Iraqi Baath Party members and Iraqi military officers were not "Nazis." Doctors and lawyers joined these organizations because that was the only path toward professional advancement. But they were not ideological. If given a second chance, they would be trustworthy Iraqi citizens in other postwar institutions.

Finally, the Palace Politics model shifts the focus from players in formal policy positions to players in proximity who attempt to influence or persuade a president or another primary decision maker. It therefore matters who in the White House or outside the government are whispering in the president's ear. They all jockey for access and try to get as close as possible to political power [9: 192–237]. In short, the palace politics model is like the three most important things in real estate: location, location, location.

3 Case Studies

In David Patrick Houghton's book called The Decision Point, the author applies these kinds of theories to six case studies in U.S. foreign policy decision-making. Similarly, in this study we will briefly apply these kinds of decision-making theories to mini-case studies in U.S. foreign policy decision-making. In addition to a deep dive into the US economic and military exits from Afghanistan, the study also looks at US decision-making debacles in Syria, Iraq, Iran, and Ukraine.

Finally, we will explain how all five case studies illustrate the broader theme (or thesis) that the credibility and resolve of U.S. foreign policy to stay the course with friends and foes has been damaged. When this occurs, friends are tempted to go their own way and to look elsewhere for more reliable and trustworthy partners. As cited earlier, adversaries like China and Russia misperceive and miscalculate U.S. intensions and conclude the U.S. is a paper tiger. Instead of the U.S. deterring Chinese and Russian aggression in great power competition, U.S. fecklessness invites aggression from these adversaries. When President Biden changes this U.S. behavior pattern and reacts more forcibly to Russian aggression in Ukraine, friends and foes are caught off guard by this erratic behavior.

President Biden then must thread the needle between strong enough conventional military assistance to Ukraine to reverse Russian aggression but not enough military power to trigger World War III and a possible nuclear holocaust. A U.S. strategy that would foster more peace and stability in the world would be one which deters Russian and Chinese aggression with stronger credibility and resolve to stay the course with friends and foes.

4 U.S. Economic and Military Exit from Afghanistan

In the years ahead, historians may well ask: How was China able to turn the tables on America? About a decade earlier, grand strategists had been bullish on America's New Silk Road strategy and plans. That perception was punctuated by US Secretary of State Hillary Clinton's "New Silk Road" speech on 20 July 2011 in Chennai, India. 2 [10].

Back then, China was on the sidelines. In contrast, General Dave Petraeus, and his interagency task force at the US Central Command (CENTCOM) were front and center, creating an inclusive New Silk Road strategy with plans for infrastructure that would go through Afghanistan and turn enemies into friends and aid into trade. All of this was intended to promote shared prosperity and collective security.

Origins of the Economic White Flag

But behind the scenes policy was starting to change. President Obama made a decision for General Petraeus to leave his coveted position as the CENTCOM Commander (responsible for Central and South Asia) and become the commander of the International Security Assistance Force (ISAF) in July 2010. General James Mattis replaced Petraeus and was the CENTCOM Commander from August 2010 until March of 2013. By 2013 Mattis had zeroed out the funding for the New Silk Road task force at CENTCOM. U.S. economic exit from the New Silk Road gave China an open playing field to promote its economic power and influence.

China turned the tables on America by filling this gap with its own version of the New Silk Road which it called the Belt and Road. In contrast to America surrendering its New Silk Road strategy, Beijing was serious. China committed itself to trillions of dollars for its version of the New Silk Road. To finance this commitment, Beijing created a New Silk Road Bank and an Asian Infrastructure Investment Bank.

How was China so successful in turning the tables on America? And why is America so reluctant to implement its version of the New Silk Road plan? The short answer is China understands how to connect economics and security in its foreign policy. In contrast, American foreign policy keeps economics and security in separate silos.

Back in the fifth century BC, Chinese General Sun Tzu wrote in the Art of War, "The supreme art of war is to subdue the enemy without firing a shot." [11] That advice still resonates in Beijing. In fact, China is competing and winning the geo-economics because the US waived the economic white flag. For instance, in 2020 China overtook the U.S. as world's leading destination for foreign direct investment. In 2020 foreign companies were optimistic about China's future domestic demand prospects. In contrast, foreign investment flows into America nearly halved in 2020.

Former Secretary of Defense Bob Gates argues that China has a strong advantage over the US. He notes that China has a more flexible system that enables it to use state-run banks and state-run enterprises more effectively than the U.S. can. For example, Gates says that nothing the US is doing compares to the scale and impact of China's Belt and Road. The World Economic Forum notes that China's Belt and

Road could see up to $8 trillion invested across 68 countries in the next two decades [1].

The state-run Belt and Road is a huge transcontinental public investment project. It includes a global network of infrastructure, railroads, highways, airports, bridges, ports, oil, and gas pipelines. President Xi calls the Belt and Road China's top foreign policy priority.

If building infrastructure is a good way for the U.S. to promote influence overseas, why is the US not more competitive with China in building infrastructure? After all, the US has impressive investment banks and amazing entrepreneurs in Silicon Valley. Gates says the problem is Silicon Valley and Wall Street are in the private sector. And the US presidents struggle to mobilize the private sector to compete overseas against China when it comes building infrastructure.

Some say the US Government's International Development Finance Corporation (or USIDFC) is a good start. This USIDFC finances private development projects with $60 Billion in capital to loan. But $60 Billion is a drop in the bucket compared China's $8 trillion.

Bob Gates criticizes the U.S. for offering China what he calls "open field running" in its pursuit of BRI. Gates says the big bureaucracies offer few incentives for people at lower levels to challenge conventional wisdom. The process is nearly always lethal for the bold idea (and sometimes for the career of the originator). Those words resonate with members CENTCOM's inter-agency task force on the New Silk Road a decade later.

5 The Rise and Fall of U.S. New Silk Road

Once every blue moon and almost by mistake someone like a George Marshall or a Dave Petraeus somehow come along who can connect the economic and military dots at a high level. Unfortunately, the economic/security connection is not institutionalized. So, when Dave Petraeus leaves the picture, foreign policy problems once again become militarized, and the U.S. looks for hammers to pound nails. In short, too many American strategists don't understand the importance of connecting economics and security.

Take Afghanistan for instance. President Obama announces in 2011 that the lion share of the 150,000 U.S. troops will come home, only leaving about 10–15% of the troops behind. A simultaneous and proportional 10–15% reduction of military contracts would also happen. U.S. military doctrine says civilians will take the lead in follow on stabilization and reconstruction operations, as the U.S. did successfully in Germany after World War II. But there is now a widening gap between the scale down of combat operations and the start of stabilization and reconstruction operations.

Bad things happen in this gap. So, when President Obama announced the U.S. drawdown in Afghanistan, the U.S. Treasury did a study that predicted how much of a negative impact would occur to the overall Afghan economy once 85–90% of the war economy went away. The best-case scenario would be a 13% cut in Afghan

GDP. That equates to the U.S. Great Depression. The worst-case scenario would be a 41% cut in Afghan GDP. Either way, jobless numbers soar.

So, what happens militarily if nothing is done to fill this gap between combat power and stabilization and reconstruction? In East Timor, violence rose when the United Nations (UN) peacekeepers played cut and run [12].

In Iraq, we saw the rise of ISIS after the US played cut and run, thus leaving the country without making any serious stabilization and reconstruction efforts [13]. But, in Iraq, at least, oil production is not located near war zones, even if low oil prices back in 2014 reduced profit margins.

In Afghanistan, the good news is that there is lots of potential wealth. In fact, the US Geological Survey says that the country has a trillion dollars of potential mineral wealth [14]. Unfortunately, Afghanistan is struggling to turn this potential wealth into actual wealth. Former MIT Professor Walt Rostow would say that it lacks preconditions for economic takeoff [15].

If land-locked Afghanistan had good infrastructure, it could still have market access. But Afghan infrastructure has been largely destroyed because of continuous wars. In this regard, only 7% of the roads are paved. Therefore, this inadequate infrastructure equates to poor market access. When it rains, the roads turn to mud or flood. So, there is little incentive to increase production.

Therefore, Afghanistan faced a double whammy. On the cyclical side, the war economy was collapsing. On the structural side, it struggled with inadequate infrastructure and poor market access. With these combined, Afghanistan faced depression economics. Therefore, it needed a Keynesian economic strategy to fill the gap in aggregate demand (or at least to soften the blow).

Former Johns Hopkins Professor Fred Starr has advocated a New Silk Road for decades. His concept is transcontinental in scope and runs from China to Europe via Central and South Asia [16]. But the U.S. government had not pushed very hard for it in the past. If the U.S. government was going to support it, conventional wisdom says the U.S. State Department is the logical place to do it. And the Policy Planning Council at STATE is the logical office at STATE to develop policy. Unfortunately, the folks at the Policy Planning Council at STATE wanted no part of it. Richard Holbrooke, the Special Representative for Afghanistan and Pakistan (SRAP), wanted no part of it either.

Since nobody else in Washington wanted to develop policy for a New Silk Road either, General Petraeus signed off on CENTCOM doing policy formulation for a New Silk Road to close the gap in aggregate demand. A US Inter-agency Task Force was created at CENTCOM to plan the New Silk Road. The CENTCOM task force developed about 20 hard and soft infrastructure projects.

The hard infrastructure included transport, mining, energy, and telecom projects.

Transport included completing the Afghan Ring Road completion, the Afghan North–South Road Corridor, the Afghan East–West Road Corridor, the Kabul-Jalalabad-Peshawar Expressway, finishing the Salang Tunnel, the Northern Rail Corridor and commercial aviation.

Mining included the Aynak Copper Mine and the Hajigak Iron Ore Mine.

Energy included Turkmenistan, Afghanistan, Pakistan, and India (TAPI) gas pipeline, the Central Asia-South Asia Electricity Project (CASA 1000) and the Sheberghan Gas Fired Thermal Power Facility.

The Fiber Optic Ring would be the telecom project.

Fortunately, interest rates were low and negative in Germany and Japan. So, the time was right for public infrastructure [17]. The task force's infrastructure projects also had demanding risk assessments. There were political, military, and economic algorithms. There were also soft infrastructure projects. These included (a) legal, policy and regulatory reforms, (b) cross-border economic zones and (c) harmonizing regional customs.

6 Decision-Making Models

In terms of decision-making models, General David Petraeus deserves kudos for building a task force that used the cost–benefit analysis of the Rational Actor Model for decisions about the 20 economic development projects to implement.

At one extreme, the task force avoided the pitfalls of rigid bureaucratic turf fights and polarization through bargaining and compromise. The task force also avoided groupthink by being inclusive. The task force believed in 360° buy-in and giving everyone in the task force a voice. Competing perspectives were heard at brainstorming session in Hyde Park offsites, and inclusive membership. The task force included 35 US government agencies and got everyone onboard.

Thanks to Marc Grossman, the new SRAP, the task force enjoyed a declaratory policy victory when U.S. Secretary of State Hillary Clinton embraced the CENTCOM New Silk Road initiative in a speech she gave in Chennai, India on 20 July 2011. With all its persuasive feasibility plans behind it, the task force was ready to execute its financing and implementation plan. General Petraeus was the ideal choice to help us get Wall Street onboard for our finance plans and the media onboard for our plan to mobilize public opinion. He holds a Ph.D. in International Relations from Princeton University and he taught economics at West Point.

Unfortunately, the New Silk Road fell apart because of dysfunctional palace politics of decision-making as well as cognitive limitations and mistakes of the leadership. The New Silk Road Initiative suffered a fatal defeat when President Obama—who never understood the strategic value of CENTCOM's New Silk Road initiative—ill-advisedly sent General Petraeus to Afghanistan. That decision meant Petraeus would get bogged down with tactical military matters and have little no time for the long view and the New Silk Road Initiative.

No Hollywood producer would ever miscast his military commanders the way Obama did. Like George Patton, General Mattis was a "legendary warrior" who should have gone to Afghanistan. Instead, Obama had Mattis replace Petraeus at CENTCOM. Like Obama, General Mattis never understood the economics or the strategic value of the New Silk Road. So Mattis zeroed out the funding for the

New Silk Road at CENTCOM and tried to eliminate the Chief Economist position that Admiral Fallon had created to link economics and security. General Lloyd Austen ignored the advice of General Mattis and kept the Chief Economist position at CENTCOM after Mattis retired. But the damage to the New Silk Road at CENTCOM lasted ten years.

China's New Silk Road

The New Silk Road task force at CENTCOM may have felt partially vindicated when Beijing announced in 2013 that it would initiate their own Belt and Road version of the New Silk Road on steroids. The good news was that China was serious. China was committed to spending trillions of dollars on its Belt and Road. The bad news is that almost all of China's New Silk Road goes around Afghanistan. CENTCOM's idea was to build infrastructure through Afghanistan, which in turn would strengthen the Afghan economy and foster transcontinental shared security. In any event, when America ill-advisedly dropped the New Silk Road ball, China happily picked it up and ran with it.

7 U.S. Military Exit from Afghanistan

Earlier we talked about President Obama's decision for a troop withdrawal in Afghanistan back in 2011. Most of the 150,000 U.S. troops would come home, only leaving about 10–15% of the troops behind. A simultaneous and proportional 10–15% reduction of military contracts would also happen. The U.S. Treasury study correctly predicted a large negative impact that would occur to the overall Afghan economy once 85–90% of the war economy went away. The study correctly predicted that jobless numbers and violence would soar.

Therefore, it's odd that the Biden administration claims that it was "caught off-guard" with the withdrawal of the rest of U.S. troops on 30 August 2021. The only possible explanation is that first President Obama and then President Biden have cognitive limitations that prevented them from understanding this critical economics/security connection. For a quarter century, the credibility and resolve of U.S. foreign policy to stay the course with friends and foes have been damaged. In Afghanistan, the rise and fall of the U.S. New Silk Road gave China an open economic playing field in great power competition. The U.S. military exit from Afghanistan triggered a humanitarian crisis and economic collapse. This pattern of lost opportunities and broken promises is also evident in Syria, Iraq, Iran, and Ukraine.

Syria: Broken Promises to Opposition Forces

In Syria President Obama was politically correct at home. He gave moralistic speeches about how President Assad was illegitimate and had to relinquish power. He also gave the opposition to Assad every reason to believe that the US military would intervene in Syria with enough military force to shove him out of power. But in the eyes of this opposition, it was a big buildup for a big letdown.

Before long, the Syrian civil war morphed into a war by proxy. Assad was receiving overwhelming military support from Russia, Iran, and Hezbollah. In contrast, the Syrian opposition was receiving much smaller amounts of military assistance from the US, Turkey, Saudi Arabia and other the Gulf Cooperation Council (GCC) states, Qatar, France, and the UK. Obama only gave the opposition enough military assistance to intensify and prolong the fighting, never enough to help the opposition compete with the Syrian military. The opposition was a victim of false expectations and ultimately felt betrayed by American's ambivalence and dithering.

Obama was fixated on the idea that the conflict could only be resolved if Assad was removed from power. The problem was that Obama grossly underestimated the strength of the Assad regime. The only way Western intervention would work would be if it put lots of American boots on the ground. Obama was not going to do that. He knew what he did not want—Assad to remain in power—but he had no idea who or what he wanted instead of Assad. Obama wanted to see a moderate, democratic, secular, pluralistic government replace Assad. The only problem was that such a possibility was not a realistic prospect [18].

In contrast, Russia, Iran, and Hezbollah knew exactly what they wanted—Assad to remain in power. This was a core interest for them. In contrast, Syria was never a core interest for Obama. Therefore, there was never any US political will to arm the political opposition to such an extent that they would have a chance to win battles against the Assad regime. There was never any clear US strategy in Syria, except that of defeating ISIS. Obama had no geopolitical clue in Syria other than counterterrorism tactics against ISIS.

What's interesting is that Robert Ford, US Ambassador to Syria, was reportedly opposed to calling for Assad's departure (Ibid). Ford was realistic. He argued that America would not be able to bring this about. In contrast, President Obama delivered stirring rhetoric calling for Assad's departure, but his rhetorical demand carried no intent to enforce that wish. The US may have been a superpower within its own borders, but it was a paper tiger in Syria.

Worst of all, Obama created false hopes among the opposition to Assad that decisive Western support was forthcoming. In the end, US support was not nearly as powerful as Obama's rhetoric. His moral high ground prevailed over pragmatism. Sadly, it took the political opposition to Assad a long time to accept the fact that they had been victims of false expectations (Ibid). In the eyes of the political opposition, Obama had reneged on his stirring commitment to help the opposition oust Assad.

So, which national security decision-making model is most relevant to the Obama administration? The cognitive model would appear to fit President Obama when it came to Syria. Leon Festinger would argue that Obama had a desire to maintain cognitive consistency and wishful thinking in the face of psychologically uncomfortable new information (or cognitive dissonance). In addition, Robert Jervis would argue that Obama's wishful thinking about "Assad must go" sent a wrong signal to the political opposition. The political opposition to Assad misperceived and miscalculated the nature and extent of Obama's security commitment [6]. As a result, the political opposition felt betrayed. They felt Obama had reneged on his stirring moral commitment. Once more, the credibility and resolve of U.S. foreign policy to stay the course with friends and allies was damaged.

8 Iraq

Back in November 2016, the Iraqi government started to celebrate its military victory over the Islamic State of Iraq and Syria (ISIS). Investors once again saw reasons for optimism about Iraq's economic prospects. The stock market was on the upswing. Fitch gave the economy another vote of confidence in March when it upgraded Iraq's economic outlook to stable, mostly based on the country's improving public finances. And its Gross Domestic Product (GDP) growth was also booming. The defeat of ISIS prompted rare optimism and a window of opportunity for social inclusion and shared prosperity among all Iraqis [19].

But before long, that window of opportunity was squandered. Just eight months after Baghdad and the Western allies celebrated a hard-won victory over ISIS, the most serious anti-government protests to hit the country in years swept across its oil-rich south. Baghdad struggled to contain the protests. Demonstrators laid siege to government buildings, ports, and oil companies.

The demonstrators faced serious electricity and water shortages as well as high unemployment. The demonstrators were fed up with their incompetent political leaders. They demanded sweeping reform to dismantle the corruption and mismanagement that crippled the Iraqi economy, despite its oil wealth. The protests undermined hopes that elections in May 2018 would be a turning point and put the embattled country on the road to social inclusion and shared prosperity after decades of conflict [20].

The Iraqi people had seen this big build-up for the big let-down movie before. Back in 2014, it also seemed like the best of times. The war was over in Iraq, and investors were pouring money into its economy. Corporate earnings were booming. International investors were confident that economic rewards in Iraq outweighed any possible risks.

Iraq's economy looked promising in so many ways. In February 2014, its oil production surged to its highest level in over 30 years, and its oil exports hit a post-2003 high. It had one of the hottest commercial markets in the world. Asia Cell Communications had the biggest initial public offering (IPO) in the Middle East since 2008 [21].

The future also looked bright. With the second-largest proven oil reserves in the Organization of the Petroleum Exporting Countries (OPEC), the International Energy Agency predicted Iraq's oil output would double by the end of the decade. Its oil output was expected to grow by 600–700%. The Economist Intelligence Unit (EIU) predicted that Iraq's GDP growth would reach 8% in 2014 and almost 9% by 2020. No other country in the world had this kind of growth trajectory [22].

But this was all a false dawn [23]. What did the optimists miss? US military doctrine in the past says that phase three in a campaign plan is civilian-led stabilization. It follows two combat arms phases (first, seize the initiative, and second, dominate the enemy). Bill Flavin at the Army War College said stabilization consolidates the gains achieved through offense and defense to achieve sustainable outcomes that align with US national interests and strategic goals. In other words, successful post-conflict stabilization assures that the US wins the war, rather than a series of battles [24]. US military doctrine also includes Reconstruction (phase four) [25].

Back in 2004 Hans Binnendijk and Stuart Johnson correctly argued that there has been a widening gap between the scale-down of combat operations and the start of stabilization and reconstruction operations in the previous US doctrine [26]. Bad things happen in this gap. The most dangerous events included ISIS filling the gap in Iraq. More recently, the Turkish military filled the gap against the Kurds in October 2019 when President Trump told US soldiers to cut and run from Syria. Once again, ISIS was free to begin another resurgence of terrorism.

Optimists thought that the fall of the divisive Prime Minister Maliki and the rise of a seemingly more inclusive Prime Minister Abadi was a hopeful sign. But it was too little, too late. Maliki's repression of the Sunnis was deep-rooted and contributed to the rapid expansion of ISIS. The initial ISIS military success and the subsequent and belated military response of the US and its allies are now well known.

However, what was not so well known at the time was the economic impact of ISIS. While the major oil fields in southern Iraq were located far away from the battle zones and not seriously affected, ISIS actions in northern Iraq occurred close to the Kurdish oil fields. In addition, ISIS-controlled areas in Iraq seriously hindered the routing of oil and gas beyond its borders. ISIS control of this area prevented Iraq's oil from being shipped via the Kirkuk-Ceyhan pipeline to Turkey and over highways to Jordan [27]. As a result, Iraq's international trade took a big hit. It lost $1.5 billion a month due to an oil export pipeline under ISIS control that ran from Kirkuk to the Turkish port of Ceylon (Ibid). And of course, low oil prices back then also hit Iraq's exports. In addition, its imports fell 60–70% [28].

Foreign reserve levels also fell. On the domestic front, two million internally displaced people (IDP) lost their jobs. Investors were then understandably scared off. Farms and factories shut down, which, in turn, caused several domestic markets to collapse (Ibid).

The Iraqi economy went from boom to bust. The International Monetary Fund (IMF) announced that the Iraqi GDP was contracting. When an economy like Iraq's gets smaller, its revenues also shrink. This, in turn, caused the budget to run a large deficit. Not surprisingly, therefore, Iraq ran a budget deficit of 4% of GDP. The Maastricht criterion from the European Union (EU) says that any budget deficit over 3% of GDP is financially unstable. In short, Iraq suddenly found itself in a fiscal crisis.

What were the drivers of this fiscal crisis? Even before the war against ISIS, Iraq was struggling to finance the reconstruction of its armed forces, but the war meant that Iraqi military spending started to skyrocket. Iraq suddenly needed to train, feed and house tens of thousands of men who volunteered to fight ISIS. The associated costs of this fighting were also rising. In addition, Iraq had no productive industry or agriculture, so, it obviously couldn't tax farms and factories that no longer existed.

That left Iraq as a "one trick pony" economy that relied on high oil prices to make ends meet. This, in turn, made it extremely vulnerable to falling oil prices which hit Iraq's economy and budget hard. Oil prices fell from over $100 a barrel in May 2014 to about $50 a barrel by early January 2015. Additionally, Iraq could not increase southern oil production and exports because of poor infrastructure and a shortage of water.

In addition to these unexpected costs, the Iraqi economy suffered from years of corruption and mismanagement, despite billions of dollars in US financing. As a result, the budget was full of structural problems. For instance, the government funded public payroll and food and energy subsidies equal to 70% of the budget. To make matters worse, the country was in arrears. Baghdad still owed the Kurdish Regional Government (KRG) billions of dollars.

9 Lessons Learned

What did we learn from the false dawn of 2014? And what should we keep in mind now that the Iraqi military has almost won the operational war against ISIS? The short answer is winning the operational war is not enough. Stabilization and reconstruction must happen immediately. If a gap widens between combat arms and stabilization/reconstruction, another insurgent group (like ISIS) will once again fill the socio-economic gap.

To avoid another false dawn and negative scenario in Iraq, the International Community (IC) should first encourage Baghdad to consider using a third party (such as the Arab League) to serve as a broker for an inclusive economic strategy that would result in shared prosperity for all Iraqis. Second, Baghdad needs to turn this economic strategy into a detailed economic plan. Key Sunni, Shia and Kurdish stakeholders need to be included in the planning process. In phase one, the plan would show how Baghdad would address Iraqi humanitarian needs in this war environment. In phase two, Baghdad should address the socio-economic needs of all Iraqis. This economic development involves upgrading basic services (power, water, and housing). In phase three, Baghdad should turn to economic reconstruction.

Since the oil industry produces very few jobs, Baghdad needs to form public–private sector partnerships (PPPs). The PPPs would help create more job opportunities for all Iraqis by rebuilding factories and the agricultural sector. Third and most importantly, there needs to be an economic roadmap. The roadmap needs to have a public awareness campaign with realistic economic assumptions. Since the fiscal crisis means Baghdad cannot do everything in one fell swoop, there needs to be (a) concrete actions, (b) specific locations and (c) achievable timelines for the economic roll-out to show progress. Public awareness of concrete progress is important because it will build mutual trust, confidence, unity, and support from all Iraqis.

That in turn will mitigate grievances, reduce the potential demand for violence and foster reconciliation and stability.

Which national security decision-making model is most relevant here? The U.S. organizational and sub-bureaucratic models seem to be most relevant here. The combat arms part of the US military kept winning its operational battles. Unfortunately, the U.S. wasted these operational victories by marginalizing what military civil affairs and the economic agencies of the US government do—stabilization and reconstruction. That in turn enabled ISIS terrorists to fill a widening gap. The point is the US makes a partial commitment and then plays cut and run, essentially wasting the operational military victory.

10 Iran

At the start of A Tale of Two Cities, Charles Dickens tells us, "It was the best of times, it was the worst of times." On 14 July 2015, the Iranian people no doubt felt it would soon be the best of times when they poured out of their houses to celebrate in the streets. After a decade of economic isolation, Iran had reached a nuclear agreement with the US, the UK, France, Germany, Russia, and China. Iran promised in the Joint Comprehensive Plan of Action (or JCPOA) that it would unwind its nuclear program in return for lifting international economic sanctions and reintegrating into the global economy. In short, the agreement was meant to turn isolation and economic coercion into Jean Monnet's vision of a web of economic interdependence and shared prosperity.

The euphoria of the Iranian people reflected both the prospect of an end to the devastating impact of the sanctions on its economy and the economic promise of the lifting of the sanctions. The JCPOA allowed Iran to export crude oil and other energy products, source foreign direct investment (FDI) in most sectors and access about $100 billion in its foreign-exchange reserves in international banks [29].

Trump Reneges on JCPOA Deal

But not so fast. President Donald Trump killed the joy and made it the worst of times for the Iranian people. On 12 October 2017, President Trump announced that he was de-certifying the nuclear agreement [30]. On 8 May 2018, he abandoned his belief in allies and withdrew from the agreement completely [31]. On 7 August 2018, he announced that "the highest level of economic sanctions" would "snap-back" again against Iran [32]. And on 12 September 2017, he slammed Iran for being "evil" in a speech he delivered at the United Nations [33].

Impact of Sanctions on Iran

What does it mean for sanctions to snap back against Iran? When severe sanctions were in place against Iran, this economic coercion targeted everything from shipping and banking to foreign investment and exports. The sanctions limited Iranian revenues and stymied industry. They had a devastating impact on the economy. Over $100 billion in financial assets were blocked. The volume of oil exports fell over 50% (from 2.5 MBD to 1.4 MBD).

Car production fell by 40%. One out of every five Iranians was jobless. The economy was 15% to 20% smaller than it would have been without the sanctions—that's the equivalent of the Great Depression in the US in the 1930s [34].

Impact on Foreign Business Community

Foreign businessmen hoped the nuclear deal would pave the way for a flood of new business deals, opening up foreign investment and international trade in crucial sectors such as oil and gas, car production, aviation, tourism, technology, mining the stock market and banking [35].

Iran's potential emergence from economic isolation could be the most significant opening of an economy since the fall of the Soviet Union and the US rapprochement with China. The potential of Iran was huge. As one of the last markets to be opened up to the world, its allure was unmistakable. Its nearly 80 million residents—60% of whom were under 30 years old—already had an affinity for Western brands, especially American ones like Coca-Cola and Chevrolet. Some shops in affluent urban areas, particularly the nation's capital, were full of Western-made products, from sunglasses and designer jeans to laptops. In addition, the Iranian population was tech-savvy. Internet penetration was 53% across the population and 77% in Tehran. About 11 million Iranians had mobile Internet access. Many senior businesspeople were educated in the US and still prized American engineering. Iran's market for technology products and services was roughly $4 billion a year. If sanctions were lifted, the market rose to $16 billion annually, which made it comparable to that of Saudi Arabia. Overall consumer expenditures were projected to be about $176.4 billion a year, with annual disposable income pegged at about $287 billion [36].

While Iran enjoyed huge oil and gas reserves, the economy was relatively diversified in a Middle Eastern context, with the oil and gas industry accounting for just 15% of the GDP [37]. In short, foreign businessmen had been eyeing a huge market in Iran for a long time.

President Rouhani deserved kudos on the foreign policy front for working with the US and the other five world powers to negotiate the nuclear deal. But he had not yet implemented significant economic reforms. In many ways, the failing economic model had not fundamentally changed from that when his predecessor, President Ahmadinejad, was in power.

10.1 Impact of Trump's Exit from Nuclear Agreement

Of course, Trump's exit from the nuclear agreement made everything even more difficult for international businesses previously trading with and/or investing in Iran. Trump was giving 3–6 months for corporations to wind down their operations with Iran.

Trump's new sanctions, which took effect on 7 August 2018, prohibited Iran from using US currency. The sanctions barred trading in cars and metals and minerals that included gold, steel, coal, and aluminum. Iran was also barred from buying US and European aircraft. The sanctions barred imports of Iranian energy and prevented financial institutions from conducting transactions with Iran's central bank [32].

10.2 Betraying US Friends and Allies

How did this economic coercion work against America's former allies? Not well. Gideon Rachman noted that America's economic power went well beyond market access. European executives who continued to do business in Iran could become subject to arrest if they travelled to the US. And European banks that did business with Iran could find themselves shut out of the US financial system, or subject to prosecution and massive fines in America [38].

All of this reflected the role of the US dollar as the world's reserve currency. It was the dollar, as much as American military might, that allowed the US to coerce its allies—as well as its adversaries (Ibid).

Trump Abandoned his Belief in Allies

Gideon Rachman, a writer for the Financial Times, predicted that 8 May 2018—the day of the US exit from the nuclear agreement—would go down in history as the day Trump abandoned his belief in allies (Ibid). Gone were the days of George Marshall and US social inclusion and shared prosperity with its allies. Donald Trump was now using economic coercion against US allies as well as against Iran.

Europeans were angry and frustrated with Trump's America Alone approach, which arbitrarily bullied the Europeans with huge opportunity costs, taking billions of dollars out of their pockets. While Iran was relatively diversified, oil and gas would arguably have been its biggest attraction for Total and other European energy companies. Few of the oil and gas industry's big players could ignore the multibillion-dollar oil and gas "candy store" that is Iran. There were lots of reasons why Iran was a golden opportunity for oil and gas companies [39].

Trump threw away a 15-year agreement that froze the Iranian nuclear program and received no viable alternative agreement in return. He was left wishing and hoping that more sanctions and economic coercion would somehow force Iran to totally give up nuclear weapons. But guess again. Wishing and hoping is not a strategy.

U.S. foreign policy kept failing in Iran because its decision-making process was seriously flawed. The psychological model in academe provides valuable theoretical insights. President Trump had acute cognitive limitations. Trump undermined Rouhani, who trusted the United States to honor its side of the agreement. Instead, Trump's withdrawal from JCPOA and his re-imposition of severe sanctions would almost certainly backfire by strengthening the hand of the hard-liners in Iran who argued that you can't trust America to honor an international commitment. America will just find another excuse to bully you with economic coercion. In short, America will betray you.

The IRGC hard-liners in Iran gained the upper hand over Rouhani and the reformers. At home, the hard-liners stifled political and economic reform. White House threats of regime change in Iran raised concerns among the Iranian hard-liners of confrontation with the United States, Israel, and Saudi Arabia. In response, the hard-liners in Tehran shored up support for its regional proxy forces as a form of deterrence.

Ukraine

Finally, let's turn to Ukraine, the elephant in the room—as an illustrative example. In 1994, three years after the collapse of the Soviet Union, Ukraine was an independent country with 1500 nuclear weapons. The problem was Ukraine had no operating control of these nuclear weapons. So, in 1994 the U.S. persuaded Ukraine to give these nuclear weapons to Russia. But there was an important quid pro quo in this Budapest Memorandum [1: 319–320].

The US guaranteed the territorial integrity and economic and political sovereignty of Ukraine. That was the only reason why Ukraine was willing to give up its nuclear weapons. Ukraine believed America had its back. Ukraine trusted the US to honor its security guarantee.

Fast forward to 2014. Russia attacks Crimea and opts for annexation as well. Russia also does a de facto land grab in Donbas. The $64,000 question was: would Obama honor the US security guarantee for Ukraine as the U.S. promised in the Budapest Memorandum of 1994? Or would he renege on that security guarantee?

Well, Obama breaks that promise just like he did the promises he made to the opposition leaders to Assad in Syria. Now imagine you are Putin in 2022. You watched Obama and Vice President Joe Biden back then and Trump and other US Presidents create a pattern of broken promises.

Most importantly, Obama reneged on Budapest Memorandum's security guarantee to Ukraine. President Biden also decided to cut and run in Afghanistan reportedly against the advice of his top generals and admirals. This pattern of broken promises sends the wrong signal to Putin. Putin sees the US as a Paper Tiger.

That misunderstanding and misperception causes Putin to miscalculate Biden's potential response to a full-blown Russian attack on Ukraine. And so, Putin assumes President Biden's response will be more of the same—more broken promises, a few economic sanctions, a little security assistance. Nothing major and all politically and economically manageable.

So, Russia invades Ukraine on 24 February 2022 expecting only a lukewarm response from the U.S. But guess again. Instead of a predictable lukewarm response, Biden reacts forcefully and in a way Putin didn't expect.

The point is strategic stability requires that the US must be more predictable and send more consistent signals to friends and foes alike. When we have a pattern of reneging on our commitments and broken promises, it makes it much harder to do a 180 and suddenly stand up to Russian aggression.

Biden is now doing the right thing. But instead of deterring Russian aggression, the U.S. must react to a full-blown Russian invasion of Ukraine. Biden now must thread the needle—which means reversing Russian aggression in Ukraine while at the same time not triggering WW3 and nuclear holocaust with Russia. Consistent, predictable strategic deterrence and stability would have been a much safer course of action.

11 Conclusion

So, what are the takeaways? US national security decisions have too often led to squandering economic opportunities and broken promises. Academe provides valuable theoretical insights. At times bureaucratic infighting theory results in polarization. At other times groupthink theory produces conformity and flawed outcomes. Most importantly, scholars talk about cognitive dissonance theory and the conceptual limitations and errors of international political leaders.

In Afghanistan, the U.S. economic exit from the New Silk Road gave China an open playing field to use its Belt and Road to promote its economic power and influence. The U.S. military exit from Afghanistan triggered a humanitarian crisis and economic collapse. This pattern of broken promises and instability is also evident in Syria, Iraq, Iran, and Ukraine.

In Syria, stirring U.S. rhetoric that "Assad must go" was followed by only enough U.S. military support to intensify and prolong the civil war.

In Iraq, the U.S. wasted its operational military success by marginalizing stabilization and reconstruction, thus enabling ISIS to fill the gap and promote terrorism and instability.

In Iran, Europeans were angry about the U.S. unilateral decision to exit from the nuclear agreement. This exit also betrayed economic reformers in Iran and strengthened Iranian hardliners who had warned that "you can't trust America."

Finally in Ukraine, the U.S. "guaranteed" Ukraine's territorial integrity with the Budapest Memorandum and then failed to reverse Russia's land grab in Crimea and Donbas. This unwillingness of the U.S. to stay the course and show resolve in Ukraine convinced Russia that it could attack Ukraine and not expect a strong pushback from the U.S.

Robert Jervis would note that this erratic pattern of weakness is dangerous and risks Moscow and Beijing misperceiving U.S. ultimate intentions. In the future, the U.S. needs to do a better job taking advantage of economic opportunities, making promises the US can keep, and sending clearer signals that reduce the risk of misperceptions, miscalculation, and great power conflict.

References

1. Gates R (2020) Exercise of Power: American Failures, Successes, and a New Path Forward in the Post-Cold War World. Alfred A. Knopf, New York
2. Allison G, Zelikow GP (1999) Essence of Decision: Explaining the Cuban Missile Crisis, 2nd edition. Longman, New York
3. Houghton P (2013) The Decision Point: Six Cases in U.S. Foreign Policy Decision Making. Oxford University Press, New York
4. Janis I (1982) Groupthink: Psychological Studies of Policy Decisions and Fiascoes. Houghton Mifflin, Boston
5. Gladwell M (2005) Blink: the Power of Thinking Without Thinking. Little Brown, New York
6. Jervis R (1976) Perception and Misperception in International Politics. Princeton, NJ, Princeton University Press
7. White R (1986) Empathizing with the Soviet Government. In: Psychology and the Prevention of Nuclear War: a Book of Readings. New York University Press, New York
8. White R (1984) Fearful Warriors: a Psychological Profile of U.S.-Soviet Relations. Free Press, New York
9. Gvosdev N, Blankshain J, Cooper D (2019) Decision-Making in American Foreign Policy: Translating Theory into Practice. Cambridge University Press, New York
10. Clinton H (2011) Remarks on India and the United States: a Vision for the 21st Century, U.S. State Department, Chennai, India, 20 July 2011
11. Tzu S (2022) Art of War. New York Union Square
12. New York Times (NYT) (2012) Peacekeepers Exit East Timor, 31 Dec 2012
13. NPR (2015) Did Obama withdraw from Iraq too soon, allowing ISIS to grow?, 19 Dec 2015
14. NBC News (2014) Rare Earth: Afghanistan Sits on $1 Trillion in Minerals, 5 Sep 2014
15. Economist (2003) Obituary: Walt Rostow, 20 Feb 2003
16. Starr F (2007) The New Silk Roads: Transport and Trade in Greater Central Asia. Central Asia Caucasus Institute Silk Road Studies Program
17. Summers L (2016) Building the case for greater infrastructure investment, 12 Sep 2016, larrysummers.com
18. Gitton P, Muzaffari M (2017) Afghanistan's integration in regional trade: a stocktaking. IMF, Dec 2017
19. Hansen A (2017) Turning fragile optimism into stability in Iraq. National Democratic Institute, 7 June 2017
20. England A (2018) Iraq's political unrest stoke instability fears. Financial Times (FT), 22 July 2018
21. DeWraver M, Albazzaz A (2013) Guest post: Iraq on the brink? Markets tell a different story. FT, 30 September 2013
22. Johnson S (2013) Iraq: can a dusty war zone become a golden opportunity? FT, 5 October 2013
23. View FT (2014) The nightmare emerging in Iraq. FT, 11 June 2014
24. Flavin B (2018) Stabilization: a New Approach to Whole of Government Operational Planning and Execution. US Army Peacekeeping and Stability Operations Institute (PKSOI), June 2018
25. Brown S (2009) Hard Lessons: the Iraq Reconstruction Experience. CreateSpace Independent Publishing Platform, 2 Feb 2009
26. Binnendijk H, Johnson S (2012) Transforming for Stabilization and Reconstruction Operations (paperback), 23 July 2012
27. Coface (2014) Iraq: Risk Assessment. www.coface.com/Economic-Studies-and-Country-Risks/Iraq
28. AINA (2014) Iraq's Economy Collapsing under Pressure of Security Crisis. http://aina.org/news/20141004193538.htm
29. Economist Intelligence Unit (EIU) (2018) Iran on a Knife's Edge: as Nuclear Deal Goes Up in Smoke. EIU Limited, London
30. Manson K (2017) Donald Trump set to repudiate Iran nuclear agreement. FT, 13 Oct 2017

31. Luce E (2018) Donald Trump Goes for Global Change. FT, 8 May 2018
32. Sevastopulo D, Khan M, Bozorgmehr N (2018) US Reimposes economic sanctions on Iran, FT, 6 August 2018
33. Rezaian J (2018) In UN speech, Trump Fails (again) to Make his Iran Case, The Washington Post, 25 September 2018
34. Glassman M (2015) What Sanctions have Done to Iran's Economy. Bloomberg, 2 March 2015
35. Bozorgmehr N, Khalaj M (2015) Businessmen eye huge opportunities in Iran. FT, 14 July 2015
36. Faucon B (2015) In Iran, Business Deals are Rarely Smooth. The Wall Street Journal, 15 July 2015
37. Johnson S (2015) Investors eye Iranian opening. FT, 7 July 2015
38. Rachman G (2018) The new World order: Donald Trump goes it alone, FT, 11 May 2018
39. Adams C, Bozorgmehr N, Crooks E (2015) Iran: the oil and gas multi-billion candy store. FT, 16 July 2015

Dr. Leif Rosenberger is a Governance Analyst at the U.S. Peacekeeping and Stability Operations Institute (PKSOI). He was previously the Distinguished Visiting Scholar of National Security at Webster University, a Visiting Scholar on the Economics Faculty at Harvard University, an Adjunct Professor of International Political Economy at the University of Pittsburgh, and an Adjunct Scholar at the Modern War Institute at West Point. In the private sector he was the Chief Economist at ACERTAS and a Co-Founder of Climate Solutions Advancement Network. On Wall Street, he was an Emerging Market Economist for Roubini Global Economics and Jim Cramer's the Street. He was also on the Board of Directors at both the Economic Innovation Institute for Africa and Think Renewables Group. Before this he had an eventful 35-year career in the U.S. government. He was a Full Professor of Economics at the U.S. Army War College, where he held the General Douglas MacArthur Academic Chair of Research in the Strategic Studies Institute. He was the Chicf Economist at PACOM and CENTCOM for almost two decades and was twice chosen as the U.S. Federal Employee of the Year. He's published twelve books and over sixty articles. He holds a BA with honors from Harvard University, a masters from Boston University and a Ph.D. in International Relations from Claremont Graduate University, where he held the Merit Fellowship as the top-ranking student.

Once and Future Great Power Competition in Afghanistan: An Interpretive History

Barnett R. Rubin

Abstract Today's state of Afghanistan owes its creation, disintegration, and periodic revival to the ebb and flow of great power competition in a region variously characterized, depending on who is doing the characterizing, as either the Heart of Asia or a remote frontier. Not just the state's external borders but also its internal governance institutions have been shaped by great powers from its origins in the eighteenth and nineteenth centuries to the 2021 U.S. military withdrawal and beyond. Afghanistan's centralized state structure is the product of the funding of that state by successive great powers to enable it to serve their security interests. During intervals of state collapse, local powerholders have emerged outside of any institutional framework, but have never established a state with decentralized institutions. The word "Afghanistan" has existed as a name for a geographic region (what we would now call the tribal areas of the Afghanistan-Pakistan border) for centuries. For just as long, "Afghans" or "Pashtuns" ruled empires and states both in and beyond the borders of today's Afghanistan, largely in India (including today's Pakistan). A state named "Afghanistan" within its current borders is a nineteenth-century creation. Afghanistan's south and east correspond to part of the old territory of Afghanistan, and its northern and western areas (Kabul, Balkh, Herat) to part of a region called Khurasan, the Persian east, which also includes Samarkand, Bukhara, and Mashhad. Today's state of Afghanistan is a sometimes-uneasy amalgam of parts of both Afghanistan and Khurasan.

Keywords Afghanistan · Great power competition · Centralized state · Khurasan

The Durrani empire, founded in 1747, rose and fell based on the fates of other pre-modern empires in Iran, India, and Central Asia. Before European land empires reached this crossroads of South, Central, and West Asia, rulership over the territory of today's Afghanistan was divided among the Mughal empire based in Delhi (and founded in Kabul), the Safavid empire based in Isfahan, and the Khanate (later Emirate) of Bukhara, a successor to the Timurid and Shaybanid empires. Bukhara was one of the Khanates that formed after the collapse of the Shaybanid empire in

---

B. R. Rubin (✉)
Center on International Cooperation, New York University, New York, NY 10003, USA
e-mail: barnettrrubin@gmail.com

A. Farhadi and A. Masys (eds.), *The Great Power Competition Volume 4*,
https://doi.org/10.1007/978-3-031-22934-3_19

the late sixteenth century. By the early to mid-eighteenth century both the Safavid and Mughal empires showed signs of impending collapse. While historians differ on the relative importance of various causes of this decline, the near simultaneity of the decline of revenue in Asian land empires in the seventeenth and early eighteenth century suggests that a principal cause was external: the rise of transcontinental European maritime trade, which put the land-based trade routes known collectively as the Silk Road under competitive pressure, eroding the tax base of the empires.

Tribes with access to mobile firepower in the form of horses took advantage of the resulting power vacuum. In 1722 the Kandahari Ghilzai tribes under Mahmud Hotaki besieged and then sacked Isfahan. The weakened Safavid capital was soon overrun by Nadir Shah Afshar, a Turkmen former ally of Hotaki, who named himself Shah in 1729. With the aid of cavalry led by Abdali (Durrani) Pashtuns of Kandahar, Nadir turned east and sacked Delhi in 1739, capturing among other booty the Mughals' Peacock Throne. When Nadir was assassinated in 1747, his Afghan cavalry commander, Ahmad Shah Abdali (later Durrani), returned to Kandahar, where he had himself named as King of the Afghans.

Like Nadir Shah and the Hotakis, Ahmad Shah then turned to raiding the neighboring areas where the empires had declined, bringing not only today's Afghanistan but parts of India, Iran, and Central Asia-Turkistan under his control in what became known as the Durrani empire. He and his descendants might have been able to turn this conquest empire into a more permanent state, relying on the revenues of the conquered provinces, but the encroachment of the Russian and British empires boxed the Durrani rulers into areas with relatively little income. The truncated Durrani empire fell into clashes among competing lineages based in different regions of the country. Due to these clashes, the kingship passed in 1818 from the Saddozai clan of the Popalzai to the Muhammadzai clan of the Barakzai tribe.

Internationally, the nineteenth century opened with clashes between the rulers of Kabul and the Sikh empire of Ranjit Singh, but it ended with Russia in Central Asia, Britain in India, and both powers competing for influence in Persia. Afghanistan, or the Kingdom of Kabul, made a transition from dealing with competition among indigenous empires in Iran, India, and Central Asia to dealing with competition within the Euro-centric imperial system.

Its foreign relations were now dominated by the competition between the UK and Russia, which were expanding into the areas surrounding the changing borders of the Kingdom of Kabul. In a familiar pattern, fear of intervention by one great power motivated intervention by another. Soon after the Muhammadzais took power, Amir Dost Muhammad Khan offered the British an alliance against the Sikhs, but his condition was the return of Peshawar (and of today's Khyber-Pakhtunkhwa province of Pakistan) to Afghan rule. When the British demurred, Dost Muhammad sent out feelers to Russia, leading to the first Anglo-Afghan War (1839–1842). The British launched what we would now call a "regime change" operation, to remove Dost Muhammad Khan because of his supposed inclination toward Russia, and reinstate Saddozai rule in the person of Shah Shuja. The British occupation in the First Anglo-Afghan War was the first time that a foreign power tried to use "traditional Afghan institutions" (Saddozai rule) to control the country. The British also launched the first attempt by a western power to build and fund Afghan security forces to empower the ruler they had placed on the throne. In a foreshadowing of the America experience 160 years later, the British overthrew Dost Muhammad Khan only to retreat leaving Dost Muhammad Khan back in power. Britain, the Soviet Union, and the U.S. replicated the experiment several times, hoping for a different outcome.

The Second Anglo-Afghan War started in 1878 when Russia sent an uninvited mission to Kabul, then ruled by Dost Muhammad's son, Sher Ali Khan. When the amir refused to accept a British mission, the British launched another "regime change" operation, replacing Sher Ali Khan with Yaqub Khan, who, under the pressure of the British occupation, signed the treaty of Gandamak. That treaty, the legitimacy of which Afghans contest, achieved the main British objectives in Afghanistan. The Amir relinquished claims to territory that later became the Northwest Frontier Province of India/Pakistan and the Tribal Agencies (together constituting today's Khyber-Pakhtunkhwa province of Pakistan). He accepted that Afghanistan's foreign policy would be subordinate to the British in Delhi. In return the Amir received a yearly subsidy in cash and weapons to enable him to control the country. The British soon replaced Yaqub Khan with Abdul Rahman Khan, who returned from exile in Tashkent. Abdul Rahman Khan reaffirmed the Treaty of Gandamak and in 1893 consented to the Durand Agreement, which demarcated the boundary between Afghanistan and the Tribal Areas of India, now incorporated into Pakistan as part of Khyber-Pakhtunkhwa province.

This arrangement with the British was intended to make Afghanistan into a stable buffer against Russian influence. This arrangement was reinforced and incorporated into an emergent global state system by the 1907 Convention between the United Kingdom and Russia relating to Persia, Afghanistan, and Tibet. This convention included measures to regulate the great-power competition between the British and Russian empires on both sides of the frontiers where they approached each other in South and Central Asia. Russia agreed that Afghanistan would be entirely within the British sphere of influence, and the British committed themselves not to intervene in the internal politics of Afghanistan.

Even without overt interference, however, the annual subsidy to the Amir of Afghanistan enabled the Amir to restructure the Afghan state in ways that were mutually beneficial to him and the British. After 1893, his agreed annual subsidy reached 1.8 million rupees per year, plus emergency payments (http://www.gutenberg-e.org/hanifi/chapter4.html) [Hanafi: 2008]. According to Ashraf Ghani (https://www.iranicaonline.org/articles/afghanistan-xi-admin), "The army was not used as a mere deterrent. It physically conquered the whole country—parts of it several times—in order to restructure the relationship of the local power holders with the central government." These coercive resources enabled the Amir to restructure the tribal-feudal kingdom of Kabul into the absolutist Emirate of Afghanistan.

The subsidies enabled Abdul Rahman Khan to transform those parts of Khurasan over which he ruled into components of an integrated state of Afghanistan, with a common administration, laws, and most important, army. He deprived some groups of access to land. He actually enslaved the Hazaras, whom his mullahs declared to be non-Muslims. Following the principle that the borders of Afghanistan should be defended by Afghans (Pashtuns), the amir combined the pacification of dissident Pashtun tribes with the establishment of rule over non-Pashtuns by exiling tribes that had resisted him to the border areas of Afghanistan with Iran and Russia (Central Asia). Others were settled on agricultural land in the north either taken from the local population or created by draining swamps. The status of these settlers, or naqelin, and their land ownership have repeatedly become points of tension during periods of conflict.

While this structure of power and ethnic relations is known to many, its relation to Afghanistan's international strategic position is less well understood. Incorporation of Afghanistan into the British sphere of influence was accomplished by strengthening a ruler from a social group (Pashtuns) extensively integrated into India and enabling that ruler to exercise absolutist power over his subjects, in particular those close to Iran and Central Asia. Without external subsidies, no ruler has ever been able to establish such centralized control over this territory and population. The internal structure of Afghanistan was in large measure a response to great power competition.

When Abdul Rahman Khan's grandson, Amir (later King) Amanullah Khan, won Afghanistan's independence after the Third Anglo-Afghan War (1919), he thereby gave up the subsidy. He attempted to extricate Afghanistan from its reliance on subsidies derived from great -power competition. He intended to build up the Afghan state and economy with intensive taxation of both agriculture and cross-border trade, combined with relations with Ataturk's Turkish Republic and Western Europe, from both of which he imported ideas of cultural and institutional reform. Amanullah also tried to make Afghanistan into a spokesman for progressive anti-colonial ideas in the Muslim world and Asia. He established what we might now call a "safe haven" for both Indian independence activists and militants fleeing Soviet repression in Central Asia.

Within a few years (by 1928) the result was a revolt, first by the border tribes whose trade Amanullah tried to tax and then by the Tajik peasants north of Kabul whose lands he tried to tax. Without the resources to field an army, Amanullah fled the country. The first political casualty was Pashtun control of the state, as Habibullah Kalakani, a Tajik from the village of Kalakan on the plains north of Kabul, conquered the capital and proclaimed himself Amir. The border tribes revolted as well. They eventually followed the leadership of another Muhammadzai sardar, Nadir Khan, who returned from exile in France to establish a base of operations in British-controlled Waziristan. With the help of weapons obtained and fighters recruited from the British-Indian side of the Durand Line, Nadir Khan's tribal army deposed and hanged Habibullah. It then waged punitive campaigns north of Kabul to reestablish Pashtun rule over the areas that had supported him.

Nader and his brothers (the Musahiban), followed by his son, Zahir Shah, and his nephew, Daoud Khan, could not return to the state-society relations of the period of Abdul Rahman Khan. The conservative elements of Afghan society had broken out of the fetters in which Abdul Rahman Khan had bound them and would not easily surrender their resources to a state bent on self-aggrandizement. Generous subsidies were not on offer from a Great Britain shaken in succession by the losses of the Great war and the global depression, while heading reluctantly into another World

War. Instead, the dynasty of the Musahiban largely left rural society and the ulama untouched and turned instead to the international market for funding.

While Afghanistan was developing fiscal and military ties to British India, its economy had remained largely autarchic. Besides foreign subsidies the main resource available to the state was land revenue, extracted from domestic production. The Musahiban, however, invested in enclaves for the export of products of commercial agriculture and pastoralism, developing a modest level of what would later be called "connectivity." Using the capital of a newly founded development bank, the regime invested in the karakul trade brought by Turkman refugees from the USSR and cotton cultivation on newly drained swamps in north Afghanistan, settled with Pashtun migrants. These enterprises, however, could fund only a weak state that, while centralized in its internal structure, depended on cooperation with tribes, dignitaries, and communities to expand its reach.

During the Great Depression and World War II, great powers were exhausted and otherwise occupied. Immediately after, the British were preoccupied with their exit from India. Stalin's Soviet Union had not yet developed a policy toward the post-colonial world—Afghanistan was still a buffer on the edge of the British Empire. All of this changed with the independence and partition of India, the launch of the Cold War, and the inauguration of Nikita Khrushchev's policy of peaceful coexistence, which supported Soviet engagement with non-aligned states like Afghanistan.

The partition of India gave Afghanistan a new neighbor, Pakistan, which inherited control of the territories across the Durand Line ceded by Yaqub Khan in the Treaty of Gandamak. Afghanistan claimed that Pakistan was not a legitimate successor state to the British Government of India and that Afghanistan's treaty obligations were null and void. It called for a plebiscite in the areas it called "Pashtunistan." Meanwhile, Pakistan had launched a war over Kashmir largely using Pashtun irregulars from that very territory. It confronted the fact that the Republic of India had inherited most of the equipment, infrastructure, and personnel of the British Indian Army, even if Pakistan's share was disproportionately large compared to its size, since northwest India had been a particular security concern.

From the beginning Muhammad Ali Jinnah and the rest of the Pakistan leadership decided that the only way they could build the army they wanted was by exploiting the opportunities offered by great power competition in the form of the Cold War. Pakistan turned to the U.S. and, in an enduring pattern, joined two anti-Soviet Pacts (CENTO and SEATO) that formed part of the containment strategy, gaining the nickname "America's most allied ally." Pakistan used this alliance to obtain arms and training that it intended primarily for use against India.

Premier Daoud Khan (cousin of King Zahir Shah) found himself facing a heavily armed Pakistan over a border Afghanistan disputed and over which periodic clashes occurred. Faced with the U.S.-supplied and supported Pakistan Army, Daoud concluded he needed a stronger army than the one left him by his uncles, who had acted as regents after the 1933 assassination of Nadir Shah. Just as Dost Muhammad had turned to the British for support against Ranjit Singh, Daoud Khan's first instinct was also to turn to the West, now led by the U.S., but Washington did not want to enter a military supply relationship with both sides of a local conflict.

Furthermore, given Afghanistan's geography and transport infrastructure, aid could reach Afghanistan only through Pakistan, which opposed strengthening its opponent.

Daoud had to look elsewhere. In November 1955 he convened a Loya Jirga to reassert Afghanistan's claims to Pashtunistan. In the context of the need to defend Pashtunistan, the delegates also approved Daoud's proposal to seek military aid from the USSR. https://media.defense.gov/2014/Apr/14/2001329862/-1/-1/0/Rouland--Great%20Game%20to%209-11.pdf When Dost Muhammad had asked Russia for aid in 1838, it took only a year before the British launched the Second Anglo-Afghan War. In this case the war for control of Afghanistan took a little more than twenty years to start and over forty years to finish.

Geopolitics were changing to Afghanistan's north as well. The new Soviet premier, Nikita Khrushchev, adopted the policy of "peaceful coexistence," which included competition for influence with the U.S. in post-colonial states. Khrushchev took this policy on the road with a December 1955 trip to Burma, India, and Afghanistan, during which he agreed to lend Afghanistan $100 million for construction of infrastructure. This was the first loan the USSR made outside the Warsaw Pact, a Soviet-led alliance all of whose members had communist governments. Khrushchev justified it by saying it was a "drop in the bucket" compared to the potential cost to the USSR of an American military base on Afghan soil. (https://media.defense.gov/2014/Apr/14/2001329862/-1/-1/0/Rouland--Great%20Game%20to%209-11.pdf, p. 43).

Afghanistan then became the second non-aligned nation (after Egypt) to receive Soviet military aid, which is estimated to have been worth $1.79 billion between1955 and 1979 [Ibid., 44]. The military aid included training for the Afghan officer corps, which was drawn from a new generation that graduated from a nationwide school system that Daoud built largely with American aid.

For about 20 years, from 1955 to 1978, Afghanistan benefited from peaceful competition between the U.S. and USSR. In addition to the army, the aid financed roads, schools, dams and powerplants. The U.S. and USSR observed a tacit agreement, which regulated great power competition as the Anglo-Russia Convention had done. Both countries supported the monarchy as the legitimate government of Afghanistan and did not support challengers to its authority. The USSR maintained fraternal relations with the People's Democratic Party of Afghanistan (PDPA) founded in 1965, and with both of the major factions into which it split in 1967: Khalq (the People) and Parcham (the Banner). The GRU also retained contact with Afghan military officers who had been trained in the USSR, some of whom were also PDPA members. The USSR also built and funded the prestigious Polytechnic and provided scholarships to study in the USSR. As long as the monarchy remained in power, however, there were no signs that the USSR was using these forces against the government.

The U.S. had ties with those Afghans who had been educated in the U.S. in a variety of fields such as engineering (in cooperation with the University of Wyoming) and education (in cooperation with Teachers' College Columbia). The U.S. also built the new campus of Kabul University and funded the expansion of primary and secondary education starting in the 1950s. European and Arab donors funded other parts off the higher education system. Most important is that the Faculty of Sharia Law at Kabul

University was twinned with al-Azhar Islamic University in Egypt. The government hoped that the students it sent would imbibe the spirit of Islamic Modernism, but by the 1960s it was the Muslim Brotherhood that inspired the youth. The U.S. had no political clients or proxies in distant Afghanistan. It would nonetheless be a mistake to infer that everyone who studied in the USSR became a militant leftist, or that those educated in the U.S. became pro-Western. Khalqi Prime Minister Hafizullah Amin studied at Columbia Teachers' College.

The U.S. and USSR observed a geographical division of labor in projects on the ground. The USSR built the northern and western part of the road network, connecting Afghanistan to the USSR and Iran, while the U.S. built the south and east, connecting to Pakistan. Khrushchev commented that the roads built by the USSR were constructed to specifications needed for Soviet tanks and military vehicles, in case they were ever needed (https://media.defense.gov/2014/Apr/14/2001329862/-1/-1/0/Rouland--Great%20Game%20to%209-11.pdf).

Until 1978 the US and USSR did not come into open conflict over Afghanistan. They managed their competition so as to avoid conflict, but they used a different arrangement from that in the 1907 agreement. Under that agreement Afghanistan was within the British sphere of influence, which Russia agreed not to challenge, and the amir resolved not to build roads or railroads that would make the country more accessible. Both parties agreed to stay out of Afghanistan except for one British Indian representative. During the first half of the Cold War, however, both parties were on the ground, interacting with different parts of the country's population and geography. The whole arrangement was predicated on the survival of the monarchy, a fixture in one form or another since 1747. Once that was gone, power struggles both domestic and international broke out.

Meanwhile, just as the subsidy agreed at Gandamak and augmented in the Durand Agreement transformed Afghan society by enabling the amir to build a highly centralized administration, the great power competition over "development" in the 1950s and 60s transformed the society by creating a new class of educated professionals, military, civilian, and Islamic. These people were drawn from all regions of the country, if mostly from the highest-ranking groups, but they tended to stay in or move to Kabul after their education or training. While the country remained overwhelmingly rural, new elites were accumulating in Kabul. They made their first move to seize power in 1973, when Parchami army officers and civilians helped Daoud overthrow the monarchy and proclaim Afghanistan a Republic. They struck out on their own in 1978, when PDPA officers killed Daoud and his family and proclaimed Afghanistan a People's Democratic Republic.

There is little or no evidence that the Soviet Union covertly organized these coups, but it acted swiftly to profit from them. The overthrow of a monarchy that had lasted 226 years—approximately nine generations—and its replacement by an untried, and, to most of the population, unknown form of government, a Republic, raised doubts about the future of the country. The stability of the monarchy, however anachronistic or unjust as it might seem, provided a focus around which great powers, neighbors, and domestic political actors could organize to limit their competition from escalating too far.

In 1978, Whitney Azoy, a former diplomat turned anthropologist, was finishing his dissertation on the Afghan game buzkashi, in which teams of (male) horsemen compete in grabbing a slaughtered goat (buz). Years earlier, as he was leaving his diplomatic post in Kabul for graduate school at the University of Virginia, an Afghan friend had advised him, "If you want to know what we're really like, go to a buzkashi game." That conversation influenced his choice of a dissertation topic. Five years later, when the news came of the 1978 coup, he asked another Afghan friend what would happen. His friend replied, "You ought to know by now. You have been studying all about it. Now the buzkashi is about to begin." Forty-four years later, it may not be over yet [1].

The buzkashi, however, did not so much reveal what Afghans are really like, as what the international system is really like. As Azoy discovered, buzkashi is much more than a game. Each team has sponsors, often very powerful ones, and the game enacts their political and social competition. Clifford Geertz uncovered the same deep structure in Balinese cockfights [2]. The political-military buzkashi that took place in Afghanistan likewise required the support of sponsors, in this case, foreign sponsors, whose competition was at least as intense and destructive as that among Afghans.

Until 1973, everyone in and outside of Afghanistan, whatever their political views, knew that the state belonged to the king and his family, and the succession was regulated by the constitution. Once Daoud became President of the Republic, a Muhammadzai was still the head of state, providing some continuity of optics, but he had come to power with the help of Parchamis and other army officers, and by the time he was overthrown and killed in April 1978 the constitution approved in 1977 had not been implemented. Daoud was born in 1909, and longevity was not a common attribute among rulers of Afghanistan.

The Soviet leadership might have been content with Daoud had he continued to rule with Parchamis and retained Afghanistan's significant dependence on aid from Moscow. Just a few months after Daoud's August 1973 coup, however, a major transformation began to upend the region: the so-called oil shock that resulted from the Israeli-Arab war in October. The price of crude oil rocketed from $2 to $11 per barrel. Even after the embargo linked to the war ended, prices remained at the same high level, thanks to OPEC production quotas. Between 1972 and 1974 the GDP of Saudi Arabia grew by a factor of 4.7, and Iran by a factor of 2.7. (https://data.worldbank.org/indicator/NY.GDP.PETR.RT.ZS?locations=IR, consulted 07/06/2022). It did not take Daoud long to devise a plan to try to lessen his dependence on the Soviet Union by tapping into the new wealth of Afghanistan's western neighbors.

At first Daoud raised the Pashtunistan issue again—it was his brand. Pakistan under Prime Minister Zulfiqar Ali Bhutto responded by sending Afghan Islamists who had fled to Pakistan to avoid arrest by Daoud's security police—notably Ahmad Shah Massoud and Gulbuddin Hikmatyar—back to Afghanistan to stage an armed uprising in 1975. Nothing happened except in Panjshir, where the government suppressed what it referred to as an "incident" in two days. Daoud went to see Bhutto in Islamabad, and the two of them agreed on a tentative plan to profit jointly from all the money flowing into the Persian Gulf, while enabling Daoud to increase his independence from the USSR and Bhutto to decrease his reliance on the U.S., which was standing in the way of Bhutto's plans to match India's 1974 nuclear test.

Bhutto and Daoud discussed what we would now call "connectivity" between Iran and Pakistan through Afghanistan, in the form of railroads and pipelines. Daoud visited the newly rich oil exporting countries, seeking aid, and signed a military training agreement with Egypt. The Soviet leaders, already wary of their Muslim clients after Egyptian President Anwar Sadat's abrupt pro-US turn in 1974, seemed to interpret this as Daoud planning to emulate the Egyptian ra'is.

In response to all these signals, in 1977 the USSR managed to reunite Khalq and Parcham in a single PDPA as a hedge against future instability. When Daoud paid a state visit to Moscow in April 1977, he raised concerns to Leonid Brezhnev about Soviet sponsorship of PDPA reunification. Brezhnev accused Daoud of allowing NATO activity in Northern Afghanistan, within the Soviet sphere of influence. Daoud bluntly replied that Afghanistan would not compromise its independence. This meeting brought to a head the suspicions by both the USSR and Afghanistan that the other side was no longer abiding by the tacit agreement on limiting great power competition. Afghanistan was still a marginal concern for the U.S., but in the context of Sadat's about-face and the U.S. defeat in Vietnam, Moscow suspected that the U.S. would try to sponsor and benefit from any rapprochement among Iran, Afghanistan, and Pakistan.

The PDPA coup that killed Daoud and his family in April 1978 (the "Sawr Revolution") seems to have been a largely unplanned reaction to an escalation of repression by Daoud, who was concerned about possible Soviet moves to subvert him. There is little evidence that Soviet officials planned the coup, but once it occurred, the West suspended aid, Moscow sent advisors to all parts of government, and it became clear that the modus vivendi that had regulated Soviet-U.S. competition in Afghanistan was now over. Soviet advisors were omnipresent in Afghan government offices, including or especially in the military. In February 1979 the Islamic Revolution in Iran overthrew a regime that had been one of the pillars of U.S. strategy in the Middle East, and Moscow was on the alert for American attempts to establish another outpost in the region. In mid-1979 the U.S. began supplying weapons through Pakistan to Afghan resistance groups.

Fearing that Amin was losing control of Afghanistan, that the Soviets were losing control of Amin, and that the Americans were investing in the mujahidin, in December 1979 the USSR sent a "limited contingent" of troops to Afghanistan. Its mission was, first, regime change, by replacing Hafizullah Amin (who was killed) with Babrak Karmal; and second, counterinsurgency against the mujahidin groups armed and funded mainly by the U.S., Saudi Arabia, and Pakistan. In the absence of the monarchy or even a Muhammadzai-led republic that the great powers might agree to recognize and support, the USSR supported the beleaguered government, and the US (in coordination with Pakistan, Saudi Arabia, and China) supported rival armed groups of various origins seeking to fight their way to power in Kabul. International cooperation over Afghanistan had resumed—the U.S. Saudi, and Chinese intelligence stations met weekly with the ISI in Islamabad to coordinate aid to the resistance –but it was solely aimed at preventing the Soviet troops from stabilizing Moscow's client regime.

The Afghan state was built with the support of international cooperation to reduce the centripetal effects of great power competition. During the 1980s instead of cooperating to aid the state, foreign powers led by the U.S. and USSR funded and armed warring parties, leading the state to lose its monopoly of legitimate violence. The armed forces were hollowed out by defections. The state was no longer able to protect the lines of communication between Kabul and many provinces and districts, leaving them cut off and vulnerable to seizure by the mujahidin. As the government lacked control or even presence in much of the "atraf," or areas outside of major cities, illegal industries such as opium production, timber smuggling, and marketing of lapis lazuli, emeralds, salt, marble, talc, and other extractive products developed outside of state control or regulation. In many areas, mujahidin groups prevented the state from functioning. Except in a few locales, they did not establish alternative administrations, but they provided channels through which people could flee to Pakistan and enjoy international protection as refugees. In a few instances resistance commanders even received Western aid to develop rudimentary public services in areas under their control.

The UN under special envoy Diego Cordovez led an effort to find a diplomatic solution. The ostensible parties, who ultimately signed the Geneva Accords of 1988, were Pakistan and the Soviet-supported government of Afghanistan. What counted most, however, were Cordovez's regular consultations with Washington and Moscow. Following the pattern of previous great power efforts to limit their competition in Afghanistan, the two sides agreed to a mutual pullback from their forward policies of intervention: the Soviet Union would withdraw its troops, and the U.S. would cease support for the mujahidin the moment the withdrawal began. Though the subject was not mentioned in the draft agreement, Moscow could continue to provide Kabul with aid, just as the British provided a subsidy to the Afghan amir within the framework of the 1907 Anglo-Russia convention. Guaranteed support or subsidy to the state, together with great-power agreement on non-interference in Afghanistan and on its strategic alignment (or non-alignment) had long been the necessary conditions for intervals of stability.

This model had proved effective in the past, but President Reagan rejected it at the last moment. He insisted that the U.S. would stop aid to the mujahidin only if Soviet military aid to the Afghan government also ceased. In the language used by the U.S., the obligations of the two parties had to be "symmetrical." In practice the continuing military and financial dependence of the Afghan government on Moscow would have virtually guaranteed a Soviet sphere of influence in Afghanistan even without the presence of Soviet troops, but a cutoff of aid to Kabul virtually guaranteed the collapse of the state, an eventuality that Washington did not consider. The U.S fully supported Pakistani and Saudi efforts to cobble together an interim government consisting of Pakistan-based-mujahidin that it hoped would replace Najibullah's Republic. The political effort, however, did not extend to direct assistance that might have enabled a government to function. Meanwhile the U.S. and Saudi Arabia continued to arm and train the Pakistan-based mujahidin, who included a growing number of foreign volunteers like Osama Bin Laden.

The USSR strenuously objected to what it (rightly) characterized as a violation of the agreement. Soviet diplomats often repeated the motto, "pacta sunt servanda" (agreements must be kept). Soviet leader Mikhail Gorbachev, however, had determined that withdrawal of troops from Afghanistan was in the Soviet national interest, and he honored the withdrawal timetable in the agreement even as the U.S. and Pakistan flagrantly violated the agreement's provisions on ending aid to the mujahidin. The U.S. agreement to withdraw in 2020 as well as the actual withdrawal in 2021 followed a similar logic. Presidents Trump and Biden had concluded that withdrawal was in the U.S. interest and went ahead despite the lack of progress in the inter-Afghan talks.

Even before the Soviet troop withdrawal concluded on February 15, 1989, Najibullah had introduced reforms in the state structure as part of his attempt to stay in power. Especially after he faced a March 1990 coup attempt by Khalqi officers supported by Hizb-i Islami, one of the mujahidin organizations, he relied less on the uniformed official armed forces and more on local militias that he funded and recognized as local authorities. He adopted but could never implement a constitution that made the empowerment of localities the focal point for national reconciliation.

The government like the mujahidin developed local armed groups based on the "andiwali" (friendship) system, in which a local leader recruited his clients and kin. Tax revenues had declined or been diverted due to the lack of territorial control, and export of natural gas to the USSR, an important source of state revenue in the 1980s, ceased when the Soviet troops left and took their engineers with them. To pay for the militias Najibullah increasingly relied on the printing of unbacked currency in Russia, leading to hyperinflation and decline of export income.

The U.S. and USSR engaged in discussions and negotiations over the future of Afghanistan, in an attempt to reinstate cooperation. The U.S called the situation where Moscow helped Kabul and the U.S. supported the mujahidin "positive symmetry." In the negotiations the two sides sought to agree on conditions for "negative symmetry," under which neither side would provide arms to its allies. Once the two sides reached an agreement, they would both terminate their core competitive behavior, the supply of weapons to Najibullah by the USSR and to the mujahidin by the U.S. Both considered placing such limits on their competition as desirable, but they differed on what was required to get there. Both agreed that an interim government without Najibullah should take over, but they disagreed on how to coordinate it with the implementation of negative symmetry. The Soviets insisted that Najibullah should remain in office during the transition even after the implementation of negative symmetry, and leave only when a new government was ready to take office. The U.S., however, wanted to eradicate all Soviet influence at the beginning of the transition and insisted that Najibullah's departure was a precondition for negotiations over a successor government.

As it turned out, the Soviet Union eradicated all Soviet influence in Afghanistan by ceasing to exist at the end of December 1991. By April 1992 Najibullah had resigned, and most of what remained of the state collapsed. With the disappearance of the USSR, great power competition over Afghanistan also disappeared, with the result that no great power perceived an interest in sustaining the Afghan state, which was many times larger, performed far more functions, employed many more people and consequently cost more than it had when Amanullah Khan fled in 1928. Great power competition provided the resources to consolidate the state and then tear it apart; the end of great power competition over Afghanistan led directly to its collapse. In the absence of great power competition, and therefore of cooperation to limit it, there was neither aid to the Afghan state nor an international effort to end the war.

Middle powers in Afghanistan's neighborhood, Pakistan and Saudi Arabia on one side and, on the other side, Iran, India, and Russia (not yet reconstituted as a great power or perceived as a threat by the U.S.) did have interests, but none of them was in a financial, military, or political position to revive the Afghan state. During 1992–1996, Afghanistan had no national security forces but rather a congeries of regional forces, one of which (Massoud's Shura-i Nazar, or Supervisory Council of the North) eventually consolidated control over most of Kabul. The country's customs posts and other sources of revenue were captured and exploited by whatever armed groups had access to them. The state was un-centralized, because the central state collapsed, but it was not decentralized, because there was no institutionalized devolution of power. Many Afghans, however, associate "decentralization" with the collapse of the state.

Mainly motivated by its competition with India, Pakistan invested what it could in the Taliban, whose discipline, hierarchical organization, and religious credentials enabled them to make more effective use of the aid they received than any other group. By 1998 they managed to gain control of the entire country except for Panjshir and Badakhshan east of the Kokcha River. They did so with aid from embedded retired Pakistani military officers, Pakistani madrasa students and other militants, and the forces of al Qaeda, organized militarily as Brigade 055. Their sources of funding included taxing the opium industry and the transit trade between the Persian Gulf and Pakistan [3]. Nonetheless, the Taliban state still had few resources. The following anecdote is, of course, anecdotal, but telling.

In June 1998 I traveled to Kabul as a consultant to the UN Office of the Coordinator for Humanitarian Affairs (UNOCHA). At that time the Afghan government still required both an entrance visa and an exit visa acquired in country. When I went to the foreign ministry to apply for my exit visa, I was directed to a room staffed by two young men, who turned out to be former members of Hizb-i Islami. There was one applicant ahead of me, an Afghan Sikh trader who was applying for an exit visa so he could go to India, and one official was fully occupied with him. I turned to the other official and asked him to process my claim. He answered with some embarrassment that he could not, because the two officers had only one pen between them. This placed me in a dilemma. I had several pens, but UN policy was to provide humanitarian assistance to Afghanistan without engaging in capacity building of the Taliban administration. Lending or even giving them one of my pens would constitute capacity building. As the case of the Sikh trader seemed to be deepening in complexity, however, I took it upon myself to make an exception to the rule and offered my pen to the official. I received the exit visa. While at least one foreign power and a transnational Salafi jihadist movement were supporting the

Taliban, virtually all the aid went to military operations, not to strengthening the state administration.

The first round of Taliban rule and the civil war that preceded it were largely started and pursued as a result of competition among regional rather than great powers, if we define great powers as those with global ambition and reach. During that "unipolar moment," The U.S. was the only great power, but it neither backed one side in a proxy war nor made a serious effort to end it.

Then came 9/11, a unique or at least rare moment in history, in which a great power was attacked and seriously damaged by a non-state actor, al Qaeda, and, even rarer, a non-state actor that was not supported by or a proxy for any government. Al Qaeda was under the protection of the Taliban in Afghanistan, which had the full support of Pakistan, however, and the U.S. had to exert extreme pressure on Pakistan to force General Pervez Musharraf to ambiguously repudiate the Taliban and grant basing and transit rights to the U.S. and its allies.

The war thus started, like the first two Anglo-Afghan wars and the Soviet wars, with a regime change operation. This had two distinct components: the military operations in Afghanistan to destroy al Qaeda bases and remove the Taliban from power (Operation Enduring Freedom); and the political work led by the United Nations to assemble a new government through the U.N. Talks on Afghanistan in Bonn. The two were linked, of course: the U.S. kept up constant pressure on the U.N. to go faster. Russia withheld agreement to participate in Bonn or to press Central Asian states to allow the U.S. to use their bases, until President Bush assured president Putin that the new government would exclude the Taliban, and that the U.S. did not intend to establish permanent bases there. Bush seems to have hinted that the U.S. would leave in five years.

Russia and Iran had long believed that the Taliban were a U.S.-sponsored project to destabilize Central Asia and to lay the groundwork for pipelines through Afghanistan from Central Asia that would marginalize Iran in global hydrocarbon markets and strengthen the sanctions regime. Seen from both Moscow and Tehran, the U.S. had now switched sides, at least temporarily, and was aligned with them rather than with Saudi Arabia and Pakistan. They moved quickly to lock in gains from that realignment.

The CIA paramilitaries and U.S. Special Operations Forces who first linked up with Ahmad Shah Massoud's successors in Panjshir (al Qaeda had assassinated Massoud on September 9) entered the country by helicopter from bases in Tajikistan, use of which was procured for them by Russia. On the ground in Afghanistan, both these U.S. forces and the diplomats who arrived later, relied on assistance they received from members of Iran's al-Qods force. Qods Force commander Qasim Suleimani, who had worked with Massoud, arrived to direct operations personally.

At Bonn, the collaboration of the U.S. delegation led by James Dobbins and Zalmay Khalilzad with the Iranian delegation led by Javad Zarif was essential to the outcome. I was there as an advisor to Lakhdar Brahimi, the chair of the UN delegation. One morning while I was having breakfast, Dobbins and Zarif approached me together. Why, they wanted to know, was there no mention of elections or counter-terrorism in the draft agreement they had seen? I asked if they were the human rights caucus, but everything they jointly wanted got into the document. On the night of December 4 and into the wee hours of December 5, Lakhdar Brahimi convened the main stakeholders in his hotel room: Yunus Qanuni, head of delegation of the United Front, whose forces already controlled Kabul; and diplomats from Iran, the U.S., Russia, India, and Germany. The last remaining item was to agree on the composition of the new cabinet, in which twenty-two out of thirty positions were held by members of one or another component of the United Front. Finally the assembled group had gotten to the point of bargaining with Qanuni over whether he would give up one ministry. He would not budge. Zarif took him off in a corner and whispered something to him. Qanuni came back and said he would give up the ministry of information and culture.

From the beginning of the talks, Burhanuddin Rabbani, president of the government overthrown by the Taliban but still recognized by the United Front, had let it be known that he did not want an agreement that would lead to his replacement. He wanted Qanuni to consult and then return to Kabul, where Russia had installed Rabbani. Qanuni worked hard to circumvent his instructions, but the breakthrough finally came when a Russian representative told Rabbani in Kabul that he would receive no support unless he accepted the Bonn agreement.

For the next several years, the international consensus on Afghanistan, with the exception of Pakistan, held. Even Pakistan did not support a full-scale insurgency by the Taliban until 2006, when the Bonn Agreement had been fully implemented without any participation by Taliban. Then gradually the consensus began to fray, and the context changed fundamentally.

Table 1 and Fig. 1 show the GDPs in 1990 (immediately after the USSR withdrew troops from Afghanistan), 2001 (when the U.S. sent troops to Afghanistan), and 2021 (when the U.S. withdrew troops from Afghanistan) of the four actual or potential great powers with stakes in Afghanistan: the U.S., China, India, and the Russian Federation. At the start of the period, approximately coinciding with the Soviet withdrawal from Afghanistan, China, Russia (USSR), and India all had GDPs about 20 percent the size of the U.S. GDP. By 2021, the U.S. GDP was still five times the size of Russia's. India was about twice the size of Russia, and China had caught up to and slightly surpassed the U.S. Of course, US GDP per capita remains five times that of China, meaning that the U.S. still has more discretionary income not tied to public services for the population and available for foreign adventures. Nonetheless, China, which was a relatively marginal player at the Bonn conference, is now perhaps the most active external state in Afghanistan and has to be involved in any international discussion.

The U.S. enjoyed such broad support starting in 2001 that great power competition over Afghanistan took a long overdue holiday. The lingering unhappiness of Pakistan and its escalating support (creating a permissive environment) for the Taliban hardly amounted to great power competition. But Russia and China supported the operation on the condition that it would be temporary: the U.S. would defeat the terrorists, the common enemy of all mankind, and then withdraw its forces from the Asian continent, where it had never established bases before.

Table 1 GDP (PPP) of great powers involved in Afghanistan in current international U.S. dollars (billions)

	1990	2001	2021
India	1048	2371	10,219
China	1228	4290	27,850
Russian Federation	1188	1075	4785
United States	5963	10,582	22,996

Source World Bank Data. China includes Hong Kong and Macao

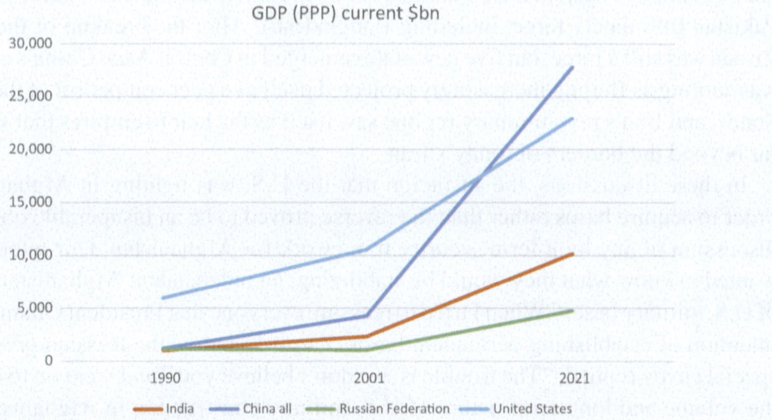

Fig. 1 GDP growth from 1990 to 2021

In 2003 the combination of the U.S. illegal invasion of Iraq and the decision to make Afghanistan into a NATO operation sparked some concern about U.S. future intentions. The May 23, 2005, U.S.-Afghan joint declaration of "strategic partnership" (https://www.govinfo.gov/content/pkg/WCPD-2005-05-30/pdf/WCPD-2005-05-30-Pg863.pdf) hardened the belief among some in Russia, China, and Iran that the U.S. was exploiting their cooperation in the name of the common objective of counter-terrorism to establish a foothold for the first time in continental Asia.

The response was not long in coming. The heads of member states of the Shanghai Cooperation Organization issued a declaration from their July 5, 2005, meeting in Astana (http://eng.sectsco.org/documents/). The declaration expressed support for counter-terrorism efforts and noted that a number of SCO member states had made infrastructure available to the "antiterrorist coalition" in Afghanistan. This referred primarily to Manas air base in Kyrgyzstan. It went on to observe:

> Given the completion of the active military phase of the antiterrorist operation in Afghanistan, the member states of the Shanghai Cooperation Organization deem it necessary for the relevant participating states of the antiterrorist coalition to set a deadline for the temporary use of said infrastructure and presence of their military contingents in the territory of the SCO member states.

As the war escalated, and the U.S. and NATO ignored the concerns expressed in the SCO declaration, suspicions likewise escalated. During 2009–2013, while I was working in the office of the U.S. Special Representative for Afghanistan and Pakistan, I organized several regional meetings in track 1.5 format (officials meeting for unofficial off the record discussions). The regional format reflected how the challenge of coordinating interstate competition over Afghanistan had become increasingly complicated. In the age of imperialism, it sufficed for Britain and Russia to reach agreement. During the Cold War both the U.S. and USSR were dominant leaders of the two alliance systems. Decolonization of India left behind two states, India and Pakistan (ultimately three, including Bangladesh). After the breakup of the USSR, Russia was still a force, but five new states emerged in Central Asia. China's economy was surging as Beijing increasingly projected itself as a peer competitor of the United States, and Iran's revolutionary regime saw itself as the heir to empires that stretched far beyond the borders of today's Iran.

In these discussions, the suspicion that the U.S. was fighting in Afghanistan in order to acquire bases rather than the reverse proved to be an insuperable obstacle to discussion of any long-term security framework for Afghanistan. Our interlocutors wanted to know what they would be stabilizing: an independent Afghanistan or a set of U.S. military bases? When I tried to reassure everyone that President Obama had no intention of establishing permanent bases, Zamir Kabulov, the Russian presidential special envoy replied, "The trouble is, we don't believe you," and went on to describe the volume and long-term nature of U.S. military construction in Afghanistan. The Iranian representative admitted that, "Perhaps Obama is a man of peace" but noted that the Republicans might come to power. He then went on to describe the three powerful groups in the U.S. who wanted to continue the war: neo-conservatives, defense contractors, and Christian extremists who believed war was a sign of the end times when only believers would be saved. I could not say he was completely wrong.

Perhaps even more than previous regime-change operations in Afghanistan, the U.S.-led intervention was so massive, not only in troops, but more important, in money, that it affected all aspects of society's relationship with the state. For better or worse, it did so in contradictory ways. The dual agendas of counter-terrorism and liberal peace were both internally incoherent and often in conflict with each other.

The first thing the U.S. did was to supercharge the forces of un-centralization. In the tradition of Afghanistan's Pashtun rulers, the Taliban had established a completely centralized Islamic Emirate in which the first principle was obedience to the amir. The Taliban army was centralized in a single command obedient to Mullah Omar. The Taliban were not believers in the second amendment. They seemed to believe that poorly regulated militias being inimical to the security of an Islamic Emirate, the right to keep and bear arms must be infringed. The first thing they did when capturing a locality was to go door to door collecting weapons from private hands to assure that the state—the Islamic Emirate—had a monopoly of the use of force.

The United Front, on the other hand, was a coalition of factional or ethnic militias with no centralized command and control, a pattern I have referred to as "un-centralization." For several years before 9/11, Massoud as defense minister of the Islamic State headed by President Rabbani, was trying to bring these various groups under a common command, in particular by seeking to persuade the UF's benefactors (mainly Iran and Russia) to provide assistance entirely through him, so he would have a strategic tool for coordination.

When the U.S. intervened, its actions on the ground destroyed not only the centralized system of the Taliban, but also the tentative steps toward command and control that Massoud had taken. The CIA decided to distribute aid directly to commanders so that the U.S. rather than any Afghans, would have strategic oversight of the campaign. The result was a pattern of warlordism or un-centralization, in which the principal commanders of each region consolidated power in their own area. The warlords or regional powerholders appropriated customs and other state revenues for themselves. Their armed forces, rather than being subordinate to a common authority, were independent of each other and sometimes engaged in armed clashes. In March 2002, I was having dinner in Kabul with Defense Minister Marshall Qasim Fahim when a message arrived that forces of Abdul Rashid Dostum (former Najibullah militia, Uzbek) and Muhammad Atta (Jamiat-I Islami, Tajik), both of whom had received millions of dollars from the U.S., were fighting over the control of a factory near Mazar-i Sharif that manufactured fertilizer from petroleum, one of the most lucrative assets in northern Afghanistan. Fahim got them on the phone and learned that they were eating watermelon together (a ritual of reconciliation) and, most likely, negotiating how to divvy up the revenues from the factory. They had previously done so with the revenue from the customs post in Hairatan, on the Uzbekistan border.

At Bonn (and on the ground in Afghanistan), the U.S. had supported the political talks convened by the U.N. under the chairmanship of Lakhdar Brahimi. This is the meeting at which the U.S. supposedly imposed a centralized system of government on "traditionally decentralized" Afghanistan. In fact, no one at Bonn ever discussed the degree of centralization of the government, except for a small delegation of Afghanistan Uzbeks from Germany who showed up at the end after all decisions had been made, talked about federalism, but had no written proposals. The centralization of the administration was implicit in the adoption of the constitution of 1964 (minus its provisions on the monarchy and parliament). That constitution codified existing practice since Abdul Rahman Khan by providing in article 108 that, "The administration of Afghanistan is based upon the principle of centralization, in accordance with the provisions of this title." Article 137 of the constitution of 2004 provided: "The government, in preserving the principle of centralization, shall transfer necessary powers, in accordance with the law to local administrations..." Various attempts at legislation to implement such devolution were either stalled in parliament or in implementation.

The main debate at Bonn, and later during the constitutional process and Loya Jirga in 2003–2004, was not over the centralization of the administration, but over the structure of the national executive. Pashtuns generally favored a pure presidential system, while non-Pashtun political leaders and groups favored either a parliamentary system or a semi-presidential system with a Prime Minister, which they saw as a mechanism for power sharing but which had nothing to do with decentralization. The U.S. may have had some influence over the rejection of the semi-presidential system in favor of the presidential system, but it also had the support of President Karzai and most Pashtuns.

The main great power influence on how Afghanistan was ruled during the U.S. occupation was not incorporated in any constitutional or legal provisions, but resulted from how the U.S. carried out military and counter-terrorist campaigns and, how it provided security for its bases. The most powerful Afghans in Afghanistan were the commanders funded by the U.S., sometimes while they served in official positions and sometimes while they did not. Over time, as the Afghan military became more capable, U.S. spending shifted from militias engaged in kinetic operations to private security companies, that were basically rebranded militias.

The priority the U.S. gave to the military campaign was felt at Bonn, where at Washington's insistence, the jurisdiction of the International Security Assistance Force was limited to "Kabul and environs" rather than including major regional and provincial centers. The presence of ISAF was supposed to replace the militia commanders as security providers, creating space for non-military politics as provided for in the Bonn agreement and then the constitution. The restriction of ISAF's geographical mandate meant that there would be international security measures making peaceful politics possible only in Kabul and environs. Even that was diluted. Annex A to the Bonn Agreement said that all armed forces and militia other than ISAF were to withdraw from Kabul, but Fahim's forces had gotten there first and controlled the streets. We felt the contradiction at Bonn. Word came both through the United Front delegation and from U.S. forces on the ground that some of the fighters armed and funded by the U.S. opposed parts of the Bonn Agreement supported by the U.S. American and UN diplomats had to negotiate agreement with U.S. funded commanders who threatened to sabotage the agreement. This was a preview of things to come. Such a hybrid system may be called many things, but "centralized" is not one of them.

The assumption of the ISAF command by NATO and the founding of provincial reconstruction teams mandated with extending the reach of the state was intended in part to redress this. PRTs, however, turned out to be a strong force for uncentralization. In many provinces the PRT was the main source of assistance, not the government. The governors had much more incentive to cooperate with the PRTs than with the distant president's office. The population learned who really had power and it was not the central government. It was the PRT, sometimes in partnership with the governor. As the U.S. drew down forces after 2012 it also began relying more on various special police or para-military forces that were barely integrated into any chain of command.

As the U.S. presence continued with no possibility of victory and no serious attempt to find a solution, the views of the neighboring states also changed. As noted, in response to 9/11, a very broad coalition supported U.S. action, including Russia, Iran, and China. The longer the war went on with no resolution in sight, and no prospect of a U.S. withdrawal, the more the neighboring states began to see the U.S. military presence itself as a threat, along with the terrorist groups whom the U.S. claimed to be fighting.

Toward the end of Karzai's administration, the U.S. negotiated a Strategic Partnership Agreement with Afghanistan, which was more detailed than the 2005 declaration and would be legally binding. President Karzai, however, refused to sign it, saying he did not want Afghanistan to be so tied to the United States. Immediately after President Ghani was inaugurated in September 2014, however, he sent his national security advisor, Haneef Atmar, to sign it, confirming the estimates of the neighbors that Ghani, a former U.S. citizen, would not deviate from U.S. plans. Those plans might include efforts to use the U.S. presence in Afghanistan for intelligence purposes, obviously against Iran, but also against Russia via Central Asia.

China however, was on a somewhat different track. China had three main concerns about Afghanistan:

- It wanted to harden the border between the two countries (in fact the border between Badakhshan and Xinjiang) and increase surveillance to prevent any cross-border activity by Uyghur Islamist separatists, some of whom had returned to the region after fighting along with the Islamic State (Da'esh) in Syria and Iraq.
- It wanted to expand the Belt and Road Initiative into and around Afghanistan. China hoped to integrate Afghanistan into the China-Pakistan Economic Corridor. This was President Xi Jinping's signature policy, and it required a secure environment.
- China also had been trying to invest in the extraction of natural resources from Afghanistan. It had signed a contract to open what would have been the world's largest copper mine in Mes Aynak, Logar province. Implementation of the project had been halted because of security concerns and the presence of valuable archeological remains.

At that time China was still following the path of "peaceful rise" and trying to establish a "new kind of great power relationship" with the United States. As part of the effort to minimize competition and enhance cooperation with the U.S., China offered to cooperate on Afghanistan. The idea originated in the Center for the Study of the U.S. in the Chinese Academy of Social Sciences, rather than in any department specifically concerned with Afghanistan.

The Obama administration responded to Chinese overtures to cooperate on Afghanistan. It agreed first to conduct joint training of Afghan diplomats. Cooperation then began on the peace process. In July 2015 Pakistan hosted the first direct talks between the Taliban and the Afghan government's High Peace Council in Murree, with both the U.S. and Chinese as observers. This process however was stalled. In July 2015 the revelation that Taliban leader Mullah Omar had died two years earlier disrupted the talks. After lengthy efforts to revive the talks, in May 2016 the U.S. assassination by drone of Taliban leader Akhtar Muhammad Mansur while he was driving to his home outside Quetta from a long visit to Iran, derailed the talks again. China concluded that the U.S. was not serious about peace talks and put an end to bilateral cooperation with the U.S. on an Afghan peace process.

Great power competition over Afghanistan had made a full comeback. Iran and Russia had reversed their longstanding opposition to negotiating with the Taliban. Russia and Pakistan also had a rapprochement. Iran and Russia both opened contact with the Taliban, Russia in the Taliban's Doha office, opened with the support of the U.S. in 2013, and Iran largely along their border. Soon U.S. military and intelligence sources reported that both were providing arms to the Taliban. Iran confirmed that to me but said it was only for border security and to fight the Islamic State. Russia denied it, but Russian military sources claimed that Russia was providing some assistance to "warlords" in Northern Afghanistan. In December 2020 Afghan intelligence in Kabul arrested 10 Chinese agents, who were allegedly contacting the Haqqani network to provide security for Chinese projects.

Seeing that efforts by the U.S. to negotiate a settlement were stalled, Kabulov started what he called the "Moscow Process." He said that peace would not come from the U.S. but from the region. The core countries in Kabulov's conception were Russia, China, Iran, and Pakistan, whom he convened in Moscow in December 2016. If the U.S. joins and supports the process, it will be more effective, but if not, the region would go ahead on its own. In subsequent meetings in 2017–2018, Russia enlarged the group to include Central Asian states, India and Afghanistan. The most important aspect of "peace" would be the complete withdrawal of U.S. troops, which Moscow considered as a potential threat to Russia by way of Central Asia. This was consistent with Russia's repeated support for the neutrality of Afghanistan. The situation was complicated, however, by the need to combat the Islamic State and militant groups from Central Asia and China's Xinjiang province. Russia claimed, however, that the U.S. was sheltering these groups in order to destabilize Afghanistan's neighbors. The Moscow format never addressed the issue of a subsidy for the Afghan government.

Donald Trump became president with a record of calls to end the Afghan war but no clear policy. At first Trump's administration viewed the Moscow process through the lens of competition with Moscow. Even when invited, the U.S refused to attend. It even dispatched an ambassador of the U.S. Embassy in Kabul (where there were four ambassadors) to the Afghan Ministry of Foreign Affairs to forbid Afghanistan from participating.

The Trump administration's initial policy on Afghanistan, issued in August 2017, was more reflective of the views of National Security advisor General H. R. McMaster than of Trump's. It did not even mention Russia and China. The policy combined counter-insurgency with diplomatic pressure on Pakistan, including by encouraging more Indian involvement in Afghanistan. This was a foretaste of the administration's national security strategy, issued in December 2017, which prioritized "great power competition" with Russia and China.

Trump and McMaster pursued this strategy while Kabulov advanced the Moscow Process. By August 2018, however, the National Intelligence Council reissued one of its greatest hits and submitted to the president a revised and updated version of "Afghanistan: Nothing You Are Doing Is Working." The military's response, of course, was that they needed more time, but Trump seized on the report as evidence that his instinct to get out was correct. On September 5, 2018, he and Secretary of State Mike Pompeo appointed veteran diplomat Zalmay Khalilzad as Special Advisor on Afghanistan with a mandate to negotiate the terms of a U.S. withdrawal, including a political settlement with the Taliban.

Khalilzad knew that a political settlement could be reached and implemented only with the backing of the other two great powers, China and Russia, who were also neighbors of Afghanistan. Over the previous several years, U.S. opposition to the Moscow process, public accusations that Russia was arming the Taliban, and the apparent determination of the U.S. to make Afghanistan a permanent military outpost had brought U.S.-Russia relations over Afghanistan to a new low. Khalilzad reversed the boycott of the Moscow Process and authorized the participation of an official of the U.S. embassy in Moscow in the next session, scheduled for November 2018, with the participation of representatives of both the Taliban and the Afghan government. Soon after, Khalilzad visited Moscow, where he met a skeptical Kabulov, who considered Khalilzad an architect of the U.S. long-term presence in Afghanistan. Both Khalilzad and Kabulov said the meeting went as well as could be expected, and they started to develop an unanticipated partnership.

The Russians held one more meeting of the Moscow process: a long-anticipated meeting in a Moscow Hotel of delegations from the Afghan Taliban and representatives of the Islamic Republic, if not directly of the Ghani government, whom the Taliban refused to meet. Khalilzad meanwhile opened direct U.S.-Taliban talks in Doha. In March, on a tour of the Middle East, Russian foreign minister Sergei Lavrov announced in Qatar that Russia that Khalilzad and Kabulov would be "closely cooperating." (https://www.mid.ru/en/press_service/minister_speeches/-/asset_publisher/7OvQR5KJWVmR/content/id/3555141). While he did not say so, he suspended the Doha process.

The next step was to create a mechanism for great power support of the Doha negotiations. Khalilzad invited the special representatives of Russia, China, and the EU to Washington for a meeting on March 22, but Russia refused to join a meeting with the EU. Russia considered that including the EU amounted to giving the U.S. an extra vote. The EU, like India, Kabulov argued, would be involved at a later stage, involving economic reconstruction. Instead, the U.S., Russia, and China formed what they named the "troika" and invited the EU to join them for lunch.

The troika met again in April, in Moscow. Following Kabulov's proposed architecture, the troika invited Pakistan and Iran to join them at the July 2019 meeting in Beijing. Pakistan accepted, but Iran, still stinging from President Trump's repudiation of the JCPOA (nuclear accord), declined. The group referred to the troika plus Pakistan as the "extended troika." The troika or extended troika held four meetings in 2019 and two meetings virtually in 2020. All members of the extended troika attended the signing of the Doha agreement between the U.S. and the Taliban on February 29, 2020.

The Doha agreement set May 1, 2021, as the deadline for the U.S. troop withdrawal, exactly one hundred days after the inauguration of President Joe Biden. On February 22, 2021, I was contacted by the aide to a very senior official of the Chinese ministry of foreign affairs. I had known the official, along with the aide for about ten years. He asked if I could arrange a meeting with a high-level State Department official "in charge of South Asia issues, including the Afghan peace process." Despite my efforts to contact everyone I could, the U.S. never responded at the appropriate level, and the meeting did not take place. Instead, the extended troika met at the level of special representative on March 18 and again on April 30. At those points the obstacles to a comprehensive settlement were primarily that the Taliban thought they had won and saw no reason to negotiate, and the Afghan government refused to believe that the U.S. would really withdraw its troops. The way the extended troika was supposed to work was:

- The four members would reach consensus on what was needed.
- Pakistan would use its influence on the Taliban.
- China would use its influence on Pakistan.
- Russia would use its influence on Iran, or at least keep it informed.
- The U.S. would use its influence on the Afghan government in Kabul.

The architecture was perfect, but the influence it presumed may have been an illusion. The government fell after the stealthy departure of President Ghani on August 15. The U.S. troop withdrawal ended on August 31 amid scenes of panicked Afghans trying to leave the country. The security forces collapsed, and the Taliban walked unopposed into Kabul.

In the aftermath, China, Iran, and Russia all accused the U.S. of a precipitous and irresponsible withdrawal. While the completion in 2021 of a withdrawal that was first announced in 2012 may not really qualify as precipitous, the great powers and other neighbors all felt they had not been genuinely consulted about the withdrawal and the consequences for them. President Biden and Secretary of State Anthony Blinken seem to have spent more time talking to NATO than to any of the countries directly affected by the U.S. decision.

Nearly a year after the fall of the Islamic Republic, there are few signs of the re-emergence of destructive competition between great and not-so great powers. The National Resistance Front led by Ahmad Massoud has been allowed to set up its headquarters in Dushanbe and travel freely. Russia may find the NRF useful from time to time to send signals to the Taliban, but there is no sign of any organized effort by any state to support and arm it. Contrary to some speculation, China is not moving in to play the role previously assumed by the U.S. or the USSR. China seems to have sent more personnel to Afghanistan, and with the temporary at least improvement of the security situation it is exploring investment opportunities, but China is still not satisfied with the efforts of the Taliban to bring the Uyghur militants under control. The Taliban have moved them away from the Xinjiang border to an area bordering on Tajikistan and Uzbekistan, but China would like them to be arrested and turned over.

The U.S. has few options. President Biden has long believed that the U.S. has put too many resources into Afghanistan. He has downgraded the office of the special representative. Since the creation of the position in the Obama administration, the SR reported directly to the Secretary of State. In the Biden administration the SR reports to the assistant secretary of state for south and central Asia. The U.S.'s freeze on most of Afghanistan's foreign exchange assets is making a palpable contribution to the humanitarian emergency in Afghanistan, but the talks with the Taliban on releasing some of them seem to be moving at a snail's pace.

The Taliban are trying to re-establish the highly centralized system of the Islamic Emirate. This has resulted in greater control over government revenues, but it has also sparked revolts among the Taliban's own Tajik, Uzbek, and Hazara commanders in North Afghanistan. So far these revolts look like a bigger threat to the Taliban than the NRF, but some former Taliban commanders have joined the NRF. The NRF says that it is not trying to overthrow the Islamic Emirate and that the only issue is decentralization. Genuine decentralization, however, would require a degree of security and administrative capacity that Afghanistan may lack for some time.

At this point no great (or not so great) power seems to consider Afghanistan valuable enough to invest in or dangerous enough to invade or even wage a proxy war over. Unfortunately, that may be the best alternative available to this long-suffering country.

References

1. Azoy GW (1982) Buzkashi, game and power in afghanistan. United States: university of Pennsylvania Press
2. Clifford Geertz (1972) Deep play: notes on the balinese cockfight daedalus. Vol 101, No 1, Myth, Symbol, and Culture (Winter, 1972), pp 1–37
3. Rubin, Barnett R (2000) The political economy of war and peace. World Development Vol 28, Issue 10, Oct 2000, pp 1789–1803

Dr. Barnett R. Rubin is a Senior Fellow at CIC, where he previously directed the Afghanistan Pakistan Regional Program. He worked at CIC from 2000 to 2020. During 1994–2000 he was Director of the Center for Preventive Action, and Director, Peace and Conflict Studies, at the Council on Foreign Relations in New York. Rubin was Associate Professor of Political Science and Director of the Center for the Study of Central Asia at Columbia University from 1990 to 1996. Previously, he was a Jennings Randolph Peace Fellow at the United States Institute of Peace and Assistant Professor of Political Science at Yale University. From April 2009 until October 2013, Dr. Rubin was the Senior Adviser to the Special Representative for Afghanistan and Pakistan in the U.S. Department of State. In November–December 2001 Rubin served as special advisor to the UN Special Representative of the Secretary General for Afghanistan, during the negotiations that produced the Bonn Agreement. He subsequently advised the United Nations Assistance Mission in Afghanistan on the drafting of the constitution of Afghanistan, the Afghanistan Compact, and the Afghanistan National Development Strategy.

Science of Decision-Making Related to Afghanistan and the CASA Region

Sean Ryan

Abstract The old adage states that managers do things the right way, and leaders do the right things. This chapter will not debate whether the U.S. did the right thing by invading Afghanistan. Rather, this chapter is dedicated to addressing the myriad of factors involved in making decisions that led to that response to the 9/11 attacks. A single chapter is not sufficient to fully address complex decision processes. However, it is my intent in this chapter to illuminate factors that influence decisions that guided the use of military force after the 9/11 attacks and highlight areas where additional research is warranted. "Lessons learned" implies that we seek knowledge that can be applied in future situations. This further implies that we had choices, hence decisions to make. This paper, therefore, attempts to address certain factors that military and strategic planners and decision makers should consider before engaging in the use of force and during extended operations. Addressing specific baseline factors related to national security level decisions in the application of the use of force will serve as a starting point in this chapter. Some decision theories and associated concepts will help to provide context for the next portion of this discussion, which offers a model for decision-making, including how temporal factors impact decision-making—or should.

Keywords Afghanistan · Decision-making · Strategic planner · CASA region

1 Brief Synopsis

Tellis and Eggers [25] provided a summary chronology of the U.S. involvement in Afghanistan. While Osama bin Laden and Al Qaida was the initial focus which brought the U.S. to Afghanistan, conditions and operational focus seemed to ebb and flow. US Forces did not take a lead role in fighting in Tora Bora in December 2001 despite intel indicating bin Laden was there. Then, Mullah Omar fled Kandahar and Afghanistan when Kandahar fell on December 9, 2001. Al Qaida remained present

S. Ryan (✉)
Arizona State University, Tempe, AZ 85281, USA
e-mail: sean1ryan@aol.com

A. Farhadi and A. Masys (eds.), *The Great Power Competition Volume 4*,
https://doi.org/10.1007/978-3-031-22934-3_20

and defiant in the mountains while tactical operations focused on ousting the Taliban. What was the primary operational objectives for the United States?

OPERATION Anaconda took the fight to about 800 Taliban and Al Qaida fighters in the Shah-i-kot vally in Paktia Province. Tellis and Eggers highlighted that Anaconda was the largest operation since Tora Bora. It was around this time in 2002 when priorities of U.S. resources started shifting to Iraq (2017). Operationally speaking this campaign against the Al Qaida terrorist network, and by extension the Taliban regime in Afghanistan, was relatively short in duration. The U.S. was involved in Afghanistan only about six months by the Spring of 2002. Political considerations coupled with flawed intelligence reporting distracted the U.S. administration from its initial objectives after 9/11. Could more clear, more concrete and achievable objectives have prevented the United States from starting a two-front conflict against enigmatic adversaries?

May 1st 2003 marked the end of major combat operations and a shift towards nation building. There was significant banter over what terminology should be used. Nation building, Capacity Building, and Stability Operations were all used at varying times. Much effort went towards veering away from images reminiscent of the U.S. experience in Vietnam. The following two years were relatively calm in Afghanistan while the Taliban reorganized across the border in Pakistan.

From 2005 to 2006, violence increased from 27 suicide attacks in 2005 to 139 in 2006 while remotely detonated explosives rose to 1677, according to Tellis and Eggers [25]. Improvised force structure in the form of Provincial Reconstruction Teams (PRT) were fielded to expand governance. PRTs were followed by Agribusiness Development Teams (ADTs) focused on assisting local farmers improve agriculture. These innovations preceded McChrystal's assumption of command on May 11, 2009. McChrystal placed constraints on air strikes to avoid prioritizing tactical success over the Taliban strategic gains with the population. Thus the focus was now on winning hearts and minds.

General David H. Petraeus highlighted that force structure and conditions drove tactical deployments during his tenure, which began in the Summer of 2010 (Petraeus, Personal Communications, 2022). To summarize force levels in Afghanistan, Tellis and Eggers [25] reported that there were 37 K troops in Afghanistan when Obama assumed the Presidency in 2009. NATO promised to add another 5000 more troops.

Then Obama shifted strategy March 27, 2009: "The core goal of the strategy, as outlined in an interagency white paper, is "to disrupt, dismantle, and defeat al Qaeda and its safe havens in Pakistan, and to prevent their return to Pakistan or Afghanistan." U.S. forces targeted to be 60–68 K by Aug 2009 ahead of the Afghanistan elections in November. Obama increased U.S. troop levels to 98,000 in December 2009 with a focus on building Afghan capacity and setting conditions for the transfer of authority to Afghanistan from U.S. and NATO forces [25].

By the tenth anniversary of the 9/11 attacks, NATO set the end of 2014 as a date for transfer of authority to Afghanistan at the Lisbon Conference. This was strategically naïve and showed the Taliban that U.S. and international commitment was waning. Osama bin Laden was killed in Pakistan. The cost of U.S. involvement in Afghanistan rose 1,800 casualties and $444 Billion. These factors all took place in the backdrop of the 2008 real estate crash, a global recession, and 9.1% unemployment in the United States. Support for protracted involvement in an increasingly violent Afghanistan was withering [25]. While these specific events may not have been foreseeable, the fact remains that Americans are not inherently patient. Prolonged conflict is not likely to retain support.

2 Foundations in Decision-Making

There are significant similarities between military planning processes and the planning processes used in business circles. A brief review of the rational decision model used in planning is useful before delving into discrete factors. Rational Decision making models are used in both business and military settings. Heraclelous, however, argues that the implicit assumptions underpinning rational decision making are unrealistic [9]. Whether military commanders or civilian leaders can be truly rational and logical may be an interesting debate, but is beyond the scope of this chapter. Focusing on the sustainability of the decision, leads to an interesting observation, however. One aspect of business that the military could benefit from is the focus on sustainable competitive advantage. Beyond tactical victories, U.S. military planners and decision makers must consider the long-term strategic advantages of specific actions.

3 Traditional Planning Process

Joint publication 5-0, Operation Planning, describes the joint planning process and the military's version of a rational decision-making methodology. Buried in Appendix G, Course of Action Comparison, JP 5-0 outlines a decision matrix that ascribes weighted quantitative comparison of the courses of action under consideration [22]. This methodology is simple enough as JP 5-0 describes in detail how to construct weighted decision matrices using criteria from the earlier planning process, assigning numerical weights, and deriving a consensus opinion across the staff. Highlighted is the need for senior leadership to approve the decision criteria and weights. This process represents an ideal situation. Political factors or a commander's preconceptions may override rational decisions. JP 5-0 also described is a non-qualitative approach that avoids using weighted values.

This military decision process is identical to what is known in business circles as a Quantitative Strategic Planning Matrix (QSPM). The QSPM uses business terminology, but is comparable in structure and content to the military equivalent [6]. This methodology is tested and reliable, even in complicated situations, where the variables are well known and criteria for success are clear or reasonably concluded. In complex situations where socio-cultural variables dominate, however, rational decision-making processes are less reliable.

JP 5-0 outlines the Joint Planning Process based on analyzing situations supported by facts and relevant assumptions. There are volumes published on various approaches to decision making. The purpose of this paper is not to espouse one method over another. Rather, we look at the strengths and weaknesses of some of these methods and highlight specific factors that are frequently ignored or overlooked.

When considering the U.S. entry into Afghanistan, we must accept that the decision to apply military force was driven by political considerations in the wake of the September 11th attacks. JP 5-0 emphasizes that national-level objectives must be clearly understood. There we must ask, "what clearly achievable military objectives supported an invasion as to a retaliatory strike against the Taliban regime that enabled and protected Osama bin Laden?" No plan can be effective if the objectives are ambiguous or are not connected decisively to the national-level objectives.

Next, we look at the nature of the operation and courses of action under consideration. Osama bin Laden and Al Qaida were the adversaries that attacked the United States. The state of Afghanistan under the Taliban regime was an enabler. While we can argue that destroying the Taliban regime was a critical path requirement to neutralizing Al Qaida, the key issue here is whether the destruction of the Taliban state was the principal target.

If destroying the Taliban regime was necessary, and we are not arguing either way, then deciding *how* is the ultimate question. Leveraging the insurgency represented by the Northern Alliance introduces complexity. Insurgency is a human endeavor driven by the passion of the insurgent leaders. Overthrowing an established government, like the Taliban, inherently results in a change between the haves and the have-nots. This aspect of Unconventional Warfare (UW) was understood—at least by the Special Forces community that initiated the invasion of Afghanistan in October 2001. A series of questions arise when considering UW:

Will the former insurgent leaders ultimately form an acceptable and sustainable government that will support long-term strategic U.S. goals? Where there are multiple insurgent elements, do they share the same goals? Can they work together?
Do we need to concern ourselves with the long-term outcomes?
Will the U.S. need to remain committed for the long term? And, if so, how long?
Will the regional environment support a long-term sustainable outcome?
Is there political will to remain engaged?
What role *should* the U.S. play post-regime change?

These questions allude to the fact that the United States leapt into a complex situation in 2001. John Boyd, author of the Boyd Cycle or OODA Loop, drew from Bertalanffy in terms of understanding the complexities of inter-connected systems [16, p. 71]. Boyd viewed things in terms of systems of systems. He advocated for a need to synthesize knowledge with intuition, which is appropriate in strategic situations. This perspective is beneficial when considering complex situations, especially those nested in the human domain.

There exists significant discussion on whether the U.S. and NATO Coalition should have tried to impose western democracy on the Afghan state. The human conditions of the prior 30 years (as of 2001) coupled with the historic cultural paradigms present in Afghanistan are generally accepted as rendering democracy unsupportable as a form of sustainable governance. Of course, the push for democratic governance was driven by ideologues, which delves into the political realm.

These complexities lead us to question whether traditional rational decision-making processes are well suited to this situation. Candidly, the lack of understanding of the complex cultural and situational factors—or the outright disregard for them by decision makers—introduced so many variables and implicit assumptions into the operational decisions that rational decisions were an illusion at best. There was no foundation for assigning weights or assessing risks rationally. Only a keen understanding of history coupled with an intimate knowledge of Afghan cultures could have informed long term perspectives.

This situation in 2001 then approaches what Mintzberg [14] contends is a conflict between planning and strategy. In the context of Afghanistan in 2001, planners were tasked with analyzing the situation which was inherently complex and devising an operational plan to remove the Taliban regime. Available intelligence was incomplete and there was hardly any foundation for making intuitive judgements. Information derived from the Northern Alliance had to be considered compromised by their own parochial agendas.

Supporting the Northern Alliance as an option to oppose the Taliban regime was an easy approach to latch onto for planners, especially in the Special Forces community. Unconventional Warfare focuses on activities to coerce, disrupt or overthrow a hostile regime as explained in Joint Publication 3-05.1. The literal definition of Unconventional Warfare evolved over time; regardless of the exact wording, the underlying truth is that leveraging the Northern Alliance to oust the Taliban regime was a natural doctrinal fit for U.S. Army Special Forces supported by joint and allied elements [28].

It is worth mentioning that Special Forces (SF) and Special Operations Forces (SOF) would serve as the lead element in this initial campaign to oust the regime that supported Al Qaida and was hiding Osama bin Laden. Historically, SF and SOF were placed in supporting roles in larger campaigns. Exceptions, like the U.S. Foreign Internal Defense (FID) mission in El Salvador was limited to 55 full-time personnel fell under the Cold War containment strategy [28]. The opportunity introduced a strong political incentive for SOF to accept the operational lead in implementing a plan that promised pay back for the September 11th attacks. The broader question of whether overthrowing the Taliban regime would produce a sustainable outcome was secondary.

Regarding the contrast between operational planning and strategy, Mintzberg [14] emphasized that strategy requires synthesis. The very nature of the complex human terrain in Afghanistan and the surrounding region undermines any chance of synthesizing facts, well-founded assumptions, and intuitive understanding. U.S. planners fell short in all three areas.

4 Foreign Policy

In our pursuit of lessons from our choices in Afghanistan, foreign policy is a central consideration. Hermann and Peterson highlighted three factors as being critical in decisions to use force in the context of foreign policy:

Political interests and priorities;
Perceived threats to those interests and perceived opportunities to advance them that national leaders see and recognize;
Ideas about "the utility, effectiveness, efficiency and acceptability at home and abroad" of using force along with the moral implications [19].

Hermann and Peterson emphasize that baseline changes in attitudes regarding the use of force can change based on context [19]. Specifically, they differentiate between the use of force involving nuclear weapons and all other cases. And, it is important to note that while there was a baseline shift from pragmatic to idealistic along with a general willingness to reduce support to NATO in the post-Cold War era, the 9/11 attacks involved people from around the globe and occurred on American soil. There was no resistance to the proposed use of force following 9/11.

A key consideration in applying the perspectives of Hermann and Peterson in the Afghanistan case relates to the implicit assumption posed in the second factor [19]. What holistic opportunities did political leaders and operational planners see *and recognize* when considering the use of force specifically to remove the Taliban regime in Afghanistan? It is easy to assume away risks and focus on the opportunities when dealing with operations in the human domain. As history demonstrated, approximately 350 U.S. Special Operations Forces personnel, working with a 100 or so interagency operatives, effectively overthrew the Taliban regime by supporting the Northern Alliance resistance in a classic UW operation [28]. The outstanding question remains, what well-informed considerations were given to what came next?

Whereas the use of force in the wake of the 9/11 attacks were supported at home and abroad, nation building was historically not supported [19]. What other options were available that could have offered the political expediency and moral acceptability while also mitigating the likelihood of 20 years and thousands of lives wasted on stabilizing a population bordered by a sanctuary in Pakistan that was unprepared to govern itself. Conventional and even SOF military planners generally do not think in those protracted terms.

5 Planning Versus Strategy

Recent research posited that planning in the near term involves different cognitive activity and long-term planning. Accepting Mintzberg's position that planning is fundamentally different than strategic thinking, an interesting question arises. Are military planners best suited for planning campaigns that will likely exceed the 12-month threshold presented in the relatively recent research? Conclusions by Wittman and Paulus [29] suggest that the estimated duration of the activity that will yield the desired outcome(s) will impact the decision.

Kirby and Marakovic [11] demonstrated that a preference for an earlier reward increases when the reward can be achieved sooner. This represents a psychological tendency among decision makers, especially those with a time preference that leans toward impulsivity. It is logical to presume that political pressures to seek justice (or revenge) for the 9/11 attacks could incentivize political leaders to seek results sooner rather than later. The natural psychology described by Kirby and Marakovic supports the conclusion that operational choices drawn from quicker results would be increasingly attractive to decision makers in the conditions following the 9/11 attacks.

Wittman and Paulus [29] take a step further in analyzing the impacts of time duration on decision making calculus. Brain activity differs between options when the expected outcomes occur before or after a 12-month horizon according to research by Wittman and Paulus. This begs several questions about how planners are screened and selected. This is especially true when considering operations, like Unconventional Warfare, where operations in the human domain may be expected to last more than a year.

While the initial UW operation to oust the Taliban regime took mere weeks, consolidating those gains and emplacing a sustainable functioning government never occurred, even after 20 years. Were the planners well suited to consider and present the risks along with the potential rewards associated with an effective UW operation? Did the military campaign plan stop short of a stable and sustainable outcome or did the objectives focus instead on the immediate results without accurately considering the longer-term results?

6 Maintaining Strategic Consistency

One lens to consider in such a protracted scenario is Social Exchange Theory (SET). SET posits that decisions are based, at least in part, on an ongoing cost–benefit analysis. Ribarsky [18] explained that people perform an ongoing cost–benefit analysis regarding relationships. It is a simple extension of that logic to conclude that the cost was out pacing the benefits of U.S. involvement in Afghanistan after 10 years in Afghanistan with mounting casualties and financial costs. This perspective introduces a key strategic question into the potential decision criteria for employing military force. Can U.S. and/or allied military power bring a conflict to culmination before the cost benefit analysis shifts away.

It is also important to note that when discussing decision making as a central function of command in operations like OPERATION ENDURING FREEDOM, that changes in commanders must be recognized. Granted, all military commanders, especially those from the same service (e.g., Army), are educated fairly similarly and are trained in the same Military Decision Making Process. Consequently, it is somewhat reasonable to assume that decision making may be more consistent when subsequent commanders come from the same service. This condition was true in Afghanistan from General McKiernan through General McChrystal, to General Petraeus. It was not until General Allen, a Marine, assumed command from General Petraeus in July 2011 that a service change occurred [25]. There is no evidence, however, that this change impacted strategic coherence.

Blackorby et al. [2] highlighted how consistent decision making over time can be problematic if the decision maker is myopic. For the purposes of this discussion, myopia alludes to a decision-maker who repeatedly pursues immediate gratification without considering the longer term effects. In the Afghanistan scenario, there was a series of commanders including McKiernan, McChrystal, Petraeus, Allen and Miller. Simultaneously, there were three different U.S. administrations (Bush, Obama, and Trump). Consequently, it is reasonable to assume that there could be significant differences in strategic direction over the 20-year period. In reality, political factors and decisions affecting force structure dominated operational considerations.

Similarly, there existed the opportunity for strategic dissonance to develop. Strategic dissonance refers to a misalignment of strategic action with the strategic intent. Burgelman and Grove [3] posited that some degree of strategic dissonance should be expected in dynamic environments over time. While Burgelman and Grove wrote specifically about strategic dissonance in the Information Technology industry, the same construct applies in the intrinsically dynamic environments experienced in Afghanistan. For the purposes of this chapter, it is important to differentiate between strategic decision making and strategic dissonance.

7 Boyd Cycle

The Boyd Cycle is one of the seminal decision-action models. What Boyd captured, whether deliberately or by accident, was that previous decisions and actions can influence subsequent decisions. In protracted situations, iterative decision-action cycles create temporal input cycles for follow-on decisions. Boyd's analysis of air-to-air combat dynamics during the Korean conflict resulted in what became known as "The Boyd Cycle" or OODA Loop. OODA is an acronym standing for Observation, Orientation, Decision, and Action [16].

Osinga [16] wrote that Boyd drew from Bertalanffy in terms of understanding the complexities of systems of systems (p. 71). Bertalanffy viewed systems from a complex biological perspective. Boyd viewed complex systems as systems of systems much the way a biologist views the human body as a group of interdependent systems (e.g., circulatory, cardio-pulmonary, neurological, and renal systems). Viewing organizations as systems of systems, Boyd concluded that people needed to combine analysis and synthesis to understand subsystems as parts of a larger whole. This perspective is consistent with strategic analysis.

Osinga, in applying Boyd's work to strategic security scenarios, referenced the Heisenberg uncertainty principle. Heisenberg posited that a person can analytically understand the location of a particle or the momentum/trajectory, but not both at the same time (2007, p. 68). Meaning that a person can understand the micro or macro level in a system of systems scenario, but not both simultaneously. If Heisenberg's analogy holds true when shifting from atomic particle analysis to complex strategic systems analysis, then temporal factors impact decisions in a significant way.

While Heisenberg's conclusion may be limited to analytical techniques or may be limited to his atomic particle scenario, the underlying premise remains valid regarding the importance of synthesizing perspectives when systems of systems are involved. The value of conducting continuing analysis as a vehicle to continuous learning, which again supports the proposition that time can impact decision making. When temporal consistency exists, time may positively impact decisions; when temporal inconsistency is present, then time may negatively impact decisions.

For context, it is worth mentioning that Boyd considered various perspectives on analyzing systems. He studied systems from Cartesian or mechanistic models to biologic systems views [16, p. 70]. The shift from considering individual parts to the whole began in the 1920s. So, this approach is not a recent evolution in theoretical thought, nor is it limited to Boyd.

In summary, Osinga makes a strong argument that Boyd's thoughts on decision-action apply at a strategic level. While this extends well above the original focus of his work, it is important to realize that strategy relies on a series of individual decisions much that same way that aerial combat relies on a series of decisions by one or more pilots. In strategy, however, the numbers of factors involved increase exponentially to the point of complexity requiring synthesis in addition to analysis.

8 Decision Model

Anecdotes from the U.S. and NATO involvement in Afghanistan provide a myriad of insights into the complexities behind decision making. Compiling factors into a more comprehensive model for decision making may inform future decision makers when they confront similar complex situations. Developing experiments to validate each of these factors, and how they potentially interact, goes beyond the scope of this chapter. The following section will address various factors and attempt to describe and depict a logical model for decision making in complex situations.

9 Decision Factors

There are a multitude of factors that influence decision making. Factors can be categorized as internal and external factors, personal and situational factors, and temporal factors. It is important to recognize that these various factors are not discreet, however.

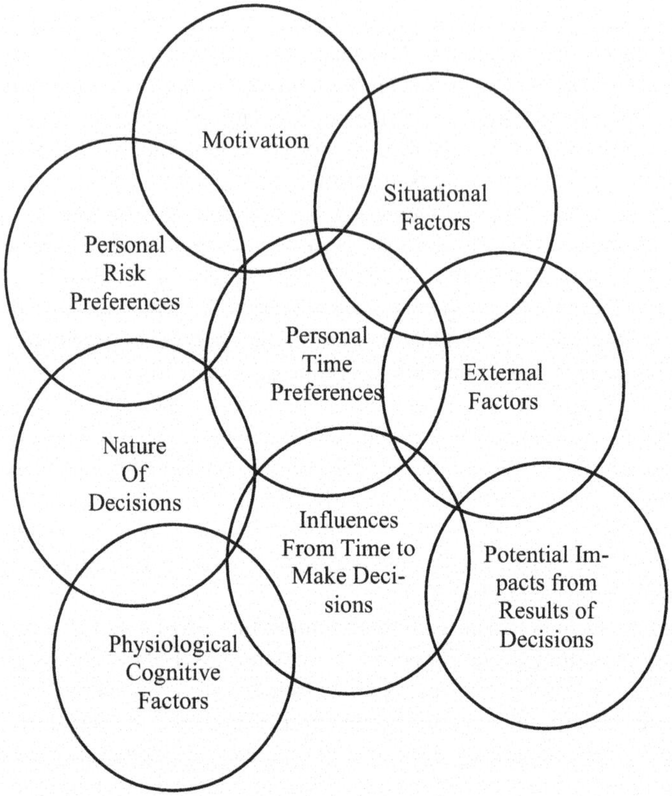

Fig. 1 Factors influencing decision making

Figure 1 depicts many of these factors in a Venn diagram structure. Most importantly, Fig. 1 is intended to demonstrate that these various factors are interdependent rather than discreet.

9.1 Personal Factors

Some factors impacting decision making are quite personal in nature. Motivational factors and personal preferences can all impact how individuals approach decisions. Whether a person is intrinsically motivated or extrinsically motivated, possesses a low time preference or is risk averse may affect decision making processes and ultimately decisions.

9.2 Motivation Factors

Ryan and Deci [20] considered intrinsic and extrinsic motivations. The authors associated intrinsic motivation with the human propensity to learn and assimilate information. In contrast, Ryan and Deci argued that extrinsic motivation can vary dramatically in terms of whether it is associated with autonomy or control. Ultimately, the authors' insights provide a foundation for understanding how complex motivations are and how they may factor into decision making.

Bénabou and Tirole [1] posited that intrinsic and extrinsic motivations can co-exist within individuals. Informed managers and leaders should understand what motivates individual workers. Parijat and Bagga [17] reviewed the value of rewards under Vroom's Expectancy Theory. Whether a reward is valuable to an individual, obviously depends on whether that individual in motivated intrinsically—by the work itself, or extrinsically—by some external reinforcement or reward. Extrinsically motivated individuals respond positively to rewards when those rewards are deemed valuable. Extrinsic rewards, however, can be detrimental in the long term when an individual is intrinsically motivated. While a sense of duty or professional responsibility may factor into how an individual tackles a task like deriving a plan for approval (i.e., decision), how that individual is motivated may well impact the level of effort that goes into informing the final decision.

Koestner et al. [12] posited that whether a person is assigned a task or undertakes the task independently and whether those individuals feel empowered and capable matters in how that task is completed. Consider a task to develop a strategic plan. Will the individual planner invest themselves more fully in a plan if they are intrinsically motivated by synthesizing vast amounts of information?

Koestner et al.'s work introduces self-efficacy as a separate variable. Koestner et al. [12] findings supported the conclusion that bolstering planners' self-efficacy is a stronger determinant of successful task completion than an assigned execution plan. Considering the spectrum of possibilities, ranging from experienced planners or decision makers to individuals who are working in totally unfamiliar territory. In the case of Afghanistan in 2001, U.S. Central Command (USCENTCOM) was the principal lead for planning military responses in the CASA region. What expertise at the USCENTCOM level possessed experience in Afghanistan or in Unconventional Warfare? The answers to these questions begs the further question, did planners derive operational plans based more on unsupported assumptions rather than synthesized understanding of the environment and tasks required?

9.3 Preferences

Frederick [8] offered insights into the time preference factor that impacts decision making. Frederick cited the plethora of previous research that focused on what is now referred to as system 1 and system 2 cognitive processes [4, 7, 10, 21, 23] as cited by Frederick [8]. In summary, system 1 processes are more intuitive whereas system 2 processes require more deliberate analysis such as a mathematical or algebraic calculation. Frederick [8] concluded that the CRT suggested a relationship, albeit undefined, between cognitive abilities and time preference.

The Cognitive Reflections Test (CRT) demonstrated the impacts of individual time and risk preferences through a series of three timed questions. Between 2012 and 2015 at Fort Bragg, NC, the CRT was administered informally in a classroom setting to 350 military students. Students ranged in rank from Staff Sergeants to Lieutenant Colonels with the preponderance being Staff Sergeants, Sergeants First Class, and Captains. Most students had served the majority of their careers up to that point in time in tactical units where immediate action drills were finely honed as part of tactical training.

Of the 350 participants, seven students answered three questions correctly in the allotted 90s. It is worth noting that I routinely instructed students to, "take your time and do your best. You have plenty of time. You have 90s; begin." When I announced the time constraint, there were obvious physical reactions observed among many students. This is consistent with the findings of Conte et al. [5] who determined that one group of participants in their experiments reacted to the imposition of time constraints while the other group did not demonstrate a time sensitivity.

Frederick [8] cites studies dating back nearly two hundred years that posit that more intelligent people tend to be more patient. Rather than delve into the debate on how intelligence is measured, this chapter simply accepts the premise that people possess varying degrees of patience. This leads to the 'so what' question.

Assessing a preference of either immediate or delayed rewards is a classic means of measuring time preference. Not all decisions are this simplistic, however. In the case of developing strategic plans a decision may be more oriented on achieving a shorter termed dramatic effect vice a longer termed, sustainable end state. Such was the case presented in Afghanistan for the United States in 2001. Of course, we are now able to look back through the perfect lens of hindsight.

Greater patience in deriving an answer to the question, what is the best course of action, may hold significant relevance to military future planners and commanders. Should planners forgo the long-term strategic advantage in favor of a more immediate win? Approaching these types of questions from a perspective of maximum sustainable advantage may offer military planners and commanders a more effective paradigm. To leverage such a paradigm, however, planners and Combatant Commanders will have to demonstrate greater patience in order to challenge intuitive answers from system 1 type thinking.

For military planners it is important to consider the relationship between a commander, the situation, and the planner when considering time preference. Time preference may not be easily isolated as a factor when a commander directs and suggests a solution. The potential impact of command influence in such a case could easily overcome greater patience in a planner seeking an optimum solution.

Risk Preferences

Frederick [8] also addressed risk preferences in the analysis of the CRT. Through his detailed findings, Frederick ascertained that more than time preference and cognitive abilities affected decisions. The CRT analysis revealed that neither the ability to calculate relative rewards, nor patience, were solely responsible for decision made. Rather, Frederick concluded that participants' willingness to gamble factored into the ultimate decision made. A willingness to gamble has direct relevance to military and national security planners, obviously.

9.4 Physical Cognitive Processes

Many directions can be taken in the discussion of intelligence or cognitive abilities. Another factor related to neuroscience, however, may provide extremely relevant insights into what impacts decisions by strategic planners. Wittman and Paulus [29] discovered that temporal horizons impact cognitive activity. Specifically, their study demonstrated that individuals possessed different levels of cognitive activity when considering issues within a 12-month horizon versus issues beyond a 12-month horizon.

10 External Factors

10.1 Situational Factors

Many situational factors can impact decision making. Situational factors represent those variables that are unique to any given situation and are beyond the control of the decision maker. There are a few common types of situational factors that may impact decisions.

Koestner [13] highlighted how the autonomy of a decision maker may impact their decisions. Koestner pointed out that individuals may advance decisions more effectively when feel more empowered and autonomous. While this factor is linked to motivational aspects of an individual's make up, it is important to recognize that an individual may not have any input into individual autonomy or team structure.

Autonomy is not binary. Meaning that an individual may possess full autonomy in one project or task and a limited amount of autonomy in another. Autonomy can occur across a wide spectrum. Consider a project team formed to make recommendations to the Secretary of the Army on ways to improve an internal process like mobilization. The team leader could be granted wide autonomy to gather recommendations and present them to the Army leadership. At the same time, the team leader could be handed a set of items to include in the recommendations. If this set of items came from the team leader's immediate superior, then autonomy just became limited.

Another situational factor may derive from an individual's prior experiences. In a case where a decision maker had a negative experience, they may possess a bias, quite possibly built on assumptions, that impact decisions made. An argument might be made that such a bias is a personal aspect attributed to the decision maker. However, such a bias is arguably associated with the specific situation and decision to be made.

The division of Active and Reserve Component Psychological Operations and Civil Affairs units between Special Operations Command and the Army General Purpose Forces might be an example of such a bias. United States Special Operations Command made that recommendation to the Army and the Secretary of Defense after frustration mounted over the need to mobilize large numbers of Reserve Component Psychological Operations and Civil Affairs units during OPERATIONS ENDURING FREEDOM and IRAQI FREEDOM. Limited force structure, most of which resided in the Army Reserves, coupled with dwell time constraints limiting individual mobilization availability generated significant frustration among Army leadership. These situational factors were viewed as having influenced decisions to divide the force structure.

Admiral Stravidis raised another factor related to the rotations of forces. Stravidis [24] highlighted how one-year tours of duty impacted strategic coherence. Discontinuity up and down the chains of command resulting from 12-month rotations contributed to the negative outcomes in Vietnam and Afghanistan. This situational factor might also find representation in temporal factors as Stravidis characterized one effect as a "sense of "I've just got to make it to my departure date"" that detracted from strategic coherence. A key point that emerges when considering some situational factors is that there exists interdependence between many factors that impact decisions.

10.2 Nature of the Decision

Decisions come in many different scenarios. The nature of the decision can directly impact the final decision. This only makes sense when we integrate other factors such as risk and time preferences. Is the nature of the decision related to what we will eat for breakfast or are we dealing with life and death ramifications? The nature of the decision can integrate with several other factors to influence the decision in complex ways.

Decisions, particularly military plans, impact peoples' lives. Deployments can last for weeks or months, even years. The very nature of military plans, which may require a commander's decision or even need Presidential approval, raises the stakes. It is only logical that such impacts will affect what decisions are made.

The nature of the decision is perhaps one of the greatest takeaways from the U.S. campaign in Afghanistan. When should the U.S. commit to the use of force? Invading Afghanistan was clearly justified on a moral and legal basis after the 9/11 attacks. But, just because the U.S. possessed the ability to invade Afghanistan, does that mean they should have invaded?

Any country has the right—and moral obligation to defend itself. Under United Nations Article 51:

> Nothing in the present Charter shall impair the inherent right of individual or collective self-defense if an armed attack occurs against a Member of the United Nations, until the Security Council has taken measures necessary to maintain international peace and security. [26]

This means that the United States and any other nation who lost citizens in the 9/11 attacks possessed the right to respond militarily. The ultimate answer to that question falls to a test of national will. The United States was strategically successful in World War II because we stayed the course. If the United States uses military force, placing blood and treasure on the line, it should be willing to fight to win. This means applying sufficient military force to break the adversaries' will and secure lasting peace and security. Fighting to a point of dominance is a tactical act without strategic advantage.

10.3 External Pressures

Stravidis [24] alludes to political pressures that were present during the campaign in Afghanistan. This is a clear example of external pressures placed on decision makers. Where Stavridis referred to fighting 20 one-year wars, he references the decisions to rotate forces every 12 months. During OPERATION IRAQI FREEDOM, senior Army leaders attempted to rotate forces every six months. It was not long before it became evident that six-month long rotations were unsupportable. The strategic dissonance incurred by yearly rotations was never dealt with due to political pressures.

Constraints placed on decision making often originate from higher up the chain of command. These represent external factors that decision-makers must accommodate. Imposing restrictions on planners before the brainstorming is done and initial courses of action are drafted can severely limit the options considered. Regardless of when constraints are imposed on decision makers, external factors clearly impact what decisions are finally made.

Koestner [13] alluded to a more subtle type of external force that can impact decision making. Koestner addressed the issue was autonomy. Autonomy can impact decision making in several, often subtle, ways. External factors come into play when autonomy is limited. This is a common situation that senior leaders should recognize and consider when delegating authority or assigning tasks.

Requirements like a need to submit a Concept of the Operation (CONOP) before executing a tactical operation are often viewed differently up and down the chain of command. Whereas higher headquarters may view a CONOP requirement as a way to ensure an operation is supported fully and synchronized with other operational activities, a subordinate element often views the need to submit a CONOP as higher headquarters looking over their shoulders—limiting their autonomy. This disparity of perspectives can be managed effectively by clear communications and explanations.

10.4 Temporal Factors

Time-related factors impact decisions in several ways. Most often people consider the amount of time available to arrive at a decision when contemplating temporal factors. This is only one aspect of temporal influence, however. The time it takes to derive results from a decision may also impact the final decision. When can results be observed and measured? This is a key question for decision makers.

10.5 Influence from Time Constraints to Make Decisions

Effects of time limitations to arrive at decisions is well documented [5, 15]. Experiments demonstrated that some people react to having time constraints imposed on them when tasked with making decisions. Other people do not react differently. Findings consistently reveal that it is not the amount of time, but rather the imposition of time limits that generate the response.

Ordonez and Benson [15] found that some people switched decision making strategies when placed under time constraints. Their experiment focused on a risk-based decision with the potential to win some amount of money. Most relevant is their finding that some people consistently reacted to the imposition of time constraints by changing to a simpler decision calculation involving less cognitive demand. Also, participants who reacted to time constraints were likely to use a similar strategy that was used previously. Ordonez and Benson grouped participants into two categories: those who reacted to the imposition of time constraints and those who did not change their approaches to analysis and decision making.

Conte et al. [5] also conducted experiments related to the imposition of time constraints with similar findings. Conte et al., found that some participants left out some information when placed under time constraints. This is a significant finding. Ultimately, Conte et al., demonstrated that one group of participants maintained a rational decision process while the other group discounted some available information when time constraints became imposed. The amount of time was not the issue, but rather the presence of time constraints.

Findings by Van der Vegt et al. [27] support conclusions by Conte et al. Van der Vegt et al., concluded that medical students without time constraints researched answers to questions more thoroughly leading to a 32% improvement in correct answers. This compared to a 6% improvement in correct answers under time constraints. Those participants who react to time constraints may lose rationality when placed under time limitations.

Collectively, it these various experiments demonstrated that certain people react to the imposition of time constraints while others do not react differently. Military and government implications should be obvious. Those in command and in planning positions would do well to remain rational and thorough under time constraints.

Much military training is placed under realistic time constraints. It would be useful to repeat experiments using trained military personnel to determine if military training desensitizes people to a natural reaction to the imposition of time constraints. Depending upon the outcome of such research, an awareness among personnel in command or planning assignments based on deliberate assessments could benefit the military services.

10.6 Potential Impacts (Positive or Negative Results from Decision)

It is important to understand when results from decisions can reasonably be expected. Tellis and Eggers [25] accurately predicted the rapid collapse of the Afghan government. Obviously, many factors contributed to the failure of the Afghan state. However, the risks to the United States of a failed state returning to Taliban rule places the entire campaign in a failed light. This highlights a key weakness in U.S. foreign policy and strategy as administrations change. Shifting priorities and strategic dissonance within the time horizons needed for a stable and sustainable regime in Afghanistan represent clear weaknesses in the U.S. approach to strategy. Accounting for this inherent intertemporal inconsistency is critical for determining the appropriate use of military force.

Aligning operational and strategic objectives with reasonable time horizons is a common expectation in setting goals. In the absence of meaningful objectives indicating the potential for a sustainable outcome, should the United States apply the use of military force? A more relevant question might be what type of military force might the United States (and allies, if appropriate) apply when regime change is unlikely to last? OPERATION EL DORADO CANYON, the bombing attack on Libya, demonstrated that alternatives to invasion do exist. The point of this section is not to argue that the decision to invade Afghanistan and overthrow the Taliban regime was right or wrong. Rather, the point here is that sufficient understanding of complex factors is important to determine when *sustainable* results should be expected. This knowledge—albeit based on rational assumptions, should inform the ultimate decision.

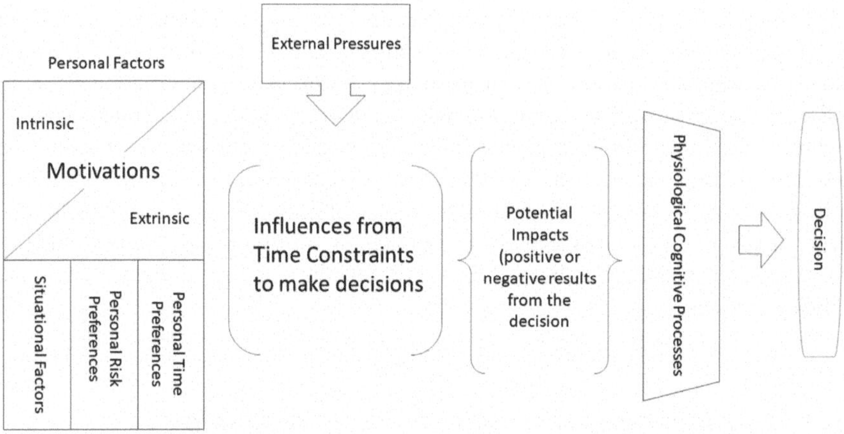

Fig. 2 Decision making model

10.7 Decision Making Model Summary

Decision making is a complex process, or set of processes, affected by many variables. More recent research suggests that physiology and temporal factors play larger roles than previously understood. Internal and external factors combine with temporal factors in complex ways. A firm understanding of how these factors interact can only benefit planners and decision makers (Fig. 2).

In the interest of improving future plans and operational decisions, the Department of Defense should invest in research into the many facets of decision making outlined above. Future research into decision making may offer insights beneficial to the military services and federal government. How does military training impact personal factors? Addressing the issue of physiological differences in cognitive processes based on temporal horizons may assist is selecting personnel who are better suited for strategic planning activities. Extensive research should be conducted to determine how temporal horizons impact cognitive activities.

References

1. Bénabou R, Tirole J (2003) Intrinsic and extrinsic motivation. Rev Econ Stud 70(3):489–520
2. Blackorby C, Nissen D, Primont D, Russell RR (1973) Consistent intertemporal decision making. Rev Econ Stud 40(2):239–248
3. Burgelman RA, Grove AS (1996) Strategic dissonance. Calif Manage Rev 38(2):8–28
4. Chaiken S, Trope Y (1999) Dual-process theories in social psychology. Guilford Press, New York
5. Conte A, Scarsini M, Sürücü O (2016) The impact of time limitation: insights from a queueing experiment. Judgm Decis Mak 11(3):260–274

6. David FR, David FR, David ME (2013) Strategic management: concepts and cases: a competitive advantage approach. Pearson, Upper Saddle River
7. Epstein S (1994) Integration of the cognitive and psychodynamic unconscious. Am Psychol 49(8):709–724
8. Frederick S (2005) Cognitive reflection and decision making. J Econ Perspect 19(4):25–42
9. Heracleous LT (1994) Rational decision making: myth or reality? Management development review
10. Kahneman D, Frederick S (2002) Representativeness revisited: attribute substitution in intuitive judgment. In: Gilovich T, Griffin D, Kahneman D (eds) Heuristics and biases: the psychology of intuitive judgment. Cambridge University Press, New York, pp 49–81
11. Kirby KN, Marakovic NN (1996) Delay discounting probabilistic rewards: rates decrease as amounts increase. Psychon Bull Rev 3:100–104
12. Koestner R, Horberg EJ, Gaudreau P, Powers T, DiDio P, Bryan C et al (2006) Bolstering implementation plans for long haul: the benefits of simultaneously boosting self-concordance or self-efficacy
13. Koestner R (2008) Reaching one's personal goals: a motivational perspective focused on autonomy. Can Psychol 49(1):60
14. Mintzberg H (1994) The rise and fall of strategic planning. The Free Press, Macmillan, Inc., New York
15. Ordonez L, Benson L III (1997) Decisions under time pressure: how time constraint affects risky decision making. Organ Behav Hum Decis Process 71(2):121–140
16. Osinga FP (2007) Science, strategy and war: the strategic theory of John Boyd. Routledge
17. Parijat P, Bagga S (2014) Victor vroom's expectancy theory of motivation—an evaluation. Int Res J Bus Manag 7(9):1–8
18. Ribarsky EN (2013) Choose your own adventure: examining social exchange theory and gendered relational choices. Activities for teaching gender and sexuality in the university classroom, 89.https://doi.org/10.1080/17404622.2012.737924
19. Ripley RB, Lindsay JM (eds) (1997) US foreign policy after the Cold War. University of Pittsburgh Pre
20. Ryan RM, Deci EL (2000) Intrinsic and extrinsic motivations: classic definitions and new directions. Contemp Educ Psychol 25(1):54–67
21. Sloman SA (1996) The empirical case for two systems of reasoning. Psychol Bull 119(1):3–22
22. Staff J (2011) Joint publication 5–0: joint operation planning. Washington, DC
23. Stanovich KE, West RF (2000) Individual differences in reasoning: implications for the rationality debate? Behav Brain Sci 22(5):645–726
24. Stravidis J (2021) I was deeply involved in war in Afghanistan for more than a decade. Here's what we must learn. Time. https://time.com/6090623/afghanistan-us-military-lessons. 16 Aug 2021 11:50 AM EDT
25. Tellis AJ, Eggers J (2017) US policy in Afghanistan: changing strategies, preserving gains. Carnegie Endowment for International Peace
26. United Nations (1984) Chapter VII—action with respect to threats to the peace, breaches of the peace, and acts of aggression. Article 51
27. Van der Vegt A, Zuccon G, Koopman B, Deacon A (2020) How searching under time pressure impacts clinical decision making. J Med Libr Assoc JMLA 108(4):564
28. Votel JL, Cleveland CT, Connett CT, Irwin W (2016) Unconventional warfare in the gray zone. Joint Forces Quart 80(1):101–109
29. Wittman M, Paulus MP (2009) Temporal horizons in decision making. J Neurosci Psychol Econ 2(1):1. https://www.sciencedirect.com/science/article/abs/pii/S1364661307002811

Dr. Sean Ryan is an Assistant Professor at West Liberty University, teaches courses in public and non-profit management, strategy, and strategic management. After graduating from West Point with a bachelor's in engineering, Dr. Ryan spent 30 years in the military. He retired in 2012 as Special Forces Colonel, with a career characterized by diverse assignments in Asia, the Middle East, Southwest Asia, and the Americas. He earned an MBA, and a master's degree in Strategic Studies and a Ph.D. in Management.

The 2021 Collapse: Lessons Learned from a Century of Upheavals and Afghanistan's Foreign Policy Vortex

Omar Samad

Abstract As a Westphalian concept, the nation-state—whether pre or post-colonial—refers to a sovereign geography whose citizens or subjects are relatively homogeneous and share a common history and communal traits. In turn, sovereignty is defined as the authority of a state to govern itself or another state. They both refer to the ability of a nation and its state to own its resources and decide on policy and systems of governance, political, economic, social, judicial and other structures. Going back more than a century, this chapter examines the meaning and application of those concepts and principles as they relate to the Afghanistan case and leading to state collapse in 2021. As a nation-state since the late nineteenth century, Afghanistan continues to be a unique case facing numerous challenges and obstacles to its sovereignty, independence and quest for peace and progress. This chapter will examine the fluctuations and nuances that drove great power politics and rivalries, regional and by-proxy pressures, and indigenous socio-political upheavals that were experienced as part of civil wars, invasions and foreign occupation, resistance and coping mechanisms. To better understand the collapse of the Western-backed Afghan republic in 2021, and the return of the Taliban's militant theocracy to power after two decades of conflict inherited from the past, one needs to look at patterns of stability and instability, state building, governance, meddling and entanglements, resistance and reconstruction. This chapter attempts to draw key lessons via the prism of foreign policy drivers, by looking at the historical record, the geopolitical context, economic dependency and the role of influential players and elites whose agendas either match or collide with indigenous cultural or ideological movements in a multi-ethnic setting. This chapter also makes use of original research involving three interviews with key leaders and a focus group survey of foreign policy practitioners.

Keywords Afghanistan · Taliban · Afghan state collapse · Afghan sovereignty · Afghan war · United States military intervention · 9/11 · Soviet invasion · Great power rivalries · Proxy war · Al Qaida

O. Samad (✉)
Atlantic Council's South Asia Center, Washington, D.C., USA
e-mail: omarsamad@gmail.com

A. Farhadi and A. Masys (eds.), *The Great Power Competition Volume 4*,
https://doi.org/10.1007/978-3-031-22934-3_21

1 An Exceptional Case

Many wonder why and how the Western-backed Afghan "Islamic Republic" collapsed like a house of cards on August 15, 2021. No doubt a casualty of geography, dependency and bad governance, however, part of the answer lies in the past century of the country's mostly turbulent history as Afghanistan underwent 14 regime changes, had 16 rulers (8 of whom died violently), and at least 18 national flag redesigns.

Few countries in the world have experienced such dramatic socio-political turnovers in such a short period of time. Despite periods of relative peace and normalcy between the 1930s and 1970s, it is estimated that in the past 50 years alone, approximately two million Afghans have lost their lives in various forms of conflict,[1] and at least five million were displaced (not counting natural disasters) as a result of coup d'états, military interventions and occupation, terrorist activities, political instability, humanitarian and economic crises and grave human rights abuses.

Unprecedented in world history, during the same period, the world's two military superpowers, the U.S.S.R (representing a Marxist-Leninist political order) and the United States (representing a liberal-capitalistic order), intervened militarily and were embroiled in prolonged and costly conflicts that involved elements of armed resistance, factional war, proxy meddling, insurgency and counter insurgency.

The Soviets intervened (1979–89 for the first and last time outside the Warsaw Pact) to secure a crumbling pro-Moscow regime and expand their geopolitical reach. Facing international condemnation, they retreated after back-breaking expenditures and more than 15,000 Red Army losses against a nationwide anti Communist Jihad[2] (Holy War) that benefitted from weaponry and funds (paid mainly by the U.S., Saudi Arabia and other allies through the covert program called Operation Cyclone),[3] using Pakistan as a conduit and staging ground. Iran and China were also contributing through their own channels.

Insofar as the U.S. intervention in reaction to the terrorist attacks on September 11, 2001 is concerned, *Operation Enduring Freedom* (OEF) signaled the start of a global war on terror,[4] as the U.S. and key allies initially used military means sanctioned by the United Nations[5] to successfully oust the Taliban regime accused of harboring al Qaida. OEF ended in 2014 after 13 years of combat operations at a cost of more than $2 trillion[6] in the Afghanistan theatre. In the process, the West got embroiled in multi-faceted "nation-building"[7] and an unfinished civil strife which paved the way by 2005 for the reemergence of a vengeful Taliban whose overtures to reconcile after

[1] Britannica [1].

[2] CIA [2].

[3] Coll [3].

[4] Naval History and Heritage Command [4].

[5] Duke University [5].

[6] Sabga [6].

[7] Acemoglu [7].

2002 were unheeded and were once again beneficiaries of Pakistani safe havens and a Jihadi support network.

Following an ebb-and-flow trajectory that lasted two decades, the U.S. and its allies finally decided at the NATO summit in Lisbon in 2010 to end their combat mission by 2015 and withdraw in stages thereafter.[8] The final drawdown was completed in August 2021 after a negotiated agreement between the U.S. and the Taliban was signed in 2020,[9] in Doha, Qatar. It is estimated that more than 240,000 people (including military personnel from all sides) have been killed in this last phase of the Afghan war.[10] Many more on all warring sides were injured and are afflicted with post-traumatic stress disorder.

As Western and Eastern (including Afghan) scholars rely on academic and theoretical tools such as geography, geopolitics, governance and ethnography to assess the causes contributing to this unique Afghan phenomenon of dissonance causing nation-state instability and human suffering, there is as of yet little consensus on any one factor or hypothesis.

What is agreed to by many Afghans though is the fact that the nexus between domestic politics and governance models, regional rivalries and geostrategic or great power dynamics have not only had a direct impact on stability and sovereignty, but also on economic development and prosperity. Today, Afghanistan is ranked as one of the poorest nations in the world, where 95% of the population lives under the poverty line,[11] and at least half the population is facing emergency levels of food insecurity.[12]

In a multi-ethnic country that prided itself half a century ago of breaking ethnic and linguistic lines under a new and modernist Constitution adopted in 1964,[13] today, the country's political class is divided between a centralized theocracy under the Taliban, who are back in power after 20 years of waging what they consider as a Jihad, while others out of power, and many across the country, are calling for a decentralized order or, at the very least, a power-sharing arrangement reflecting the country's socio-political diversity.

Many Afghans attribute today's fragmentation to years of conflict and several rounds of purges and brain-drains, while others point to under-development and foreign meddling as key causes. However, the last century demonstrates that the country experienced periods of relative peace, stability and development when sovereignty was least undermined, and the sense of nationhood and ownership were strongest in an environment where the geopolitical context enabled the country's foreign, security and development policies to maintain a functional equilibrium and avoid severe entanglements.

---

[8] Council on Foreign Relations [8].

[9] US State Department [9].

[10] Myers [10].

[11] Human Rights Watch [11].

[12] Human Rights Watch [11].

[13] Constitution of Afghanistan [12].

Conversely, Afghanistan has been vulnerable and unstable whenever the level of dependency has spiked and/or foreign meddling has undermined growth, sovereignty, ownership and political participation.

For many today, the trauma of war, displacement, corruption, injustice and uncertainty on one hand, and the painful soul-searching on the other, have resulted in disbelief and trauma bordering on extreme tendencies that have diluted older concepts of nationalism and patriotism. While the majority in Afghanistan's multi-ethnic society is silent and feels marginalized by events beyond their control, a "caretaker" Taliban government, facing many challenges at home and abroad, is embroiled in a veiled internal debate between pragmatists and hardliners[14] over issues such as female education, gender employment rights, foreign policy engagement, counter terrorism and political participation.

Meanwhile, as the Taliban consolidate power in a country that is no longer at war but facing sporadic Daesh-type (instigated by the Islamic State Khurassan Province (ISKP) in this region) attacks and opportunistic targetting by remants of the old regime and local opposition group fighters, dispersed groupings of old guard followers, once-privileged exiled elites and "secular" remnants of past regimes scattered outside the country are struggling to regroup and challenge the Taliban hold and vision.[15]

A third category of non-partisan mediation groups have also emerged since the collapse of the Republic, with the aim to rekindle intra-Afghan talks and seek non-violent solutions to the country's political inclusivity and gender rights issues. Taking historic and current trends into account, this chapter dwells on Afghanistan's foreign policy challenges as the country moved from being in subordinate isolation at the turn of the twentieth century to passive neutrality, bipolar balancing acts, occupation and resistance, the Jihadi resurgence, and global re-engagement before reverting into forced isolation under severe economic sanctions[16] imposed by the donor community once the Taliban regained control in 2021.

Making use of three exclusive conversations (one with former Afghan President, Hamid Karzai, the second with former U.N. envoy, Ambassador Lakhdar Brahimi, and the third with Prof. Ishaq Nadiri, former Senior Economic Advisor to the Afghan President), and a focus group survey of practitioners, this chapter will also shed light on the challenges that Afghan leaders and key stakeholders faced during these two decades in a violent, highly volatile and demanding geopolitical environment, putting under stress both sovereign rights—as expressed through foreign policy, reconstruction and reconciliation efforts—and the search for a functional equilibrium.

---

[14] Mohseni [13].

[15] BBC Persian [14].

[16] U.S. Department of the Treasury [15].

2 Afghan Foreign Policy 1893–1978: From Isolation to Independence

In their quest for dominance and expansion in South-Central Asia, nineteenth century imperial rivals, Russia and Great Britain, came to agree that instead of fighting over the rugged terrain and its traditionalist inhabitants, their interests would be better served if the country became a "buffer state" where the British Raj influenced foreign policy lines, thus containing the threat to Russia. British officials first used the buffer state term in 1883, before it was officially coined as such after the Durand Line agreement of 1893[17] between Emir Abdul Rahman and Sir Mortimer Durand. Later, the Anglo-Russian Convention of 1907[18] once again delineated the current boundaries as a geographical buffer between the two powers. The Durand Line, as we will see later, remains a divisive and unresolved issue to this day, despite the fact that the international community at-large has accepted it as a de jure border between Pakistan and Afghanistan since 1947.

In 1919, under King Amanullah, after the third Anglo-Afghan War, Britain agrees to grant Afghanistan full independence and end the country's semi protectorate status. But independence and sovereignty during the subsequent decade, tied to a rushed modernization program at home and warm relations with Moscow's new Soviet rulers were met with skepticism and pushback in London.

Emboldened and proud of its new independent status, Amanullah took Afghanistan out of isolation. After touring Russia, Turkey and Europe, he opened up embassies in key capitals, established cultural, educational and economic ties and borrowed modern governance practices. Turkey's secular leader, Kamal Ataturk, became a role model, while upper-class emancipation and westernization were encouraged in Kabul. However, the liberalization agenda proved to be excessive and implemented too hastily for a religiously conservative, tribal and agrarian society.

The first jolt came in 1929 as King Amanullah, hailing from a Pashtun dynasty, was deposed by Habibullah Kalakani (known as Bacha-e Saqao or the water carrier's son), a Tajik commoner, who led an anti-reformist rebellion that many attribute to some level of British prodding caused by displeasure with a monarch whose foreign policy was viewed as too independent.

Kalakani ended up being a short-lived distraction. Less than a year later, Mohammad Nader Khan, a cousin of the former King, with British backing and Russian acquiescence, violently overthrew the Tajik ruler and reestablished the urbanized Pashtun dynasty. Although Nader Shah was assassinated by a disgruntled student in 1933, the monarchy remained stable till the end of his son Zaher Shah's reign in 1973—the longest period of uninterrupted peace and development in the country's modern history. During these 40 years, the country initially went through slower paced reform to maintain stability and avoid a social backlash, and then by the

[17] Institute for Conflict Management [16].

[18] Department of British Foreign and Commonwealth Office [17].

mid-50s pursued higher-speed modernization till the Communist coup and start of protracted conflict in 1978.

On the foreign relations front, mindful of great power intentions, the country avoided getting embroiled in regional or great power rivalries. Afghanistan remained neutral[19] (or *betaraf* as in not taking sides) during World War II, and became a member of the non-aligned movement in the emerging post-war bipolar world.

But relations with Pakistan soured from the start over the Durand Line as Afghanistan launched the Pashtunistan campaign[20] calling for self-determination of the semi-autonomous Pashtun and Baluch tribal regions. Concerned about domestic unrest in its own tribal regions, Afghanistan has yet to recognize the Line as the international border. On its side, Pakistan has used the issue to foment instability, crack down on irredentist and nationalist elements, and align itself with religious movements affiliated to its own Deobandi establishment.[21]

While Iran and Pakistan became members of Western military alliances after WWII, the Cold War rivalries between the U.S. and the USSR gave Afghanistan the opportunity to carry favors with both sides and become the recipient of economic aid from both East and West. Kabul turned to Moscow after an unsuccessful attempt in the 1950s to acquire American military aid.[22] The Soviets offered an initial $100 million military modernization and infrastructure package. Between 1954 and 1978,[23] Afghanistan received more than $1 billion in Soviet aid, including substantial military assistance. The United States, in contrast, provided about $500 million in development and humanitarian aid to the country prior to the 1978 coup. Both blocs contributed to the funding of different stretches of the critical ring road connecting Eastern Afghanistan via Kabul to the West, and other key infrastructural projects. It was an era of stability, growth, balancing acts and a high level of Afghan sovereignty.

In 1964, King Zaher Shah, prodded by intellectuals and modernists, introduced far-reaching political reforms by limiting the royal family's influence and introducing democratic changes by promulgating a new Constitution that maintained the State's Islamic foundation, but also gave all citizens—men and women—equal rights and basic freedoms. An elected parliament and press freedom were enshrined in the document, but the system failed to detect the counter currents, including one led by a disgruntled ex-Prime Minister and the King's cousin, Mohammad Daoud, who colluded with pro-Moscow leftists to undo the monarchy and the democratic experiment. This period also saw the rise of leftist and right-wing Islamic parties that would play a decisive role throughout the 1980s and 90s during different phases of conflict.

The monarchy was toppled by Daoud and his leftist associates within the Soviet-trained military in a bloodless coup in 1973. The countries of the region did not show much concern because stability was maintained and Daoud was a known entity, but Pakistan started to host anti regime Islamist elements as a counterbalance.

---

[19] Dupree [18].

[20] Ahmad et al. [19].

[21] Global Security [20].

[22] Samad Ghaus [21], See also Barfield [22].

[23] Oliker [23].

Proclaiming a republic and shunning the pro Soviet elements over time, Daoud made overtures to Pakistan[24] (including talks on the thorny Durand Line issue), the Shah of Iran, Saudi Arabia and Egypt as a means to gradually shift away from the Soviet Union and normalize relations with Western and Islamic nations. The balancing act failed to keep Afghanistan steady. A second military-led coup toppled Daoud in a bloody attempt at pushing Afghanistan deeper into the Soviet orbit.

In short, During the 40-year long peace stretch, despite a high degree of sovereignty rights and socio-economic progress, three key game-changing factors impacted equilibrium and future stability:

1. The Pashtunistan policy tied to the colonial legacy of the Durand Line agreement.
2. The tilt toward the Soviet Union as the main supplier of military hardware and training.
3. The rift within the royal family that undermined democracy-building and resulted in a coup d'état and subsequent unraveling.

All three were the result of a mix of domestic, regional and great power dynamics impacting Afghan stability, growth and sovereignty.

3 The Unraveling Between 1978–2001

The bloody April 1978 military coup instigated by the twin pro-Moscow factions of Khalq and Parcham under the umbrella of the People's Democratic Party of Afghanistan (PDPA) not only overthrew a dynasty that cost the lives of Daoud Khan and his immediate family, but also put the country on a slippery track that is impacting Afghan society to this day, and also took the lives of tens hundreds of thousands of Afghans. Immediately followed by a systemic purge of regime opponents (real and imaginary), the reaction to this purported Marxist Leninist takeover was swift and widespread. Within weeks, clusters of anti PDPA armed insurgency and mutiny within the military were visible across the country, which pushed the so-called "revolutionary" followers of PDPA to unleash an indiscriminate reign of terror that further outraged both urban and rural communities.

Internal divisions and rivalries within the PDPA also further inflamed the situation and made it more difficult for their patron saints in Moscow to maintain order in the cities and prevent the spread of rebellion in the countryside. Not only were forcibly implemented land reforms and socialist style socio-economic changes facing resistance from all segments of society, but both city dwellers and villagers facing oppression started to head out of the country and establish refugee and anti-regime "Mujahedeen" bases in Pakistan and Iran or resettle in the West.

The party experienced three turnovers and four leaders in the ensuing decade. After a short-lived honeymoon, the PDPA splintered again as the predominantly Pashtun Khalq wing headed by Nur Mohamad Taraki and Hafizulah Amin ousted

---

24 Arwin [24].

the non Pashtun Parcham faction headed by Babrak Karmal by the end of 1978. Another internal row in 1979 led to the killing of Taraki by Amin supporters.

The octogenarian Politburo in Moscow assessed its options and decided to break with Warsaw Pact traditions and deploy more than 100,000 Red Army troops backed by airborne division into the non-aligned Islamic country to prevent a regime collapse and expand its zone of influence into South Central Asia. Not only was the region and Islamic world viewing this Christmas eve 1979 invasion as a major geopolitical shift, but the United States and China were also closely watching developments and assessing their options. Calling it "a serious threat to peace,"[25] the Carter Administration boycotted the 1980 Moscow Olympics and postponed SALT-II negotiations. Some saw it as the end of the age of détente between East and West.

With Khalq defanged, the Parcham wing of the PDPA, advised and backed by the USSR, was now facing a growing insurgency that used bases in Pakistan and Iran. For the next decade, key events shaped Afghanistan, regional dynamics and great power relations, ultimately resulting in a multibillion-dollar covert program to defeat the Soviets—mainly using a pipeline that was managed by Pakistan's intelligence services. During this period, Muslims from other nations joined the "Holy War," and some, like Osama Bin Laden and other militants, built local relationships and established organizations in Pakistan that would later morph into anti-Western terror groups like al Qaida.

As the United States started supplying the Afghan Mujahedeen with sophisticated anti air weapons like the Stinger missile, Soviet air vulnerabilities and losses also started to increase. With the Kremlin's old guard out of the way and a new crop of leaders experimenting with Perestroika and Glasnost in the mid-1980s,[26] Karmal was replaced by Mohammad Najibullah, a former intel chief, in Kabul in 1986. Najibullah's mission was to retreat from a hardline pro-Moscow communist line and prepare the ground for national reconciliation and a Soviet retreat. The United Nations started mediation talks between Kabul, Russian, American and Pakistani diplomats to discuss the terms for an asymmetrical disengagement. The Geneva accords[27] were signed in 1988 between Kabul and Islamabad, with the U.S. and USSR as guarantors. It paved the way for a full Soviet retreat by February 1989. Meanwhile, in Moscow, as the situation unfolded swiftly, the collapse of the Soviet Union and break up of Eastern Europe took place between 1989 and 1991.

Since the Mujahedeen were not a party to the accord, they continued to fight the Kabul government. Najibullah's beleaguered government eventually collapsed in 1992 as rival Mujahedeen factions entered Kabul, and he took refuge at the UN compound. Unable to cobble together a unity government, the country fell into a new round of factional warfare with the main focus on control of the capital.

Forces belonging to veteran Mujahedeen commander, Ahmad Shah Massoud, and his allies inside and outside the crumbling regime captured Kabul from the North while forces belonging to his rival, Gulbudin Hekmatyar, who had close ties to the

---

[25] The History Channel [25].

[26] Reuveny [26].

[27] Wikipedia [27].

Pakistani establishment and Najibullah factional supporters within the regime, dug in on the Southern and Eastern flanks of the city. Most other armed groups allied themselves with either side as the fighting increasingly became ethnic and splintered along North–South alignments. All regional countries with a stake in the Afghan conflict used hedging strategies that further weakened Afghan sovereign rights and introduced a new level of proxy warfare. As a result, thousands were killed in Kabul, while a new wave of refugees headed out of the country.

As Western engagement with the country reached new lows, a new Islamic militant group called the Taliban, most of whom were former Mujahedeen affiliated to more extreme Pashtun Islamist factions based in Pakistan and recruited from among refugee and Deobandi religious schools (Madrassa) emerged in Southern Afghanistan by 1994.[28] Their motto was to end warlordism, impose peace and Sharia or Islamic law as they saw fit given their village and refugee madrassa experience in Pakistan.

Swiftly, the Taliban, seen as a fresh force with the potential to grow, received backing by religious and intelligence circles in Pakistan and Gulf states, and advanced rapidly across Afghanistan. They captured Kabul in September 1996, as Massoud forces retreated North and Hekmatyar fighters lost their bases and most of their support network shifted to the Taliban side.

Taliban's control of Kabul started a new phase of fighting dubbed *moqawemat* or the Afghan resistance era and pitted Taliban fighters against Massoud's seasoned multi-ethnic force that controlled no more than 20% of the country and received scant assistance from India, Russia and Iran. The Taliban recognized by Pakistan, Saudi Arabia and the UAE, also agreed to extend refuge to foreign fighters who were roaming between the Af-Pak tribal regions seeking a base in secure territories under their control. Al Qaida and other non-state and terror organizations took advantage of the Taliban sanctuary and in return helped them fight other Afghan resistance forces.

With the country's economy at a virtual standstill and relying primarily on humanitarian aid, the cultivation, production and export of narcotics increased dramatically (with the exception of 2000 caused by a tactical crackdown), mainly in Taliban territories. The presence of terrorist groups, especially after the bombings of US embassies in East Africa and on the USS Cole, had gradually regenerated Western interest in Afghanistan and the adjoining region, but despite warnings by Massoud[29] and others of an imminent threat directed at U.S. interests, there was no clearcut policy to pursue a more forceful engagement or counter terrorism agenda. All of that changed on September 9, when two foreign terrorists masquerading as journalists assassinated Massoud in Northern Afghanistan. 48 h later, the U.S. homeland was attacked on September 11.

The main takeaways form this segment are as follows:

1. Foreign military interventions, proxy war interferences and even non-state actor activities—as in militancy and terrorism—do not happen without local alliance

---

[28] Congressional Research Service [28].

[29] SBS [29].

and agency. They are highly disruptive and can undermine key principles such as the right to self-determination and territorial integrity.
2. Forceful imposition of an imported ideology or system on a traditional society can result in violent backlashes and chaos.
3. A comprehensive and timely political settlement is the best option to resolve a conflict when the opportunity presents itself. Failing to do so can lead to more instability and the emergence of a failed state.
4. A failed state is not only a threat to social cohesion, basic rights and development, but can also have repercussions that can reach other shores.

4 The Western Intervention 2001–2021

In the aftermath of the terrorist attacks of September 11, 2001, until the collapse of the Islamic Republic on August 15, 2021, the Afghanistan mission pursued two key strategic objectives that were presented as one but were not fully integrated. More influential countries involved on the military and donor sides pursued their programs in a manner that was not always in synch with Afghan priorities and at times lacked sufficient donor coordination and oversight.

4.1 The External Pillar

This dimension reflected Western interests and was managed by the United States and key Western donors and contributors. That strategy was initially driven by the intent to take revenge for the tragedies of 9/11, mobilize nations against a new global enemy and bring about regime change in Afghanistan with the explicit use of military means and foreign aid. That initiative was supported and welcomed by many in the international community as well as most Afghans who were victimized by more than two decades of continued warfare.

Over a 20-year span, Western policy and objectives centered on military/counter-terrorism/counter-insurgency/counter-narcotics, regional connectivity[30] (the New Silk Road that failed to materialize), nation building and economic development as part of a costly war economy using modernized financial management systems that lacked transparency in the procurement and contracting domains.[31]

As illustrated in the three exclusive interviews (with former President Karzai, Amb. Lakhdar Brahimi and Prof. Ishan Nadiri) and a focus group survey that are part of this chapter, major war and peace as well as rebuilding decisions and benchmarks were defined by the major donors unless Afghan input was needed or requested.

Consequently, the diplomatic and international community presence in Afghanistan gained more, not less, influence over time as dependency grew, the

[30] US State Department, Bureau of Public Affairs [30].

[31] Alexander [31].

war economy drove a large portion of the national economy, and a new generation of foreign educated/trained young professionals rose to prominence. While many of them lacked work experience or an appreciation for difficulties faced by the majority of Afghans beyond large cities, they rose to prominence during Ashraf Ghani's seven-year tenure, at a time when corruption levels[32] reached new highs, governance faced new hurdles and human losses[33] mounted as government forces faced a growing insurgency, impacting morale and civilian support.

The continuation of conflict created a stalemate as the two sides dug in for a protracted war. After the signing of the *Enduring Strategic Partnership Agreement Bilateral*[34] in 2012 and the *Bilateral Strategic Agreement* (BSA)[35] in 2014 between the Ghani government and the U.S., the Taliban strategy changed from one of engaging in three-way peace talks to aiming to force an American withdrawal and deal directly with Washington on a timetable, shunning Ghani's government, seen as a "client" administration—in the same manner in which the Mujahedeen had dealt with the Najibullah regime in the early 1990s.

Meanwhile, the growth of Western influence was also being watched with mounting apprehension by most regional capitals, especially the near and far neighbors. Countries across the region started new hedging strategies as they realized that the Western military presence in Afghanistan was stalemated and there was no clear end-state for the NATO presence in the country.

As part of the gradual shift in the Afghan balance of power, Russia, China and Iran started direct contacts with the Taliban during Ghani's tenure.[36] The signing of the BSA was probably a turning point for key regional powers that had previously accepted at face value American pledges to leave Afghanistan after Bin Laden was found and killed by an American special ops team in Abbottabad, Pakistan in May 2011.

As the US and China moved deeper into a new great powers' competition arena, Beijing was no longer just concerned about the US War on Terror or Afghan governance. Mindful of security and state control in the majority Muslim province of Xinjiang, where separatist Uyghur militants are said to pose a threat, the Chinese government expanded its monitoring of Jihadi groups sympathetic to the Uyghur cause. China also aggressively promoted its own version of the new Silk Road, called the Belt and Road Initiative (BRI), avoiding Afghanistan in the first phase, but sending signals to the Taliban that it is considering connecting a corollary of the China-Pakistan Economic Corridor (CPEC) to Afghanistan[37] once the country is stable enough.

Ironically, since the US withdrawal and return of the Taliban back to power, both sides have expanded direct talks and even though China has yet to recognize the

[32] UK Department for International Development (DFID) [32].

[33] Wikipedia [33].

[34] US State Department [34].

[35] Just Security [35].

[36] NBC News [36].

[37] Greenfield [37].

Taliban government, bilateral relations are warmer and focused on security,[38] trade, investment and connectivity.

4.2 *The Internal Pillar*

This pillar involved mainly Afghan powerbrokers and beneficiaries of the trillion-dollar war economy. Some were participants at the U.N. sponsored Bonn Conference of 2001[39] that drew a new political roadmap for the post-Taliban Afghanistan and selected a moderate political activist from Kandahar, Hamid Karzai, with ties to the exiled monarch Zahir Shah, to head the interim government made up of a mix of anti-Talib Mujahedeen, ex technocrats and NGO practitioners.

Although the Afghans were put in the driver's seat, but the vehicle's fuel and maintenance were being supplied key foreign backers—the U.S. being "the elephant in the room." Karzai faced two immediate challenges: solidify his base by engaging in a balancing act at home with fractious and influential factions, and balancing Afghanistan's complex new set of relations among a diverse group of countries that did not always see eye-to-eye on issues pertaining to a Western military presence and regional rivalries. Although he acknowledged in his talk (in this chapter) that during the initial stage after 2002, "the entire international community came together to support Afghanistan, friends and foes, opponents and those with differences… backed by the UNSC and other powers," but he admitted that later during the Ghani Administration, "clear signs emerged that the regional consensus on Afghanistan was weakening, and some former government officials talked to me about this unravelling taking place around us."

Bringing the old monarch back to Kabul as the Father of the Nation, Karzai pushed for unity as a priority after two decades of fragmentation. A new Islamic Republic Constitution was promulgated in 2004[40] and elections were held the following year, giving Karzai the legitimacy needed to govern more effectively.

5 The Reconstruction Bonanza

Mindful of Afghanistan's dire needs for foreign aid to rebuild the dilapidated infrastructure and institutions and offer vital public services, the international community started to organize donor conferences[41] to channel resources under international

---

[38] Murtazashvili [38].

[39] United Nations Peacemaker [39].

[40] The Constitution of Islamic Republic of Afghanistan [40].

[41] Wikipedia [41].

supervision to the Afghan government. The reconstruction strategy and fund allocation were driven mainly by donor priorities while tactical decisions about coordination and implementation were increasingly handled by a small cadre of trusted and connected Afghans with past work experiences with the foreign NGO and aid communities. Ashraf Ghani, an ex-World Bank anthropologist and new Finance Minister in the Karzai Government, who had lived overseas for more than 30 years, was a key interlocutor for Western donors during the initial stages.

On the regional front, the Western-backed Karzai, and later Ghani, administrations used new initiatives to engage the neighboring countries, resolve outstanding issues, aim for regional connectivity, expand trade and transit agreements, attract foreign and Afghan investment, open land and air corridors to link up to strategic transport arteries, connect the country to regional railroad networks, act as a power and energy import and transit hub and start preliminary feasibility of the country's trillion dollar unexploited mineral wealth.

With few exceptions, most initiatives remained plans and feasibility studies on paper.

Adopting an open market economy, primarily funded by the donors and Afghan business interests, the private sector started playing an increasingly important role in the implementation of projects. Subcontractors from Turkey, India, Pakistan, Central Asian and European nations took advantage of the opportunities by gaining a large share of the reconstruction pie.

In 2008, the Afghan National Development Strategy (ANDS) became the government's "overarching strategy for promoting growth, generating wealth and reducing poverty and vulnerability." Adopted as a 5-year plan, it provided the framework for the development of government policies but was subject to U.N. and World Bank scrutiny and donor approval. As Dr. Nadiri said in the interview below: "basic long term investment programs to change the structure of the Afghan economy and to achieve long term and comprehensive sustainability unfortunately did not take place."

Furthermore, Dr. Adib Farhadi, the former Executive Director of the ANDS said "the strategy was a serious attempt to 'Afghanize' the country's development priorities. However, this Afghanization was a miserable failure largely due to donors' unwillingness to relinquish any control of their funds or priorities".

On the domestic front, aid money was primarily allocated to the road building, education, healthcare, institution and capacity building, gender empowerment, media and civil society development, rural development, refugee resettlement and other sectors deemed as important to the donors and to the government at the national and subnational levels.

This necessitated new legal frameworks and an administrative and judicial system that could handle the workload. International and multilateral institutions such as the United Nations' specialized agencies, the World Bank, the IMF, Asian Development Bank, foreign aid agencies and others played a critical role during this period. The need and scope of work was so extensive that hundreds of new local and foreign contractors and NGOs also got involved in the reconstruction effort.

As a flagship development program and darling of Western aid agencies, the National Solidarity Program (NSP)[42] was launched in 2003 to empower rural Afghanistan's grassroots through elected Community Development Councils (CDC) that included women, using grants to restore rural infrastructure and create employment. Financed by a consortium of international donors led by the World Bank, more than 14 million villagers in some 20,000 communities benefited from the NSP for a total cost of $2.3 billion. By 2015, the program was replaced by the follow-on Citizens' Charter NSP at a time when rural inhabitants made up about 74%[43] of the total population (up from 92% in the 1960s).

Meanwhile, critical fields that were neglected for a long time included the power generation, agribusiness, urban planning and water management sectors. Key challenges facing development and reconstruction included: corruption and waste, a bloated bureaucracy, security threats, criminality and the illicit economy fueled by the narcotics business.

Given the gargantuan levels of aid that poured into Afghanistan between 2004–2021, the country's GNP climbed from an estimated $4 billion in 2002 to more than $19 billion in 2020, while per capita income went from $180 in 2002 to more than $500 in 2020 (capping at $650 in 2013).[44] These figures have dropped significantly since the fall of Kabul and the imposition of financial and economic sanctions on the Taliban regime. Despite the rise in income, more than 50% of the country's population still lived below the national poverty-line in 2020. That number has sharply risen to about 97% in 2022.[45]

The fleeting successes experienced primarily in the education, healthcare and capacity building fields were offset by the erratic strategic decision-making processes at the top, as well as the high level of corruption across national and subnational institutions.

The justice system lost credibility, and in some regions, people relied on the Taliban court system to settle cases.[46] These governance deficiencies helped fuel disenchantment and a mix of religious nationalistic sentiments that bolstered the Taliban anti-occupation narratives. This, in turn, alienated large pockets of the population that felt disconnected or victimized in regions and enclaves that were seen as either sympathetic to the Taliban or were contested.

[42] The World Bank [42].

[43] The World Bank [43].

[44] The World Bank [44].

[45] United Nations Development Program [45].

[46] Baczko [46].

5.1 *The Regional Balancing Act*

In order to address regional security, rivalries and contentious issues, Afghanistan launched several initiatives to build goodwill and expand cooperation and connectivity. In December 2002, mindful of meddling and undue interferences, the *Kabul Declaration on Good Neighborly Relations*[47] was signed with all six neighbors. The aim was to stress on good relations and non-interference in each other's affairs.

Another initiative to reduce tensions between Afghanistan and Pakistan was launched in 2003 as part of the Tripartite Commission on security and involving U.S./NATO military representatives as well. The aim was to address border security and management issues. Nothing substantial was achieved because of Pakistan's unwillingness to admit to the presence of Taliban sanctuaries on its soil.

Other trilateral and multilateral fora were established to reduce tensions and engage in trust-building, especially as the Taliban insurgency picked up speed after 2005. Chinese, Turkish, Iranian, Russian and Central Asian diplomats joined Pakistani and Afghan officials to expand the discussion beyond security and the nagging insurgency, to also include regional cooperation and connectivity.

The largest initiative, The Heart of Asia—Istanbul Process,[48] was launched with Turkish partnership in 2011 in Kabul. It included 15 countries: Afghanistan, Azerbaijan, China, India, Iran, Kazakhstan, Kyrgyzstan, Pakistan, Russia, Saudi Arabia, Tajikistan, Turkey, Turkmenistan, Uzbekistan and the United Arab Emirates. With few accomplishments, it held its last session in 2021 in Dushanbe, a few months before the fall of Kabul.

6 Centralization Versus Decentralization

Afghans warmly embraced the democratic shift in 2004. However, many were dismayed by the level of internal and exteranl corruption, political fraud and foriegn manipulation witnessed in subsequent presidential and parliamentary elections and rubber-stamp Loya Jirgas held between 2005 and 2020. Public support for the shaky centralized democracy, viewed by the public as marred in fraud,[49] was waning after 2009. Presidential and parliamentary Elections held in 2009, 2010, 2014, 2018 and 2019 were highly controversial and deepened the socio-political divides.

Aside from Taliban threats, the number of illegible voters who cast ballots went from close to 9 million in 2004 to less than 1.8 million (an inflated figure contested by opposition political parties) in 2019.[50]

---

[47] United Nations Peacemaker [47].

[48] Nadirova [48].

[49] Callen and Long [49].

[50] Wikipedia [50].

Some attributed this to structural deficiencies and constitutional over-reach where power was centralized in the office of the President. Others saw a historical trend where Pashtun elites, opposed to a national census that could potentially alter the perceived demographic landscape, were in favor of a political system that gave them a monopoly of power at the top.

Although some voices during the Constitutional Loya Jirga of 2004 favored a decentralized parliamentary system over the centralized presidential order[51] that was adopted, another counter movement emerged after the controversial elections of 2014 when non-Pashtun elites lobbied for constitutional amendments authorizing more provincial autonomy as part of a decentralization process. Some elements within parliament advocated for a federal system. A fringe group, accusing Ghani of handing power[52] by default to the predominantly Pashtun Taliban as part of a conspiracy in 2021, are now in support of the country's North–South partition along perceived ethnic and linguistic lines that are blurred in many districts where a patchwork of villages belonging to different ethnicites live a short distance from each other.

Meanwhile, the Taliban are consolidating a Taliban style centralized theocracy[53] where mostly Pashtuns belonging to a handful of pro Talib tribes share power under an all-powerful "Emir of the faithful" in Kandahar. The political battle lines are drawn between those favoring a centralized versus decentralized system, even though pragmatists[54] within Taliban ranks are in favor of a light power-sharing political order, as was discussed during the Doha talks, but not pursued once Kabul fell into Taliban hands. By the end of 2022, two other issues of great importance to most Afghans, as well as to the U.N. and donor community, also revealed rifts within the Taliban on policy and ideological tendencies: girls access to secondary and higher education and a ban on women employment with domestic and foreign NGOs [54].

7 The Peacebuilding Conundrum

As President Karzai points out (in his interview below), his government did its best under the circumstances to address all governance issues but failed to convince its key allies to pursue intra-Afghan reconciliation talks aimed at ending a war that was part civil, part proxy and part ideological. He said "Sovereignty is vitally important for any country. One of the issues regarding foreign policy in the past eight years (prior to the fall of Kabul) was the extent of influence on decision-making."

Only two factions, the Taliban and the Hezb-i Islami of Hekmatyar, acted as an armed opposition to the government during the U.S. occupation period. A peace

[51] International Crisis Group [51].

[52] Nasar [52].

[53] Aikins [53].

[54] Mohseni [13].

accord[55] was eventually signed with Hezb in 2016 which allowed for the safe return of its leaders to Kabul. The Taliban, however, refrained from engaging Kabul as long as foreign troops remained in the country.

After an initial unsuccessful attempt to open a liaison office in Qatar in 2013, reconciliation talks were put on hold until a new American envoy, Amb. Zalmay Khalilzad, was asked by the Trump Administration in 2018, to reengage the Taliban on seeking a withdrawal and a settlement of the Afghan war. Since the Taliban were adamantly opposed to three-way talks with Kabul, US and Taliban negotiations led to an agreement that included a timetable, a roadmap, counter terrorism commitments and a secret annex with technical details. The document was signed in Doha on February 29, 2020.[56]

It was stipulated that intra-Afghan talks would be pursued until an inclusive political settlement and a permanent ceasefire were reached. Intra-Afghan talks between a Kabul negotiating team and Taliban representatives kicked off after staged delays and attempts by Ghani team members at sabotaging the process[57] in order to gain time till American elections were held at the end of 2020, in the hopes that a Biden Administration would overturn the Trump deal and prevent an American withdrawal.[58] That hope was shattered when Biden, after a thorough review, decided to change the date but, nonetheless, push for a full pullout by September 2021.[59]

Instead of taking the political track as was discussed in Doha, Ghani and his support base in Kabul and Washington shunned advice from different quarters, including U.S. Secretary of State Anthony Blinken, who proposed the formation of a "transitional Peace Government of Afghanistan" that would eventually transfer power to a permanent government following the adoption of a new constitution and national elections.[60]" Paradoxically, while the demoralized Afghan army was on the back foot, he called for controversial elections and, eventually, for a Jirga to endorse a power transfer, promising his small base of continuity.

Meanwhile, the Taliban took advantage of the political chaos and pushed hard to advance on large cities and provincial capitals between June and August 2021. Days before the fall, while corps commanders and security chiefs were left in a state of confusion, Ghani played a double game of pretending to agree to a power transfer, but in reality, making arrangements with his closest allies to flee the country and leave Kabul without a political transition plan or effective security blanket. With the Americans unwilling and unable[61] to step in and provide security during a transition, the vacuum was filled by the Taliban who had, by then, advanced to the gates of Kabul. The city fell to the Taliban on August 15, for the second time in 25 years.

---

55 Ludin [55].

56 US State Department [9].

57 Feroz [56].

58 Lawler [57].

59 The White House [58].

60 Tanzem [59].

61 Lawler [57].

Post facto, lamenting about the inability of the UN to prioritize peace, the former UN Secretary General's Special Envoy, Amb. Lakhdar Brahimi, said "I tried very hard to help the Afghans take the lead in every field. Those attempts did not really succeed. No peace building can fully succeed if the locals are not fully in charge." (see interview in this chapter).

8 Conclusion

After a costly 20-year long American-led mission came to an end in August 2021, historians and scholars probing the main causes of state collapse in Afghanistan will undoubtedly focus more on the military-insurgency and state-building aspects of the mission and less on other factors such as Afghan sovereignty woes and regional geopolitical challenges facing its foreign policy, both in terms of balancing competing interests and implementation as our focus survey (detailed below) of Afghan foreign policy practitioners reflects in this chapter.

Many theorists and practitioners of foreign and security policies agree with Prussian military theorist Carl Von Clausewitz's famous dictum that "war is the continuation of politics by other means," in the same manner that many believe that "foreign policy is also the continuation of domestic politics outside national boundaries." In other words, in a normal bilateral or multilateral setting, foreign and security policy arenas are supposed to be the reflection of a country's domestic policy and national will as represented by its leadership, with all the nuances, nexuses and complexities that differentiate the two.

That has clearly not been the case for most of Afghanistan's past four decades of conflict. The democracy-building and state-building experiences of the last 20 years, supposedly a period of resurgent sovereignty and independence, faced serious challenges when key strategic decisions or policy formulations required input or advice from key foreign military powers and donors who also provided most of the funding. In some cases, the external powers exercised a right to veto and had the last word. The veto to engage the Taliban early on in peace talks or the decision to use air and ground forces during night raids into suspect villages, fall into those categories.

As a result, the U.S. military intervention went beyond the immediate scope of neutralizing terror outfits or denying them a foothold, and instead, turned into a costly state and nation-building enterprise with inconsistent strategic goals, commitments and timelines.

A new urban Afghan elite addicted to power and resources lost touch with the rest of the country and assumed that progress and social engineering could only be achieved by forcibly imposing a Western modernization model on the rest. The same mistake that Afghan communist apparatchiks had made in the 1980s by using military means and violence to engage in social engineering, was repeated under the banners of democracy, female emancipation and a free market economy. Government policies further divided an already war-afflicted society, weakened ethno-political bonds, undermined sovereign rights and led to an even more violent armed insurrection with local and regional ties to an established religious infrastructure, and by default in some cases, to terrorist networks and state and non-state actors.

As our focus group survey demonstrates, Afghan leaders were selectively only involved in some aspects of strategic planning, policy making and resource prioritization. Almost 90% believe that decisions were "influenced by a mix of foreign and external factors and interests." Almost the same percentage point to the U.S., EU and NATO as yielding the most influence on Afghan affairs during these two decades.

On the domestic front, foreign affairs was seen as being the domain of the President, as 90% consider the presidential palace as exerting more influence than all other institutions or agencies. While about half saw the success rate in achieving foreign policy objectives as mixed, 90% identified weak leadership, low sovereignty, inexperienced management/low capacity, excessive external dependency and regional counter measures as factors contributing to regime collapse in 2021.

Dissent or outright opposition to decisions that were deemed as hurtful to Afghan interests were rare and at times came with political cost. Many Afghan observers consider the U.S. military establishment's volte face against then sitting President Hamid Karzai, as a sign of discontent caused by his opposition to night raids and aerial bombardments that did not take into account civilians on the ground. Conversely, Ghani was seen as more accommodating and in line with the military's designs.

As the Afghanistan case demonstrated, conflict was not only driven by political will and intent, but also by a dependent war economy where after 20 years, foreign funding still accounted more than 75% of the operational and development budget.[62]

Looking back at Afghanistan's century-long journey through instability, the ebb and flows of sovereignty, the harm caused by dependency, the confluence of geopolitical and security interests, mixed with domestic power plays and governance factors, it becomes apparent that the country can easily toggle between becoming a kleptocracy, a failed state or a rentier state.

To avoid such scenarios, we can conclude that the degree of sovereignty and independence of a nation-state like Afghanistan that is geographically challenged, yet has the opportunity to balance domestic and foreign policy interests and priorities, correlates directly with the degree of stability and development experienced by that nation.

Excerpts from a Conversation with Former President Hamed Karzai (in Kabul) on July 30, 2022

SAMAD: Mr. President, what did we Afghans do right or failed to accomplish in the post 9/11 period that ended with the 2021 collapse?

KARZAI: Post the 9/11 tragedies, we welcomed the international assistance so that with their help, we could free our country from the grips of al-Qaida, Bin Laden, terrorist networks and the creeping invasion of a neighboring country. My focus before the US arrival was also to make sure that Afghanistan ends the internal conflict, which I believe was driven by outside interests—then and now. I consider the Taliban and those called the Muqawemat (resistance) both as Afghans… and

[62] The World Bank [60].

wanted a settlement of issues… through the traditional mechanism of a Loya Jirga, which had taken us (including yourself) to meet with others inside and outside the country, and under the 3-point formula by HM the late King of Afghanistan, as part of an honorable Afghan effort for a solution with our own means and initiative… and the backing of the international community was necessary.

Following the tragedy of 9/11, there was sudden attention which led to the arrival of the international community. For the good luck of our country, the entire international community came together to support Afghanistan, friends and foes, opponents and those with differences… backed by the UNSC and other powers…

Our foreign policy of that time was a great success… In my view, Afghans did all they could to prosper, democracy was embraced… people came together from different backgrounds. However, there was one serious shortcoming from the beginning… the absence of Taliban in the Bonn talks (2001). Not having them represented there meant that we were denying an important segment of Afghan society the right to participate, and that was more of an international decision than an Afghan decision.

SAMAD: Major decisions were taken in the first few years like the deployment of NATO forces and overtures for reconciliation. How much did Afghan input and leadership views from you and your advisers matter during the first few years? And how did the new government balance domestic versus foreign policy dynamics?

KARZAI: The USA aspect of our relationship went very well for a long time until the incidents of house searches, night raids and bombardments began.

On overtures for peace, as soon as I was declared head of the Interim Government, while I was in Northern Kandahar, my first statement was about reconciliation and general amnesty for all Taliban. This was not liked at that time by the U.S. and their military establishment. But I did not notice at that time the seriousness of their opposition to this. I thought it was just a temporary negative reaction to my statement for peacebuilding and reconciliation, and that as we establish the government it would go away. They even came to Kandahar to arrest some former Mujahedeen with ties to the Talibs, but I did not allow that.

In Kabul, as I started to talk more and more about peace, I began to see very slowly though that there was some undercurrent moving against the peace process… but I did not visibly see the hands that were at play in it. This became visible around 2007–08. When I began calling on the late King of Saudi Arabia, King Abdullah, to help us talk to the Taliban to reconcile and bring peace, there I noticed that there was some visible opposition by the United States. They would not say it to me, but I noticed that they were not happy with this. It was the beginning of the difficulties regarding the efforts for peace.

SAMAD: Was it just the US Mr. President?

KARZAI: I think that Pakistan played a very important role in either opposing the peace process or actively encouraging conflict in Afghanistan… What I did not see clearly was US opposition… this became visible when in 2007, we got word from some of the Taliban in Helmand through a letter… asking me whether the international community was friends with us or with them (the Taliban)? I asked the then governor (Mr. Wafa) to find out what does this mean… questioning NATO forces' intentions? … soon he caught two gentlemen, one British, the other Irish at the Bost Hotel in Lashkargah, working undercover, one for the UN and the other for the European Union, with documents and whole evidence meeting with Taliban members… when we questioned them, they said they were working for peace. That was the first time we noticed that there was something deeper going on to undermine our peace efforts and to add to conflict. That is why they were persona non grata and expelled… Mr. Semple and Mr. Marvin I think (Marvin Patterson). This was the first indication that made me doubt the intentions… not so much the US at that time as I thought it was more a British thing.

As we moved forward, it became clear to me, and I began to believe that Afghans were seeking peace but some of our international partners were not… and Pakistan was actively promoting conflict.

SAMAD: Mr. President, under your leadership, facing such issues and revelations, and looking back at history as a student of history… 200 years of geopolitical challenges, how did you attempt and manage this balancing act between domestic needs for peace and stability and versus these foreign policy dynamics and agendas? How difficult was it and how did you manage it?

KARZAI: Look, there was without a doubt a tremendous balancing act, not only by me but by the whole country. As an example, at the height of my differences with the United States in 2010–11, when I was pressing for the closure of the Bagram prison and was at loggerheads in trying to shut down the private security firms that the US, Britain and others had created…

One day I met a large group of elders from Zurmat or Hesarak district, I think. That day, I was very dismayed and disappointed. I told them that the US is making it impossible to work for a decent relationship with them by undermining the peace process by having prisons and promoting security firms that are the biggest source of lawlessness in the country… and I said, I want to get them out of here. An elder got up and said we understand your frustration and anger… but please, we never want you to go that far… that will get Afghanistan in a greater mess. Another said, even now as so-called friends and allies, they are hobnobbing and supporting Pakistan against us… imagine if they turn against us… they will get us hurt…

You may have noticed, during my time in office, I never called for the withdrawal of the United States and their allies from Afghanistan. I kept asking them for correcting their behavior. It was precisely because of this balancing that we needed to do as a nation… While I did not ask them to leave, I also did not give them the bilateral security agreement (BSA). I conditioned that upon the US guaranteeing peace in Afghanistan and explain the extent of their relations with Pakistan vis-à-vis Afghanistan, and on our own relations with Pakistan…

So that's the kind of balancing I did, mainly as a result of consultations with community leaders… and of course within the government as well, there were voices clearly in support of the US continuing in Afghanistan.

SAMAD: What kind of foreign and regional policy approaches plus lessons can we draw after 20 years, in terms of balance between sovereignty and stabilization?

KARZAI: That's a very important question that I have confronted many times while in office. Involved India-Pakistan, Saudi-Iran, Russia-US, China-US relations, some easier than others to deal with. The Russians understood the presence of the

US in Afghanistan and supported us regardless… and we too proved in many ways that we were an independent country and that they did not need to worry about the US presence. For example, we took side on the Crimea issue…

Also, Iran approached us with tremendous wisdom despite immense difficulties with America, but understood that the US was here and it was a necessity of the time, keeping in view the events of September 11, and that Afghanistan needed international backing and support… I also made sure to be clear with the US administration… that Iran was a grand old neighbor of ours with whom we share language, culture… and that under no circumstances would we not side with Iran if they had difficulties… or that we would allow Afghan territory to be used against them… It was understood. The same with the Saudis and the Iranians… The challenging part of our relations though was with India and Pakistan.

I tried my best during my tenure to do all we could to improve relations with Pakistan. I visited them 20 times. Found the civilian leadership, especially PM Nawaz Sharif to be inclined to improving relations. I had a basic proposal: Pakistan, you do not need to use violence and extremism in order to gain friendship with Afghanistan… you can get all you want through positive and civilized engagement… the difficulty constantly was in the Pakistani military, who were for their own reasons seeking a weakened Afghanistan and for reasons of their own alliance with other powers and interests.

Recently, even given the friendly relations between Pakistanis and the Taliban, you saw them bomb Khost… this shows that they do not want a stable Afghanistan no matter who is in power. The Durand Line, the legacy of the British and Pakistan's own internal insecurities… they have difficult ties with India, but it fears Afghanistan because of historical reasons and composition of its own ethnic structures. On our side, we Afghans have to be wiser and do all we can to improve relations with Pakistan and keep ourselves strong… as for India, the Afghan people gave immense friendship to India and considered it a strategic ally. We expected India to continue to engage with Afghanistan and remain committed to the Afghan people. As an example, they could have allowed the return of Afghan students back to India to pursue their education. However, they did facilitate some trade to take place, but more could be done.

SAMAD: Twenty years later, do you now see Afghanistan entering another phase of great power competition or proxy rivalries, and given our experiences, what practical and policy steps do you propose at this stage?

KARZAI: I see very clear signs of great power rivalry in Afghanistan. Our difficulties today are the consequence of that rivalry. Even going back 4–5 years ago, clear signs emerged that the regional consensus on Afghanistan was weakening, and some former government officials talked to me about this unravelling taking place around us… any Afghan government here would have to correct that through friendship and trust with all our neighbors. We need national strength to handle relations with them. Regarding the US and its allies, we gave them tremendous support and respect, expecting a reliable ally to do well for itself and us as well. The events last year at Kabul Airport was dishonorable to both our nations.

SAMAD: What would you have done differently from a strategic perspective given Afghanistan's situation?

KARZAI: I think that I managed the balance of relations between us and great powers pretty well… it was challenging to manage the India-Pakistan relations on Afghanistan. At times I wonder whether the US failed to understand Afghanistan, or we failed to understand them? That question lingers on. Some of us think that what happened to us is exactly what the US wanted… that the way they behaved, especially from a military perspective, was not a repetition of mistakes but how they wanted to behave. I have my own judgment about it, but I will keep it to myself right now.

SAMAD: Are there lessons to be drawn from the last eight years, and what opportunities did we squander?

KARZAI: I will give you two examples. You remember that I was, and still am, calling Taliban "brothers". I tell them that all Afghans who are opposing them today are also our brothers, and they must reconcile and bring them home, so that this country becomes a country for all Afghans again. Having said this, I was in favor of peace, but I did not accept the way the Qatar office for negotiations was initially handled (in 2013). I asked the US to shut it down because it was indicative of a splintered Afghanistan with two governments.

I would have not conducted the Doha talks the way they were run between 2018–2021. I would not have agreed to the US talking to the Taliban alone… They proposed that to me too in 2011. Their Special Envoy (Mark Grossman) proposed a three-way talk with the Americans in the middle, but I said no way. That meant undermining Afghanistan's unity and stability. It should have been Afghans (Taliban and the Republic) on one side and the Americans on the other side of the table.

Sovereignty is vitally important for any country. One of the issues regarding foreign policy in the past eight years was the extent of influence on decision-making.

Also, when the mother of all bombs (MOAB) was dropped on us, that was a shame for us. Firstly, the government should not have allowed it, and should have protested it.

SAMAD: Was it a sovereignty or political will problem?

KARZAI: It could be both. One leads to the other, if you don't have the political will, then that may lead to the question of sovereignty… One creates the other.

My foreign policy objectives were friendship with the world, especially those who were there to work with us. I compartmentalized our relations with America into two spheres: one, the military and their activities that I opposed strongly. Two, the American people, who helped us with their money, which I supported strongly. We did not see what the military was doing as the work of the American people. We still want relations with America and appreciate all their and others' contributions… But we also wanted and proved that we are a sovereign country and that our lack of means does not reduce our sovereignty.

Excerpts from a Conversation with Ambassador Lakhdar Brahim, Former Srsg, July 9, 2022 In Paris, France

SAMAD: Tell us about the UN role since the Bonn conference, what was done right and where the UN and others failed to have accomplishments in the first few years of the Republic.

BRAHIMI: Even if the UN was seemingly given the lead at Bonn and in Kabul, the US was very much in the lead. The US was from day one—meaning even before they started the bombing campaign in Afghanistan in October 2001—interested primarily in Iraq, not Afghanistan.

Bonn was representative of the rich variety of Afghanistan. People say today we should have invited the Taliban. Wrong! The Talibs would not have come even if they had been invited. But we should have reached out to them immediately after Bonn. I think many would have joined the Bonn process.

The first Loya Jirga was as representative as the situation of the day permitted. Frankly, I think the selection process of participants was better than the very expensive elections which were organised after that.

Far too little was done to improve the Justice system. I tried very hard to help the Afghans to take the lead in every field. Those attempts did not really succeed. No peace building can fully succeed if the locals are not fully in charge.

SAMAD: In terms of major decisions, like the NATO deployment or reconciliation overtures, how much did Afghan input or leadership views matter at the end of the day?

BRAHIMI: the ISAF concept was a good one. The US and the rest of the Western world refused to expand it outside Kabul first. Looking back the NATO deployment was not a good idea. It was presented to us as a purely bureaucratic arrangement. It was finally a way of absorbing ISAF into a fully integrated NATO Operation. That ended up—among other things—blurring the independence of the UN. After 9/11, common wisdom was "forget about the Taliban, they are finished."

SAMAD: What foreign policy lesson do we draw after 20 years for a geopolitically challenged country like AFG, about sovereignty and stabilization?

BRAHIMI: The locals have to be in the lead effectively and all the time. That is a hell of a lot easier said than done. But everyone should try much harder than it is generally done to make that a reality. Lip service is paid to the idea of 'national ownership" but nothing more than that—lip service.

It was an uphill struggle all the way to create a national commission to draft the Constitution and an almost impossible task to make that work.

Everyone is to blame—first of all the nationals themselves with whom the national interest far too often comes second to other interests. But also, foreigners who use the power of money to impose their own choices and priorities. An issue that needs to be discussed internationally much more than it is.

SAMAD: Do you see AFG entering another phase of great power competition or by proxy conflict? What practical steps or policies do you propose to avoid another round of suffering and disruption?

BRAHIMI: the Talibs are in total charge. Understandably, they feel all powerful and invulnerable. At this stage, all those who can have access to them should advise them to reach out to all other parties. Of course, the Talibs are divided and their supreme leader seems to be inaccessible.

The neighbors and near neighbors are all important: Pakistan, Iran, India and China and also Turkey and Qatar are more Important than the rest of the world put together. If they all understand and accept that peace is to their benefit, not proxy war, the Afghan problem will be resolved.

A Discussion (August 2022) with Prof. Ishaq Nadiri, Former Jay Gould Professor of Economics at NYU and senior economic advisor to president Hamid Karzai (2004–2010)

SAMAD: An important part of Afghan reconstruction and development post 9/11 involved decisions about economic recovery strategy and prioritization of development and investment sectors and funding. To what extent was the Afghan government independent and sovereign in these decisions and resource allocation?

Dr. NADIRI: Yes, the size and frequency of aid from many countries were substantial and the basic analytical framework for receiving the aid and allocation was the Afghan National Development Strategy (ANDS). The strategy was based on ideas of the Afghan government through the proposals and requests of the various ministries. The role of foreign experts in formulating some of the basic policies was significant. The Afghan government was not in position to specify the amount and delivery of the aid resources. Most of these decisions were made by each donor. The aid from various donors were not pooled into a large endowment fund for well managed and systematic development purposes over time. The Donor countries concentrated on sectors like security, infrastructure and education but not as much on rural development and skills development.

SAMAD: What were the main obstacles or challenges Afghans faced in terms of reconstruction strategy and resource allocation internally and externally with key donors? Were we able to balance aid and investment between great power and regional rivals?

NADIRI: The main problem was the Afghan government was being established at the time in a broken society with multiple concentration of centers of power. It did not have the capacity to develop and execute efficient programs. The donors hired many consultants to advise the ministries and ANDS (National Development Strategy). The issue of balancing investments by major powers and regional rivals were beyond the power and capacity of the Afghans. The questions were discussed occasionally, but not seriously.

SAMAD: Given the unprecedented sums pledged over 20 years, what would you have done differently if you were able to have the last word?

NADIRI: The best way for the enormous size and variety of multi-country foreign aid to the country would have been to set up an Afghanistan endowment fund. A much careful expenditure and allocation strategies could have developed. Selection of multi policies and major projects would have been systematically carried out methodically over time under the control of a combined Afghan-International administration. The funds would have been spent with careful development of the Afghan absorptive capacity overtime and on more essential sectors like regional integration, rural agriculture, employment and training skilled labor and management capacities.

There were many impressive accomplishments that did take place in many fields and areas that we all know. But basic long term investment programs to change the structure of the Afghan economy and to achieve long term and compressive sustainability unfortunately did not take place.

Focus Group Survey

Conducted during July 2022. Respondents: 20 out of 24 responses received. 80% male, 20% female. Includes: Former diplomats, government officials and Parliamentary members.

Questions:

1. How would you characterize AFG foreign policy formulation and application during the Republic's 2002–2021 period:
A. Stable and steady.
B. Erratic and disoriented.
C. A mix of steady and erratic.
Result: 50%—C and 50%—B
2. Agree mostly that it was:
A. Driven by AFG leadership and elite priorities thru consensus.
B. Influenced by a mix of foreign and external factors and interests.
C. A mix of both elite and foreign agendas intersecting.
Result: 90%—C and 10%—B
3. Which foreign entities were the most influential.
A. United States, EU and NATO nations.
B. Regional powers.
C. UN and other multilateral orgs.
Result: 85%—A then B then C
4. Which Afghan institutions were the most influential:
A. The president and ARG.
B. Ministry of Foreign Affairs.
C. Political/jihadi elites.
D. Other institutions.
Result: 90%—A and 10%—C
5. Were Afghan strategic priorities and national interests taken into account:
A. To a large degree.
B. Intermittently at times.
C. Seldom.
Result: 60%—C and 30%—B and 10%—A
6. To what degree was AFG foreign policy calibrated/coordinated effectively with security and development/economic policies?
A. High level.

B. Mid level.
C. Low level.
Result: 60%—C and 30%—A and 10%—B
7. How would you rate AFG success and failure over 20 years:
A. 80% success.
B. 50%
C. 30% or less.
D. Other.
Result: 80%—B and 10%—C and 10%—D
8. Main foreign policy factors of regime collapse in 2021:
A. Weak leadership.
B. Low sovereignty.
C. Inexperienced management/low capacity.
D. Excessive external dependency.
E. Regional counter measures.
F. A mix of most or all of the above.
Result: 90%—F and 10%—A

https://ctc.usma.edu/lessons-from-the-collapse-of-afghanistans-security-forces/.
https://www.sigar.mil/pdf/lessonslearned/SIGAR-21-46-LL.pdf.

References

1. Britannica T (2021) Editors of Encyclopedia. Afghan War. Encyclopedia Britannica. https://www.britannica.com/event/Afghan-War
2. CIA (unclassified 2000) The cost of Soviet involvement in the Afghanistan. https://www.cia.gov/readingroom/docs/DOC_0000499320.pdf
3. Coll S (1992) Anatomy of a victory: CIA's covert Afghan War. The Washington post. https://www.washingtonpost.com/archive/politics/1992/07/19/anatomy-of-a-victory-cias-covert-afghan-war/1bd10b14-a0cc-441c-99cc-d2b5d1ba6e2d/
4. Naval History and Heritage Command (2020) Operation enduring freedom. https://www.history.navy.mil/browse-by-topic/wars-conflicts-and-operations/middle-east/operation-enduring-freedom.html
5. Duke University (2002) Terrorism, the use of force and international law after 11 September. https://scholarship.law.duke.edu/cgi/viewcontent.cgi?referer=, https://www.google.com/&httpsredir=1&article=5575&context=faculty_scholarship
6. Sabga P (2021) The US spent $2 trillion in Afghanistan—and for what? https://www.aljazeera.com/amp/economy/2021/8/16/the-us-spent-2-trillion-in-afghanistan-and-for-what
7. Acemoglu D (2021) Why nation-building failed in Afghanistan. Project syndicate. https://www.project-syndicate.org/commentary/afghanistan-top-down-state-building-failed-again-by-daron-acemoglu-2021-08?barrier=accesspaylog
8. Council on Foreign Relations (2021) The U.S. war in Afghanistan. https://www.cfr.org/timeline/us-war-afghanistan
9. US State Department (2020) Agreement for bringing peace to Afghanistan between the Islamic Emirate of Afghanistan which is not recognized by the United States as a state and is known as the Taliban and the United States of America. https://www.state.gov/wp-content/uploads/2020/02/Agreement-For-Bringing-Peace-to-Afghanistan-02.29.20.pdf

10. Myers M (2021) Afghanistan war cost more than $2T and 240,000 lives, report finds. Military times. https://www.militarytimes.com/news/your-military/2021/04/16/afghanistan-war-cost-more-than-2t-and-240000-lives-report-finds/
11. Human Rights Watch (2022) Economic causes of Afghanistan's Humanitarian Crisis. https://www.hrw.org/news/2022/08/04/economic-causes-afghanistans-humanitarian-crisis
12. Constitution of Afghanistan. https://www.constituteproject.org/constitution/Afghanistan_1964.pdf?lang=en
13. Mohseni S (2022) Can the Taliban be contained? Foreign Affairs. https://www.foreignaffairs.com/afghanistan/can-taliban-be-contained
14. BBC Persian (2022) Interview with the Mohammad Younesi. https://www.facebook.com/watch/?extid=NS-UNK-UNK-UNK-AN_GK0T-GK1C-GK2C-GK3C&v=1016153759072196
15. U.S. Department of the Treasury (2022) Afghanistan-related sanctions. https://home.treasury.gov/policy-issues/financial-sanctions/sanctions-programs-and-country-information/afghanistan-related-sanctions
16. Institute for Conflict Management. Durand line agreement. https://www.satp.org/satporgtp/countries/pakistan/document/papers/durandlineagrrement.htm
17. Department of British Foreign and Commonwealth Office (2012) FCO properties in the UK and overseas. www.fco.gov.uk/en/about-us/our-history/historic-views-kabul/learning-from-history/
18. Dupree (1988) Myth and reality in Afghan 'Neutralism'. Central Asian Survey 7, no. 2–3, p:1
19. Ahmad R, Waseem R, Ahmed AS (2022) Pashtunistan: a construction of an imaginary entity. Multicul Educ 8(6). http://ijdri.com/me/wp-content/uploads/2022/06/22.pdf
20. Global Security. Deobandi Islam: The Religion of the Taliban. https://www.globalsecurity.org/military/library/report/2001/Deobandi_Islam.pdf
21. Samad Ghaus A (1988) The Fall of Afghanistan: an insider's account. Pergamon-Brassey's International Defense Publishers, Washington, p 209
22. Barfield T (2010) Afghanistan: a cultural and political history. Princeton University Press, Princeton, NJ, p 238; Freedman RO (1991) Moscow and the middle east: soviet policy since the invasion of Afghanistan. Cambridge University Press, Cambridge
23. Oliker O (2011) Historical overview: 20th-century security aid to Afghanistan before the Soviet invasion. In: Building Afghanistan's security forces in wartime: the Soviet experience. RAND Corporation, pp 3–18. http://www.jstor.org/stable/10.7249/mg1078a.10
24. Arwin R (2020) Afghanistan and Pakistan's oft-ignored history—1947–1978. The tribune. https://tribune.com.pk/article/97165/afghanistan-and-pakistans-oft-ignored-history-1947-1978
25. The History Channel (2022) Carter reacts to Soviet intervention in Afghanistan. https://www.history.com/.amp/this-day-in-history/carter-reacts-to-soviet-intervention-in-afghanistan
26. Reuveny R, Prakash A (1999) The Afghanistan war and the breakdown of the Soviet Union. Rev Int Stud 25:693–708. https://faculty.washington.edu/aseem/afganwar.pdf
27. Wikipedia. Geneva Accords (1988) https://en.m.wikipedia.org/wiki/Geneva_Accords_(1988)#:~:text=The%20Geneva%20Accords%2C%20known%20formally,Soviet%20Union%20serving%20as%20guarantors
28. Congressional Research Service (2021) Taliban Government in Afghanistan: background and issues for congress. https://crsreports.congress.gov/product/pdf/R/R46955
29. SBS (2021) 9/11: The unheeded warning. https://www.sbs.com.au/ondemand/video/1935245891546/911-the-unheeded-warning
30. US State Department, Bureau of Public Affairs. U.S. Support for the New Silk Road. https://2009-2017.state.gov/p/sca/ci/af/newsilkroad/index.htm
31. Alexander K, Croslin C, Moktar J, Weyman F (2012) Overview of political economy, contracting and corruption in Afghanistan. UK Department for International Development (DFID). http://curbingcorruption.com/wp-content/uploads/2018/07/Kyle-et-al-2012-Overview-of-political-economy-contracting-and-corruption-in-Afghanistan-A-case-study-in-Kandahar.pdf

32. UK Department for International Development (DFID) Our activity in Afghanistan. https://www.transparency.org/en/countries/afghanistan
33. Wikipedia. List of Afghan security forces fatality reports in Afghanistan. https://en.wikipedia.org/wiki/List_of_Afghan_security_forces_fatality_reports_in_Afghanistan
34. US State Department (2012) Agreement between the United States of America and Afghanistan. https://www.state.gov/wp-content/uploads/2019/02/12-704-Afghanistan-Relations.pdf
35. Just Security. Security and defense cooperation agreement between the Islamic Republic of Afghanistan and the United States of America. https://www.justsecurity.org/wp-content/uploads/2014/10/BSA-ENGLISH-AFG.pdf
36. NBC News (2021) Iran, China and Russia make PR hay from U.S. Afghanistan withdrawal. https://www.nbcnews.com/news/world/iran-china-russia-make-pr-hay-u-s-afghanistan-withdrawal-n1278697
37. Greenfield C (2021) Pakistan discussing expansion of CPEC to Afghanistan-ambassador. Reuters. https://www.reuters.com/world/asia-pacific/pakistan-discussing-expansion-cpec-afghanistan-ambassador-2021-09-27/
38. Murtazashvili J (2022) China's activities and influence in South and Central Asia. Carnegie Endowment for International Peace. https://carnegieendowment.org/2022/05/17/china-s-activities-and-influence-in-south-and-central-asia-pub-87146
39. United Nations Peacemaker. Agreement on provisional arrangements in Afghanistan pending the re-establishment of Permanent Government Institutions (Bonn Agreement). https://peacemaker.un.org/afghanistan-bonnagreement2001
40. The Constitution of Islamic Republic of Afghanistan. https://www.constituteproject.org/constitution/Afghanistan_2004.pdf?lang=en
41. Wikipedia. List of International Conferences on Afghanistan. https://en.wikipedia.org/wiki/List_of_international_conferences_on_Afghanistan
42. The World Bank (2018) Flagship Afghan rural program lays strong foundation for the future. https://www.worldbank.org/en/news/feature/2018/01/16/flagship-afghan-rural-program-lays-strong-foundation-for-future
43. The World Bank (2018) Rural population (% of total population)—Afghanistan. https://data.worldbank.org/indicator/SP.RUR.TOTL.ZS
44. The World Bank (2021) Afghanistan GDP: USD: gross national income. https://www.ceicdata.com/en/afghanistan/gross-domestic-product-nominal/af-gdp-usd-gross-national-income
45. United Nations Development Program (2021) 97 percent of Afghans could plunge into poverty by mid 2022, says UNDP. https://www.undp.org/press-releases/97-percent-afghans-could-plunge-poverty-mid-2022-says-undp
46. Baczko A (2021) How the Taliban justice system contributed to their victory in Afghanistan. Social Science Research Council. https://items.ssrc.org/insights/how-the-taliban-justice-system-contributed-to-their-victory-in-afghanistan/
47. United Nations Peacemaker: Kabul declaration on good neighborly relations. https://peacemaker.un.org/afghanistan-neighbourly-relations2002
48. Nadirova G (2021) Heart of Asia—Istanbul process. Eurasian Research Institute. https://www.eurasian-research.org/publication/heart-of-asia-istanbul-process/
49. Callen M, Long JD (2015) Institutional corruption and election fraud: evidence from a field experiment in Afghanistan. Am Econ Rev 105(1):354–381. http://www.jstor.org/stable/43497063
50. Wikipedia. 2019 Afghanistan Presidential Election. https://en.wikipedia.org/wiki/2019_Afghan_presidential_election
51. International Crisis Group (2003) Afghanistan: The Constitutional Loya Jirga. https://www.crisisgroup.org/asia/south-asia/afghanistan/afghanistan-constitutional-loya-jirga
52. Nasar K (2021) Afghanistan: Taliban leaders in bust-up at presidential palace, sources say. BBC news. https://www.bbc.com/news/world-asia-58560923
53. Aikins M (2022) The Taliban's dangerous collision course with the west. The New York Times. https://www.nytimes.com/2022/08/08/magazine/taliban-afghanistan.html
54. Afghanistan: Taliban ban women from working for NGOs - BBC News

55. Ludin Q (2019) How peace was made: an inside account of talks between the Afghan Government and Hezb-e Islami. United States Institute of Peace. https://www.usip.org/publications/2019/03/how-peace-was-made-inside-account-talks-between-afghan-government-and-hezb-e
56. Feroz E (2019) How the Afghan president became irrelevant. TRT News. https://www.trtworld.com/opinion/how-the-afghan-president-became-irrelevant-31940
57. Lawler D (2022) Interview: former U.S. envoy Khalilzad on leaving Afghanistan to the Taliban. AXIOS. https://www.axios.com/2022/08/15/zalmay-khalilzad-interview-fall-kabul-taliban
58. The White House: Remarks by President Biden on the Drawdown of U.S. Forces in Afghanistan. https://www.whitehouse.gov/briefing-room/speeches-remarks/2021/07/08/remarks-by-president-biden-on-the-drawdown-of-u-s-forces-in-afghanistan/
59. Tanzem A (2021) Blinken Warns Afghanistan's Ghani of dire consequences without urgent changes. VOA NEWS. https://www.voanews.com/a/south-central-asia_blinken-warns-afghanistans-ghani-dire-consequences-without-urgent-changes/6202999.html
60. The World Bank (2019) Afghanistan: public expenditure update. https://www.worldbank.org/en/country/afghanistan/publication/afghanistan-public-expenditure-update

Omar Samad is Afghanistan's former Ambassador to Canada and France, and nonresident senior fellow with the Atlantic Council's South Asia Center.

Learning from Afghanistan and Beyond: Recent Developments in Operation Assessment

Jonathan Schroden

Abstract In 2011, criticisms of operation assessment—a practice aimed at determining progress of a military operation toward the accomplishment of its desired goals—identified a systematic cycle of failure with its practice in Afghanistan and other theaters in which the United States and its coalition partners were at war with insurgent and terrorist groups. This paper examines activities since 2011 to address the five elements of that failure cycle—lack of advocacy, poor doctrine, inadequate training, poor processes and products, and commander disinterest—to identify where progress has or has not been made. It finds notable progress with regards to advocacy, doctrine, and processes and products, but little activity or advancement in the areas of training and commander interest. From this examination, the paper offers a set of key lessons for the continued future improvement of operation assessment.

Keywords Assessment · Operation assessment · Afghanistan · Lessons learned · Operations research

1 Introduction

In 2011, the author published a paper in the Naval War College Review titled "Why Operations Assessments Fail: It's Not Just the Metrics" [1]. In it, he argued that a key component of the counterinsurgency campaign in Afghanistan—operation assessment—was failing to help commanders understand whether progress was being made and to make better decisions aimed at generating desired outcomes.[1] Counter to common belief at the time, he argued further that the cause of this failure was not that we lacked proper metrics to assess the campaign. Rather, the author posited that there was a systemic cycle of failure behind the inability of operation assessment to deliver on its theoretical utility to operational commanders.

[1] Current usage of the term is "operation assessment." Prior to codification of it in U.S. joint doctrine, both this and the term "operations assessment" were used to describe the same concept.

J. Schroden (✉)
The Center for Naval Analyses, Arlington, VA 22201, USA
e-mail: Schrodj@cna.org

A. Farhadi and A. Masys (eds.), *The Great Power Competition Volume 4*,
https://doi.org/10.1007/978-3-031-22934-3_22

Just over a decade later, President Joe Biden decided to withdraw all U.S. forces from Afghanistan and the North Atlantic Treaty Organization (NATO) followed suit. The result of that withdrawal was the collapse of the government of the Islamic Republic of Afghanistan and its replacement by the Taliban's Islamic Emirate. With the U.S. and NATO war in Afghanistan now over, it is worth reflecting on what, if any, progress was made in addressing the operation assessment "failure cycle" and what lessons we should learn from our experience with attempts to assess progress in that war and others like it (e.g., Iraq). The remainder of this paper will do so, in four major parts. First, it will summarize the operation assessment failure cycle for those not previously familiar with it. Second, it will identify attempts that were made to address the cycle's components and to what extent those efforts were or were not successful. Third, it will derive lessons from those attempts and offer implications of those lessons for future conflicts. Last, it will offer some brief concluding remarks.

2 Operation Assessment and the Failure Cycle

Assessment is currently defined by the U.S. military as "determination of the progress toward accomplishing a task, creating a condition, or achieving an objective. A continuous process that measures the overall effectiveness of employing capabilities during military operations" [2]. Operation assessment is further described as "an integral part of planning and execution of any operation, fulfilling the requirement to identify and analyze changes in the [operational environment] and to determine the progress of the operation" [3].[2]

Prior to the issuance of revised joint doctrinal publications within the past seven years, the definitions for operation assessment varied slightly across U.S. military doctrine, but the stated theoretical purpose for conducting this activity varied greatly. A list of purposes compiled by the author in 2011 included the following: informing commanders' decision making; completing the planning or design cycle; recognizing changing conditions in the environment; stimulating and informing adaptation and innovation; reducing uncertainty and bounding risk; showing causal linkages between actions and the achievement of objectives; documenting the commander's decision-making process; and evaluating the performance of subordinate units [1]. While this list is wide ranging in scope, a common feature among these purposes is that they are actions most military commanders would find helpful, if they were done well. The issue identified by the author in 2011, however, is that operation assessments being conducted in Afghanistan at that time were not being done well.

There was no single reason for this situation. Rather, there were three.

The first was deficiencies in U.S. military doctrine. In 2011, the major doctrinal publications guiding U.S. activities in Afghanistan were Joint Publication (JP) 3-0

[2] Operation assessment is thus distinct from strategic assessment, which is typically defined as attempts by a state to align its political and military goals, and those goals with its military capabilities, to achieve a desired political outcome [4].

(Operations) and JP 5-0 (Planning), and the Army's field manual (FM) for counterinsurgency (FM 3-24). In addition to offering a diverging set of purposes for operation assessment, these sources focused primarily on drawing distinctions between terms such as measures of performance ("doing things right") and measures of effectiveness ("doing the right things"). They offered little in the way of guiding principles or specific models and processes for the conduct of operation assessment.

The second was lack of training for practitioners. In 2011, there were no formal training courses for members of the Department of Defense tasked with the conduct of operation assessment. Nor was there any professional designation for people with experience in operation assessment. Rather, operation assessments were typically conducted by untrained staff officers or those trained in operations research and systems analysis (colloquially called "ORSAs").[3] Contrary to popular belief, ORSAs are not, in fact, trained or educated to conduct operation assessment. Rather, typical curricula for ORSA programs focus on mathematical modeling and optimization, and statistical analysis. While these are sometimes useful for the conduct of operation assessment, the latter can rarely be reduced to only these approaches, especially for human-centric conflicts like counterinsurgencies.

The third was the failure of operation assessments to deliver to commanders practical utility that matched their expectations. Relative to the list of purposes of operation assessment above, there were few instances in Afghanistan in which practitioners of assessment were able to deliver on one or more of them. Anecdotally, this often manifested in commanders losing interest in operation assessment, observable through such indicators as commanders ceasing to attend assessment updates, delegating oversight and participation in the assessment process to low-level subordinates, or asking for wholesale revamps of assessment products, processes, or organizational structures. That lack of interest further tended to result in a lack of advocacy for improving assessment practice.

In looking holistically at these three reasons why operation assessments were failing to deliver on their theoretical utility in Afghanistan, the author was able to identify a systematic "failure cycle" at work (Fig. 1). In short, the cycle ran as follows: "poor and confusing doctrine leads (in part) to inadequate (or no) training of assessment practitioners, which leads to poor assessment processes and products, which leads to commanders who are uninterested in assessment, which leads to a lack of advocacy for fixing assessment, which leads to a perpetuation of poor doctrine—and the cycle continues" [1].

In seeking to address the failure cycle, one could, in principle, start at any point along it. But the author argued that it would be best to first gain an advocate. Doing so would help drive change and offer an entity around which to build institutional knowledge that would be necessary to identify what changes to make and how best to implement them. Once an advocate was gained, efforts to improve doctrine would

[3] Operations research is defined by the Institute for Operations Research and the Management Sciences as "a discipline that deals with the application of advanced analytical methods to help make better decisions" (see: https://www.informs.org/Explore/What-is-O.R.-Analytics/What-is-O.R).

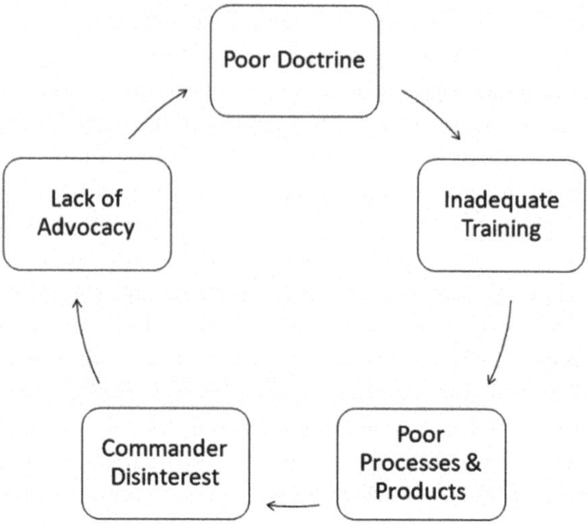

Fig. 1 Failure cycle for operation assessment (adapted from Schroden [1])

have the most value because such changes could then be used to drive the development of training courses and the processes and products on which to train practitioners. If those changes could be made, presumably operation assessment practice and products would also improve over time, and commanders would increasingly see the benefit of assessment in both theory and practice.

With that summary of the operation assessment failure cycle in hand, a reasonable question to ask is: "How much progress has been made in addressing the failure cycle since it was identified in 2011?" The next section of the paper will address that question.

3 Breaking the Failure Cycle of Operation Assessment: An Assessment

Assessing the extent to which the failure cycle of operation assessment has been adequately addressed requires first examining efforts that were undertaken between 2011 and the end of the U.S. war in Afghanistan in 2021. In turn, we will do so for each of the components of the failure cycle shown in Fig. 1, beginning with the lack of advocacy.

3.1 Lack of Advocacy

In November 2012, U.S. Central Command (CENTCOM)—spurred on by the Analysis and Requirements Branch of its Resources Division (CCJ8-ARB) and supported by members of the Center for Army Analysis (CAA)—hosted a special meeting of the Military Operations Research Society (MORS) at MacDill Air Force Base in Tampa, Florida. The stated purpose of the meeting was to "develop information that can help inform doctrine, policy, and methods for organizations and countries performing assessments" [5]. The special meeting, which was attended by 142 individuals spanning five countries and all the major components of the U.S. Department of Defense (DOD), featured a plenary session and seven working groups focused on the following topics: principles of assessment; operational environment; strategic theater assessments; campaign assessments; operational and tactical assessments; operations analysis tools for assessments; and data and knowledge management.[4] Key themes that were identified in the conference report included the following: assessment should be a co-equal staff function to planning and operations; assessment doctrine represented a key area for improvement (e.g., in its definitions and principles); the importance of educating commanders on their key role in shaping and guiding the assessment process; and the necessity of integrating coalition partners into assessment processes for multinational operations [5].

In the immediate wake of this special meeting, the CENTCOM CCJ8-ARB team began a series of engagements with the writers of U.S. joint doctrine on the Joint Staff while members of CAA engaged doctrine writers at the Army's Training and Doctrine Command. CAA also submitted a proposal to the NATO Systems Analysis and Studies Panel (SAS) for an update to the NATO Assessment Handbook as part of an overall study on irregular warfare [5]. The results of these engagements as they pertain to changes in doctrine will be discussed in more detail below.

In the years following the meeting, CENTCOM's CCJ8-ARB team continued to advocate for improvement to the ideas and practice of operation assessment. Two additional examples of this advocacy are worth highlighting. The first was their advocacy for the Air Land Sea Application Center (ALSA) to develop a multi-service tactics, techniques, and procedures (MTTP) document to serve as a doctrinal guide for the practice of operation assessment for tactical to operational level commands. The second was their sponsorship of a study called the Joint Assessment Doctrine Evaluation (JADE) Quick Reaction Test Project, which was overseen by the U.S. Army's Test and Evaluation Command. While the results of this study were not released to the public, they nonetheless proved instrumental in contributing to the significant revisions to U.S. military doctrine on operation assessment that occurred between 2013 and 2017 [6].

In short, during the first six years (2011–2017) of the decade of interest for this chapter (2011–2021), CENTCOM's CCJ8-ARB team, in conjunction with CAA, took on the role of change advocate. In 2018, MORS (spurred on by a former member of the CCJ8-ARB team and a faculty member from the Army War College) held a second special session to address operation assessment, which continued to generate momentum for positive change. And in the years since then, several former members of CENTCOM, CAA, and operation assessment cells in Afghanistan continued to

[4] In full disclosure, the author was a co-chair for the working group on campaign assessments and one of several panelist speakers at the event.

play advocacy roles, for example by organizing a special issue of the Journal of Defense Modeling and Simulation dedicated to assessment [7]. As we will see, the combined efforts of these individuals resulted in numerous substantive and positive changes to the ideas behind, and practice of, operation assessment.

3.2 Poor Doctrine

In June 2013, prompted by the MORS special meeting in Tampa and a series of published works that had been critical of doctrine on assessment (e.g., Downes-Martin [8], Schroden [1], Connable [9]), the Joint Staff's Joint Doctrine Analysis Division (JDAD) conducted a special study "to determine whether current assessment doctrine provides sufficient guidance and procedures on how to evaluate progress toward achieving objectives, creating desired conditions, and accomplishing tasks during joint operations" [10].

This study resulted in the following findings: the term "assessment" was present in all 77 joint doctrinal publications; these usages comprised 19 distinct types of assessment; 15 of these types utilized a common dictionary definition of "assessment" as opposed to a specific military meaning; and the remaining types related to the four different military meanings of assessment that were codified in JP 3-0 (Operations). The study described this situation as "confusing" [10]. The study also concluded that "current joint doctrine does not provide a useable description or format for development of an assessment plan and its integration/synchronization with joint planning and execution" nor does it "provide useful vignettes and examples…that could clarify various aspects of assessment" [10].

The study's first recommendation was to specifically define the term "operation assessment" in joint doctrine as "a continuous process that determines the overall effectiveness of employing joint force capabilities during military operations by measuring the progress toward accomplishing a task, creating a condition, or achieving an objective." Additional recommendations included adding to doctrine instructions for developing an assessment plan—as well as vignettes and real or notional operation assessment frameworks and plans—and developing a Joint Doctrine Note (JDN) to serve as a bridging source of guidance on operation assessment until the next regularly scheduled updates of the capstone joint publications (e.g., those on planning and operations) were completed [10].

The first major change in doctrine that emerged in the wake of this study was the Joint Staff's publication of JP 3-24, Counterinsurgency, in 2013. This document was the first joint doctrinal publication to include an entire chapter on assessment and, following recommendations from advocates like the CENTCOM J8 and the assessments community writ large, it advanced a set of key tenets about operation assessment. These were as follows: it is a commander-centric function; it requires linking multiple echelons of command; it requires a whole-of-staff effort to be effective; it should be integrated into planning and operations cycles; it should include information and intelligence internal and external to the staff; assessment reports must be transparent and credible; and assessment is a continuous process [11]. For the first time, JP 3-24 also identified and discussed an "assessment process" that illustrated how assessments should be integrated into staff planning and operations cycles, and it included limited discussion of methods and organizing for the conduct of operation assessment.

The inclusion of a chapter on operation assessment in JP 3-24 influenced the Army to include a chapter on assessment in its revision of FM 3-24, Insurgencies and Countering Insurgencies, which it published in 2014.[5] This chapter contained less emphasis on assessment purpose, process, and practice than JP 3-24 and more on the identification of specific measures of effectiveness and performance [12]. Nonetheless, it represented a step forward from the Army's first iteration of FM 3-24, which contained only three pages of superficial discussion on assessment [13].

In early 2015, the Joint Staff acted more fulsomely on one of the key recommendations from the JDAD study by publishing JDN 1-19, Operation Assessment. The stated purpose of the JDN was "to elaborate on specific assessment fundamentals and articulate how operation assessment is initiated during joint planning and executed throughout operations" [14]. Further, the authors of the JDN sought to "define the term 'operation assessment' in joint doctrine and to standardize a notional operation assessment framework" [14]. In this way, the JDN was meant to apply not just to countering insurgencies, but to all types of joint DOD operations. To those ends, the JDN formalized the definition for operation assessment quoted above, discussed its purpose and key tenets, provided organizational design options for conducting assessment, and identified the fundamental steps (with limited examples) of creating an assessment framework and developing an assessment plan.

Later in 2015, ALSA—in response to advocacy from the CENTCOM CCJ8-ARB team—published a multi-service doctrinal publication titled Operation Assessment that spanned "the tactical and operational levels of war for the Army division, Marine expeditionary force, and joint task force and below" [15]. This document built on JDN 1-19 by providing significantly more, and more detailed, examples of assessment frameworks that had been developed by practitioners (mostly during the wars in Iraq and Afghanistan) as well as a detailed discussion on metrics, standards, and indicators. The document also provided a novel section on how to integrate assessment into the operations process.

The fulsome inclusion of operation assessment in JP 3-24, development of JDN 1-19, and publication of the ALSA multi-service manual paved the way for the inclusion of substantially more content on this topic in the Joint Staff's subsequent routine update to JP 5-0, Joint Planning. This update was published in June 2017 and for the first time it included an entire chapter dedicated to assessment. This chapter opened with the following two paragraphs, which codified key principles that the operation assessment community had been advancing for years [16].[6]

[5] This field manual doubles as a warfighting doctrinal publication for the U.S. Marine Corps.

[6] In the interest of full disclosure, the author was one of a small group of individuals who were asked to contribute ideas and written sections to this chapter, as well as to review various drafts before publication.

Operation assessments are an integral part of planning and execution of any operation, fulfilling the requirement to identify and analyze changes in the [operational environment] and to determine the progress of the operation. Assessments involve the entire staff and other sources such as higher and subordinate headquarters, interagency and multinational partners, and other stakeholders. They provide perspective, insight, and the opportunity to correct, adapt, and refine planning and execution to make military operations more effective. Operation assessment applies to all levels of warfare and during all military operations.

Commanders maintain a personal sense of the progress of the operation or campaign, shaped by conversations with senior and subordinate commanders, key leader engagements, and battlefield circulation. Operation assessment complements the commander's awareness by methodically identifying changes in the [operational environment], identifying and analyzing risks and opportunities, and formally providing recommendations to improve progress towards mission accomplishment. Assessment should be integrated into the organization's planning (beginning in the plan initiation step) and operations battle rhythm to best support the commander's decision cycle.

The updated JP 5-0 also set forth a more nuanced definition of assessment, as "a continuous activity that supports decision making by ascertaining progress toward accomplishing a task, creating an effect, achieving an objective, or attaining an end state for the purpose of developing, adapting, and refining plans and for making campaigns and operations more effective" [16]. The chapter contained detailed sections on how to conduct operation assessment and the publication included an appendix with detailed examples of assessment and data collection plans to support their execution.

In 2019, the Army issued an update to Army Doctrinal Publication (ADP) 5-0, The Operations Process, which was its third update within a decade. The 2010 version of this publication (then called FM 5-0) included a designated chapter on assessment, but its emphasis on flawed, quantitative models formed a primary basis of criticism for numerous scholars (e.g., Downes-Martin [8], Schroden [1], Connable [9]). The Army ditched the chapter on assessment in its 2012 update and slimmed the discussion down to two pages, though the content of those two pages still represented a major step forward from 2010. For example, the section on "guides to effective assessment" outlined four principles ("commanders prioritize the assessment effort," "incorporate the logic of the plan," "use caution when establishing cause and effect," and "combine quantitative and qualitative indicators" [17]) which aligned with the recommendations of advocates for change in the assessment community (e.g., MORS [5]). The Army's 2019 update again contained a dedicated chapter for assessment, which provided an overview of key assessment ideas and directed the reader to the ALSA publication for details on key implementation aspects such as the assessment process [18].

In 2020, the Joint Staff again updated its capstone doctrinal publication on planning, JP 5-0. This newest version again includes a dedicated chapter to operation assessment with much of the same content that existed in the 2017 version. However, the 2020 version took two notable steps backward relative to its predecessor.

First, its definition of assessment was revised to be "determination of the progress toward accomplishing a task, creating a condition, or achieving an objective. A continuous process that measures the overall effectiveness of employing capabilities during military operations" [3]. This definition omits the critical purposes of assessment as supporting decisions and making operations more effective and more narrowly defines it as a measurement activity. In so doing, it tempts the revival of criticisms of "junk arithmetic" and overemphasis on quantitative approaches made by scholars such as Downes-Martin. It also places the definition out of step with the remainder of the chapter, which largely adheres to the same commander- and decision-centric purpose, and mixed methods approaches, to assessment that existed in the 2017 version.

Second, the 2020 version relegates substantially more of its discussion on assessment to an appendix and removes all the notional examples of assessment plans and sample assessment report formats that were present in the 2017 version. As such, the newer version reverts to providing less information to practitioners on how to conduct operation assessment than did the prior publication.

4 Inadequate Training

Unlike the degree of development that has occurred with respect to key U.S. doctrinal publications on the topic of operation assessment, DOD has still not created or implemented any formal training programs for practitioners.

The closest thing to an operation assessment training course inside of DOD was the Deploying Analysts Course that CAA used to run from its headquarters in Fort Belvoir, Virginia. CAA created and designed this course to prepare its ORSA personnel (as well as analysts from other organizations who chose to participate) to deploy to support U.S. combat operations in Iraq, Afghanistan, the Horn of Africa, the Philippines, and elsewhere. This course, which ran for two weeks, typically included at least one or two days of instruction on how to conduct operation assessment. The content of those sessions initially hued closely to the prior experience of CAA analysts as well as those of other practitioners (including the author, who occasionally provided instruction for the course) but became more formally aligned with emergent joint doctrine after the Joint Staff's publication of JDN 1-19. As the U.S. wars in Iraq and Afghanistan wound down, CAA stopped running the course, though it captured the content of it in a Deployed Analyst Handbook and it continues to train its own staff in the conduct of operation assessment [19].

While not a DOD program, the NATO School at Oberammergau created a formal training course for operation assessment in 2012. The goal of the course is to "explain operation assessment theory and the application of operation assessment processes to individuals assigned to a NATO operation assessment or related position" [20]. Some U.S. personnel have served in such positions or been allowed to attend the course, so in that way it has offered a degree of training to the U.S. military.[7] Overall, however, DOD remains without a formal training course for operation assessment.

4.1 Poor Processes and Products

When the author published his criticism of operation assessment in 2011, many practitioners were struggling to apply the purely mathematical approaches that were highlighted in doctrine at that time. Since then, however, substantial progress has been made in identifying, developing, and refining a variety of new tools and approaches for operation assessment, including mixed quantitative–qualitative methods and interdisciplinary ones drawing on established fields of practice in similar areas (e.g., monitoring and evaluation).

One of the earliest of these was published in 2013, in which five members of the Afghan Assessment Group (AAG) described a new framework for strategic and operation assessment that they created for the International Security Assistance Force (ISAF) in Afghanistan. In this article, the authors described key issues with the previous ISAF assessment framework, which included the inability of subordinate or supporting commands to provide input to, or challenge the results of, ISAF's assessment. The focus of the framework on assessing progress along lines of operation, as opposed to assessing progress toward ISAF's desired objectives, was also deemed problematic.

To rectify these and other key issues, the AAG created a two-tier assessment framework, featuring a strategic assessment and a campaign assessment. The strategic assessment centered on answering a set of seven "strategic questions," which were designed to be "high-level, focused, unemotional, enduring, and most importantly, critical to senior decision-makers" [23]. These questions were deliberately designed to "cut across lines of operation and staff organization (i.e., such that they would require elements from several staff sections to answer), to help break down artificial stovepipes of information and force staffs and commands to communicate and coordinate with each other" [23]. The responses to these questions were used in conjunction with a regular rhythm of staff meetings and engagements to compile a narrative assessment report that included the answers to the strategic questions, along with identified obstacles to future progress and risks to the accomplishment of ISAF's strategic objectives if those obstacles were not overcome [23].

The campaign assessment comprised two parts. First, ISAF's subordinate and supporting commands were required to report their progress against ISAF's eight essential campaign tasks across four domains (security, governance, economics, and regional relations), using a set of five standards per domain. Their inputs were then plotted on a spider chart to illustrate areas of agreement or disagreement among

[7] This course was one part of a three-pronged effort by NATO to improve its own practice of operation assessment. The other two parts were a formal experiment to identify best practices in operation assessment [21] and a book on innovations in the field [22].

the commands, as well as areas of progress or regress since the prior assessment. Each command was also allowed a narrative explanation of its ratings, which was used alongside the spider charts to provide necessary nuance to the ratings. Second, ISAF's subordinate and supporting commanders were required to submit a personal assessment of progress directly to the ISAF commander, which allowed them a channel through which to provide their unfiltered views of progress and challenges. The results of the campaign assessment were discussed at a quarterly meeting of the ISAF commander, his senior staff, and his subordinate and supporting commanders [23]. The design elements of this effort were subsequently deemed of sufficient utility that they were written into joint doctrine as examples of how to conduct strategic and campaign level assessments [16].

Another example of a new approach for operation assessment is that of a horizon framework, which was advanced by Larsen and Tygesen [24]. In this approach, rational-expectation theory is applied in the context of a counterinsurgency campaign to identify indicators of population behavior with varying time (expectation) horizons. This is done to gauge trends toward a decisive majority of the population supporting the state (thereby depriving insurgents of the popular support they require). According to the authors, a horizon framework is useful for operation assessment in counterinsurgency because it "privilege[es] behavioral indicators, emphasiz[es] the need to differentiate between time horizons, and enforce[es] a focus on capturing the behavioral shift among the population necessary for counterinsurgency success" [24].

A third example is the use of an explicit theory of change—which describes the underlying logic that planners use to link activities to desired results—to guide the development of assessment indicators. Drawing from the robust literature on monitoring and evaluation, Christopher Paul argued that having a clearly articulated theory of change allows assessors to go beyond the typical steps of identifying indicators pertaining to activities (measures of performance) and outcomes (measures of effectiveness), by enabling them to also identify indicators for the intermediate linkages between activities and outcomes. Identifying a theory of change also makes clear the broad assumptions that planners have made in their logic, which opens the possibility of continuously monitoring the veracity of those assumptions via additional indicators. As Paul states, "if one of those measurements does not yield the desired results, assessors will have a pretty good idea of where in the chain the logic is breaking down… They can then make changes to the theory of change and to the activities being conducted, reconnecting the logical pathway and continuing to push toward the objectives" [25]. In his article, Paul also articulates a "hierarchy of evaluation" model for assessment, in which assessors first start at a base level of "assessment of the need for an effort," and then work up through four additional hierarchies: assessment of design and theory, assessment of process and implementation, assessment of outcome/impact, and assessment of cost-effectiveness [25].

These techniques have been further advanced by other practitioners, who have applied and refined them in the context of assessment beyond counterinsurgency campaigns. In one example, a group of authors married theories for why people engage in terrorist activities with a questions-based assessment approach to create a novel template for assessing counterterrorism operations [26]. In another example, Thurston demonstrated the use of theories of change as a tool for strategic assessment of national-level strategies [27]. In another, a group of researchers provide substantial detail on the application of evaluation theory to operation assessment. They explore the application of expertise-oriented, program-oriented, decision-oriented, and participant-oriented evaluation models to security-related applications. They also identify and examine the relationship between three specific types of monitoring and evaluation designs and operation assessment: "descriptive designs (case study, cross-sectional, time-series), quasi-experimental designs (interrupted time-series, comparison group, case study), and experimental designs (posttest-only, pre-post)" [28].[8]

Authors have also examined some of the tools used by assessors to identify best practices for their employment. One group of authors reviewed common issues that arise with attempts to use public perception polling—and especially polls conducted by third-party contractors with no direct military oversight. This group of authors provided tips for practitioners on the use of active polling, as well as the latent use of social media, for gauging popular attitudes in support of operation assessment [30]. In two additional examples, practitioners reviewed the use of common presentation tools. The author discussed the use of colored maps in Vietnam and Afghanistan [31], and Arnhart and King discussed thermographs, stoplight charts, so-called "color math," arrows, indices, and 100-point scales. After identifying pros and cons of these methods, Arnhart and King argued that written assessments—especially those that include written evaluations of risk—are the most fulsome means of presenting assessment results. The authors identify additional best practices as "using objective development criteria [for assessment indicators], such as the acronym SMART (specific, measurable, achievable, relevant, and time bound) and the similar initialism RMRR (relevant, measurable, responsive, and resourced)" [32].

Last, authors have also sought to capture practical lessons from their own experiences. In one example, three authors with separate stints at Army and joint commands discussed three elements they saw as common to their experiences: "assessors must understand context and should often use reason and qualitative data more than rely on numerical metrics...assessors can help staffs and leaders accurately identify trends amidst noisy and chaotic operational environments...[and] the assessment team can leverage and assist the staff to enable better assessments" [33]. In another example, a group of authors with shared experiences with ISAF's AAG wrote about challenges they had encountered when trying to use trends in violence as an indicator for assessment of counterinsurgency operations [34].

[8] Of note, DOD formally enshrined the use of monitoring and evaluation techniques for the assessment of major security cooperation initiatives in 2017 [29].

5 Commander Disinterest

In contrast to developments in doctrine, processes, and products, there has been comparatively little written by commanders regarding their level of interest or disinterest in operation assessment. Anecdotally, the author can state that there was commander-level interest in assessment at ISAF during the tenure of Generals Stanley McChrystal (who directed the establishment of the AAG and a formal assessment process at ISAF), David Petraeus (who hosted a quarterly assessment conference), and John Allen (who requested the author deploy to ISAF to redesign the command's assessment processes) [23]. Commander level interest in operation assessment continuously waned, however, during the subsequent tenures of Generals Joseph Dunford, John Campbell, Mick Nicholson, and Scott Miller. The extent to which this was a result of assessment products and processes not proving valuable to these commanders or the result of other factors (e.g., waning US resources for, and interest in, the war) remains unclear.

6 Lessons Learned

As described above, there have been significant and substantial efforts to improve the theory and practice of operation assessment in the wake of critics concluding it was "broken" in Afghanistan and elsewhere circa 2011 [1, 8, 9]. From these developments, two positive and two negative key points emerge.

6.1 Positive Lessons

First, the example of operation assessment is one among many that illustrate DOD as a learning organization.[9] In response to operation assessment failures, as well as external and internal stimuli—criticisms from independent scholars and advocacy from service members and DOD civilians, respectively—elements of DOD (e.g., the Joint Staff, Department of the Army, ALSA) took substantive steps to improve the doctrinal publications that guided the conduct of operation assessment alongside planning and operations. In addition, practitioners of operation assessment within DOD took it upon themselves to further develop, articulate, and share ideas for the improvement of this function and best practices for its conduct. And CAA created a training course for analysts being deployed to various war zones to be assessors. In this way, the story of operation assessment since 2011 parallels other stories of military adaptation in Afghanistan [36]. It also exhibits some key aspects of general arguments about the sources of military adaptation in war [37]. The role

[9] Another prominent example from the wars in Afghanistan and Iraq was attempts to counter improvised explosive devices over time [35].

of the CENTCOM CCJ8-ARB team in this regard should be highlighted and held in esteem—were it not for the tireless advocacy of key members of that team, it is doubtful that nearly as much progress would have been made by DOD in advancing the ideas behind operation assessment.

Two key lessons should be learned from this experience. The first is that DOD can adapt in wartime, and on timescales that are relevant to the conduct of irregular conflicts. While DOD's refusal to transition itself to a full "war stance" for the counterinsurgencies in Iraq and Afghanistan meant that it remained fixed on routine update cycles for capstone doctrinal publications, it was nonetheless able to use internal studies and temporary fixes (such as the issuance of JDN 1-19) to address short-term gaps and shortfalls in doctrine. The second lesson is that this type of adaptation can be greatly accelerated by the combined actions of strong internal advocates and pressure from independent sources. The MORS special meeting in 2012 served as a major catalyst for the cascading set of changes to doctrine that occurred in the following seven years. That event was the product of both external criticisms from scholars such as Downes-Martin, Connable, and the author and laudable internal advocacy by members of CENTCOM and CAA.

Second, not only has operation assessment doctrine—which captures best practices from recent military operations—improved markedly over the past ten years; the theory of operation assessment has also improved. This has been the result of two primary lines of effort. The first has entailed a mix of practitioners, researchers, and experts in adjacent fields publishing scholarly articles seeking to advance ideas for the improvement of operation assessment theories, practice, presentation, and organizational design. The second has consisted of at first episodic, but by now, routine meetings and conferences to convene those with experience and interest in operation assessment to further discuss and debate ways to continuously improve. Ten years after a slate of articles that described operation assessment as at best out of date and at worst intellectually bankrupt, it now appears to be an expanding area of academic and practical study. The level of discourse has expanded to the point where it is now possible for scholars to pen review articles to summarize developments in the field [32, 38] and for organizations to publish edited volumes of contributions from experts on assessment [22]. This development bodes well for the future relevance and continued improvement of operation assessment.

A lesson to be learned from this experience is the positive role that academia (e.g., university researchers), learning organizations (e.g., MORS), and government sponsored research entities—such as the federally funded research and development centers (e.g., Center for Naval Analyses, RAND) and analytic agencies like CAA—can play together in rapidly advancing a particular field of study and practice in support of military operations.

6.2 Negative Lessons

First, while the doctrine and field of study pertaining to operation assessment have markedly advanced over the past ten years, there has been little progress in developing formal education and training opportunities for practitioners. DOD has yet to establish any formal pathways for those tasked with the conduct of operation assessment to become proficient in it, and the closest thing that existed—the Deployed Analyst Course run by CAA—ended along with the US wars in Iraq and Afghanistan. While NATO developed a formal course of instruction that some US military personnel have been able to attend, US practitioners are overwhelmingly left to their own devices when it comes to learning how to conduct operation assessment—a function that US doctrine now describes as "an integral part of planning and execution of any operation" [3]. This remains DOD's biggest failure in addressing the shortfalls with operation assessment over the past decade.

It also stands as the biggest risk to the future effective practice of operation assessment. This is the result of three factors. First, the bulk of the learning on this topic that has occurred over the past decade has been in the context of countering insurgencies and terrorist groups—operations that the United States' 2018 National Defense Strategy (NDS) made clear are no longer the priority of DOD [39]. Second, the priorities that the 2018 NDS and its 2022 update have set—deterrence, preparation for high-end conflict, competition with other states, and crisis response [40]—will likely require continued adaptations in operation assessment relative to approaches used for irregular wars. Third, the absence of any formal training courses for assessors removes a critical cog in the learning and adaptation loop, insomuch as there is no "schoolhouse" to which practitioners of assessment in the current and future security environment can provide their gained knowledge to be acted and built upon. While the expanding field of study of operation assessment noted above helps mitigate this risk, it would be notably further reduced via the creation of a formal DOD course of instruction for operation assessment.

Second, while much has been written by practitioners and scholars of operation assessment over the past ten years, there is precious little that has been written by the audiences of such assessment. Most notably, we have virtually no collective understanding of how commanders and senior policy makers viewed assessments during their tenure, beyond fragmented bits of information (e.g., comments by Secretary of Defense Donald Rumsfeld in his infamous "snowflake" memos) and anecdotes included in the experiential writings of practitioners (e.g., Schroden et al. [23]). The absence of documented feedback from commanders and senior officials on what they found most and least useful about assessment remains the biggest gap in the literature on this topic from the past ten years. Systematic studies of the views of these audiences pertaining to assessment—spanning echelons from operational to strategic levels—would be of enormous help to furthering this field of study and continuing to improve and adapt operation assessment in the future.

7 Conclusion

Ten years after the author described the "failure cycle" of operation assessment at work in Afghanistan and beyond, there have been notable advances by DOD, and a community of scholars and practitioners, in addressing three of the cycle's five components. First, notable groups of advocates—both internal and external to DOD—have emerged as champions for the improvement of operation assessment. Second, major improvements have occurred in the quality and utility of US (and NATO) doctrine on operation assessment. Third, significant advances have been made in the theories, approaches, processes, products, and organizational design pertaining to assessment, largely as the result of a growing field of academic and practical study on this topic.

These positive developments notwithstanding, the other two elements of the failure cycle—inadequate training and commander disinterest—remain almost completely unaddressed. These, combined with the continuously changing nature of the security environment in which assessors will find themselves in the years to come, represent the largest risks to the future adaptation and successful practice of operation assessment. The community that has emerged around this field is a dedicated and professional one that has learned significant lessons from time spent struggling to conduct assessment in the face of the failure cycle's headwinds. Honoring the legacy of their wartime efforts requires us—and especially DOD—to act on the lessons learned from their experiences. Operation assessment has come a long way in the past ten years; it would bode ill for the effectiveness of future military operations if we allowed that momentum to subside.

References

1. Schroden J (2011) Why operations assessments fail: it's not just the metrics. Navy War College Rev 64(4):89–102
2. Joint Chiefs of Staff (2021) DOD dictionary of military and associated terms. U.S. Department of Defense, Nov 2021
3. Joint Chiefs of Staff (2020) Joint publication 5-0: joint planning. U.S. Department of Defense, 1 Dec 2020
4. Brooks R, Strategy S (2008) The civil-military politics of strategic assessment. Princeton University Press, Princeton, NJ
5. Military Operations Research Society (2012) Assessments of multinational operations: from analysis to doctrine and policy. MORS report on proceedings
6. Joint Test and Evaluation (2015) Joint assessment doctrine evaluation (JADE) quick reaction test project: final report. U.S. Department of Defense, June 2015
7. Arnhart L, King M, Schroden J (2019) Operation assessment: fundamentals and best practices to create a baseline for assessors. J Defense Model Simul 16:4
8. Downes-Martin S (2011) Operations assessment in Afghanistan is broken—what is to be done? Naval War Coll Rev 64(4):103–125
9. Connable B (2012) Embracing the fog of war: assessment and metrics in counterinsurgency. RAND Corporation, Santa Monica, CA

10. Joint Doctrine Analysis Division (2013) Assessment special study report. Joint Chiefs of Staff J-7, 20 Oct 2013
11. Joint Chiefs of Staff (2013) Joint publication 3-24: counterinsurgency. U.S. Department of Defense, 22 Nov 2013
12. Headquarters, Department of the Army (2014) FM 3-24: insurgencies and countering insurgencies. U.S. Department of Defense, May 2014
13. Headquarters, Department of the Army (2006) FM 3-24: counterinsurgency. U.S. Department of Defense, Dec 2006
14. Joint Chiefs of Staff (2015) Joint doctrine note 1-15: operation assessment. U.S. Department of Defense, 15 Jan 2015
15. Air Land Sea Application Center (2015) Operation assessment: multi-service tactics, techniques, and procedures for operation assessment. U.S. Department of Defense, 18 Aug 2015
16. Joint Chief of Staff (2017) Joint publication 5-0: joint planning. U.S. Department of Defense, 16 June 2017
17. Headquarters, Department of the Army (2012) ADP 5-0: the operations process. U.S. Department of Defense, May 2012
18. Headquarters, Department of the Army (2019) ADP 5-0: the operations process. U.S. Department of Defense, July 2019
19. Center for Army Analysis (2016) Deployed analyst handbook. U.S. Department of Defense, June 2016
20. NATO School Oberammergau (2022) Academic course guide 2022. North Atlantic Treaty Organization
21. NATO Headquarters Supreme Allied Commander Transformation (2013) NATO operations assessment experiment final report, 11 July 2013
22. Williams A, Bexfield J, Farina FF, de Nijs J (eds) (2013) Innovation in operations assessment: recent developments in measuring results in conflict environments. NATO Headquarters Supreme Allied Commander Transformation, Norfolk, VA
23. Schroden J et al (2013) A new paradigm for assessment in counterinsurgency. Mil Oper Res 18:3:5–20
24. Larsen KK, Tygesen CB (2013) The horizon framework. Small Wars J
25. Paul C (2013) Foundations for assessment: the hierarchy of evaluation and the importance of articulating a theory of change. Small Wars Journal, 30 July 2013
26. Schroden J, Rosenau W, Warner E (2016) Asking the right question: a framework for assessing counterterrorism actions, CNA DRM-2015-U-012261-Final, Feb 2016
27. Thurston CQ (2019) Assessing the theory of change in national strategies. J Defense Model Simul 16:4
28. Keith A, Ahner D, Curtis N (2019) Evaluation theory and its application to military assessments. J Defense Model Simul 16:4
29. Office of the Under Secretary of Defense for Policy (2017) DOD instruction 5132.14: assessment, monitoring, and evaluation policy for the security cooperation enterprise. Department of Defense, 13 Jan 2017
30. Lindquist J, Peng D, Rochelle S, Turner D, Warshaw M (2019) Effective use of public perceptions in assessments. J Defense Model Simul 16:4
31. Schroden J (2021) The challenges of mapping Taliban control in Afghanistan. Lawfare, 1 Aug 2021
32. Arnhart L, King M (2018) Are we there yet? Implementing best practices in assessments. Mil Rev, May-June 2018
33. Jablonski J, Wade B, Alt J (2019) Operation assessment: lessons learned across Echelons. J Defense Model Simul 16:4
34. Gons E, Schroden J, McAlinden R, Gaul M, Van Poppel B (2012) Challenges of measuring progress in Afghanistan using violence trends: the effects of aggregation, military operations, seasonality, weather, and other causal factors. Defense Secur Anal 28:2:100–113

35. Martin B, Manacapilli T, Crowley JC, Adams J, Shanley MG, Steinberg P, Stebbins D (2013) Assessment of joint improvised explosive device defeat organization (JIEDDO) training activity. RAND Corporation RR421
36. Farrell T (2013) Introduction: military adaptation in war. In: Farrell T, Osinga F, Russell JA (eds) Military adaptation in Afghanistan. Stanford University Press, Stanford, pp 1–23
37. Posen BR (1984) The sources of military doctrine: France, Britain, and Germany between the world wars. Cornell University Press, Ithaca, NY
38. MacLeod M, Banko K (2021) A Canadian perspective on military assessments. Defense Research and Development Canada, DRDC-RDDC-2021-R030, March 2021
39. "Summary of the 2018 National Defense Strategy of the United States of America: Sharpening the American Military's Competitive Edge," DOD, 2018
40. "Fact Sheet: 2022 National Defense Strategy," DOD, 2022
41. Headquarters, Department of the Army (2010) FM 5-0: the operations process. U.S. Department of Defense, March 2010

Dr. Jonathan Schroden is the Director of CNA's Countering Threats and Challenges Program (CTCP), whose mission is to support US government efforts to better understand and counter state and non-state threats and challenges. CTCP includes geographic research portfolios focused on the Middle East, Africa, South and Central Asia, Latin America, and North Korea. Dr. Schroden also directs CNA's Special Operations Program, which focuses on bringing CNA's "full spectrum" research and analysis capabilities to bear on the most complex and challenging issues facing special operations forces (SOF) today and in the future.

The manufacturer's authorised representative in the EU is Springer Nature Customer Service Centre GmbH, Europaplatz 3, 69115 Heidelberg, Germany. If you have any concerns regarding our products, please contact ProductSafety@springernature.com

Printed and bound by CPI Group (UK) Ltd, Croydon, CR0 4YY
01/10/2026
02225892-0016